Drift Exploration in Glaciated Terrain

Geological Society Special Publications
Series Editors
P. DOYLE
A. J. HARTLEY
R. E. HOLDSWORTH
A. C. MORTON
M. S. STOKER
J. TURNER

Special Publication reviewing procedures

The Society makes every effort to ensure that the scientific and production quality of its books matches that of its journals. Since 1997, all book proposals have been refereed by specialist reviewers as well as by the Society's Publications Committee. If the referees identify weaknesses in the proposal, these must be addressed before the proposal is accepted.

Once the book is accepted, the Society has a team of series editors (listed above) who ensure that the volume editors follow strict guidelines on refereeing and quality control. We insist that individual papers can only be accepted after satisfactory review by two independent referees. The questions on the review forms are similar to those for *Journal of the Geological Society*. The referees' forms and comments must be available to the Society's series editors on request.

Although many of the books result from meetings, the editors are expected to commission papers that were not presented at the meeting to ensure that the book provides a balanced coverage of the subject. Being accepted for presentation at the meeting does not guarantee inclusion in the book.

Geological Society Special Publications are included in the ISI Science Citation Index, but they do not have an impact factor, the latter being applicable only to journals.

More information about submitting a proposal and producing a Special Publication can be found on the Society's web site: www.geolsoc.org.uk.

GEOLOGICAL SOCIETY SPECIAL PUBLICATION NO. 185

Drift Exploration in Glaciated Terrain

EDITED BY

M. B. McCLENAGHAN
Geological Survey of Canada, Canada

P. T. BOBROWSKY
British Columbia Geological Survey, Canada

G. E. M. HALL
Geological Survey of Canada, Canada

&

S. J. COOK
Hudson Bay Exploration and Development Co. Ltd., Canada

2001
Published by
The Geological Society
London

THE GEOLOGICAL SOCIETY

The Geological Society of London was founded in 1807 and is the oldest geological society in the world. It received its Royal Charter in 1825 for the purpose of 'investigating the mineral structure of the Earth' and is now Britain's national society for geology.

Both a learned society and a professional body, the Geological Society is recognized by the Department of Trade and Industry (DTI) as the chartering authority for geoscience, able to award Chartered Geologist status upon appropriately qualified Fellows. The Society has a membership of 9099, of whom about 1500 live outside the UK.

Fellowship of the Society is open to persons holding a recognized honours degree in geology or a cognate subject and who have at least two years' relevant postgraduate experience, or not less than six years' relevant experience in geology or a cognate subject. A Fellow with a minimum of five years' relevant postgraduate experience in the practice of geology may apply for chartered status. Successful applicants are entitled to use the designatory postnominal CGeol (Chartered Geologist). Fellows of the Society may use the letters FGS. Other grades of membership are available to members not yet qualifying for Fellowship.

The Society has its own Publishing House based in Bath, UK. It produces the Society's international journals, books and maps, and is the European distributor for publications of the American Association of Petroleum Geologists (AAPG), the Society for Sedimentary Geology (SEPM) and the Geological Society of America (GSA). Members of the Society can buy books at considerable discounts. The Publishing House has an online bookshop (*http://bookshop.geolsoc.org.uk*).

Further information on Society membership may be obtained from the Membership Services Manager, The Geological Society, Burlington House, Piccadilly, London W1V 0JU (E-mail: *enquiries@geolsoc.org.uk;* tel: +44 (0)207 434 9944).

The Society's Web Site can be found at *http://www.geolsoc.org.uk/*.The Society is a Registered Charity, number 210161.

Published by The Geological Society from:
The Geological Society Publishing House
Unit 7, Brassmill Enterprise Centre
Brassmill Lane
Bath BA1 3JN, UK

(*Orders*: Tel. +44 (0)1225 445046
 Fax +44 (0)1225 442836)
Online bookshop: *http://bookshop.geolsoc.org.uk*

The publishers make no representation, express or implied, with regard to the accuracy of the information contained in this book and cannot accept any legal responsibility for any errors or omissions that may be made.

© The Geological Society of London 2001. All rights reserved. No reproduction, copy or transmission of this publication may be made without written permission. No paragraph of this publication may be reproduced, copied or transmitted save with the provisions of the Copyright Licensing Agency, 90 Tottenham Court Road, London W1P 9HE. Users registered with the Copyright Clearance Center, 27 Congress Street, Salem, MA 01970, USA: the item-fee code for this publication is 0305-8719/01/$15.00.

British Library Cataloguing in Publication Data
A catalogue record for this book is available from the British Library.

ISBN 1-86239-082-7
ISSN 0305-8719

Typeset by Bath Typesetting, Bath, UK
Printed by Alden Press, Oxford, UK

Distributors

USA
AAPG Bookstore
PO Box 979
Tulsa
OK 74101-0979
USA
Orders: Tel. + 1 918 584-2555
 Fax + 1 918 560-2652
 E-mail *bookstore@aapg.org*

Australia
Australian Mineral Foundation Bookshop
63 Conyngham Street
Glenside
South Australia 5065 Australia
Orders: Tel. +61 88 379-0444
 Fax +61 88 379-4634
 E-mail *bookshop@amf.com.au*

India
Affiliated East-West Press PVT Ltd
G-1/16 Ansari Road, Daryaganj,
New Delhi 110 002 India
Orders: Tel. +91 11 327-9113
 Fax +91 11 326-0538
 E-mail *affiliat@nda.vsnl.net.in*

Japan
Kanda Book Trading Co.
Cityhouse Tama 204
Tsurumaki 1-3-10
Tama-shi
Tokyo 206-0034 Japan
Orders: Tel. +81 (0)423 57-7650
 Fax +81 (0)423 57-7651

Contents

Introduction

KLASSEN, R. A. A Quaternary geological perspective on geochemical exploration in glaciated terrain   1

Sampling techniques

MCMARTIN, I. & MCCLENAGHAN, M. B. Till geochemistry and sampling techniques in glaciated shield terrain: a review   19

LEVSON, V. M. Regional till geochemical surveys in the Canadian Cordillera: sample media, methods and anomaly evaluation   45

Heavy minerals in mineral exploration

AVERILL, S. A. The application of heavy indicator mineralogy in mineral exploration, with emphasis on base metal indicators in glaciated metamorphic and plutonic terrain   69

MCCLENAGHAN, M. B. & KJARSGAARD, B. A. Indicator mineral and geochemical methods for diamond exploration in the glaciated terrain of Canada   83

Lake sediment geochemistry

COOK, S. J. & MCCONNELL, J. W. Lake sediment geochemical methods in the Canadian Shield, Cordillera and Appalachia   125

Biogeochemistry

DUNN, C. E. Biogeochemical exploration methods in the Canadian Shield and Cordillera   151

Data interpretation

HARRIS, J. R., WILKINSON, L. & BERNIER, M. Analysis of geochemical data for mineral exploration using a GIS - A case study from the Swayze greenstone belt, northern Ontario, Canada   165

Case Histories I - Geochemical exploration in Shield Terrain

MCCLENAGHAN, M. B. Regional and local-scale gold grain and till geochemical signatures of lode Au deposits in the western Abitibi Greenstone Belt, central Canada   201

EARLE, S. Application of composite glacial boulder geochemistry to exploration for unconformity-type uranium deposits in the Athabasca Basin, Saskatchewan, Canada   225

Case Histories II Geochemical exploration in Appalachia

STEA, R. R. & FINCK, P. W. An evolutionary model of glacial dispersal and till genesis in Maritime Canada   237

BATTERSON, M. J. & LIVERMAN, D. G. E. Contrasting styles of glacial dispersal in Newfoundland and Labrador: methods and case studies — 267

Case Histories III - Geochemical exploration in the Cordillera

PLOUFFE, A. The glacial transport and physical partitioning of mercury and gold in till: implications for mineral exploration with examples from central British Columbia, Canada — 287

LETT, R. E. Geochemical signatures around massive sulphide deposits in Southern British Columbia, Canada — 301

PAULEN, R. C. Glacial transport and secondary hydromorphic metal mobilization: examples from the southern interior of British Columbia, Canada — 323

Index

It is recommended that reference to all or part of this book should be made in one of the following ways:

MCCLENAGHAN, M. B., BOBROWSKY, P. T., HALL, G. E. M. & COOK, S. J. (eds) 2001. *Drift Exploration in Glaciated Terrain*. Geological Society, London, Special Publications, **185**.

DUNN, C. E. Biogeochemical exploration methods in the Canadian Shield and Cordillera. *In*: MCCLENAGHAN, M. B., BOBROWSKY, P. T., HALL, G. E. M. & COOK, S. J. (eds). *Drift Exploration in Glaciated Terrain*. Geological Society, London, Special Publications, **185**, 151–164.

Preface

This special publication is a compilation of papers presented at the Drift Exploration in Glaciated Terrain Short Course held in conjunction with the 19th International Geochemical Exploration Symposium in Vancouver, British Columbia, Canada in April, 1999. The short course was sponsored by the Association of Exploration Geochemists.

The volume focuses on the application of till geochemical and indicator mineral methods to mineral exploration in the glaciated terrain of Canada. The principles and examples described, however, have direct applications for explorationists working in glaciated parts of North America, northern Europe and Asia, as well as mountainous regions of South America. Mineral exploration in glaciated terrain requires an appreciation and understanding of glacial processes, surficial sediments, glacial history, and soil formation in addition to economic geology. The following papers address these issues and are organized to lead the reader from the general to the specific.

The first half of the volume is an introduction to glaciated terrain. Sampling techniques are described, followed by reviews of indicator mineral methods used for diamond, gold, and base-metal exploration. Lake sediment and biogeochemical methods are included to complement geochemical and indicator mineral methods. A paper describing the application of GIS methods to till geochemical data has also been included, reflecting the importance of data interpretation and display as essential parts of regional geochemical surveys. The second half of the volume consists of a series of case studies addressing each of the three major glaciated terrains of Canada: flat lying Shield terrain of central and northern Canada, rugged mountainous terrain of the western Canadian Cordillera and the rounded mountains of Appalachia on the east coast.

The editors wish to acknowledge their appreciation of the many hours the authors have devoted to preparing presentations for the short course and to modifying the course notes for subsequent publication in this special publication. The editors would also like to thank the following dedicated colleagues for their comprehensive and thoughtful reviews of the manuscripts: A. Brooks, J. J. Clague, L. Clark, W. B. Coker, A. Dixon-Warren, M. Fedikow, M. Fenton, J. Franklin, P. Friske, E. Grunsky, S. M. Hamilton, L. Hulbert, L. Jackson, B. Janse, R. Lett, A. A. Levinson, V. Levson, E. Nielsen, R. C. Paulen, A. Plouffe, T. Pronk, B. Schreiner, S. Sibbick, R. R. Stea, P. Taufen, I. Thomson, L. H. Thorleifson, B. C. Ward and S. Williams.

A Quaternary geological perspective on geochemical exploration in glaciated terrain

RODNEY A. KLASSEN

Geological Survey of Canada, 601 Booth Street, Ottawa, Ontario K1A 0E8, Canada (e-mail: klassen@nrcan.gc.ca)

Abstract: The application of Quaternary geology and glacial sedimentology is given as a broad guide for geochemical exploration in glaciated terrain. Predictive models of glacial dispersal provide an important basis for tailoring drift prospecting methods to suit regional variations in ice flow history and dynamics. The models relate compositional variations in glacial dispersal trains to ice flow direction, glacial history and subglacial processes. They are continually refined with reference to the geological and physical properties of the ice bed; new empirical field evidence constraining particle trajectories; and knowledge of subglacial processes affecting glacial erosion, transport and deposition. Transport at the ice bed leads to an exponential decrease in indicator concentrations with increasing distance of glacial transport, whereas linear decrease is associated with englacial transport, and may be characteristic of ice streams. The partitioning of rock and mineral fragments through subglacial comminution leads to compositional differences among size fractions that can reflect intensity of subglacial process, distance of transport, and provenance; hence, the choice of size fraction is important to drift prospecting by geochemical methods.

Drift prospecting is based on the premise that lithological, mineralogical or geochemical indicators of economic mineralization can be traced in glacial deposits to locate their bedrock source. Of the varied types of glacial sediment, till is most directly related to bedrock composition and has been successfully used as a sample medium for mineral exploration. Till is mechanically derived from bedrock and preglacial sediments along the path of ice flow, and modified by glacial processes of erosion, transport and deposition. Till geochemistry is determined by its constituent rock fragments and mineral grains, including both common rock-forming minerals and economic indicators such as sulphides. Hence, in glaciated terrain the practice of geochemical exploration is firmly linked to our understanding of clastic sedimentology and glacial process. As such, it may be considered as '... a micro-variation of boulder tracing – as a method of searching for infinitesimal 'boulders'' (Kauranne 1959).

Drift prospecting is seldom straightforward, reflecting the complexities of ice flow and subglacial processes, and their variation in time and space at scales of metres to thousands of kilometres. Furthermore, in Canada it is typically carried out in a remote setting where little or no surficial geological information is available, and predictive models of glacial dispersal constitute important guides for exploration practice (e.g. Coker & DiLabio 1989; Shilts 1993, 1996). Over the past decade the basis for drift prospecting has been significantly improved through ongoing refinement of ice sheet models, continued acquisition of geological field evidence, and knowledge of glacial processes acquired through study of modern ice sheets. The ice sheet and glacial dispersal models provide an evolving context for interpreting till provenance and geochemistry in terms of subglacial thermal regimes, ice flow velocities, and glacial processes.

The principal constraints for ice sheet and glacial dispersal modelling are provided by the geological record, especially empirical field evidence for ice flow and drift composition. It has been largely acquired through federal and provincial surveys, and includes hundreds of thousands of geochemical, mineralogical, and lithological till analyses. Maps of surficial geology and drift composition portray variation at local to regional scales, tens to hundreds of kilometres, whereas most ice sheet models are

From: McClenaghan, M. B., Bobrowsky, P. T., Hall, G. E. M. & Cook, S. J. (eds) 2001. *Drift Exploration in Glaciated Terrain*. Geological Society, London, Special Publications, **185**, 1–17.
0305-8719/01/$15.00 © The Geological Society of London 2001.

continental scale, 1000's of kilometres. They aid exploration by establishing a context for investigations at property-scale (hundreds of metres to kilometres). The ongoing developments in the related fields of Quaternary geology, glacial sedimentology, and ice sheet modelling provide important new tools for tailoring drift prospecting methods to suit regional variations in ice flow history and dynamics.

This paper illustrates applications of glacial dispersal and ice sheet models to geochemical exploration in Canada. It is intended as a broad guide for explorationists unfamiliar with studies in the related fields of glaciology, glacial sedimentology, and Quaternary geology. It describes key aspects of ice sheet modelling, glacial record, and dispersal models, emphasizing their evolving nature and increasingly accurate portrayals of debris transport. It also illustrates the effects of ice flow history and glacial process on sediment composition. The focus is on the Laurentide Ice Sheet, although Fennoscandian research is reported because it has direct application to Canadian exploration practice. The text and references, however, are not comprehensive, and many deserving studies have been omitted in respect of space.

Ice sheet models

Early models of the Laurentide Ice Sheet assumed a rigid, unyielding bed, and plastic ice flow, and the ice sheet was portrayed as a stable monolith centred on Hudson Bay with broadly radial flow lines outward across Canada (e.g. Flint 1971). The landforms shown by the Glacial Map of Canada (Prest *et al.* 1968) were interpreted to reflect a uniform application of subglacial process and simplistic relations among ice temperature, crustal heat flow, and ice flux (Sugden 1977, 1978). The monolithic model was pre-eminent from the 1940s to the early 1970s, when geological fieldwork demonstrated that glacial flow lines described by Flint's model were not consistent with glacial dispersal patterns (Shilts 1979). Through the 1980s, ice sheet models radically changed, incorporating physical properties of the ice bed, topography, and subglacial hydrology as controls on ice flow dynamics (Boulton *et al.* 1985; Fisher *et al.* 1985; Boulton & Hindmarsh 1987; Budd & Smith 1987). They were further refined through acquisition of geological data, including striations, indicator erratics, stratigraphy, and radiocarbon dates, as well as through glaciological studies of the modern Antarctic and Greenland ice sheets. The accuracy of palaeoflow models is principally based on the field evidence of indicator erratics

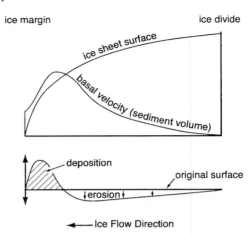

Fig. 1. Schematic profiles of ice flow velocity from ice divide to ice margin, and corresponding areas of erosion and deposition. (After Boulton & Clark 1990*b*).

hundreds to thousands of kilometres from their source (e.g. Boulton *et al.* 1985; Fisher *et al.* 1985; Clark 1987; Prest 1990).

The intensity of subglacial processes, and hence the formation of till and glacial dispersal trains, relates to subglacial ice flow velocity, temperature, and water, as well as to the mechanical properties of the ice bed. Spatial and temporal variations in subglacial environments shown by ice sheet models indicate: (1) near ice divides, basal ice flow velocity and subglacial erosion are minimal, and glacial history and debris transport is complex; (2) erosional processes are predominant in the outer margin of an expanding ice sheet where flow velocities are greatest; and (3) depositional processes are predominant in the outer margin of an ice sheet at its maximum and during deglaciation with decrease in flow velocities (Boulton 1984, 1996*a*,*b*) (Fig. 1). The evolution from erosional to depositional regimes in an expanding ice sheet and areal differences in ice flow duration are thereby linked to regional variations in glacial transport distances and deposit thickness. The proportions of far-travelled to local debris increase upwards in till, and with distance outward from the ice divides (e.g. Boulton 1984, 1996*a*,*b*; Clark 1987).

Glacial deposits and landforms of Canada, shown on the Surficial Materials Map of Canada (Fulton 1995) at 1:5 000 000 (Fig. 2), are the product of ice flow dynamics associated with contiguous, semi-independent ice sheet sectors (e.g. Sugden 1977; Boulton 1984, 1996*a*,*b*; Dyke & Prest 1987; Aylsworth & Shilts 1989*a*;

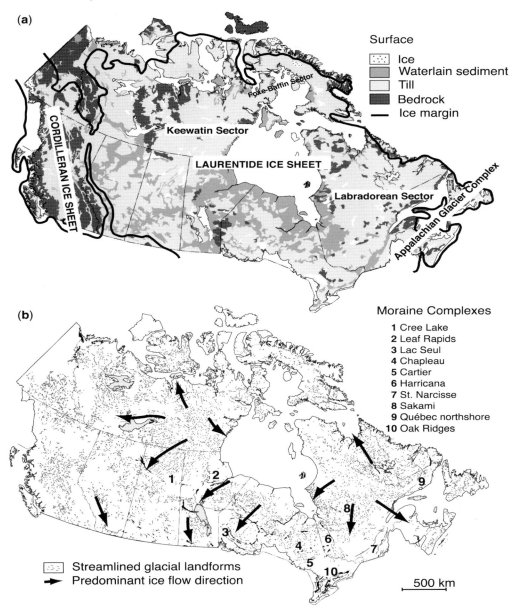

Fig. 2. (a) Regional distribution of bedrock, till, and waterlain sediments in Canada. (b) Trend and orientation of streamlined glacial landforms and morainic complexes. (Fulton 1995).

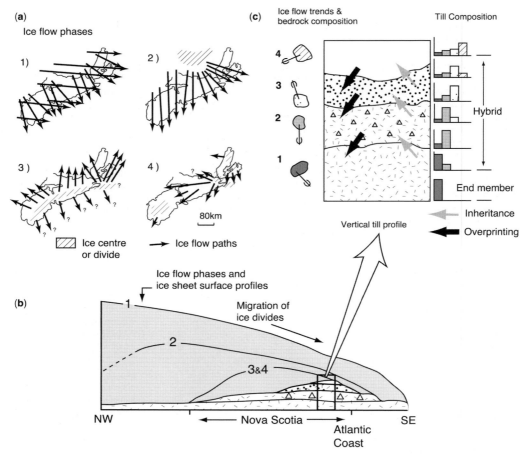

Fig. 3. In Nova Scotia, spatial and temporal variations in ice flow dynamics are related to interaction between the Laurentide Ice Sheet and the Appalachian Glacier Complex. Till there is described as: (1) end member, comprising detritus soley derived from bedrock along the path of ice flow; and (2) hybrid, which includes a component either derived from till overridden along the path of ice flow (inheritance), or introduced by overridding ice (overprinting). (Modified from Finck & Stea 1995; Stea *et al.* 1989; also R. R. Stea, pers. comm. 1997)

Bouchard & Salonen 1989, 1990; Boulton & Clark 1990*a,b*), and bedrock geology (Aylsworth & Shilts 1989*b*). From glaciology, they are inferred to record concentric successions of marginal ice flow regimes, characterized by an outermost zone of strongly developed depositional forms, an intermediate zone with both strong erosional and depositional forms, and central zones dominated by erosional landforms and superimposed late glacial depositional forms (Fulton 1989, 1995; Boulton 1996*b*). In contrast to early portrayals, the central ice divides were active throughout glaciation, although changing in shape, size, and location (Shilts 1979; Boulton *et al.* 1985; Boulton & Clark 1990*a,b*; Clark 1993; Dyke & Prest 1987).

In the Surficial Materials Map (Fig. 2), narrow to lobate ice flow patterns extend hundreds to thousands of kilometres that commonly terminate at major morainic complexes and have lateral margins marked by discontinuities in regional ice flow trends (Dyke 1984; Dyke & Morris 1988; Hicock *et al.* 1989; Dyke *et al.* 1992; Thorleifson & Kristjansson 1993; Dredge 1995). These features are associated with ice streams, which are zones in an ice sheet that flow more rapidly than the surrounding ice (Bentley 1987). Ice streams occur where shear stress at the ice bed is low as the result of either subglacial sediment deformation or meltwater between the ice and its bed, or both. They are inferred to have been prominent features of the Laurentide Ice Sheet, playing a vital role in its flow dynamics and mass balance (Marshall *et al.* 1996), and serving as important agents of glacial dispersal and till formation. Their distribution reflects bed topography, subglacial sediment properties, and subglacial thermal and hydrological regimes (Boulton 1996*a*, *b*; Marshall *et al.* 1996; Shaw *et al.* 1996).

Ice flow trends of the Appalachian glacier complex reflect topographic effects and interaction with the Laurentide Ice Sheet. There, drift composition reflects flow: (1) in the outer margins of a continental ice sheet, with high ice velocities and long-continued directions of flow; and (2) near smaller ice divides of the maritime complex (Pronk *et al.* 1989; Rappol 1989; Shilts & Smith 1989; Stea *et al.* 1989; Charbonneau & David 1993). Interaction among ice sheets and changing ice divides is described by Stea *et al.* (1989) in a zonal context (Fig. 3). Dispersal associated with the Laurentide Ice Sheet and with adjacent ice centres outside the Province (phase 1) is along regionally prominent flow paths and the debris includes a significant far-travelled component. In contrast, greater areal variation in ice flow direction and glacial transport distance is evident from local dispersal centres (phases 2, 3, 4).

Geology and sedimentology

Empirical field evidence is used to establish particle trajectories in the ice sheet, to model glacial dispersal, and to infer subglacial processes active in the formation of glacial deposits. It also constrains theoretical models derived from principles of glaciology and palaeoclimatology. The evidence includes: indicator erratic trains that define net ice flow paths, including multi-cycle transport; ice flow trends shown by striations and streamlined landforms; and the properties and provenance of glacial sediments. Although much of Canada remains unmapped, field-based evidence acquired in the past two decades has provided a basis for reconstructing a comprehensive history for much of the last glaciation, not only the late-glacial. Advances in glacial sedimentology have further refined our ability to interpret the geological record in terms of glacial process and depositional environment, and to distinguish natural background variations in till geochemistry from the often subtle signals of exploration interest. The following text briefly describes developments related to the geological record and glacial sedimentology, and their implications for drift prospecting.

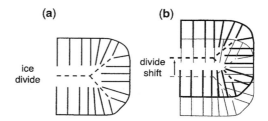

Fig. 4. Complex ice flow trends can result from simple shift in ice divide location. (**a**) Ice flow pattern associated with a principal ice divide and subsidiary divides. (**b**) Flow pattern resulting from a shift in the ice divide, showing crosscutting trends in subsidiary divide areas; line thicknesses reflect the relative prominence of striations (Boulton & Clark 1990*b*).

Ice flow record

The ice flow record is both erosional, which includes striations, grooves, whalebacks and roche moutonées, and depositional, which includes drumlins and flutes. The varied components represent different stages in ice sheet growth and decay, rather than a single temporal flow regime. Striations and grooves are lineations created by debris dragged by ice across the glacier bed. Although they define the trend and relative age of ice flow events, they do not record the duration of ice flow and cannot be used to establish distance of glacial transport (e.g. Prest 1983; Kleman 1990). Striations can provide a detailed and comprehensive record of ice flow throughout the last glaciation (Veillette 1986; Kleman 1990; Klassen & Thompson 1993; Stea 1994; Parent *et al.* 1996; Veillette *et al.* 1999). Near ice divides, striation trends are typically complex, reflecting the evolution in shape, extent, and location of the divides, as well as outward migration of the zone of basal erosion (Boulton & Clark 1990*a,b*) (Fig. 4). Near the outer margins trends are less variable, reflecting greater distance from ice divides and corresponding stability in ice flow direction. Discontinuities in regional ice flow trends can record the overprint of a localized glacial event, such as ice streaming.

Striations pre-dating the last glacial maximum occur on exposed outcrop surfaces as crosscutting sets, and on surfaces sheltered from later ice flow (e.g. Bouchard & Martineau 1985; Veillette 1986; Kleman 1990; Lundqvist 1990; Klassen & Thompson 1993; Stea 1994; Veillette & Roy 1995; Parent *et al.* 1996; Veillette & McClenaghan 1996). Their preservation, despite subsequent glaciation, indicates: (1) limited, differential glacial erosion of bedrock; (2) the potential importance of topography and bed roughness to both the formation and preservation of dispersal trains; and (3) the significance of ice flow direction relative to the aspect of bedrock slopes. By inference, outcrop surfaces of different aspect are not equally represented in the till composition because of differential erosion. The growing body of striation evidence has been the key to establishing a comprehensive ice flow record for the last glaciation.

On the assumption that streamlined glacial landforms become younger with distance from the ice margin, late glacial ice flow is characterized by complex, linear ice divides that evolved in size, shape and relative importance (Dyke & Prest 1987). Crosscutting relations among landforms, however, indicates that a simple interpretation of progressive change in relative age with distance from the late glacial maximum ice margin may not be correct (Boulton & Clark 1990a, b).

Glacial erosion

Glacial erosion relates to basal ice flow velocities. Net depths of bedrock erosion are estimated to be metres to tens of metres, determined from debris volumes in glacial dispersal trains (Kaszycki & Shilts 1980; Lundqvist 1990; Charbonneau & David 1993), and by preservation of older glacial and interglacial deposits and landforms (DiLabio et al. 1988; Thorleifson et al. 1993; Kleman 1994) and preglacially weathered bedrock beneath till (Boyle 1996; Hirvas 1991). Buried preglacial deposits are widespread across the Prairies (Fenton 1984) and the Hudson Bay Lowlands (Thorleifson et al. 1993); occurring in bedrock depressions in northern Ontario and Quebec (DiLabio et al. 1988; McClenaghan et al. 1992; Smith 1992), and the southernmost margins of the ice sheet (Hansel et al. 1987). They indicate where glacial erosion was incomplete and that underlying bedrock occurs in surface till only as redeposited debris. In the arctic and along the eastern and northern continental margins, the ice sheet was cold-based and protective of its bed (Dyke & Morris 1988; Dyke et al. 1992; Dyke 1993; Dredge 1995). There, glacial deposits can be difficult to distinguish from bedrock rubble, and major landform elements could be Tertiary in age (Dyke et al. 1992; Charbonneau & David 1993; Dredge 1995).

Streamlined glacial landforms, such as crag-and-tail hills, drumlins and flutings, have also been described as erosional, created by subglacial sediment deformation (Boyce & Eyles 1991) or subglacial meltwater (Shaw 1990). Large-scale, catastrophic subglacial meltwater floods have been interpreted to have formed significant parts of the landform record attributed to ice (Shaw 1990; Shaw et al. 1996). Meltwater features include varied erosional marks and forms developed in bedrock (e.g. Kor et al. 1991; Kor & Cowell 1998; Rampton 2000). The meltwater discharge model interprets the streamlined landforms to have been created by vast subglacial outburst floods of short duration early in deglacial time. The landforms define coherent linear belts extending hundreds of kilometres from the ice sheet margins that, in some cases, have also been attributed to ice streams. In contrast to transport in glacier ice, the role of subglacial meltwater on particle transport, distinct from glacial, is not known. The effects of subglacial meltwater processes are potentially important to drift prospecting, especially in terms of property scale exploration where facies differences in glacial sediments can reflect marked difference in indicator concentrations (Rampton 2000).

Glacial sedimentology

The sophistication of glacial sedimentology and the criteria to distinguish sediment facies and depositional environments are reflected by evolution in the definition and usage of terms describing till (e.g. Dreimanis 1989, 1990). The terms encompass glacial processes (e.g. lodgment, deformation), position of transport in the ice (e.g. basal, supraglacial), and depositional environment, especially related to subglacial hydrological conditions (e.g. meltout). For indicator tracing, recognition of depositional environments and debris transport processes, including topographically directed ice flow, glacial surging into glacial lakes, debris flow and iceberg rafting, furthers our capability to trace indicator debris (e.g. DiLabio 1979; Salonen 1988; Klassen & Murton 1996; Veillette & McClenaghan 1996). Sedimentological evidence can be obtained from natural exposures, including stream-cut sections, and excavations. Most exploration samples, however, originate from shallow pits and drill core, where structures and stratigraphic relations required for sediment classification can rarely be established. Where sediment facies cannot be distinguished, compositional differences among sites may record differences in subglacial process as well as provenance, and the interpretation of geochemical survey data is rendered more difficult.

An example of till facies recognition is illustrated by deformation till, a sediment type formed by shearing of unconsolidated sediment at the base of a warm-based ice sheet, which may be characteristic of ice streams (e.g. Benn & Evans 1996). Although subglacial shearing can affect a wide range of materials in the ice bed, it most likely occurs in fine-grained sediment having low porosity and high porewater pressure, such as glaciolacustrine and glaciomarine sediments overridden by ice, and in the fine-textured till derived from them (Boulton & Hindmarsh 1987; Hicock et al. 1989; Boulton 1996a, b). Deformation till is characterized by: (1) a fine-grained matrix and faceted clasts, reflecting intense abrasion and the incorporation of fine-grained older sediment by ice; (2) lack of structure, either due to shear attenuation and homogenization of sediments overridden by the ice, or to complex folding (Hicock & Dreimanis 1992); (3) a well-defined lower erosional contact reflecting the limit of shear deformation and contact with undeformed glacial sediment (decollement surface) (Clark 1991); and (4) over-consolidation (Alley 1991). Compared to lodgement till, deformation till is relatively thick, compositionally homogenous, and is characterized by significant glacial transport distance.

A terminology of relevance to the exploration community has been proposed by Finck & Stea (1995) to describe tills in terms of glacial process (Fig. 3). End-member till results from subglacial processes associated with a single glacial event, whereas hybrid till is a product of multiple events. Hybrid till composition results from either inheritance, where older glacial deposits are incorporated in younger till, or from overprinting, where later glacial events impress a compositional record on underlying, older glacial deposits, possibly through penetrative shear deformation. Palimpsest dispersal trains (Parent et al. 1996) occur where older trains are sources of indicator debris to later glacial events (e.g. Charbonneau & David 1993; Klassen & Thompson 1993; Parent et al. 1996). Multi-cycle dispersal trains comprise hybrid tills with a component inherited from an earlier advance and modified by later flow.

Glacial dispersal models

Basal transport

Indicator concentrations in till commonly decrease exponentially with distance down-ice from a source (e.g. Gillberg 1965; Shilts 1976; Clark 1987; Salonen 1987, 1992; Puranen 1990; Finck & Stea 1995; Parent et al. 1996; Klassen 1999) (Fig. 5a). Exponential dispersal profiles

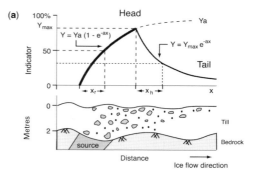

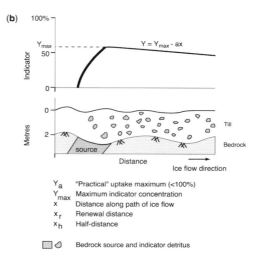

Fig. 5. Schematic profiles illustrating compositional variations of indicator erratics with distance of glacial transport and their expected distribution in till. (a) Exponential uptake and decay curves reflect erosion, modification, and deposition of debris transported at the base of the ice. (b) Linear decay reflects englacial transport with little or no modification of debris during transport. This profile may be characteristic of dispersal by ice streams where flow occurs by deformation in the ice bed (see Fig. 9, profile A–B).

are associated with transport either at or near the ice bed, where abrasion and comminution of clasts is most intense, and where there is ongoing erosion and deposition. The relation is expressed in the form:

$$y = y_0 \, e^{-a \, x}$$

where, y is the debris concentration at a point along the flow path, y_0 is the maximum concentration achieved either at or down-ice of the source ($\leq 100\%$), x is the distance of transport down-ice of the maximum, e is the base of the natural logarithm, and a is a constant reflecting the rate of decay (Gillberg 1965). Conversely, where ice flows across a source, indicator concentrations increase exponentially as the result of uptake by the ice (uptake curve; Peltoniemi 1985; Finck & Stea 1995). The width of the bedrock source, and the concentration of the indicator within it, both determine the maximum concentration of indicator debris in till, and its contrast with background (Puranen 1990).

The uptake and deposition curves give a net dispersal profile which rises rapidly to a head either over or directly down-ice of the source, and decreases in the direction of ice flow in a tail where indicator concentrations approach background concentrations (e.g. Shilts 1976). As noted by Shilts (1976), till in the tail segment is most likely to be sampled because it constitutes the most extensive part of the dispersal train. To maximize the spacing of samples in geochemical exploration, the faint signal of mineralization in the tail must be reliably distinguished from background noise related to geological and analytical variability.

The shape of the exponential curve, reflected by the value for constant a, is controlled by: (1) the velocity and duration of ice flow (Dyke 1984; Clark 1987; Bouchard & Salonen 1990; Aario & Peuraniemi 1992); (2) the physical properties of the source, including its areal extent, topographic exposure (Clark 1987; Salonen 1992) and its susceptibility to glacial erosion and comminution (Gillberg 1965); and (3) the balance between re-entrainment of older glacial debris and addition of new detritus from bedrock at the base of the ice sheet (Parent *et al.* 1996). It also reflects the intensity of glacial process associated with clast comminution, and hence ice flow velocity, sediment deformation, and water content.

From the exponential relation, glacial transport distances can be characterized in terms of a geometric mean or a half-distance (Perttunen 1977). The half-distance is the distance for maximum concentrations to decrease to half their value (Gillberg 1965) (Fig. 5a). Where peak concentrations are 100%, the term is synonymous with geometric mean (Bouchard & Salonen 1990). Half-distances range widely, but numerous reports indicate glacial debris has typically undergone limited transport, on the order of hundreds of metres to several kilometres (Clark 1987; Puranen 1988, 1990; Bouchard & Salonen 1989, 1990; Coker & DiLabio 1989; Charbonneau & David 1993; Finck & Stea 1995).

Ice sheet constructs (e.g. Boulton 1984,

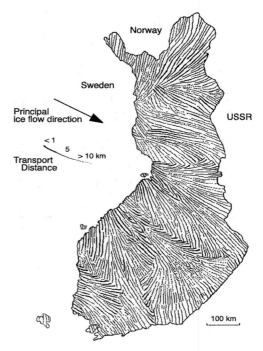

Fig. 6. Schematic characterization of boulder transport distances in Finland shown in the context of glacier flowlines. Glacial transport distance is reflected by line thickness. (Bouchard & Salonen 1990).

1996*b*) and dispersal models (e.g. Puranen 1988; Klassen & Thompson 1993; Finck & Stea 1995; Stea *et al.* 1989; Veillette & McClenaghan 1996) link drift composition, glacial transport, and landform-sediment associations to spatial and temporal variations in ice flow dynamics. Through field studies, glacial transport distances can differ between the distal and proximal sides of end moraines, the core and surface of landforms, and the marginal regions of the ice sheet and ice divides (Graves & Finck 1988; Puranen 1988, 1990; Bouchard & Salonen 1990; Aario & Peuraniemi 1992). In Finland, the bulk of till (>70 %) is transported 5–17 km in drumlins, 0.4–3 km in hummocky moraine, and 0.8–10 km for surface cover (Salonen 1987, 1988, 1992). Along flow lines in ice lobes, mean transport distances increase with increasing net distance down-ice, possibly reflecting change in transport from basal to englacial (Fig. 6).

Where an indicator source cannot be used to model glacial dispersal, the transport distance distribution (TDD) can be used to establish half-distance (Salonen 1988; Fig. 7). TDD is determined from a log-normal plot of the concentration of the lithological types present at a site

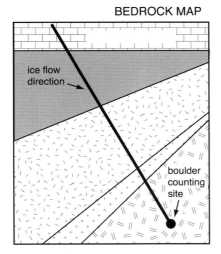

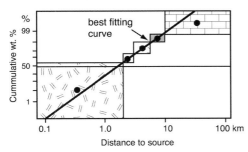

Fig. 7. Transport distance distribution (TDD) applied to the determination of the geometric mean of the glacial transport distance (half-distance). For each rock type along the path of flow, the concentration and distance to the source is plotted on a cumulative frequency graph. The estimate of transport distance can be used as a predictive tool for mineral exploration. (Bouchard & Salonen 1990).

against the distance to their bedrock source in the direction of ice flow. From a linear approximation, a half-distance characteristic of till in the region can be determined, and the transport distance for mineralized indicator erratics at the site estimated (Bouchard & Salonen 1989, 1990). To model glacial dispersal independent of outcrop width, the renewal distance, or distance of ice flow across a source required for an indicator concentration to reach 50%, can also be determined (Peltoniemi 1985).

Englacial transport

A second type of glacial dispersal profile shows linear decrease in indicator debris concentrations with increasing distance of glacial transport (Fig. 5b). The dispersal profile is flat or gently sloping with minimal change in indicator debris concentration with increasing distance of transport (e.g. Dyke 1984; Dyke & Prest 1987; Kaszycki 1989; Dyke *et al.* 1992; Thorleifson & Kristjansson 1993; Dredge 1995; Finck & Stea 1995). It is the product of englacial transport in which there is minimal to no modification of detritus during transport. The dispersal process can be linked to a conveyor belt, with no mixing between the ice bed and debris in transport, and may be characteristic of ice streams where ice flow is accommodated by deformation of the ice bed.

Ice flow velocity relates to the effectiveness of ice as an agent of dispersal, and ice streams are associated with well-defined plumes of far-travelled (> 50 km) debris. They are the dispersal mechanism responsible for carbonate-rich drift across large parts of northern Ontario and Manitoba, hundreds of kilometres from the bedrock source (e.g. Dredge 1988; Hicock *et al.* 1989; Thorleifson & Kristjansson 1993), and in arctic regions (Dyke *et al.* 1992). Marked compositional contrasts between areas affected by ice streams and adjacent areas where till is more closely linked to underlying bedrock (Fig. 8) can affect the interpretation of geochemical exploration at property scales.

Dispersal train shapes

The size, shape, orientation, and internal compositional variations of glacial dispersal trains reflect: (1) change in ice flow direction and provenance; and (2) distance of glacial transport related to either change in flow velocity or to debris position in the ice (i.e. englacial v. basal) (e.g. Boulton 1984, 1996*a, b*; Hansel *et al.* 1987). Dispersal trains have been described as ribbon, fan, and amoeboid (Shilts 1976; DiLabio 1990) (Fig. 9). Ribbons are aligned in a single ice flow direction, with a width comparable to the outcrop source, measured perpendicular to ice flow. Fans broaden down-ice; their outer margins are aligned in the two most divergent directions of ice flow. The maximum angle of fan divergence defines a probability sector for tracing indicators in an up-ice direction (Hirvas 1989). Near ice divides, where flow variation is marked and reversals of flow are possible, amoeboid dispersal trains of irregular shape extend in all directions about their source (e.g. Shilts 1976; Stea *et al.* 1989; Parent *et al.* 1996). Skip zones are areas of non-deposition along the path of ice flow (Finck & Stea 1995). Where debris is introduced from a topographic promi-

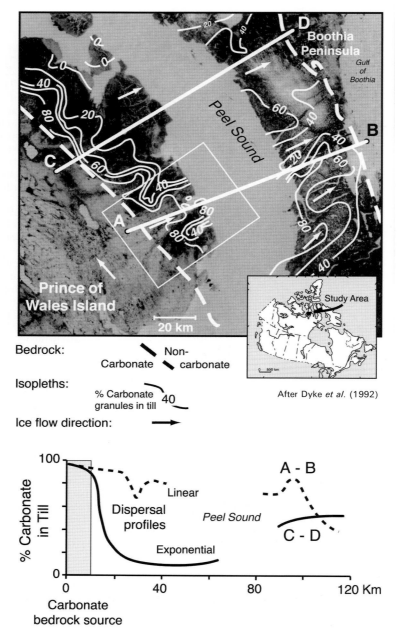

Fig. 8. Aerial photograph of part of Prince of Wales Island and Boothia Peninsula showing the distribution of carbonate-rich till (light-toned areas; carbonate isopleths shown). The location of a fomer ice stream extending eastward across Peel Sound is marked by a conspicuous plume of carbonate debris (line A–B). Profile A–B, at the bottom, shows a linear decrease in carbonate debris with increasing distance of glacial transport. Outside the inferred ice stream, carbonate concentrations decrease exponentially (profile C–D). (Dyke et al. 1992).

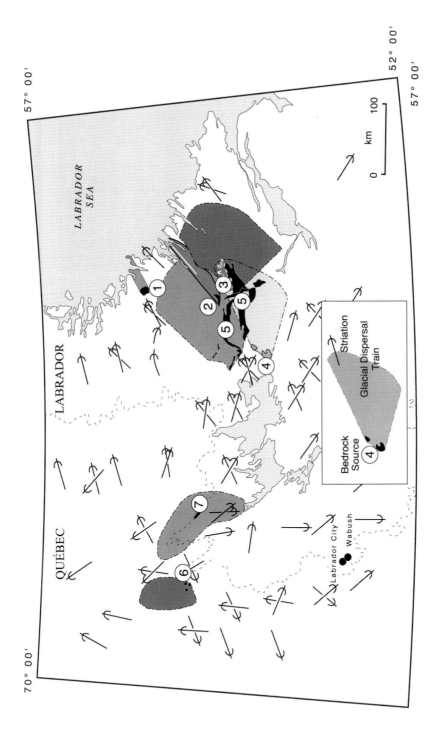

Fig. 9. The shape and orientation of glacial dispersal trains reflects their context in the ice sheet and ice flow history. Near the outer margins, where there was little change in ice flow direction, they are narrow ribbons (source 1). Where ice flow varied, they are fans whose lateral margins trend in the two most divergent ice flow directions (sources 2–6); across ice divide locations, fans open in opposing directions (compare sources 1–5 v. 6). In areas of ice divides where ice flow directions are complex, including reversal in flow, they are amoeboid, with debris distributed in all directions about the source (source 7) (Klassen & Thompson 1993).

nence, it is transported englacially and deposited farther down-ice at a point where it contacts the bed, for example at a topographic obstruction (Batterson 1989; Puranen 1990). Separation between the head of a train and its source also occurs by glacial erosion and removal of debris during later events (Stea *et al.* 1989; Parent *et al.* 1996), and where the train is covered by other glacial sediment. Topographic relief can also affect ice flow, especially during ablation as the ice sheet thins, resulting in local ice flow that is either around or otherwise constrained by obstacles (Gillberg 1965; Shilts 1976).

Vertical profiles

Indicator concentrations are typically greatest at the base of till over their source; in till sheets dispersal trains commonly rise in a down-ice direction at a low angle of inclination (DiLabio 1990) (Fig. 5a). The vertical rise is attributed to continued erosion of the source and addition of debris to the base of the ice sheet along the flowpath (Puranen 1988, 1990), and to upward shear diffusion in a deforming bed (Charbonneau & David 1995). For exploration, the horizontal separation between the bedrock source and the first occurrence of mineralized debris at the surface represents a gap that must be prospected in the subsurface by trenching or drilling. The separation distance tends to increase with till thickness.

Glacial process and partitioning

In addition to distance and direction of glacial transport, glacial processes affect the concentration and distribution of indicators among size fractions. Rocks and minerals are partitioned to varying degrees among size fractions, with the degree of partitioning dependent on the duration and intensity of subglacial processes associated with abrasion and crushing, and the mechanical properties of the detritus (Shilts 1971, 1993, 1995; Haldorsen 1978, 1983; Nevalainen 1989; DiLabio 1995). Hence, the definition of glacial dispersal trains depends on the physical and geochemical properties of the indicator, glacial processes, and the size fraction chosen for analyses. Net transport distances can also differ among size fractions. Outside the source terrain, milling of coarse glacial detritus to finer fractions alters the relative concentrations of indicator debris among fractions, lengthening the tail segment of dispersal trains in finer fractions (Peltoniemi 1985; Klassen 1999). Where indicator minerals or elements concentrate in a particular size fraction, the probability of identifying them at low overall concentration is enhanced by separation and analyses of that fraction. Hence, the choice of size fraction is important to drift prospecting and the interpretation of geochemical analyses.

In Shield-derived tills, rock fragments occur as pebble and coarser-sized fractions (>2 mm), mineral grains occur as sand (2–0.063 mm), and both mineral grains and grain fragments occur as silt and clay (<0.063 mm) (Haldorsen 1983; Shilts 1993, 1995; DiLabio 1995). Where detritus is derived from sedimentary or volcanic rock, which tend to be softer and finer-grained than plutonic and metamorphic rock, rock fragments comprise a greater component of the sand fraction and partitioning of minerals among finer sizes is less pronounced because they are silt-sized and finer in the bedrock source. With increasing transport distance, the minerals achieve a terminal grade size resistant to further comminution that depends on their original size and physical properties affecting their resistance to glacial comminution (Dreimanis & Vagners 1971; Haldorsen 1983). Quartz and feldspar, for example, typically concentrate in the sand and silt-sized fractions, whereas phyllosilicates and amphiboles, which are softer and more readily cleaved, concentrate in the silt and clay-sized fractions (Nevalainen 1989).

As the result of mineralogical partitioning, some metals, including copper, lead, and zinc, preferentially concentrate in the clay-sized fraction to levels that can be one or more orders of magnitude greater than in coarser fractions (Nikkarinen *et al.* 1984; Salminen 1992; DiLabio 1995; Shilts 1995). The geochemical differences reflect either the preferential concentration of metal-rich phyllosilicate minerals in the clay-sized fraction, or of metal-poor quartz and feldspar in silt-sized and coarser fractions, or both. They have also been attributed to differences in mineral surface areas among fractions, assuming that metals are bound to surface sites (Horowitz 1991)

Strong geochemical correlations between coarse (<2 mm) and fine (<0.063 mm) fractions of till can indicate a common provenance (McClenaghan 1992; Tarvainen 1995). In addition to glacial partitioning, compositional differences among size fractions can also reflect provenance, and the incorporation of preglacial deposits, including glaciolacustrine sediments and weathered regolith (Nikkarinen *et al.* 1984; Peuraniemi 1991; Räisänen *et al.* 1992; Salminen 1992; Finck & Stea 1995; Mäkinen 1995). Provenance determined from gravel lithology can aid the interpretation of geochemical and mineralogical analyses of finer fractions (Taipale et al. 1986; Minell 1990; Hornibrook et al. 1991; McClenaghan et al. 1992).

Where a source preferentially contributes to the fine fraction, or has distinctive geochemical properties, or both, it can have a disproportionate effect on till geochemistry (Vallius 1993). Till derived from soft ultramafic rock leads to prominent geochemical anomalies for chromium and nickel extending tens to hundreds of kilometres down-ice from their source (Shilts 1976, 1984). In contrast, till derived from rock impoverished in trace metals, such as carbonate and quartzite, can lead to negative geochemical dispersal trains (Klassen & Shilts 1977; Hicock et al. 1989; Kaszycki 1989; Mäkinen 1992; Vallius 1993; Dredge 1995; Eden & Bjorklund 1995). In both cases the trains mask the compositional expression of underlying bedrock and pose a challenge for geochemical exploration. In general, geochemically-defined dispersal trains are typically more limited (hundreds of metres) in their down-ice extent than those indicated by heavy mineral or erratic indicators (Lehmuspelto 1987; Kokkola 1989; Balzer & Broster 1994).

Two approaches distinguish geochemical variations related to texture and mineral partitioning from those related to provenance. The Geological Survey of Canada has used the clay-sized fraction for geochemical analysis (e.g. Shilts 1971, 1973, 1977, 1984; Lindsay & Shilts 1995). Due to its fine grain size, the mineralogy of the clay-sized minerals is least likely to change through either subglacial or postglacial winnowing (Shilts 1971, 1973; Haldorson 1983; Rampton 2000), and its geochemical properties are less likely to be affected by change in diluent minerals like quartz and feldspar which have a silt-sized terminal grade. Another approach uses geochemical ratios (Mäkinen 1995). Where elements share a common mineralogical control, their ratios can be independent of total concentrations, and hence unaffected by mineral partitioning among size fractions.

Concluding remarks

For mineral exploration, ice sheet and glacial dispersal models provide a predictive basis for drift prospecting and a context for interpreting analytical results. They serve as guides to establish effective exploration strategies in the context of sediment-landform associations and subglacial process. The models are continually revised by reference to the physical properties of the ice bed, and to the empirical field evidence for glacial dispersal, including indicator erratics and the ice flow record of striations and streamlined landforms. Glacial processes and depositional environments inferred from principles of glacial sedimentology enhance our ability to trace particle distributions in glacial sediments, and to distinguish the signal of economic indicators in the context of natural background variations.

The author has drawn freely from the works of numerous authors, and acknowledges the significant contributions of past and present coworkers at the Geological Survey of Canada, notably W. W. Shilts and R. N. W. DiLabio. The manuscript has benefited from critical reviews by B. McClenaghan, and by J. Clague and A. Dixon-Warren whose thoughtful comments have led to major revision and significant improvement. Geological Survey of Canada Contribution No. 1999204.

References

AARIO, R. & PEURANIEMI, V. 1992. Glacial dispersal of till constituents in morainic landforms of different types. *Geomorphology*, **6**, 9–25.

ALLEY, R. B. 1991. Deforming-bed origin for the southern Laurentide till sheets? *Journal of Glaciology*, **37**, 67–76.

AYLSWORTH, J. A. & SHILTS, W. W. 1989a. *Glacial features around the Keewatin Ice Divide: Districts of Mackenzie and Keewatin*. Geological Survey of Canada, Paper **88-24**.

AYLSWORTH, J. A. & Shilts, W. W. 1989b. Bedforms of the Keewatin Ice Sheet, Canada. *Sedimentary Geology*, **62**, 407–428.

BALZER, S. A. & BROSTER, B. E. 1994. Comparison of clast and matrix dispersal in till: Canterbury area, New Brunswick. *Atlantic Geology*, **30**, 9–17.

BATTERSON, M. J. 1989. Glacial dispersal from the Strange Lake alkalic complex, northern Labrador. *In:* DILABIO, R. N. W. & COKER, W. B. (eds) *Drift Prospecting*. Geological Survey of Canada, Paper **89-20**, 31–39.

BENN, D. I. & EVANS, D. J. A. 1996. The interpretation and classification of subglacially deformed materials. *Quaternary Science Reviews*, **15**, 22–52.

BENTLEY, C. R. 1987. Antarctic ice streams: a review. *Journal of Geophysical Research*, **92**, 8843–8858.

BOUCHARD, M. A. & MARTINEAU, G. 1985. Southeastward ice flow in central Quebec and its paleogeographic significance. *Canadian Journal of Earth Sciences*, **22**, 1536–1541.

BOUCHARD, M. A. & SALONEN, V.-P. 1989. Glacial dispersal of boulders in the James Bay lowlands of Québec, Canada. *Boreas*, **18**, 189–199.

BOUCHARD, M. A. & SALONEN, V.-P. 1990. Boulder transport in shield areas. *In:* KUJANSUU, R. & SAARNISTO, M. (eds) *Glacial Indicator Tracing*. A. A. Balkema, Rotterdam, 87–107.

BOULTON, G. S. 1984. Development of a theoretical model of sediment dispersal by ice sheets. *In:* GALLAGHER, M. J. (ed.) *Prospecting in Areas of Glaciated Terrain 1984*. Institution of Mining and

Metallurgy, 213–223.

BOULTON, G. S. 1996a. The origin of till sequences by subglacial sediment deformation beneath mid-latitude ice sheets. *Annals of Glaciology*, **22**, 75–84.

BOULTON, G. S. 1996b. Theory of glacial erosion, transport and deposition as a consequence of subglacial sediment deformation. *Journal of Glaciology*, **42**, 43–62.

BOULTON, G. S. & CLARK, C. D. 1990a. A highly mobile Laurentide Ice Sheet revealed by satellite images of glacial lineations. *Nature*, **346**, 813–817.

BOULTON, G. S. & CLARK, C. D. 1990b. The Laurentide ice sheet through the last glacial cycle: the topology of drift lineations as a key to the dynamic behaviour of former ice sheets. *Transactions of the Royal Society of Edinburgh: Earth Sciences*, **81**, 327–347.

BOULTON, G. S. & HINDMARSH, R. C. A. 1987. Sediment deformation beneath glaciers: rheology and geological consequences. *Journal of Geophysical Research*, **92**, 9059–9082.

BOULTON, G. S., SMITH, G. D., JOHN, A. S. & NEWSOME, J. 1985. Glacial geology and glaciology of the last mid-latitude ice sheets. *Journal of the Geological Society of London*, **142**, 447–474.

BOYCE, J. I. & EYLES, N. 1991. Drumlins carved by deforming ice streams below the Laurentide ice sheet. *Geology*, **19**, 787–790.

BOYLE, D. R. 1996. Supergene base metals and precious metals. *In*: ECKSTRAND, O. R., SINCLAIR, W. D. & THORPE, R. I. (eds) *Geology of Canadian Mineral Deposit Types*. Geological Survey of Canada, Geology of Canada, **8**, 92–108.

BUDD, W. F. & SMITH, I. N. 1987. Conditions for growth and retreat of the Laurentide Ice Sheet. *Géographie physique et Quaternaire*, **41**, 279–290.

CHARBONNEAU, R. & DAVID, P. P. 1993. Glacial dispersal of rock debris in central Gaspésie, Quebec, Canada. *Canadian Journal of Earth Sciences*, **30**, 1697–1707.

CHARBONNEAU, R. & DAVID, P. P. 1995. A shear-diffusion model of till genesis based on the dispersal pattern of indicator rocks in the Grand-Volume Till of central Gaspésie, Québec, Canada. *Boreas*, **24**, 281–292.

CLARK, C. D. 1993. Mega-scale glacial lineations and cross-cutting ice-flow landforms. *Earth Surface Processes and Landforms*, **18**, 1–29.

CLARK, P. U. 1987. Subglacial sediment dispersal and till composition. *Journal of Geology*, **95**, 527–541.

CLARK, P. U. 1991. Striated clast pavements: Product of deforming subglacial sediment? *Geology*, **9**, 530–533.

COKER, W. B. & DILABIO, R. N. W. 1989. Geochemical exploration in glaciated terrain: geochemical responses. *In*: GARLAND, G. D. (ed.) *Proceedings of Exploration '87; Third Decennial International Conference on Geophysical and Geochemical Exploration for Minerals and Groundwater*, Ontario Geological Survey, Toronto, 336–383.

DILABIO, R. N. W. 1979. Drift prospecting in uranium and base-metal mineralization sites, District of Keewatin, Northwest Territories, Canada. *In*: *Prospecting in Areas of Glaciated Terrain*. Institution of Mining and Metallurgy, Dublin, 91–100.

DILABIO, R. N. W. 1990. Glacial dispersal trains. *In*: KUJANSUU, R. & SAARNISTO, M. (eds) *Glacial Indicator Tracing*. A. A. Balkema, Rotterdam, 109–122.

DILABIO, R. N. W. 1995. Lithological analysis in drift prospecting studies. *In*: BOBROWSKY, P. T., SIBBICK, S. J., NEWELL, J. H. & MATYSEK, P. F. (eds) *Drift Exploration in the Canadian Cordillera*. British Columbia Ministry of Energy, Mines and Petroleum Resources, Paper **1995-2**, 139–148.

DILABIO, R. N. W., MILLER, R. F., MOTT, R. J. & COKER, W. B. 1988. *The Quaternary stratigraphy of the Timmins area, Ontario, as an aid to mineral exploration by drift prospecting*. Geological Survey of Canada, Paper **88-1C**, 61–65.

DREDGE, L. A. 1988. Drift carbonate on the Canadian Shield. II: Carbonate dispersal and ice-flow patterns in northern Manitoba. *Canadian Journal of Earth Sciences*, **25**, 783–787.

DREDGE, L. A. 1995. *Quaternary geology, of northern Melville Peninsula, District of Franklin; Northwest Territories: surface deposits, glacial history, environmental geology, and till geochemistry*. Geological Survey of Canada, Bulletin **484**.

DREIMANIS, A. 1989. Tills: their genetic terminology and classification. *In*: Goldthwait, R. P. & Matsch, C. L. (eds) *Genetic Classification of Glacigenic Deposits*. A. A. Balkema, Rotterdam, 17–83.

DREIMANIS, A. 1990. Formation, deposition, and identification of subglacial and supraglacial tills. *In*: KUJANSUU, R. & SAARNISTO, M. (eds) *Glacial Indicator Tracing*. A. A. Balkema, Rotterdam, 35–59.

DREIMANIS, A. & VAGNERS, U. 1971. The dependence of the composition of till upon the rule of bimodal composition. *In*: *Études sur le Quaternaire dans le Monde 2*. Seventh Congress, International Conference for Quaternary Research, Paris, 787–789.

DYKE, A. S. 1984. *Quaternary geology of Boothia Peninsula and northern District of Mackenzie, central Canadian Arctic*. Geological Survey of Canada, Bulletin **407**.

DYKE, A. S. 1993. Landscapes of cold-centred Late Wisconsinan ice caps, Arctic Canada. *Progress in Physical Geography*, **17**, 223–247.

DYKE, A. S. & MORRIS, T. F. 1988. Drumlin fields, dispersal trains, and ice streams in Arctic Canada. *The Canadian Geographer*, **32**, 86–90.

DYKE, A. S. & PREST, V. K. 1987. Late Wisconsinan and Holocene history of the Laurentide Ice Sheet. *Géographie physique et Quaternaire*, **41**, 237–263.

DYKE, A. S., MORRIS, T. F., GREEN, D. E. C. & ENGLAND, J. 1992. *Quaternary geology of Prince of Wales Island, Arctic Canada*. Geological Survey of Canada, Memoir **433**.

EDEN, P. & BJORKLUND, A. 1995. Geochemistry of till in Fennoscandia from ultra-low density sampling. *Journal of Geochemical Exploration*, **52**, 285–302.

FENTON, M. M. 1984. Quaternary stratigraphy of the Canadian Prairies. *In*: FULTON, R. J. (ed.) *Quaternary Stratigraphy of Canada*. Geological Survey of Canada, Paper **84-10**, 57–68.

FINCK, P. W. & Stea, R. R. 1995. *The compositional*

development of tills overlying the South Mountain Batholith, Nova Scotia. Nova Scotia Department of Natural Resources, Paper **95-1**.

FISHER, D. A., REEH, N. & LANGLEY, K. 1985. Objective reconstructions of the Late Wisconsinan Laurentide Ice Sheet and the significance of deformable beds. *Géographie physique et Quaternaire*, **39**, 229–238.

FLINT, R. F. 1971. *Glacial and Quaternary Geology*. John Wiley & Sons, New York.

FULTON, R. J. 1989. Quaternary geology of the Canadian Shield. *In*: FULTON, R. J. (ed.) *Quaternary Geology of Canada and Greenland*, Geological Survey of Canada Geology of Canada, **1**, 177–317.

FULTON, R. J. (compiler) 1995. *Surficial Materials of Canada*. Geological Survey of Canada Map **1880A**.

GILLBERG, G. 1965. Till distribution and ice movements on the northern slopes of the south Swedish highlands. *Geologiska Foreningens i Stockholm Förhandlingar*, **86**, 433–484.

GRAVES, R. M. & FINCK, P. W. 1988. The provenance of tills overlying the eastern part of the South Mountain batholith, Nova Scotia. *Maritime Sediments and Atlantic Geology*, **24**, 61–70.

HALDORSEN, S. 1978. Glacial comminution of mineral grains. *Norsk Geologisk Tidsskrift*, **58**, 241–243.

HALDORSEN, S. 1983. Mineralogy and geochemistry of basal till and their relationship to till-forming processes. *Norsk Geologisk Tidsskrift*, **63**, 15–25.

HANSEL, A. K., JOHNSON, W. H. & SOCHA, B. J. 1987. Sedimentological characteristics and genesis of basal tills at Wedron, Illinois. *In*: KUJANSUU, R. & SAARNISTO, M. (eds) *INQUA Till Symposium, Finland 1989*. Geological Survey of Finland, Special Paper **3**, 11–21.

HICOCK, S. R. & DREIMANIS, A. 1992. Deformation till in the Great Lakes region: implications for rapid flow along the south-central margin of the Laurentide Ice Sheet. *Canadian Journal of Earth Sciences*, **29**, 1565–1579.

HICOCK, S. R., KRISTJANSSON, F. J. & SHARPE, D. R. 1989. Carbonate till as a soft bed for Pleistocene ice streams on the Canadian Shield north of Lake Superior. *Canadian Journal of Earth Sciences*, **26**, 2249–2254.

HIRVAS, H. 1989. Application of glacial geological studies in prospecting, Finland. *In*: DILABIO, R. N. W. & COKER, W. B. (eds) *Drift Prospecting*. Geological Survey of Canada, Paper **89-20**, 1–6.

HIRVAS, H. 1991. *Pleistocene stratigraphy of Finnish Lapland*. Geological Survey of Finland, Bulletin **354**.

HORNIBROOK, E. R. C., BROSTER, B. E., GARDINER, W. W. & PRONK, A. G. 1991. Glacial dispersal of heavy minerals in Late Wisconsinan till, central New Brunswick. *Atlantic Geology*, **27**, 199–208.

HOROWITZ, A. J. 1991. *A Primer on Sediment-Trace Element Chemistry*. Lewis Publishers, Chelsea, Michigan.

KASZYCKI, C. A. 1989. *Surficial geology and till composition, northwestern Manitoba*. Geological Survey of Canada, Open File **2118**.

KASZYCKI, C. A. & SHILTS, W. W. 1980. *Glacial erosion of the Canadian Shield – calculation of average depths*. Atomic Energy of Canada, Technical Record **TR-106**.

KAURANNE, L. K. 1959. Pedogeochemical prospecting in glaciated terrain. Geological Survey of Finland, Bulletin **184**, 1–10.

KLASSEN, R. A. 1999. The application of glacial dispersal models to till geochemistry in Labrador, Canada. *Journal of Geochemical Exploration*, **67**, 245–269.

KLASSEN, R. A. & MURTON, J. B. 1996. Quaternary geology of the Buchans area, Newfoundland: implications for mineral exploration. *Canadian Journal of Earth Sciences*, **33**, 363–377.

KLASSEN, R. A. & SHILTS, W. W. 1977. Glacial dispersal of uranium in the District of Keewatin, Canada. *In*: JONES, M. J. (ed.) *Prospecting in Areas of Glaciated Terrain*. Institution of Mining and Metallurgy, Espôo, Finland, 80–88.

KLASSEN, R. A. & THOMPSON, F. J. 1993. *Glacial history, drift composition, and mineral exploration, central Labrador*. Geological Survey of Canada, Bulletin **435**.

KLEMAN, J. 1990. On the use of glacial striae for reconstruction of paleo-ice sheet flow patterns – with application to the Scandinavian ice sheet. *Geografiska Annaler*, **72A**, 217–236.

KLEMAN, J. 1994. Preservation of landforms under ice sheets and ice caps. *Geomorphology*, **9**, 19–32.

KOKKOLA, M. 1989. Is till matrix transported-or is the way to its study wrong? *In*: PERTTUNEN, M. (ed.) *Transport of Glacial Drift in Finland*. Geological Survey of Finland, Special Paper **7**, 55–58.

KOR, P. S. G. & COWELL, D. W. 1998. Evidence for catastrophic subglacial meltwater sheet flood events on the Bruce Peninsula. *Canadian Journal of Earth Sciences*, **35**, 1180–1202.

KOR, P. S. G., SHAW, J. & SHARPE, D. R. 1991. Erosion of bedrock by subglacial meltwater, Georgian Bay, Ontario: a regional view. *Canadian Journal of Earth Sciences*, **28**, 623–642.

LEHMUSPELTO, P. 1987. Some case histories of the till transport distances recognized in geochemical studies in northern Finland. *In*: KUJANSUU, R. & SAARNISTO, M. (eds) *INQUA Till Symposium, Finland 1985*. Geological Survey of Finland, Special Paper **3**, 163–168.

LINDSAY, P. J. & SHILTS, W. W. 1995. A standard laboratory procedure for separating clay-sized detritus from unconsolidated glacial sediments and their derivatives. *In*: BOBROWSKY, P. T., SIBBICK, S. J., NEWELL, J. M. & MATYSEK, P. F. (eds) *Drift Exploration in the Canadian Cordillera*. B.C. Ministry of Energy, Mines and Petroleum Resources, Paper **1995-2**, 165–166.

LUNDQVIST, J. 1990. Glacial morphology as an indicator of the direction of ice flow. *In*: KUJANSUU, R. & SAARNISTO, M. (eds) *Glacial Indicator Tracing*. A. A. Balkema, Rotterdam, 61–70.

MÄKINEN, J. 1992. *The relation between unit weight and geochemical and mineralogical compositions in the fine fraction of till*. Geological Society of Finland, Bulletin **64**, 59–74.

MÄKINEN, J. 1995. Effects of grinding and chemical

factors on the generation and composition of the till fine fraction: an experimental study. *Journal of Geochemical Exploration*, **54**, 49–62.

MARSHALL, S. J., CLARKE, G. K., DYKE, A. S. & FISHER, D. A. 1996. Geologic and topographic controls on fast flow in the Laurentide and Cordilleran ice sheets. *Journal of Geophysical Research*, **101**, 17,827–17,839.

MCCLENAGHAN, M. B. 1992. Surface till geochemistry and implications for exploration, Black River-Matheson area, northeastern Ontario. *Exploration Mining Geology*, **1**, 327–337.

MCCLENAGHAN, M. B., LAVIN, O. P., NICHOL, I. & SHAW, J. 1992. Geochemistry and clast lithology as an aid to till classification, Matheson, Ontario, Canada. *Journal of Geochemical Exploration*, **42**, 237–260.

MINELL, H. 1990. The part of Quaternary geology in uranium prospecting. *Striae*, **29**, 79–83.

NEVALAINEN, R. 1989. Lithology of fine till fractions in the Kuhmo greenstone belt area, eastern Finland. *In*: PERTTUNEN, M. (ed.) *Transport of glacial drift in Finland*. Geological Survey of Finland, Special Paper, **7**, 59–65.

NICHOL, I., LAVIN, O. P., MCCLENAGHAN, M. B. & STANLEY, C. R. 1992. The optimization of geochemical exploration for gold using glacial till. *Exploration Mining Geology*, **1**, 305–326.

NIKKARINEN, M., KALLIO, E., LESTINEN, P. & AYRAS, M. 1984. Mode of occurrence of Cu and Zn in till over three mineralized areas in Finland. *Journal of Geochemical Exploration*, **21**, 239–247.

PARENT, M., PARADIS, S. J. & DOIRON, A. 1996. Palimpsest glacial dispersal trains and their significance for drift prospecting. *Journal of Geochemical Exploration*, **56**, 123–140.

PELTONIEMI, H. 1985. Till lithology and glacial transport in Kuhmo, eastern Finland. *Boreas*, **14**, 67–74.

PERTTUNEN, M. 1977. *The lithologic relation between till and bedrock in the region of Hameenlinna, southern Finland*. Geological Survey of Finland, Bulletin **291**.

PEURANIEMI, V. 1991. Geochemistry of till and humus in the Kotkajärvi Cu-Co-Au prospect, Kalvola, southern Finland. *In*: BJÖRKLUND, A. J. (ed.) Gold Geochemistry in Finland. *Journal of Geochemical Exploration*, **39**, 363–378.

PREST, V. K. 1983. *Canada's Heritage of GlacialFeatures*. Geological Survey of Canada, Miscellaneous Report **28**.

PREST, V. K. 1990. Laurentide ice-flow patterns: a historical review, and implications of the dispersal of Belcher Island erratics. *Géographie physique et Quaternaire*, **44**, 113–136.

PREST, V. K., GRANT, D. R. & RAMPTON, V. N. 1968. *Glacial Map of Canada*. Geological Survey of Canada Map **1253A**.

PRONK, A. G., BOBROWSKY, P. T. & PARKHILL, M. A. 1989. An interpretation of Late Quaternary glacial flow indicators in the Baie des Chaleurs region, northern New Brunswick. *Géographie physique et Quaternaire*, **43**, 179–190.

PURANEN, R. 1988. *Modelling of glacial transport of basal tills in Finland*. Geological Survey of Finland, Report of Investigation **81**.

PURANEN, R. 1990. Modelling of glacial transport of tills. *In*: KUJANSUU, R. & SAARNISTO, M. (eds) *Glacial Indicator Tracing*. A. A. Balkema, Rotterdam, 15–34.

RÄISÄNEN, M. L., TENHOLA, M. & MÄKINEN, J. 1992. Relationship between mineralogy and the physicochemical properties of till in central Finland. Geological Society of Finland, Bulletin **64**, 35–58.

RAMPTON, V. N. 2000. Large-scale meltwater effects of subglacial meltwater flow in the southern Slave Province, Northwest Territories, Canada. *Canadian Journal of Earth Sciences*, **37**, 81–93.

RAPPOL, M. 1989. Glacial history and stratigraphy of northwestern New Brunswick. *Géographie physique et Quaternaire*, **43**, 191–206.

SALMINEN, R. 1992. Geochemical Dispersion in the Secondary Environment. *In*: KAURANNE, L. K., SALMINEN, R. & ERIKSSON, K. (eds) *Regolith Exploration Geochemistry in Arctic and Temperate Terrains*. Handbook of Exploration Geochemistry, **5**. Elsevier, Amsterdam, 93–125.

SALONEN, V.-P. 1987. Observations on boulder transport in Finland. *In*: KUJANSUU, R. & SAARNISTO, M. (eds) *INQUA Till Symposium*. Geological Survey of Finland, Special Paper **3**, 103–110.

SALONEN, V.-P. 1988. Application of glacial dynamics, genetic differentiation of glacigenic deposits and their landforms to indicator tracing in the search for ore deposits. *In*: GOLDTHWAIT, R. P. & MATSCH, C. L. (eds) *Genetic Classification of Glacigenic Deposits*. A. A. Balkema, Rotterdam, 183–190.

SALONEN, V.-P. 1992. Glacigenic dispersion of coarse till fragments. *In*: KAURANNE, L. K., SALMINEN, R. & ERIKSSON, K. (eds) *Regolith Exploration Geochemistry in Arctic and Temperate Terrains*. Handbook of Exploration Geochemistry, **5**. Elsevier, Amsterdam, 127–142.

SHAW, J. 1990. A qualitative view of sub-ice-sheet landscape evolution. *Progress in Physical Geography*, **18**, 159–184.

SHAW, J., RAINS, B., EYTON, R. & WEISSLING, L. 1996. Laurentide subglacial outburst floods: landform evidence from digital elevation models. *Canadian Journal of Earth Sciences*, **33**, 1154–1168.

SHILTS, W. W. 1971. Till studies and their application to regional drift prospecting. *Canadian Mining Journal*, **92**, 45–50.

SHILTS, W. W. 1973. *Drift prospecting, geochemistry of eskers and till in permanently frozen terrain: District of Keewatin, Northwest Territories*. Geological Survey of Canada, Paper **72-45**.

SHILTS, W. W. 1976. Glacial till and mineral exploration. *In*: LEGGET, R. F. (ed.) *Glacial Till: A Symposium*. Royal Society of Canada, Special Publication **12**, 205–224.

SHILTS, W. W. 1977. Geochemistry of till in perennially frozen terrain of the Canadian Shield – application to prospecting. *Boreas*, **5**, 203–212.

SHILTS, W. W. 1979. Flow patterns in the central North American ice sheet. *Nature*, **286**, 213–218.

SHILTS, W. W. 1984. Till geochemistry in Finland and Canada. *Journal of Geochemical Exploration*, **21**, 95–117.

SHILTS, W. W. 1993. Geological Survey of Canada's contributions to understanding the composition of glacial sediments. *Canadian Journal of Earth Sciences*, **30**, 333–353.

SHILTS, W. W. 1995. Geochemical partitioning in till. *In*: BOBROWSKY, P. T., SIBBICK, S. J., NEWELL, J. M. & MATYSEK, P. F. (eds) *Drift Exploration in the Canadian Cordillera*. B.C. Ministry of Energy, Mines and Petroleum Resources, Paper **1995-2**, 149–166.

SHILTS, W. W. 1996. Drift exploration. *In*: MENZIES, J. (ed.) *Past Glacial environments, sediment forms, and techniques*. Butterworth Heinemann, Toronto, 411–439.

SHILTS, W. W. SMITH, S. L. 1989. Drift prospecting in the Appalachians of Estrie-Beauce, Quebec. *In*: DILABIO, R. N. W. & COKER, W. B. (eds) *Drift Prospecting*. Geological Survey of Canada, Paper **89-20**, 41–59.

SMITH, S. L. 1992. *Quaternary stratigraphic drilling transect, Timmins to the Moose River Basin, Ontario*. Geological Survey of Canada, Bulletin **415**.

STEA, R. R. 1994. Relict and palimpsest glacial landforms in Nova Scotia, Canada. *In*: WARREN, W. P. & CROOT, D. G. (eds) *Formation and Deformation of Glacial Deposits*. A. A. Balkema, Rotterdam, 141–158.

STEA, R. R., TURNER, R. G., FINCK, P. W. GRAVES, R. M. 1989. Glacial dispersal in Nova Scotia: a zonal concept. *In*: DILABIO, R. N. W. & COKER, W. B. (eds) *Drift Prospecting*. Geological Survey of Canada, Paper **89-20**, 155–169.

SUGDEN, D. E. 1977. Reconstruction of the morphology, dynamics, and thermal characteristics of the Laurentide Ice Sheet at its maximum. *Arctic and Alpine Research*, **9**, 21–47.

SUGDEN, D. E. 1978. Glacial erosion by the Laurentide Ice Sheet. *Journal of Glaciology*, **20**, 367–391.

TAIPALE, K., NEVALAINEN, R. & SAARNISTO, M. 1986. Silicate analysis and normative compositions of the fine fraction of till: examples from eastern Finland. *Journal of Sedimentary Petrology*, **56**, 370–378.

TARVAINEN, T. 1995. The geochemical correlation between coarse and fine fractions of till in southern Finland. *Journal of Geochemical Exploration*, **54**, 187–198.

THORLEIFSON, L. H. & KRISTJANSSON, F. J. 1993. *Quaternary geology and drift prospecting, Beardmore-Geralton area, Ontario*. Geological Survey of Canada, Memoir **435**.

THORLEIFSON, L. H., WYATT, P. H. & WARMAN, T. A. 1993. *Quaternary stratigraphy of the Severn and Winisk drainage basins, northern Ontario*. Geological Survey of Canada, Bulletin **442**.

VALLIUS, H. 1993. Technical notes: Regional geochemical relation between bedrock and till in Lake Päijänne area, southern Finland. *Institution of Mining and Metallurgy, Transactions, Section B; Applied Earth Science*, **102**, B48–B49.

VEILLETTE, J. J. 1986. Former southwesterly ice flows in the Abitibi-Timiskaming region: implications for the configuration of the late Wisconsinan ice sheet. *Canadian Journal of Earth Sciences*, **23**, 1724–1741.

VEILLETTE, J. J. & MCCLENAGHAN, M. B. 1996. *Sequence of glacial ice flows in Abitibi-Timiskamming: implications for mineral exploration and dispersal of calcareous rocks from the Hudson bay Basin, Quebec and Ontario*. Geological Survey of Canada, Open File **3033**.

VEILLETTE, J. J. ROY, M. 1995. *The spectacular cross-striated outcrops of James Bay, Quebec*. Geological Survey of Canada, Paper **1995-C**, 243–248.

VEILLETTE, J. J., DYKE, A. S., & ROY, M. 1999. Ice-flow evolution of the Labrador Sector of the Laurentide Ice Sheet: a review, with new evidence from northern Quebec. *Quaternary Science Reviews*, **18**, 993–1019.

Till geochemistry and sampling techniques in glaciated shield terrain: a review

ISABELLE McMARTIN & M. BETH McCLENAGHAN

Geological Survey of Canada, 601 Booth Street, Ottawa, Ontario K1A 0E8, Canada
(e-mail: imcmarti@nrcan.gc.ca)

Abstract: Till is a favoured sample medium for locating mineral deposits in glaciated shield terrains of Canada and Fennoscandia because it best reflects the primary composition of the bedrock source area. In the sampling phase, an important and costly component of till surveys, sample density, sample depth and sample method must be chosen according to the needs of the exploration program. Surface till sampling methods in forested areas differ from those used in permafrost terrain. However, in both areas, concentrations of labile ore minerals and their products of decomposition can be detected in the fine fraction (<2 mm) of weakly oxidized till. In thin drift-covered areas, till samples are collected by hand excavation or trenching at <5 m depth. In areas of thicker drift, more expensive methods such as reverse circulation rotary drills, rotasonic drills and portable drills are used to collect till samples at depth and to determine lateral and vertical variations in till geochemistry. Laboratory methods are an essential part of till geochemical surveys. The choices of the size fraction and analytical methods are determined by the nature and composition of the expected bedrock target, and by costs.

Materials having an intimate relation to their source are the most suitable for geochemical exploration (Rose *et al.* 1979; Levinson 1980). In glaciated terrain, drift prospecting is an integral part of mineral exploration because of the widespread cover of till which is a first-cycle sediment directly deposited by glacier ice (e.g. Shilts 1984; Coker & DiLabio 1989; Kujansuu & Saarnisto 1990; Klassen 1997). Till is composed of freshly crushed bedrock blended with re-worked older sediments, and can be transported from a few metres up to many kilometres along a preferred orientation related to the ice flow history. Thus, drift sampling, combined with striation mapping and boulder tracing, has become an important tool for tracing base and precious metal deposits and, more recently, diamonds on the Canadian and Fennoscandian Shields.

Mineral exploration in drift covered terrain is often driven by the need to locate the bedrock source of an indicator erratic. In till, the particle size of the glacially eroded ore material ranges from fine powder to boulders, and hence the composition of the fine fractions can be used to define the shape, size and location of the bedrock source of geochemical dispersal trains (Shilts 1976; DiLabio 1989; McClenaghan *et al.* 1997, 2000). Till geochemistry, the subject of this paper, refers to the sampling of C-horizon soil material deposited by clastic dispersal, and the processing and geochemical analysis of the finest fractions of till to detect minerals and their weathering products. The paper presents an overview of field and laboratory methods used in temperate and arctic climates underlain by shield terrains in Canada and Fennoscandia. An outline of the basic concepts of till geochemistry is presented first to provide a background for drift prospecting.

Principles of till geochemistry

Canada and Fennoscandia have been periodically covered by glacial ice during the last million years of geological time. The surficial sediments deposited by glaciers have been exposed to weathering for only 8000 to 12 000 years, i.e. since deglaciation. As a result, the processes of geochemical dispersion in sediments transported by glacier ice are different from those in non-glaciated areas of residual weathering (Rose et al. 1979; Shilts 1993). In glaciated terrains, deposits are composed largely of unweathered, crushed bedrock detritus, the result of glacial dispersal processes. Weathering processes are active mainly in the uppermost part (0 to 2 m depth) of the glacial sediments and are important factors to consider when till geochemical samples are collected in the near-surface environment.

Glacial dispersal processes

Erosion, transport and deposition by glacial ice basically reflect glacial history and ice flow dynamics. These processes affect the provenance and the geochemistry of glacial deposits. In till, the fine fraction (<2 mm) is produced largely by the physical crushing and abrasion of primary minerals. Fine-grained metal-bearing particles and grains of ore minerals are glacially comminuted and the elements redistributed into different grain sizes (Dreimanis & Vagners 1971; Shilts 1995). During weathering, these elements may preferentially accumulate in the clay-sized fraction (<2 μm) because clay minerals in this fraction have a large surface area, high cation exchange capacity and can accommodate wide ranges of ionic radii (Shilts 1995). Hence, geochemical variations in the fine fraction of till may also reflect textural differences to some degree. Orientation studies are commonly carried out to determine the relationship between the size fraction of ore minerals in the host rocks, the size fractions to which they are glacially comminuted and the size fraction where elements have accumulated during weathering (DiLabio 1995).

A geochemical anomaly in till derived from a discrete bedrock source is three-dimensional. In the ideal case, the anomaly pattern resembles a thin plume extending in the direction of ice movement from the subcropping ore zone towards the surface of the glacial overburden (Drake 1983; Miller 1984). Ore-rich debris dispersed by a glacier usually occurs with maximum frequency close to its origin and in exponentially decreasing amounts in the direction of glacial transport (Krumbein 1937; Shilts 1976; Salonen 1987). The distance of glacial transport and shape of the decay curve have been shown to vary with size of transported clasts or particles, nature of bedrock types, width of outcrop, topography, and mode of glacial behaviour (Kauranne et al. 1992; Shilts 1996; Klassen 1997). A second type of dispersal profile has also been described in which debris concentrations decrease linearly with increasing distance of glacial transport, leading to a flat and gently sloping profile (e.g. Dyke 1984; Thorleifson & Kristjansson 1993; Dredge 1995). Linear decay reflects englacial transport with minimal to no deposition during transport, which may be characteristic of ice streams (Klassen 2001).

Glacial dispersal trains have been widely described in the literature (e.g. Shilts 1976; Klassen & Thompson 1989; DiLabio 1990; Puranen 1990). They may have ribbon, fan or amoeboid shapes. Palimpset dispersal trains have been described where older trains served as sources of indicator debris to later glacial events (e.g. Bird & Coker 1987; Stea 1994; Parent et al. 1996). In areas with relatively little relief such as shield terrains, components of glacial sediments can be transported over topographic barriers along relatively straight lines, and may cross one or many drainage basins. Therefore, the shape and extent of a dispersal train are largely controlled by the location of the source outcrops relative to former centres of outflow and by flow characteristics within former ice sheets (Boulton 1984; Hicock 1988; Dredge 1995).

Weathering processes

Postglacial weathering can have a significant effect on the geochemistry of till in the zone of oxidation above the groundwater or permafrost table, commonly to a depth of a few metres (e.g. Shilts 1975, 1984; Peuraniemi 1984; Saarnisto 1990; Shilts & Kettles 1990). During postglacial weathering and subsequent soil formation, clastically dispersed labile minerals such as sulphides and carbonates are progressively decomposed in the oxidation zone, releasing elements which are dissolved and transported by percolating rain water. Some of the elements are adsorbed or precipitated almost immediately, depending on the element and the local geochemical environment, whereas others are trapped in organic matter, clay-sized phyllosilicates or secondary oxides/hydroxides, or are completely leached out of the soil profile by groundwater. The destruction of labile minerals under surface conditions has a significant impact on mineral exploration because many of these minerals are glacially dispersed ore minerals. Variations in trace metal values related to natural soil-forming processes, such as biogeochemical enrichment in the surface organic layer and post-depositional mobilization of elements, may overshadow primary geochemical variations due to glacial transport and depositional history. Secondary weathering processes may result in 'mineralized' horizons whereas primary glacial processes may result in multiple stratigraphic units and multiple till facies. Deeper, in the non-oxidizing groundwater regime, or in permafrost, ore minerals are preserved. Thus, an understanding of vertical geochemical variations within a till unit is essential when evaluating geochemical patterns.

Field methods in shield terrain

The Canadian and Fennoscandian Shields are geologically complex terrains that have been subjected to continental glaciation of a more widespread and continuous nature than the glaciated valley areas of the Canadian Cordillera or Norwegian Fjordlands. As a result, glacial drift is more extensive, and very thick in places. Furthermore, overlapping sequences of proglacial lake clays and marine sediments may conceal till and bedrock. Although these conditions have created the greatest challenge for mineral exploration, successful exploration methods using till geochemistry, heavy minerals and boulder tracing have been established in Canada and Fennoscandia.

Typical shield terrain is a region of low, rolling hills, rounded by glacial erosion of hard crystalline bedrock. Drainage has been extensively disrupted by glacial processes and most of the lake basins occupy depressions scoured by glacial erosion or formed by melting of ice blocks in areas of thicker drift (Shilts *et al.* 1987). In Canada, the climatic zones of the Precambrian Shield can be divided into two major areas characterized by vegetation differences: (1) tundra, north of the southern limit of continuous permafrost, and (2) forest, boreal for the most part up to the tree-line, mixed with deciduous in the southern fringe of the Shield.

Since deglaciation, soils in these two climatic zones have been subjected to different geomorphological and weathering processes which affect drift prospecting methods.

Preliminary work

When undertaking detailed and regional till sampling programs, a Quaternary geological framework is needed prior to any sampling, analysis and interpretation (Hirvas & Nenonen 1990; Coker 1991). Since sampling is the most important aspect of data collection, and accounts for about 50 to 80% of geochemical survey costs (Kauranne *et al.* 1992), the correct identification of types of glacial sediments is critical (e.g. Dreimanis 1990). Proper sampling sites can often be derived from surficial geology maps that show the distribution, thickness and types of all glacial deposits. Till should be examined in terms of glacial processes and glacial history as they relate to the type and orientation of glacial landforms, and stratigraphy if present. Bedrock geology, topography and soils maps are also used to assess the regional context for drift prospecting. In the field, the mapping of distinct indicator bedrock units and associated erratics, as well as detailed mapping of ice-inscribed features on bedrock (glacial striations) (e.g. Veillette 1989; Lundqvist 1990), can be used for reconstructing directions and distances of glacial transport in thin drift areas. Till pebble fabrics are measured in areas of thick overburden to evaluate glacial transport directions (e.g. Kujansuu 1990).

Sampling design

Till geochemical surveys vary tremendously in terms of scale, methodology and objectives. They range from reconnaissance-scale surveys to local-scale studies near potential mineraliza-

Table 1. *Scales of till geochemical surveys*

Scale	Sampling density	Sample spacing	Length of dispersal trains detected	Objectives, target
Reconnaissance	one/100 to 500 km^2	10 to 25 km	500 to 1000 km	geochemical provinces, continental glacial dynamics
Regional	one/10 to 100 km^2	4 to 10 km	10 to 100 km	mineral belt, kimberlite cluster
Local	one/1 to 4 km^2	1 to 2 km	1 to 5 km	mineralized ground, tails of dispersal trains from orebodies
Detailed	100 to 1000 samp./km^2	25 to 100 m	100 m to 1 km	ore body

tion (Shilts 1984, 1993). The choice of the sampling scale is dictated by the primary objective of the exploration program, the size and accessibility of the study area and, of course, available funds (Kauranne et al. 1992). Table 1 summarizes the sampling density typically chosen for the different scales and purposes of geochemical programs. Samples are collected at very low density (one sample/100 to 500 km^2) as a first step in a geochemical study, to derive regional information about geochemical provinces and dispersal trains over distinctive provenance areas, and to answer continental and regional scale questions of glacial dynamics (Shilts 1984; Prest & Nielsen 1987). This low density sampling is the reconnaissance scale of exploration surveys where thousands of square kilometres are evaluated. Regional scale surveys in promising zones (one sample/10 to 100 km^2) may detect a mineral belt, or large anomalies of mineralized drift transported tens to hundreds of kilometres. Higher density till sampling surveys are used to outline mineralized ground (one sample/1 to 4 km^2), or to pinpoint buried ore bodies with the greatest possible precision (100 to 1000 samples/km^2), prior to drilling.

In glaciated shield terrain, sampling grids are commonly designed according to the scale of study, the predominant direction of glacial transport, the expected character of the anomaly and the predicted size of the bedrock target. Topography, which is generally subdued, has little influence on the design of the sampling grid and spacing. A regular sampling grid can be of two types: a grid regular in every direction, and a grid where the sampling sites are situated along lines transverse to the direction of ice movement, with sample spacing along lines much shorter than the space between lines (Kauranne et al. 1992; McClenaghan et al. 1997). The sampling grid that is regular in every direction is usually preferred in areas where anomalies have no clearly defined length or where the character of the buried sources are not well known, which is the case in most regional and reconnaissance scale sampling surveys. However, the sampling sites are sometimes constrained by the extent and distribution of the surface till units, and for reasons of economy by accessibility, such that the sampling grid is commonly an irregular one with variable distances between sites. In areas of extensive glacial transport where elongate dispersal trains are expected, the use of sampling lines may be more economical to delineate anomalies at all scales (Kauranne 1975; McClenaghan et al. 1997). With increasing sampling density, such as in local and detailed scale surveys, the sampling grid should be more regular, either in every direction, or along sampling lines if the direction of ice movement and the strike of the bedrock are known.

Sampling depth

The proportions of far-travelled to local debris typically increase upward in till deposits (Sauramo 1924) so that concentrations of indicator minerals derived from a buried source rock increase with depth towards the source. Therefore, as the surface part of the till blanket represents a wider source area, sampling should be close to the *till surface* (0.5–1.0 m depth) in reconnaissance and regional scale geochemical surveys, in order to intersect the tail of dispersal trains. On the other hand, in local and detailed surveys, till sampling close to the *bedrock surface* is most effective because the composition of the till more closely resembles the underlying bedrock. Dispersal trains commonly rise in a down-ice direction at a low angle of inclination so that the distance between the bedrock source and the first appearance of the dispersal train at the surface tends to increase with till thickness (Miller 1984). In order to detect the continuity of a dispersal train in areas of thick glacial sediments, excavation or overburden drilling may be necessary. If separate till beds have been deposited by different ice flow events, sampling should be done at different depths to characterize till stratigraphy and determine vertical variations in till geochemistry. In permanently frozen terrain, the depth of sampling is often restricted by the thickness of the active layer which commonly reaches 1–2 m in till during the maximum summer thaw period (Shilts 1977; Klassen 1995).

Sampling in the near-surface environment

Secondary trace metal variations related to surface weathering can overshadow the primary variations related to till provenance. Hence, samples that are collected at constant depth in the zone of oxidation, and without proper identification and description, may fail to reliably define anomalous element concentrations related to mineralization (Hoffman 1986). For example, false geochemical patterns (both highs and lows) in surface till samples may reflect unusually high levels of specific soil components, such as organic matter or Fe–Mn hydroxides. Because the different horizons in a soil profile are distinctively depleted or enriched in some of these components, geochemical anomalies related to provenance can only be identified by samples taken from the same

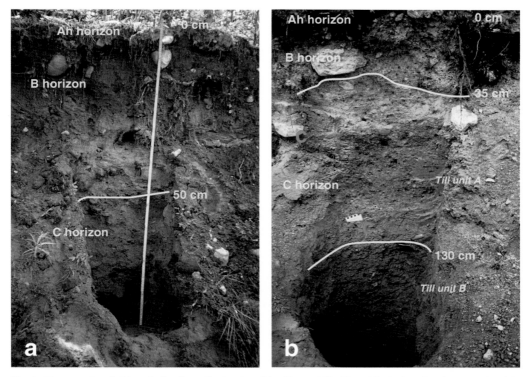

Fig. 1. Photographs of: (**a**) typical soil profile developed in till near Flin Flon, Manitoba, showing organic- rich, mineral horizon (Ah), rusty B-horizon and relatively unweathered upper C-horizon; and, (**b**) soil profile developed in strongly calcareous till near The Pas, Manitoba, showing Ah horizon and leaching of carbonates in the B-horizon, secondary carbonate layers (C) precipitated in the upper C-horizon, and colour difference between two stratigraphic till units in the C-horizon.

material type (e.g. till), and in the case of surface sampling, from the same soil horizon developed in till. Weathering processes are significantly different in forested areas than those acting in the tundra, and hence an understanding of soil formation processes is necessary for optimizing surface till sampling methods.

Forested areas. Chemical weathering is the dominant factor controlling the nature of the weathering products above the zone of oxidation in forested areas. Biological activity is also important, but prevails only in the near-surface organic-rich zone of soils. In non-glaciated terrain with residual overburden, geochemical exploration based on B-horizon sampling is commonly used because many mobile metals tend to become enriched in the ferruginous B-horizon, where contrasts between background and anomalies are enhanced. In glaciated terrain, the interpretation of geochemical data from B-horizon material is complicated because of the difficulty in distinguishing clastic (glacial)

dispersal from soil formation and hydromorphic patterns. Furthermore, pollution from activities such as mining and smelting sometimes contaminates the B-horizon material (e.g. Kaszycki *et al.* 1996; Henderson *et al.* 1998; McMartin *et al.* 1999). The B-horizon of forest soils developed on till is usually recognized by a rusty-brown colour (Fig. 1a), and characterized by an enrichment in Fe, Mn, Al, and organic matter. Considerable secondary, low-temperature chemical alteration of all mineral fractions can be expected, such that labile minerals including sulphides and carbonates are absent from this horizon. The B-horizon typically has a transitional boundary with the underlying C-horizon, a boundary which varies in depth from about 30 to 100 cm, depending on till texture and local drainage conditions.

In the C-horizon soil developed on till, minerals are fresh, or weakly oxidized above the water table (Peuraniemi 1984). Therefore, geochemical anomalies found in the C-horizon are easiest to interpret because the material more

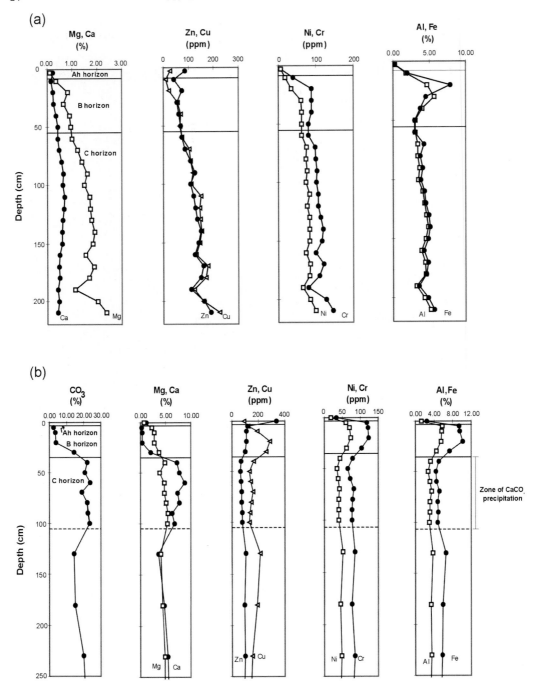

Fig. 2. Geochemical profiles in: (**a**) non-calcareous till (shield derived); and, (**b**) calcareous till (from McMartin *et al.* 1996). Trace and major elements in the < 0.002 mm fraction of till were determined by ICP-ES, following aqua regia digestion. Total carbonate was analyzed in the < 0.063 mm fraction of till by AAS, following digestion in an 1:1 HCl solution. Uppermost sample in each section consists of organic rich, mineral material (< 0.425 mm fraction).

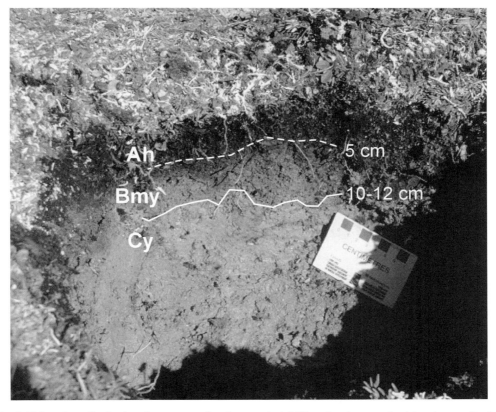

Fig. 3. Thin soil profile developed on vegetated and cryoturbated till surface in permafrost terrain near Rankin Inlet, Nunavut. Organic-rich, mineral horizon (Ah) is 4 to 6 cm thick and the B-horizon (Bmy) extends to a maximum of 12 cm depth from the surface.

closely reflects the primary composition of the bedrock from a source area. C-horizon till is widely used in glaciated terrain of Canada and Fennoscandia to define geochemical dispersal trains formed by mechanical processes (e.g. DiLabio 1989; Kujansuu & Saarnisto 1990; Shilts 1993). In unoxidized till, grains of ore minerals are directly responsible for anomalous geochemical trace metal concentrations. In weakly oxidized C-horizon material, metals originally present in labile minerals such as sulphides have been redistributed to a certain degree and can be detected by geochemical analysis. C-horizon till samples are often collected at depths of ≥75 cm, preferably above the water table, where trace metal concentrations are stabilized (Fig. 2a). In calcareous till, the carbonate minerals have been leached from the B-horizon and re-precipitate deeper in the upper C-horizon as secondary carbonate layers (Fig. 1b). Trace metal concentrations in these discrete carbonate layers can vary significantly from the concentrations in the enclosing C-horizon material (Fig. 2b), and hence till samples should be taken well below the leached B-horizon and preferably below the secondary carbonate layers (Kaszycki et al. 1996; McMartin et al. 1996).

Permafrost areas. North of the tree line, physical weathering is by far the dominant process in the near-surface zone. Soils developed on till are generally thin and immature because of relatively little chemical weathering (Fig. 3). Some leaching of Ca, Na and K, and enrichment in Fe, Mg, Mn and organic matter may occur below thin organic mats to a depth of a few centimeters in till. However, in areas underlain by permafrost, frost action constitutes a major factor in disrupting and destroying the poorly developed soil profiles (Shilts 1973). The net result is a vertical mixing of the soil, where surface organic horizons and weathered material can be redistributed at depth in the active layer. At the maximum depth of summer thaw (1 to 2 m in

Fig. 4. Active mudboil showing freshly extruded till in the centre surrounded by vegetation rim. Shovel is 15 cm wide.

till), the impermeable frozen-sediment surface restricts the normal downward groundwater circulation. As a result, mudboils may form on poorly-sorted, silt- or clay-rich deposits such as till or fine-grained marine or lacustrine sediments. Mudboils, or non-sorted circles (Washburn 1973), can be recognized in the field by bare to lichen-covered, round to oval patches, commonly surrounded by ridges of rocks and vegetation (Fig. 4). They form in plastic sediments with low liquid limits ($<20\%$) and appreciable amounts of silt and clay. Material is extruded to the surface because of high water pressures built up in the active layer, above the permafrost table (e.g. Shilts 1974, 1978; Egginton 1979). These processes are important because they bring relatively unweathered till to surface for easy sampling. Mudboils have been used extensively in permafrost terrain of the Canadian Shield for geochemical sampling (e.g. Shilts 1973, 1977; Klassen 1995; Dredge et al. 1997).

In areas of continuous permafrost, repeated exposure of till to oxidation at the surface through frost-heaving and frost-stirring causes most labile minerals such as sulphides to be destroyed in the active zone (Shilts 1977; Klassen 1995). However, near or directly down-ice from mineralization, fresh sulphides such as pyrite grains, and pseudomorphs of goethite after pyrite, have been observed in surface till collected near gold-bearing iron formations of Nunavut (McMartin 2000). In the active layer, most of the cations released by the destruction of the labile minerals are carried away in solution through surface runoff and/or are scavenged by primary or secondary minerals that exist mainly in the finest fractions of till (Shilts 1973). Therefore, the seasonally thawed layer is analogous to the upper C-horizon layer above the water table in forested areas. In mudboils, the material is fairly homogenized and trace metal concentrations do not vary considerably with depth (Fig. 5), hence samples may be collected at shallow depth (>30 cm). However, Shilts (1975) observed a slight decrease in trace metal concentrations with depth in mudboils, and DiLabio (1979) reported significant U variation down profile near U mineralization in Keewatin. Where shallow soils are developed, trace metals adsorbed or complexed with fine-grained organic matter recycled from the surface and redistributed throughout the active layer by cryoturbation as discrete organic-rich layers can result in abrupt geochemical variations (McMartin et al. 2000). On the other hand, Laurus & Fletcher (1999) observed a decrease in Au concentrations with depth in profiles of arctic soils. In any case, special care must be taken not to include layers of organic material or heavily oxidized clasts. Interpretation of geochemical data in areas covered by a thin (<1 m) veneer of postglacial marine or lacustrine sediments can be further

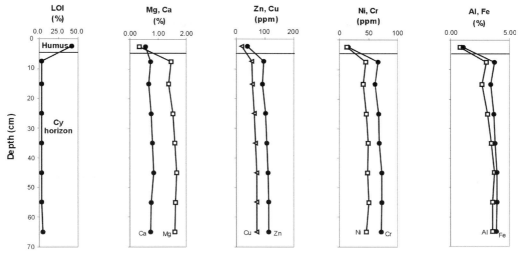

Fig. 5. Geochemical profile in till collected in mudboil from Kaminak Lake area, Nunavut. Trace and major elements in the <0.002 mm fraction of till were determined by ICP-ES, following aqua regia digestion. LOI (loss-on-ignition) gives an estimate of total organic content (by weight). Top sample consists of humus collected adjacent to the mudboil (<2 mm fraction).

complicated by circulation in the active layer that results in mixing of till and younger sediments. In these areas, mudboils with little or no clasts on the surface, abundant shells, or exotic debris from offshore areas, should be avoided.

Sampling methods in thin drift areas

Once the decisions concerning sample density and sample depth are made, the choice of sampling equipment becomes important. Plouffe (1995) summarized the advantages and disadvantages of common sampling methods, which are listed in Table 2. Methods used depend on glacial sediment thickness as well as cost, and most importantly project objectives.

Hand excavation. In relatively thin drift-covered shield terrain or in areas of thick till where shallow sampling (<1 m) is acceptable (regional and reconnaissance scales), the most cost-effective procedure in many cases is to take samples from pits dug with a shovel or pick. In upland terrain, till commonly forms a discontinuous veneer with thicker accumulations occurring in depressions or as tails in the down-ice side of bedrock knobs (Fig. 6). In topographically lower terrain that was inundated by a proglacial lake or a post-glacial sea, till is commonly capped by a bouldery mantle of unsorted debris, or a veneer of clayey to sandy glaciolacustrine or marine sediments. In these areas, acceptable sampling sites are usually found on the down-ice side of bedrock highs where the offshore sediment cover thins (Fig. 6). Several regional and detailed till geochemical surveys have been recently completed with hand-dug holes in drift-covered areas of the Canadian Shield (e.g. McClenaghan *et al.* 1998; Thorleifson & Kristjansson 1993; Bajc 1996; McMartin *et al.* 1996; Kerr *et al.* 1998). Hand augers can be used before trenching with a shovel in permafrost and non-permafrost terrains, in order to identify the proper sampling material and facilitate the choice of the sample site. Till samples can be collected easily with a shovel from road cuts or borrow pits where till has been removed for road construction, and from natural sections along river or lake shorelines.

In forested areas, pits can be dug to a depth of about 1 m to reach the relatively unoxidized till with limited environmental damage. However, the difficulty to extract large boulders and penetrate a compact forest root system to reach the till below may result, in some cases, in abandoning the sample site. Depending on accessibility and spacing between sample sites, about 8 to 15 holes can be hand-dug in a 10 h day by a two-person crew. The hand excavation method has the advantages of low cost, minimal environmental damage, no limitation in the size of the samples, excellent sample recovery, and low sample contamination. In permafrost terrain, till samples collected from hand-dug pits over wide areas accessed by helicopter is

Table 2. Features of till sampling methods

	Hand digging	Trenching (backhoe)	Small Percussion and Vibrasonic Drills (various)	Auger Drills (Various)	Reverse Circulation Drills (Longyear or Acker) (Nodwell Mounted)	Rotasonic Drills (Nodwell or Truck Mounted)
1. Production cost estimate per: - day (10 hrs) - metre	$100 per sample	$500–$1000	$500–$1000 $20–$40	$800–$1500 $25–$50	$1800–$2000 $25–$40	$3000–$4000 $50–$80
2. Penetration depth	1 to 1.5 m	3 to 5 m	10–20 m (greater ?)	15–30 m (boulder free)	Unlimited (125 m ?)	Unlimited (125 m ?)
3. Environmental damage	minor	5 m wide cut trails	nil	2–3 m wide cut trails (Nodwell, muskeg, all terrain vehicle mounted quite manoeuverable)	5 m wide cut trails (may have to be cut in areas of larger trees)	5 m wide cut trails
4. Size of sample	no limitation	no limitation	300 g (dry), or continuous core	3–6 kg (dry or wet)	5 kg (wet)	Continuous core (9 cm)
5. Sample of bedrock	Yes (chips), if bedrock reached	Yes (chips), if bedrock reached	Yes (chips) if bedrock reached	Unlikely, if hollow auger, split spock sampler can be used for chips	Yes (chips)	Yes (core)
6. Sample recovery a) till b) stratified drift	excellent excellent	excellent excellent	good good	good poor to moderate	good moderate	excellent excellent
7. Holes per day (10 hrs)	10 to 15 (1 m deep, close spacing)	10 to 15 trench (short and close spacing)	5 @ 6 to 10 m	1 to 3 @ 15 to 20m 1 @ 60 to 80 m	4 @ 15 to 20 m 1 @ 60 to 80 m	4 @ 15 to 20 m 1 @ 60 to 80 m
8. Meters per day	10 to 15 m	30 to 50 m	30 to 50 m	20 to 60 m	60 to 80 m	60 to 80 m
9. Time to pull rods	N/A	10–20 min	30–60 min @ 15m	20–40 min @ 15m	10 min @ 15 m	10 min @ 15 m
10. Time to move	5–10 min		30 min	15–60 min		15–30 min
11. Negotiability	good	good	good (poor if manually carried on wet terrain)	good to reasonable	good	moderate
12. Trails required	no	yes, may have to be cut in areas of larger trees	no	yes and no	yes, may have to be cut in areas of larger trees	yes†, must be cut
13. Ease in collecting sample	excellent	good	sometimes difficult to extract from sampler flow through, split spoon, or piston samplers	good (contamination?)	good	excellent, continuous core
14. Type of bit	N/A	N/A	hydraulic percussion (gas engine percussion, vibrasonic)	auger with tungsten carbide teeth	milltooth or tungsten carbide tricone	tungsten carbide ring bits
15. Type of power	human	tractor	hydraulic jack, hand	hydraulic-rotary	hydraulic-rotary	hydraulic-rotasonic
16. Method of pulling rods	N/A	N/A jack or winch		winch or hydraulics	hydraulic	hydraulic
17. Ability to penetrate or move boulders	very poor	excellent	poor	poor to moderate	excellent	excellent, cores bedrock
18. Texture of sample	original texture	original texture	original texture	original texture (dry) to slurry (wet)	slurry (disturbed sample)	original texture (core can be shortened, lengthened and/or contorted
19. Contamination of sample	nil	nil	nil (tungsten)	nil to high (tungsten)	nil, fines lost (tungsten)	nil (tungsten)

From Plouffe (1995). Drilling costs based on 1985 data (Canadian $).

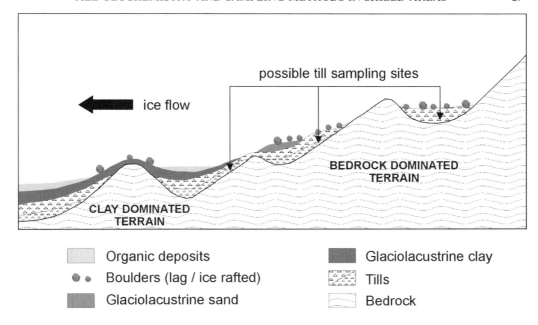

Fig. 6. Sampling strategies in shield terrain covered by relatively thin drift: upland bedrock-dominated terrain and topographically lower areas dominated by thick clay deposits (Henderson 1995a).

Fig. 7. Backhoe digging a trench in till on top of The Pas Moraine, central Manitoba.

extremely efficient for reconnaissance and regional surveys because of the shallow depth of sampling (< 50 cm) and the ability to recognize mudboils (potential sample sites) from air photo interpretation or directly from the aircraft (Shilts & Wyatt 1989; Klassen 1995). As many as 30 to 50 samples can be collected by pitting during a day with helicopter support. This

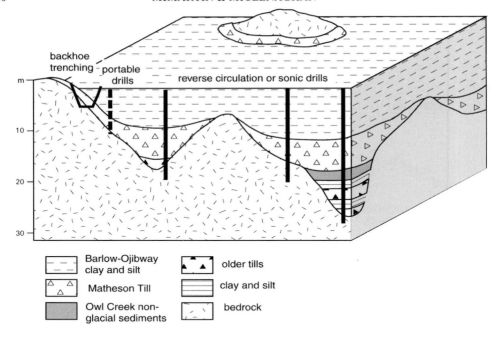

Fig. 8. Block diagram showing till sampling strategies in shield terrain covered by thick drift. This example is for the Abitibi region of central Canada where multiple till units are capped by thick glaciolacustrine sediments (McClenaghan 2001).

method has been used extensively in the search for diamonds in the Canadian tundra.

Trenching. Terrain and access permitting, the use of a backhoe excavator for digging trenches is economical in areas of thicker drift as a typical machine can dig trenches from 3 to 5 m deep (Fig. 7). In Finland, the backhoe has been used extensively over past decades for sampling till in the up-ice ends of known ore-boulder trains or over geochemical or geophysical anomalies (e.g. Often & Olsen 1986; Saarnisto *et al.* 1991; Huhta 1993). In Canada, the use of the backhoe has been more restricted because of poor access and the significant cover of late- or postglacial lacustrine or marine sediments. Nevertheless, some regional and detailed surveys have used a backhoe primarily to document till stratigraphy and vertical geochemical variations (e.g. Snow & Coker 1987; Ford *et al.* 1988; McClenaghan 1992; Henderson 1995b; McMartin *et al.* 1996). Bedrock is commonly exposed at the bottom of the trench and direct observations of the bedrock and till features are possible. Large and representative till samples can be obtained along a profile at regular intervals. The best sampling results will be achieved in areas where surficial sediment cover does not greatly exceed 5 m.

Advantages of this method over hand excavation are the ability to reach a greater sampling depth (hence interval sampling is facilitated), larger exposure of sediments, and ease of sampling in bouldery till. However, environmental impact is greater, particularly if trails have to be cut in areas of larger trees, and the hazards of working in a trench are significant (Hirvas & Nenonen 1990; Plouffe 1995). Costs average $30 to $75 (U.S.) per hour, and 10 to 15 short and closely spaced trenches can be excavated in one day (Coker & Dilabio 1989). The pits should be filled with excavated material after the till sampling is completed in order to avoid any hazards to people or wildlife.

Sampling methods in thick drift areas

Where glacial sediment thickness exceeds 5 m, drilling is required to collect till samples at depth or below the surface cover of other sediments, in order to characterize the till stratigraphy and to determine lateral and vertical variations in till geochemistry (Fig. 8). Availability of equipment and cost are major factors in the choice of drilling method (Coker & DiLabio 1989; Coker 1991). For all drilling methods, till samples should be collected continuously over depth

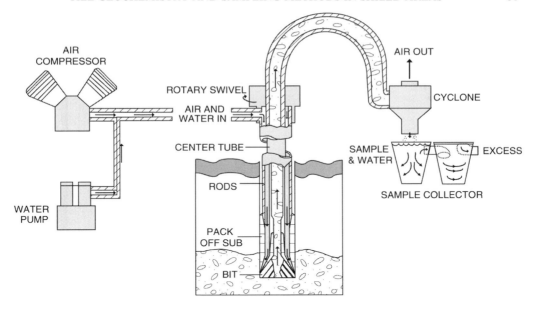

Fig. 9. Schematic cross-section of a reverse circulation rotary drill (modified from Averill 1990).

intervals of 1 to 2 m through entire till units because: (1) dispersal trains can be thin, forming discrete horizons, and till units can vary from one to tens of metres in thickness; (2) dispersal trains are not always in contact with the bedrock surface; and (3) corroboration by vertically adjacent samples is a key to confirming a geochemical anomaly (Geddes 1982; Averill 1990; McClenaghan 1994). Brass fittings, diamond drill bits, and tungsten carbide drill bits can contaminate heavy mineral fractions, and drilling grease used on drill rods can contaminate geochemical samples with Zn, Pb and Mo (Proudfoot *et al.* 1975; Averill *et al.* 1986; Averill 1990). Plouffe (1995) has described the advantages and disadvantages of several drilling methods, which are listed in Table 2. The most common drilling methods are summarized below from Averill (1990), Hirvas & Nenonen (1990), Kauranne *et al.* (1992), and Plouffe (1995).

Reverse circulation rotary drilling. Reverse circulation rotary drilling was first used in the early 1970s in the Timmins area of the Archean Abitibi Greenstone Belt in central Canada (Skinner 1972; Thompson 1979). Since that time, it has been used to explore across the shield terrain of central Canada for a variety of commodities, including Au, base metals, U and diamonds (e.g. Geddes 1982; Harron *et al.* 1987; Sauerbrei *et al.* 1987; Brereton *et al.* 1988). This method is used in areas of extremely thick glacial sediments (10 m to 125 m), such as the Abitibi Greenstone Belt, and where till is stony and bouldery. Dual-tube rods are employed (Fig. 9) to drill a continuous 7 cm diameter hole through glacial sediments and into bedrock. Air and water are injected at high pressure down the outer tubes of the drill rods to a tricone bit at the bottom of the hole. The tricone bit (Fig. 10a), fitted with tungsten carbide buttons, directs the compressed air and water mixture onto the bit as it cuts. Drill cuttings are carried up to the surface through the inner tube as a continuous slurry of <1 cm diameter chips and mud. The material is delivered to the surface quickly, where it is passed through a cyclone to reduce the slurry velocity. As the slurry exits the cyclone, it passes through a 4.0 mm (ASTM 5-mesh) screen and then into a two-bucket system to allow the sediment to settle and excess water to flow off (Fig. 10b). A 10 kg till sample is collected from material in the first bucket approximately for every 1 to 2 m drilled. Most clay-sized material, and approximately 30% of the silt-sized material in till, is lost by this drilling method. This loss is an important consideration when using reverse circulation drilling to collect till samples in areas where a significant amount of Au is known to occur in the fine fraction of till (Shelp & Nichol 1987; Plouffe 1995). Also, till samples can become cross-contaminated by the recirculating water (Coker & Shilts 1991). Holes usually extend

Fig. 10. Reverse circulation drill: (**a**) tricone with carbide buttons; (**b**) cyclone used to decrease the velocity of the slurry returned from drill and the two-bucket system used to allow excess water to flow off and to collect the till sample in the first bucket; and, (**c**) drill mounted on a Nodwell tracked vehicle drilling in a swampy area in the summer time. Note man on roof of drill shack for scale.

1.5 m into bedrock and bedrock cuttings are collected to determine lithology, and for geochemical analysis.

Reverse circulation drills weigh approximately 15 tonnes and are usually mounted on all-terrain tracked vehicles (Fig. 10c) that can travel on frozen lakes and wetlands in the winter. Holes can be drilled relatively quickly because the process does not involve coring, and casing does not have to be used to keep the hole open. Recovery is generally good in all sediment types and the drill can penetrate boulders and bedrock easily and quickly. The quality of the stratigraphic interpretation is limited because the sample is a disturbed slurry of mud and chips and the geologist has only one chance to describe and sample the material as the drill rapidly penetrates the ground. Matrix grain size is difficult to determine because of the loss of fine-grained material during drilling. Quality of interpretation also depends on the expertise of the on-site Quaternary geologist and the driller. Clast lithology type and percentage content, matrix particle size (granules to clay), matrix colour of coarser material retained on the screen (Fig. 11) as well as behaviour of the drill rods all can provide important stratigraphic information (Averill 1990; Plouffe 1995).

Rotasonic drilling. Rotasonic drilling is an ideal drilling method for areas of extremely thick glacial sediments (10 m to 125 m) where natural exposures are rare or absent, and where till is stony and bouldery. It is used instead of reverse

Fig. 11. Observing material retained on screen after it exits the cyclone can indicate the type of glacial sediment being drilled: (a) chunks of sticky clay indicate fine-grained glaciolacustrine sediments; (b) rock chips of variable lithologies indicate glaciofluvial gravel or till; and (c) rocks fragments of only one lithology indicate a boulder or bedrock.

circulation drilling when more detailed stratigraphic information is needed (e.g. DiLabio et al. 1988; Shilts & Smith 1988). Since the early 1980s, most rotasonic drilling has been in support of exploration for Au, base metals, and diamonds in thick drift-covered parts of the shield terrain in central Canada (e.g. Averill & Zimmerman 1986; Bird & Coker 1987; Bajc 1991; Thorleifson & Kristjansson 1993; McClenaghan 1994; McClenaghan et al. 1996, 1999).

Rotasonic drilling uses a combination of high frequency (averaging 5000 cycles per minute) resonant vibration and rotation to drill through glacial sediments, boulders and bedrock with minimal compaction or disturbance to recover a continuous core 9 cm in diameter. No drilling fluids are required except when coring large boulders or bedrock. Core barrels with tungsten carbide button drill bits (Fig. 12a) are resonated/rotated down first, casing is then advanced to the same level using water to clear the cuttings and sand from between the core barrel and the casing (Fig. 13). Casing is used to prevent collapse of the borehole when the rods and core barrels are pulled out of the hole to retrieve the core. Core is resonated from the core barrels into plastic sleeves usually in 1.5 m increments (Fig. 12b) and then rods and core barrels are replaced down the hole. These steps are repeated after each 3 to 9 m (1 to 3 rods) of drilling, using an efficient hydraulically operated rod handling system, until the desired depth is reached. Usually 1.5 m of bedrock are cored at the bottom of every hole; however, softer bedrock lithologies can be drilled to greater depths. For example, McClenaghan et al. (1996, 1998, 1999) used a rotasonic drill to core 10 m of kimberlite in several drill holes in the Kirkland Lake area. Core is stored in 1.5 m length wooden core boxes and transported to a core logging and sampling site. Averill et al. (1986) and Averill (1990) describe in detail the procedures to achieve the best production rates and sample recovery in a variety of sediment types. A rotasonic drill

Fig. 12. Rotasonic drill: (a) bit with tungsten carbide buttons; (b) close-up of rotasonic drill operations. A core barrel is pulled to surface and core is resonated out of the barrel into plastic sleeves in 1.5 m segments; and (c) rotasonic drill set-up in the winter time with the drill mounted on a tracked Nodwell and backed up to a second tracked vehicle carrying the drill rods.

weighs about 14 tonnes and can be mounted on a truck, skid, Nodwell tracked vehicle (Fig. 12c) or barge so that it can be used in different terrains and during different seasons.

The major advantage of rotasonic drilling over reverse circulation drilling is that it recovers high quality drill core (Fig. 14). This core: (1) can be logged and sampled any time after the hole is drilled, and can be archived for future reference; (2) provides more detailed information on sediment stratigraphy and genesis; (3) has not lost any fine-grained material during drilling, and thus the fine till fraction is also available for geochemical analysis; and, (4) contains intact pebbles that can be collected for detailed pebble lithology studies. The major disadvantage of this method over reverse circulation drilling is the higher cost (Table 2). Also core shortening or lengthening can occur with this drilling method and the presence of an on-site geologist is essential to monitor actual depths from which the core was obtained and determine the reason for core lengthening or shortening. Some soft sediment deformation can occur during drilling and Smith & Rainbird (1987) describe criteria for identifying these phenomena. On a hourly basis, rotasonic and reverse circulation drilling costs are approximately equal. However, the need to use casing, pull core barrels and rods to recover core and re-enter the hole reduces the productivity of rotasonic drilling.

Portable drills. Descriptions and overviews of portable drilling equipment that can be used for drift exploration in permafrost and non-permafrost terrains are presented by Veillette & Nixon (1980), Hirvas & Nenonen (1990) and Salminen (1992). Low frequency hammer or percussion drills, such as the Pionjar and Cobra (Fig. 15a), are amongst the most common portable drills used for till geochemical surveys in Finland, and to a lesser extent, in Canada (e.g. Hartikainen & Nurmi, 1993; Gustavsson *et al.* 1994; Gleeson &

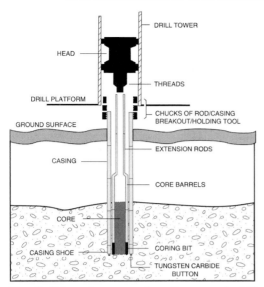

Fig. 13. Schematic cross-section of a rotasonic drill (McClenaghan 1991).

Cormier 1971; Gleeson & Sheehan 1987). Percussion drills use short (100 cm), small diameter rods and a short (0.3 m) flow-through ejector bit of 2 to 3 cm in diameter. A single till sample is collected at the bottom of a drill hole, to a maximum of 8 to 10 m depth in sandy till with few boulders (Hirvas & Nenonen 1990). Till samples weigh between 150 and 300 g and small 'buttons' of bedrock (2 to 3 cm) can be collected where bedrock is crushed, loosened or weathered (Plouffe 1995). Several holes are drilled at one location to collect till samples from varying depths within a thick till unit. Rods are hammered into the ground and extracted using a jack. The lightest of the percussion drills weighs 10 to 25 kg, which can be carried into the field along with the rods. The advantages of portable drills are that they are portable and inexpensive to operate. However, they cannot drill deep holes or penetrate compact material such as dry sand, overconsolidated interglacial clay or very stony and bouldery till (Averill 1990). The sample collected in the flow-through

Fig. 14. Drill core recovered using a rotasonic drill: (**a**) varved clay and silt, (**b**) silty-sand till, (**c**) well sorted glaciofluvial sand, (**d**) mafic volcanic bedrock. Scale card is 8.5 cm in length.

Fig. 15. (a) A portable percussion drill used to collect till samples in Finland; and, (b) small 150 to 300 g samples collected using the flow through sampler on a portable percussion drill.

sampler (Fig. 15b) is small (150 to 300 g). It can be used for fine fraction till geochemistry but is not large enough for heavy mineral geochemical or Au grain studies.

Laboratory methods

Sample preparation

Till samples for geochemical, mineralogical and lithological studies are usually 8 to 10 kg in size. Small subsamples (500 g) may be set aside for geochemical analysis and archiving (Fig. 16, sections B & C). Geochemical subsamples are dried at temperatures below 40°C to prevent the loss of volatile elements such as Hg, and sieved using stainless steel or plastic screens to prevent contamination and recover specific size fractions for geochemical analysis. Centrifuging is used to recover the clay-sized fraction (< 0.002 mm) of till samples for geochemical analysis (Lindsay & Shilts 1995).

Geochemical analysis

Geochemical exploration involves detecting anomalous concentrations of ore-indicator elements, or elements associated with the ore or host rocks, called pathfinder elements. Table 3 gives a summary of ore-indicator and pathfinder elements in major types of mineral deposits. The choice of size fraction and analytical methods used to detect elements depends on the commodity sought, the mineral form, and the weathering history of the bedrock source and till.

Size fraction. In the last decade, many studies have used the < 0.063 mm till fraction for Au exploration (e.g. Pronk & Burton 1988; Toverud 1989; Äyräs 1991; Lestinen et al. 1991; Koljonen 1992; Hartikainen & Nurmi 1993; Cook et al. 1995; Sibbick & Kerr 1995). Other studies have recommended analysis of both the heavy mineral and fine till fractions (Campbell, 1986, 1987; Makela et al. 1988; Bloom & Steele 1989; Campbell & Schreiner 1989; Chapman et al. 1990; McClenaghan 1992, 1994; McMartin 2000). This is done in order to detect both fine- or coarse-grained Au and because the heavy mineral fraction may provide better contrast between background and anomalous samples (Brereton et al. 1988; Bernier & Webber 1989; Gleeson et al. 1989). Although expensive to recover, several geochemical surveys for base metal exploration analysed the < 0.002 mm fraction of till (e.g. Steele 1988; Kettles 1992; Kaszycki et al. 1996; McMartin et al. 1996) because of its greater capacity to retain elements released during weathering, and to avoid textural bias on geochemistry. Other fractions less commonly analysed geochemically include heavy minerals (e.g. Campbell & Schreiner, 1989; Glumoff & Nikkarinen 1991; McClenaghan 1992, 1994; Thorleifson & Kristjansson 1993) and the < 2 mm bulk till fraction (e.g. Bloom & Steele 1989; Tarvainen 1995).

Analytical methods. Geochemical methods commonly used to analyse till are summarized by Hall (1991, 1997), Koljonen (1992), Kauranne et al. (1992), and Lett (1995). Most geochemical surveys use partial digestions such as aqua regia to decompose samples (Koljonen & Malisa 1991). The elements mostly bound to sulphide, oxide and clay minerals, organic matter and secondary precipitates are almost completely dissolved with aqua regia (Niskavaara 1995), and these components are the major sources of trace metals in till (Räisänen et al. 1997). However, some regional surveys in Fennoscan-

Table 3. *Association of elements in major types of mineral deposits*

Type of deposit	Major components	Associated elements
Hydrothermal deposits		
Porphyry copper and/or molybdenum	Cu, Mo, S	Fe, Ag, Au, REE
Ni-Cu-Fe sulphides	Ni, Cu, S, Fe	Pt, Co, As, Au
VMS	Fe, S, Cu, Zn	Cd, Hg, Au, Pb, As, Sb, Ba, Bi
Native copper	Cu	Ag, As, S
Precious metal	Au, Ag	Sb, As, Hg, Te, Se, S, U
Native silver and Ni-Co arsenides	Co, Ni, Ag, S	As, Sb, Bi, U
Polymetallic veins	Pb, Zn, Cu, S	Ag, Au, As, Sb, Mn
Iron formations	Au, As, S	Sb, Fe
Mercury	Hg, S	Sb, As
Mississipi Valley Pb-Zn	Zn, Pb, S	Ba, F, Cd, Cu, Ni, Co, Hg
Skarn-iron (magnetite) deposits	Fe	Ti, Cu, Co, S
Skarn Cu-Zn-Pb deposits	Fe, Cu, Pb, Zn	Cd, S, Au, Ag, Cu, Co
Skarn W-Mo-Sn deposits	W, Mo, Sn	F, S, Cu, Be, Bi
Magmatic deposits		
Pt-Ni-Cu rich intrusions	Pt, Ni, Cu	Cr, Co, S
Chromite ores	Cr	Ni, Fe, Mg
Carbonatites	K, Na, Fe, P, F, Cl, Zr, Nb	Sr, Ti, Ta, U, REE
Pegmatites	Be, Li, Cs, Rb	B, U, Th, RE, Nb, Sc
Kimberlites (diamonds)	Sr, Nb, Ba, Cr, Ni, Mg, Ca	Al, K, Ta, Th, REE
Sedimentary types		
Copper shales	Cu, S	Ag, Zn, Pb, Co
Uranium vein	U	Mo, Pb, F
Sandstone-type U	U	Se, Mo, V, Cu, Pb

Modified from Rose *et al.* (1979) and Levinson (1980).

dia have used both total and partial digestions to investigate variable element solubility in samples as it relates to rock type and degree of weathering (Koljonen 1992; Lahtinen *et al.* 1993; Tarvainen 1995). Regional surveys use instrumental neutron activation analysis (INAA) and inductively coupled plasma emission spectrometry (ICP-ES) for >30 elements to provide extensive and complementary analytical data for the fine till fractions (<0.063 mm and <0.002 mm). Certain elements, however, require specialized techniques. Mercury is commonly determined using cold vapour-atomic absorption spectrometry (CV-AAS). Platinum group elements (PGE) and Au are determined by fire assay followed by graphite furnace-AAS, ICP-ES or ICP-MS (Hoffman *et al.* 1999), or by INAA. Major oxides are determined by x-ray fluorescence spectrometry (XRF) or ICP-ES. Heavy mineral concentrates are analysed best by non-destructive INAA and, in the case of unoxidized sediments, a small representative split may be analysed for base metals by aqua regia/ICP-ES (Fig. 16, section D). The use of destructive analytical methods (e.g. fire assay) for heavy mineral concentrates is strongly discouraged because no sample can be retained for future mineralogical analysis. Ten to 15% of each analytical batch should comprise in-house and certified geological reference materials, as well as field and post-preparation duplicates to monitor analytical accuracy and precision (Garrett 1991; Smee 1998).

For the <0.063 mm and the <0.002 mm fraction, geochemical analysis of as little as 0.5 g can be used to identify Au-pathfinder elements, such as Ag, As, Sb, S, Te, Mo, B, Cu, Bi, and Co, which may be more evenly distributed and more abundant than Au itself (e.g. Brereton *et al.* 1988; Saarnisto *et al.* 1991; Hartikainen & Nurmi 1993; Sibbick & Kerr 1995). In analysis of Au, a minimum of 30 to 50 g of the fine fraction is required because of the particulate nature in which Au occurs coupled with sample inhomogeneity (the 'nugget' effect). Unusually high or low levels of major elements in till also may act as pathfinders because variations can indicate bedrock alteration related to Au mineralization (Hartikainen & Damsten 1991). Whole rock geochemistry (major and trace) by XRF and ICP-MS of the till matrix, till clasts or indicator bedrock can also be used for till provenance studies (e.g. Campbell *et al.* 1999).

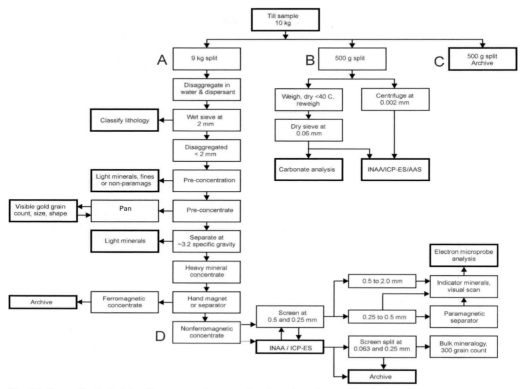

Fig. 16. Generalized glacial sediment sample processing flow chart for geochemical analyses.

Summary

Till collected from the C-horizon is the preferred sampling medium for mineral exploration in glaciated shield terrains. It is deposited directly by glacial ice, and as such, can be used for tracing mineral deposits. The effects of weathering in the uppermost 1 to 2 m of till at surface are a concern for mineral exploration because many of the labile ore minerals are oxidized under surface conditions. However, in the upper C-horizon where surface till samples are collected, the products of this decomposition are scavenged by primary clay-size minerals and secondary minerals, and can be detected by geochemical analysis of the finer till fraction. In addition, C-horizon soil has become an important sampling medium for environmental studies because it is largely unaffected by anthropogenic and biological factors.

In a till geochemical survey, sampling is the most important aspect of data collection, and the most expensive. Therefore, it is essential to design a sampling survey that is appropriate to the needs of the exploration program. In areas covered by relatively thin drift (< 5 m), till samples are commonly collected from hand-dug pits or backhoe excavations, which are two relatively inexpensive sampling methods. These methods are not appropriate in shield terrain covered by thick drift. In these areas, overburden drilling methods must be used to collect subsurface till samples, characterize till stratigraphy and identify vertical and lateral geochemical variations. Laboratory methods are driven by the nature and composition of the expected ore deposit, and by costs. An understanding of mineral partitioning, both physical and chemical, is essential for selecting the proper size fraction to be analysed, and the proper analytical method.

J. Campbell, B. Schreiner and R. Lett are thanked for their careful reviews. Suggestions by R. DiLabio, P.

Henderson and D. Kerr (GSC) helped to improve the manuscript. Descriptions of laboratory methods were taken largely from McClenaghan et al. (1997). Geological Survey of Canada Contribution No. 1999197.

References

AVERILL, S. A. 1990. Drilling and sample processing methods for deep till geochemistry surveys: making the right choices. In: AVERILL, S. A., BOLDUC, A., COKER, W. B., DILABIO, R. N. W., MAURICE, Y., PARENT, M., VEILLETTE J. & LASALLE, P. (eds) Application de la géologie du Quaternaire à l'exploration minérale. Association professionelle des géologues et des géophysiciens du Québec, 139–173.

AVERILL, S. A. & ZIMMERMAN, J. R. 1986. The Riddle resolved: the discovery of the Patridge gold zone using sonic drilling in glacial overburden at Waddy Lake, Saskatchewan. Canadian Geology Journal of CIM, **1**, 14–20.

AVERILL, S. A., MACNEIL, K. A., HUNEAULT, R. G. & BAKER, C. L. 1986. Rotasonic Drilling Operations (1984) and Overburden Heavy Mineral Studies, Matheson Area, District of Cochrane. Ontario Geological Survey, Open File Report **5569**.

ÄYRÄS, M. 1991. Geochemical gold prospecting at Vinsanmaa, northern Finland. Journal of Geochemical Exploration, **39**, 379–386.

BAJC, A. F. 1991. Till Sampling Survey, Fort Frances Area, Results and Interpretation. Ontario Geological Survey, Study **56**.

BAJC, A. F. 1996. Regional Distribution of Gold in Till in the Peterlong Lake-Radisson Lake Area, Southern Abitibi Subprovince; Potential Exploration Targets. Ontario Geological Survey, Open File **5941**.

BERNIER, M. A. & WEBBER, G. R. 1989. Mineralogical and geochemical analysis of shallow overburden as an aid to gold exploration in southwestern Gaspésie, Québec, Canada. Journal of Geochemical Exploration, **34**, 115–145.

BIRD, D. J. & COKER, W. B. 1987. Quaternary stratigraphy and geochemistry at the Owl Creek Gold Mine, Timmins, Ontario, Canada. Journal of Geochemical Exploration, **28**, 267–284.

BLOOM, L. B. & STEELE, K. G. 1989. Gold in till: preliminary results from the Matheson area, Ontario. In: DILABIO, R. N. W. & COKER, W. B. (eds) Drift Prospecting. Geological Survey of Canada, Paper **89-20**, 61–70.

BRERETON, W. E., BRIGGS, D. N. & ROLLINSON, J. P. 1988. Till prospecting in the area of the Farley Lake gold deposits, northwestern Manitoba, Canada. In: MACDONALD, D. R. & MILLS, K. A. (eds) Prospecting in Areas of Glaciated Terrain. Canadian Institute of Mining and Metallurgy, Halifax, 225–239.

BOULTON, G. S. 1984. Development of a theoretical model of sediment dispersal by ice sheets. In: Prospecting in Areas of Glaciated Terrain. Institution of Mining and Metallurgy, London, 213–223.

CAMPBELL, J. E. 1986. Quaternary Geology of the Waddy Lake Area Applied to Prospecting for Gold. Saskatchewan Research Council, Report **R-842-1-E-86**.

CAMPBELL, J. E. 1987. Quaternary Geology and Till Geochemistry of the Sulphide-Hebden Lakes Area. Saskatchewan Research Council, Report **R-842-4-E-87**.

CAMPBELL, J. E. & SCHREINER, B. T. 1989. Quaternary geology and its implications to gold exploration in the La Ronge and Flin Flon domains Saskatchewan. In: DILABIO, R. N. W. & COKER, W. B. (eds) Drift Prospecting. Geological Survey of Canada, Paper **89-20**, 113–126.

CAMPBELL, J. E., QUIRT, D. H. & MILLARD, M. J. 1999. The application of till geochemistry to mineral exploration in Northern Saskatchewan, Canada. In: FLETCHER, W. K. & ELLIOTT, I. L. (eds) Symposium Program and Abstracts Volume. 19th International Geochemical Exploration Symposium, The Association of Exploration Geochemists, Vancouver, B.C., 8–9.

CHAPMAN, R., CURRY, G. & SOPUCK, V. 1990. The Bakos deposit discovery – a case history. In: BECK, L. S. & HARPER, C. T. (eds) Modern Exploration Techniques. Saskatchewan Geological Society, Special Publication, **10**, 195–212.

COKER, W. B. 1991. Overburden geochemistry in mineral exploration. In: Exploration Geochemistry Workshop. Geological Survey of Canada, Open File **2390**, 3-1 – 3-60.

COKER, W. B. & DILABIO, R. N. W. 1989. Geochemical exploration in glaciated terrain, geochemical responses. In: GARLAND, G. D. (ed.) Proceedings of Exploration' 87. Ontario Geological Survey, Special Volume, **3**, 336–383.

COKER, W. B. & SHILTS, W. W. 1991. Geochemical exploration for gold in glaciated terrain. In: FOSTER, R. P. (ed.) Gold Metallogeny and Exploration. Blackie, London, 336–359.

COOK, S. J., LEVSON, V. M., GILES, T. R. & JACKAMAN, W. 1995. A comparison of regional lake sediment and till geochemistry surveys: a case study from the Fawnie Creek area, central British Columbia. Exploration and Mining Geology, **4**, 93–110.

DILABIO, R. N. W. 1979. Drift prospecting in uranium and base-metal mineralization sites, District of Keewatin, Northwest Territories, Canada. In: Prospecting in Areas of Glaciated Terrain. Institution of Mining and Metallurgy, London, 91–100.

DILABIO, R. N. W. 1989. Chapter 10: Terrain geochemistry. In: Fulton R.J. (ed.) Quaternary Geology of Canada and Greenland. Geological Survey of Canada, Geology of Canada, **1**, 645–663.

DILABIO, R. N. W. 1990. Glacial dispersal trains. In: KUJANSUU, R. & SAARNISTO, M. (eds) Glacial Indicator Tracing. A.A. Balkema, Rotterdam, 109–122.

DILABIO, R. N. W. 1995. Residence sites of trace elements in oxidized tills. In: BOBROWSKY, P. T., SIBBICK, S. J., NEWELL, J. M. & MATYSEK, P. F. (eds) Drift Exploration in the Canadian Cordillera. British Columbia Ministry of Energy, Mines and Petroleum Resources, Paper **1995-2**, 139–148.

DILABIO, R. N. W., MILLER, R. F., MOTT, R. J. & COKER, W. B. 1988. The Quaternary stratigraphy of the Timmins area, Ontario as an aid to mineral exploration by drift prospecting. *In: Current Research*. Part C, Geological Survey of Canada, Paper **88-1C**, 61–65.

DRAKE, L. D. 1983. Ore plumes in till. *Journal of Geology*, **91**, 707–713.

DREDGE, L. A. 1995. *Quaternary Geology of Northern Melville Peninsula, District of Franklin; Northwest Territories: Surface Deposits, Glacial History, Environmental Geology, and Till Geochemistry*. Geological Survey of Canada, Bulletin **484**.

DREDGE, L. A., KERR, D. E., KJARSGAARD, I. M., KNIGHT, R. D. & WARD, B. C. 1997. *Kimberlite Indicator Minerals in Till, Central Slave Province, Northwest Territories*. Geological Survey of Canada, Open File **3426**.

DREIMANIS, A. 1990. Formation, deposition, and identification of subglacial and superglacial tills. *In:* KUJANSUU, R. & SAARNISTO, M. (eds) *Glacial Indicator Tracing*, A.A. Balkema, Rotterdam, 35–59.

DREIMANIS, A. & VAGNERS, U. J. 1971. Bimodal distribution of rock and mineral fragments in basal tills. *In*: GOLDTHWAIT, R. P. (ed.) *Till – A Symposium*. Ohio State University Press, Columbus, Ohio, 237–250.

DYKE, A. S. 1984. *Quaternary Geology of Boothia Peninsula and Northern District of Mackenzie, Central Canadian Arctic*. Geological Survey of Canada, Bulletin **407**.

EGGINTON, P. A. 1979. Mudboil activity, central District of Keewatin. *In: Current Research*. Part B, Geological Survey of Canada, Paper **79-1B**, 349–356.

FORD, K. L., DILABIO, R. N. W. & RENCZ, A. N. 1988. Geological, geophysical, and geochemical studies around the Allan Lake carbonatite, Algonquin Park, Ontario. *Journal of Geochemical Exploration*, **30**, 99–121.

GARRETT, R. G. 1991. The management, analysis and display of exploration geochemical data. *In*: *Exploration Geochemistry Workshop*. Geological Survey of Canada, Open File **2390**, 9-1–9-41.

GEDDES, R. S. 1982. The Vixen Lake indicator train, northeastern Saskatchewan. *In:* DAVENPORT, P. H. (ed.) *Prospecting in Areas of Glaciated Terrain*. Canadian Institute of Mining and Metallurgy, St. Johns, 264–283.

GLEESON, C. F. & CORMIER, R. 1971. Evaluation by geochemistry of geophysical anomalies and geological targets using overburden sampling at depth. Canadian Institute of Mining and Metallurgy, Special Volume **11**, 159–165.

GLEESON, C. F. & SHEEHAN, D. G. 1987. *Humus and till geochemistry over the Doyon, Bosquet and Williams gold deposits*. Canadian Institute of Mining and Metallurgy Bulletin, **80**, 58–66.

GLEESON, C. F., RAMPTON, V. N., THOMAS, R. D. & PARADIS, S. 1989. Effective mineral exploration for gold using geology, Quaternary geology and exploration geochemistry in areas of shallow till. *In*: DILABIO, R. N. W. & COKER, W. B. (eds) *Drift Prospecting*. Geological Survey of Canada, Paper **89-20**, 71–96.

GLUMOFF, S. & NIKKARINEN, M. 1991. *Regional Geochemical Mapping Based on the Heavy Fraction of till in Central Finland*. Geological Survey of Finland, Report of Investigation **96**.

GUSTAVSSON, N., LAMPIO, E., NILSSON, B., NORBLAD, G., ROS, F. & SALMINEN, R. 1994. Geochemical maps of Finland and Sweden. *Journal of Geochemical Exploration*, **51**, 143–160.

HALL, G. E. M. 1991. Analytical methods used in exploration geochemistry. *In: Exploration Geochemistry Workshop*. Geological Survey of Canada, Open File **2390**, 8-1 – 8–90.

HALL, G. E. M. 1997. Recent advances in geoanalysis and their implications. *In*: GUBINS, A. G. (ed.) *Proceedings of Exploration 1997*. Fourth Decennial International Conference on Mineral Exploration, 293–294.

HARRON, G. A., MIDDLETON, R. S., DURHAM, R. B. & PHILIPP, A. 1987. *Geochemical and geophysical gold exploration in the Timmins area, Ontario; a case history*. Canadian Institute of Mining and Metallurgy, Bulletin **80**, 52–57.

HARTIKAINEN, A. & DAMSTEN, M. 1991. Application of till geochemistry to gold exploration, Ilomantsi, Finland. *Journal of Geochemical Exploration*, **39**, 303–322.

HARTIKAINEN, A. & NURMI, P. A., 1993. Till geochemistry in gold exploration in the Late Archean Hattu schist belt, Ilomantsi, eastern Finland. *Geological Survey of Finland, Special Paper* **17**, 323–352.

HENDERSON, P. J. 1995a. Summary report on the surficial geology and drift composition in the Annabel Lake-Amisk Lake area, Saskatchewan (NTS 63L/9 and 16, and part of 63K/12 and 13). *In*: RICHARDSON, D. G. (ed.) *Investigations Completed by the Saskatchewan Geological Survey and the Geological Survey of Canada under the Geoscience Program of the Canada-Saskatchewan Partnership Agreement on Mineral Development (1990-1995)*. Geological Survey of Canada, Open File **3119**, 121–130.

HENDERSON, P. J. 1995b. *Surficial Geology and Drift Composition of the Annabel Lake-Amisk Lake Area, Saskatchewan (NTS 63L/9, L/16, and Part of 63K/12 and K/13)*. Geological Survey of Canada, Open File **3026**.

HENDERSON, P. J., MCMARTIN, I., HALL, G. E., PERCIVAL, J. B. & WALKER, D. A. 1998. The chemical and physical characteristics of heavy metals in humus and till in the vicinity of the base metal smelter at Flin Flon, Manitoba, Canada. *Environmental Geology*, **34**, 39–58.

HICOCK, S. R. 1988. Calcareous till facies north of Lake Superior, Ontario: implications for Laurentide ice streaming. *Géographie Physique et Quaternaire*, **42**, 120–135.

HIRVAS, H. & NENONEN, K. 1990. Field methods for glacial indicator tracing. *In:* KUJANSUU, R. & SAARNISTO, M. (eds) *Glacial Indicator Tracing*. A.A. Balkema, Rotterdam, 217–248.

HOFFMAN, S. J. 1986. Soil sampling. Exploration geochemistry: design and interpretation of soil surveys. Society of Economic Geologists, *Reviews in Economic Geology*, **3**, 39–77.

HOFFMAN, E. L., CLARK, J. R. & YEAGER, J. R. 1999. Gold analysis-fire assaying and alternative methods. *Exploration and Mining Geology*, **7**, 155–160.

HUHTA, P. 1993. *The use of heavy mineral concentrates from till in gold exploration in the late Archean Hattu Schist belt, Ilomantsi, Eastern Finland*. Geological Survey of Finland, Special Paper **17**, 363–372.

KASZYCKI, C. A., NIELSEN, E. & GOBERT, G. 1996. Surficial geochemistry and response to volcanic-hosted massive sulphide mineralization in the Snow Lake region. *In:* BONHAM-CARTER, G. F., GALLEY, A. G. & HALL, G. E. M. (eds) *Extech I: a Multidisciplinary Approach to Massive Sulphide Research in the Rusty Lake-Snow Lake Greenstone Belts, Manitoba*. Geological Survey of Canada, Bulletin **426**, 139–154.

KAURANNE, K. 1975. Regional geochemical mapping in Finland. *In:* JONES, M. J. (ed.) *Prospecting in Areas of Glaciated Terrain*. Institution of Mining and Metallurgy, London, 128–137.

KAURANNE, K., SALMINEN, R. & ERIKSSON, K. (eds) 1992. *Regolith Exploration Geochemistry in Arctic and Temperate Terrains. Handbook of Exploration Geochemistry* **5**. Elsevier, Amsterdam.

KERR, D. E., KNIGHT, R. D. & DREDGE, L. A. 1998. *Till Geochemistry and Gold Grain Results, Contwoyto Lake Map Area, Northwest Territories (NTS 76E, North Half)*. Geological Survey of Canada, Open File **3654**.

KETTLES, I. M. 1992. *Glacial Geology and Glacial Sediment Geochemistry in the Clyde Forks-Westport Area of Ontario*. Geological Survey of Canada, Paper **91-17**.

KLASSEN, R. A. 1995. *Drift Composition and Glacial Dispersal Trains, Baker Lake Area, District of Keewatin, Northwest Territories*. Geological Survey of Canada, Bulletin **485**.

KLASSEN, R. A. 1997. Glacial history and ice flow dynamics applied to drift prospecting and geochemical exploration. *In:* GUBINS, A. G. (ed.) *Proceedings of Exploration 1997. Fourth Decennial International Conference on Mineral Exploration*, 221–232.

KLASSEN, R. A. 2001. A Quaternary geological perspective on geochemical exploration in glaciated terrain. *In:* MCCLENAGHAN, M. B., BOBROWSKY, P. T., HALL, G. E. M. & COOK, S. J. (eds) *Drift Exploration in Glaciated Terrain*. Geological Society, London, Special Publications, **185**, 1–18.

KLASSEN, R. A. & THOMPSON, F. J. 1989. Ice flow history and glacial dispersal patterns, Labrador. *In:* DILABIO R. N. W. & COKER, W. B. (eds) *Drift Prospecting*. Geological Survey of Canada, Paper **89-20**, 21–29.

KOLJONEN, T. (ed.) 1992. *Geochemical Atlas of Finland, Part 2: Tills*. Geological Survey of Finland.

KOLJONEN, T. & MALISA, E. 1991. Solubility in aqua regia of selected chemical elements occurring in the fine fraction of till. *In:* PULKKINEN, E. (ed.) *Environmental Geochemistry in Northern Europe*. Geological Survey of Finland, Special Paper **9**, 49–52.

KRUMBEIN, W. C. 1937. Sediments and exponential curves. *Journal of Geology*, **45**, 577–601.

KUJANSUU, R. 1990. Glacial flow indicators in air photographs. *In:* KUJANSUU, R. & SAARNISTO, M. (eds) *Glacial Indicator Tracing*. A.A. Balkema, Rotterdam, 71–86.

KUJANSUU, R. & SAARNISTO, M. (eds) 1990. *Glacial Indicator Tracing*. A.A. Balkema, Rotterdam.

LAHTINEN, R., LESTINEN, P. & SAVOLAINEN, H. 1993. *The use of total and partial dissolution till geochemical data in delineating favourable areas for Ni prospects: an example from the Tampere-Hmeenlinna area, southern Finland*. Geological Survey of Finland, Special Paper **18**, 101–111.

LAURUS, K. A. & FLETCHER, W. K. 1999. Gold distribution in glacial sediments and soils at Boston Property, Nunavut, Canada. *Journal of Geochemical Exploration*, **67**, 271–285.

LESTINEN, P., KONTAS, E., NISKAVAARA, H. & VIRTASALO, J. 1991. Till geochemistry of gold, arsenic, and antimony in the Seinajoki district, western Finland. *Journal of Geochemical Exploration*, **39**, 343–361.

LETT, R. E. 1995. Analytical methods for drift samples. *In:* BOBROWSKY, P. T., SIBBICK, S. J., NEWELL, J. M. & MATYSEK, P. F. (eds) *Drift Exploration in the Canadian Cordillera*. British Columbia Ministry of Energy, Mines and Petroleum Resources, Paper **1995-2**, 215–228.

LEVINSON, A. A. 1980. *Introduction to Exploration Geochemistry. The 1980 Supplement*. Applied Publishing, Chicago, Illinois.

LINDSAY, P. J. & SHILTS, W. W. 1995. A standard laboratory procedure for separating clay-sized detritus from unconsolidated glacial sediments and their derivatives. *In:* BOBROWSKY, P. T., SIBBICK, S. J., NEWELL, J. M. & MATYSEK, P. F. (eds) *Drift Exploration in the Canadian Cordillera*. British Columbia Ministry of Energy, Mines and Petroleum Resources, Paper **1995-2**, 165–166.

LUNDQVIST, J. 1990. Glacial morphology as an indicator of the direction of glacial transport. *In:* KUJANSUU, R. & SAARNISTO, M. (eds) *Glacial Indicator Tracing*. A.A. Balkema, Rotterdam, 61–70.

MAKELA, M., SANDBERG, E. & RANTALA, O. 1988. Geochemical exploration of gold-bearing veins associated with granitoids in western Finland. *In:* MACDONALD, D. R. & MILLS, K. A. (eds) *Prospecting in Areas of Glaciated Terrain*. Canadian Institute of Mining and Metallurgy, Halifax, 255–270.

MCCLENAGHAN, M. B. 1991. *Geochemistry of Tills from the Black River-Matheson (Brim) Sonic Overburden Drilling Program and Implications for Exploration*. Ontario Geological Survey, Open File Report **5800**.

MCCLENAGHAN, M. B. 1992. Surface till geochemistry and implications for exploration, Black River-Matheson area, northeastern Ontario. *Exploration and Mining Geology*, **1**, 327–337.

MCCLENAGHAN, M. B. 1994. Till geochemistry in areas of thick drift and its application to gold exploration, Matheson area, northeastern Ontario. *Exploration and Mining Geology*, **3**, 17–30.

MCCLENAGHAN, M. B. 2001. Regional and local-scale gold grain and till geochemical signatures of lode Au deposits in the Western Abitibi Greenstone Belt, Central Canada. *In:* MCCLENAGHAN, M. B.,

Bobrowsky, P. T., Hall, G. E. M. & Cook, S. J. (eds) *Drift Exploration in Glaciated Terrain*. Geological Society, London, Special Publications, **185**, 201–224.

McClenaghan, M. B., Thorleifson, L. H. & DiLabio, R. N. W. 1997. Till geochemical and indicator mineral methods in mineral exploration. *In:* Gubins, A. G. (ed.) *Proceedings of Exploration 1997. Fourth Decennial International Conference on Mineral Exploration*, 233–248.

McClenaghan, M. B., Thorleifson, L. H. & DiLabio, R. N. W. 2000. Till geochemical and indicator mineral methods in mineral exploration. *Ore Geology Reviews*, **16**, 145–166.

McClenaghan, M. B., Kjarsgaard, I. M., Schulze, D. J., Stirling, J. A. R., Pringle, G. & Berger, B. R. 1996. *Mineralogy and Geochemistry of the B30 Kimberlite and Overlying Glacial Sediments, Kirkland Lake, Ontario*. Geological Survey of Canada, Open File **3295**.

McClenaghan, M. B., Paulen, R. C., Ayer, J. A., Trowell, N. F. & Bauke, S. 1998. *Regional Till and Humus Geochemistry of the Timmins-Kamiskotia (NTS 42A/11, 12, 13, 14) Area, Northeastern Ontario*. Geological Survey of Canada, Open File **3675**.

McClenaghan, M. B., Kjarsgaard, I. M., Schulze, D. J., Stirling, J. A. R., Pringle, G., Kjarsgaard, B. A. & Berger, B. R. 1999. *Mineralogy and Geochemistry of the C14 Kimberlite and Associated Glacial Sediments, Kirkland Lake, Ontario*. Geological Survey of Canada, Open File **3295**.

McMartin, I. 2000. *Till composition across the Meliadine Trend, Rankin Inlet area, Kiralliq Region, Nunavut*. Geological Survey of Canada, Open File **3747**.

McMartin, I., Hall, G. E. M., Kerswill, J. A., Sangster, A. L. & Vaive, J. 2000. Mercury cycling in perennially frozen soils of Arctic Canada, Kaminak Lake area, Nunavut. *In: Proceedings of the 25th Anniversary International Conference on Heavy Metals in the Environment*. University of Michigan, USA.

McMartin, I. Henderson, P. J., Nielsen, E. & Campbell, J. E. 1996. *Surficial Geology, Till and Humus Composition Across the Shield Margin, North-central Manitoba and Saskatchewan: Geospatial Analysis of a Glaciated Environment*. Geological Survey of Canada, Open File **3277**.

McMartin, I., Henderson, P. J. & Nielsen, E. 1999. Impact of a base metal smelter on the geochemistry of soils from the Flin Flon region, Manitoba and Saskatchewan. *Canadian Journal of Earth Sciences*, **36**, 141–160.

Miller, J. K. 1984. Model for clastic indicator trains in till. *In: Prospecting in Areas of Glaciated Terrain*. Institution of Mining and Metallurgy, London, 69–77.

Niskavaara, H. 1995. A comprehensive scheme of analysis for soils, sediments, humus and plant samples using inductively coupled plasma atomic emission spectrometry (ICP-AES). *In: Current Research*. Geological Survey of Finland, Special Paper **20**, 167–175.

Often, M. & Olsen, L. 1986. Gold transport in till in the complex glaciated Karajok greenstone belt area, Finnmark, Norway. *In: Prospecting in Areas of Glaciated Terrain*. Institution of Mining and Metallurgy, London, 83–94.

Parent, M., Paradis, S. J. & Doiron, A. 1996. Palimpset glacial dispersal trains and their significance for drift prospecting. *Journal of Geochemical Exploration*, **56**, 123–140.

Peuraniemi, V. 1984. Weathering of sulphide minerals in till in some mineralized areas of Finland. *In: Prospecting in Areas of Glaciated Terrain*. Institution of Mining and Metallurgy, London, 127–135.

Plouffe, A. 1995. Drift prospecting sampling methods. *In:* Bobrowsky, P. T., Sibbick, S. J., Newell, J. M. & Matysek, P. F. (eds) *Drift Exploration in the Canadian Cordillera*. British Columbia Ministry of Energy, Mines and Petroleum Resources Paper **1995-2**, 43–52.

Prest, V. K. & Nielsen, E. 1987. *The Laurentide Ice Sheet and Long-distance Transport*. Geological Survey of Finland, Special Paper **3**, 91–101.

Pronk, A. G. & Burton, D. M. 1988. Till geochemistry as a technique for gold exploration in northern New Brunswick. *Canadian Institute of Mining and Metallurgy*, Bulletin **81**, 90–98.

Proudfoot, D. A., Skinner, R. G. & Shilts, W. W. 1975. *Contamination in Overburden Samples Obtained by the Rotary, Dual-tube Drilling Technique*. Geological Survey of Canada, Open File **277**.

Puranen, R. 1990. Modelling of glacial transport in tills. *In:* Kujansuu, R. & Saarnisto, M. (eds) *Glacial Indicator Tracing*. A.A. Balkema, Rotterdam, 15–34.

Räisänen, M. L., Kashulina, G. & Bogatyrev, I. 1997. Mobility and retention of heavy metals, arsenic and sulphur in podzols at eight locations in northern Finland and Norway and the western half of the Russian Kola Peninsula. *Journal of Geochemical Exploration*, **59**, 175–195.

Rose, A. W., Hawkes, H. E. & Webb, J. S. 1979. *Geochemistry in Mineral Exploration*. 2nd Edition. A.W. Rose, State College, PA, USA.

Saarnisto, M. 1990. An outline of glacial indicator tracing. *In:* Kujansuu, R. & Saarnisto, M. (eds) *Glacial Indicator Tracing*. A.A. Balkema, Rotterdam, 1–13.

Saarnisto, M., Tamminen, E. & Vaasjoki, M. 1991. Gold in bedrock and glacial deposits in the Ivalojoki area, Finnish Lapland. *Journal of Geochemical Exploration*, **39**, 303–322.

Salminen, R. 1992. Field methods. *In:* Kauranne, K., Salminen, R. & Eriksson, K. (eds) *Handbook of Exploration Geochemistry*. Volume **5**: Regolith Exploration Geochemistry in Arctic and Temperate Terrains. Elsevier, Amsterdam, 165–184.

Salonen, V.-P. 1987. *Observations on boulder transport in Finland*. Geological Survey of Finland, Special Paper **3**, 103–110.

Sauerbrei, J. A., Pattison, E. F. & Averill, S. A. 1987. Till sampling in the Casa-Berardi gold area, Quebec: a case history in orientation and discovery. *Journal of Geochemical Exploration*, **28**, 297–314.

Sauramo, M. 1924. *Tracing of Glacial Boulders and its Application in Prospecting*. Bulletin de la Commission Géologique de Finlande, **67**.

SHELP, G. S. & NICHOL, I. 1987. Distribution and dispersion of gold in glacial till associated with gold mineralization in the Canadian Shield. *Journal of Geochemical Exploration*, **27**, 315–336.

SHILTS, W. W. 1973. *Drift Prospecting; Geochemistry of Eskers and Till in Permanently Frozen Terrain, District of Keewatin, Northwest Territories*. Geological Survey of Canada, Paper **72-45**.

SHILTS, W. W. 1974. Physical and chemical properties of unconsolidated sediments in permanently frozen terrain, District of Keewatin. *In*: *Current Research*. Part A, Geological Survey of Canada, Paper **74-1**, 229–235.

SHILTS, W. W. 1975. Principles of geochemical exploration for sulphide deposits using shallow samples of glacial drift. *Canadian Institute of Mining and Metallurgy*, Bulletin **68**, 73–80.

SHILTS, W. W. 1976. Glacial till and mineral exploration. *In*: LEGGE, R. F. (ed.) *Glacial Till*. Royal Society of Canada, Special Publication **12**, 205–223.

SHILTS, W. W. 1977. Geochemistry of till in perennially frozen terrain of the Canadian Shield – application to prospecting. *Boreas*, **6**, 203–212.

SHILTS, W. W. 1978. Nature and genesis of mudboils, central Keewatin, Canada. *Canadian Journal of Earth Sciences*, **15**, 1053–1068.

SHILTS, W. W. 1984. Till geochemistry in Finland and Canada. *Journal of Geochemical Exploration*, **21**, 95–117.

SHILTS, W. W. 1993. Geological Survey of Canada's contributions to understanding the composition of glacial sediments. *Canadian Journal of Earth Sciences*, **30**, 333–353.

SHILTS, W. W. 1995. Geochemical partitioning in till. *In*: BOBROWSKY, P. T., SIBBICK, S. J., NEWELL, J. M. & MATYSEK, P. F. (eds) *Drift Exploration in the Canadian Cordillera*. British Columbia Ministry of Energy, Mines and Petroleum Resources, Paper **1995-2**, 149–163.

SHILTS, W. W. 1996. Drift exploration. *In*: MENZIES, J. (ed.) *Glacial Environments, Sediment Forms and Techniques*. Butterworth Heinemann Ltd., Toronto, 411–439.

SHILTS, W. W., AYLSWORTH, J. M., KASZYCKI, C. A. & KLASSEN, R. A. 1987. Canadian Shield. *In*: GRAFF, W. L. (ed.) *Geomorphic Systems of North America*. Geological Society of America, Centennial Special Volume **2**, 119–161.

SHILTS, W. W. & WYATT, P. H. 1989. *Gold and Base Metal Exploration Using Drift as a Sample Medium, Kaminak Lake – Turquetil Lake Area, District of Keewatin*. Geological Survey of Canada, Open File **2132**.

SHILTS, W. W. & KETTLES, I. M. 1990. Geochemical-mineralogical profiles through fresh and weathered till. *In:* KUJANSUU, R. & SAARNISTO, M. (eds) *Glacial Indicator Tracing*. A.A. Balkema, Rotterdam, 187–216.

SHILTS, W. W. & SMITH, S. L. 1988. Glacial geology and overburden drilling in prospecting for buried gold deposits, southeastern Quebec. *In:* MACDONALD, D. R. (ed.) *Prospecting in Areas of Glaciated Terrain*. Nova Scotia Department of Mines and Energy, Halifax, 141–169.

SIBBICK, S. J. & KERR, D. E. 1995. Till geochemistry of the Mount Milligan area, north-central British Columbia; recommendations for drift exploration for Porphyry Cu-Au mineralization. *In*: BOBROWSKY, P. T., SIBBICK, S. J., NEWELL, J. M. & MATYSEK, P. F. (eds) *Drift Exploration in the Canadian Cordillera*. British Columbia Ministry of Energy, Mines and Petroleum Resources, Paper **1995-2**, 167–180.

SKINNER, R. G. 1972. *Drift Prospecting in the Abitibi Clay Belt: Overburden Drilling Program Methods and Costs*. Geological Survey of Canada, Open File **116**.

SMEE, B. W. 1998. Overview of quality control procedures required by mineral exploration companies. *In*: *Workshop on Quality Control Methods in Mineral Exploration*. January 25–26, 1998, Association of Exploration Geochemists.

SMITH, S. L. & RAINBIRD, R. H. 1987. Soft sediment deformation structures in overburden drill core, Quebec. *In*: *Current Research*. Part A. Geological Survey of Canada, Paper **87-1A**, 53–60.

SNOW, R. J. & COKER, W. B. 1987. Overburden geochemistry related to W-Cu-Mo mineralization at Sisson Brook, New Brunswick, Canada: an example of short and long distance glacial dispersal. *Journal of Geochemical Exploration*, **28**, 353–368.

STEA, R. R. 1994. Relict and palimpsest glacial landforms in Nova Scotia, Canada. *In:* WARREN, W. P. & CROOT, D. G. (eds) *Formation and Deformation of Glacial Deposits*. A.A. Balkema, Rotterdam, 141–158.

STEELE, K. G. 1988. Utilizing glacial geology in uranium exploration, Dismal Lakes, Northwest Territories, Canada. *Boreas*, **17**, 183–194.

TARVAINEN, T. 1995. The geochemical correlation between coarse and fine fractions of till in southern Finland. *Journal of Geochemical Exploration*, **54**, 187–198.

THOMPSON, I. S. 1979. *Till prospecting for sulphide ores in the Abitibi Clay Belt of Ontario*. Canadian Institute of Mining and Metallurgy, Bulletin **72**, 65–72.

THORLEIFSON, L. H. & KRISTJANSSON, F. J. 1993. *Quaternary Geology and Drift Prospecting, Beardmore-Geraldton Area, Ontario*. Geological Survey of Canada, Memoir **435**.

TOVERUD, Ö. 1989. Geochemical prospecting for gold in the county of Jämtland, upper central Sweden. *Journal of Geochemical Exploration*, **32**, 61–63.

VEILLETTE, J. J. 1989. Ice movements, till sheets and glacial transport in Abitibi-Timiskaming, Quebec and Ontario. *In*: DILABIO, R. N. W. & COKER, W. B. (eds) *Drift Prospecting*. Geological Survey of Canada, Paper **89-20**, 139–154.

VEILLETTE, J. J. & NIXON, F. M. 1980. *Portable Drilling Equipment for Shallow Permafrost Sampling*. Geological Survey of Canada, Paper **79-21**.

WASHBURN, A. L. 1973. *Periglacial Processes and Environments*. Edward Arnold Ltd., London.

Regional till geochemical surveys in the Canadian Cordillera: sample media, methods and anomaly evaluation

VICTOR M. LEVSON

British Columbia Geological Survey, 1810 Blanshard Street, Victoria, British Columbia V8V 1X4, Canada (e-mail: Vic.Levson@gems9.gov.bc.ca)

Abstract: Basal tills have become a widely used regional geochemical sampling medium in recent years in the Canadian Cordillera. They reflect the primary composition of the source bedrock and contrast with B-horizon soil that can be developed on a variety of glacial and non-glacial surficial sediment types. Detailed sedimentological data are critical to collect and they are used to differentiate basal tills from other visually similar sediments including englacial and supraglacial tills, colluvial debris flow deposits, and very poorly sorted, glaciofluvial or glaciolacustrine sediments (e.g. diamictons or gravelly muds). The variable transport and depositional processes that form these different sediments make interpretation of geochemical data difficult. Deep (usually > 0.75 m) C-horizon sampling of basal till minimizes the complicating effects of pedogenesis, weathering, surface washing and gravity remobilization of the tills. The latter processes, particularly pronounced in the wet, steep terrain, typical of much of the Canadian Cordillera, lead to depleted concentrations of heavy minerals (notably Au) and hydromorphic dispersion of mobile elements in the near surface sediments. Also, elements that are preferentially concentrated in the fine fraction can be selectively removed by surface waters.

Offset sampling lines, oriented perpendicular to the dominant ice-flow direction, are most effective for detecting regional geochemical anomalies which are typically narrow and elongated parallel to ice-flow. Erratics trains and till anomalies are usually a few to several kilometres long and up to one or more kilometres wide. For some metals such as Au, anomalies are generally larger and more readily detected in till than in B-horizon soil. Surface till anomalies reflect up-ice sources and not the immediately underlying bedrock; down-ice displacements of > 500 m often occur in areas of thick till. Basal till anomalies usually can be traced to source along linear transport paths reflecting topographically controlled valley-glacier flow in mountainous areas and unidirectional ice-sheet flow in many plateau areas, chiefly representative of the last glacial event. Interpretations of till geochemical data are enhanced with a clear understanding of the surficial and bedrock geology, Quaternary stratigraphy, ice-flow history and down-ice dispersal characteristics around known mineral deposits.

One of the primary objectives of till geochemistry is to identify areas where glaciers eroded mineralized bedrock, transported and dispersed the debris in the direction of glacial flow, and redeposited the mineral-rich sediment over relatively large areas. The resultant deposits, containing elevated levels of elements that were concentrated in the source bedrock, are commonly referred to as till geochemical anomalies or glacial dispersal trains. Although a specific definition is preferred, 'geochemical anomaly' is commonly used to refer to areas where geochemical concentrations are elevated above background values. In this paper, specific definitions are used where possible, but otherwise this general meaning applies. Further discussion on this subject is provided later under the title 'Evaluation of till geochemical anomalies'. Since these dispersal trains may be hundreds of times larger in area than their original bedrock source, they provide a cost effective target for mineral exploration programs in drift-covered terrains (c.f. Shilts 1976; DiLabio 1990; Levson & Giles 1995). In addition, tills are 'first-derivative' products of bedrock (Shilts 1993) and, having been transported to their present

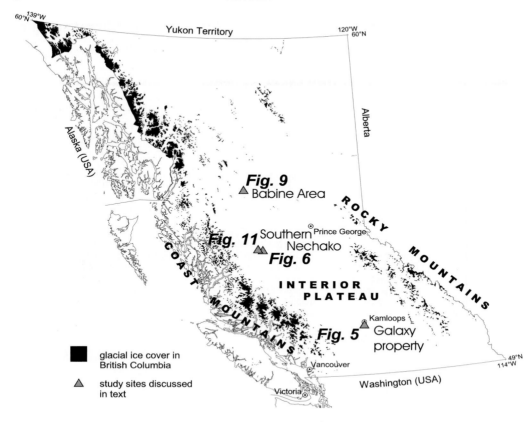

Fig. 1. Location map of properties discussed in detail in the text: Galaxy Cu–Au deposit (Fig. 5); Laidman Lake Au–Ag property (Fig. 6); Old Fort Mountain regional till geochemical survey (Fig. 9); and the Wolf Au–Ag prospect (Fig. 11). The current distribution of alpine glaciers in the British Columbia Cordillera is also shown in relation to the Coast Mountains, Rocky Mountains and the relatively low-lying Interior Plateau.

location mainly by relatively linear glacier flow during one or more glacial episodes, they are more readily traced to source than higher order derivative products such as glaciofluvial or glaciolacustrine sediments. In the simplest case of unidirectional ice flow, mineralized material at a point source is eroded, transported and redeposited to produce a narrow dispersal train, elongated parallel to the direction of ice flow. These trains are well documented for shield terrains, especially in Scandinavia and Canada (e.g. Kauranne 1959; Dreimanis 1960; Shilts 1971, 1973*a,b*; 1976; DiLabio 1981; Salonen 1989; Kauranne *et al*. 1992; Klassen 1997), and recent examples also have been described from the Canadian Cordillera (e.g. Fox *et al*. 1987; Kerr *et al*. 1992; Levson & Giles 1995, 1997).

Media-specific geochemical exploration techniques, such as till geochemistry, have been increasingly employed in the Canadian Cordil-lera in recent years. In particular, the British Columbia Geological Survey and the Geological Survey of Canada began publishing regional till geochemical survey results in the early 1990s (e.g. Kerr *et al*. 1992; Plouffe & Ballantyne 1993, 1994; Levson *et al*. 1994). The objectives of this paper are to discuss the use of till as a geochemical sample media, to describe methods of till geochemistry now being used in the Cordillera, and to provide information that can be used to evaluate the significance of till anomalies. Although mineralized erratics and geochemically anomalous surficial materials are commonly identified by traditional exploration methods in the Cordillera such as B-horizon soil sampling (using a variety of sediment types), the locations of their buried, mineralized bedrock sources often elude explorationists. Some of the reasons why bedrock sources are not found are discussed and examples are provided to illustrate

Fig. 2. Typical basal till deposit displaying poor sorting (diamicton), massive structure, matrix-supported framework, subhorizontal fissility and striated, subrounded to subangular clasts. Scale markings along top of card are in cm.

how till geochemistry can help resolve these difficulties. The location of examples discussed in detail in this paper are provided on Figure 1 along with some major physiographic elements of the region (specifically the Coast Mountains, Interior Plateau and Rocky Mountains) in relation to the current extent of alpine glaciers in the British Columbia Cordillera.

Previous work

Papers relating to till geochemical studies in the Canadian Cordillera are too numerous to describe here. A review of 85 published (pre-1994) studies on drift prospecting in the British Columbia Cordillera was provided by Kerr & Levson (1995). Since that review, many more studies and regional till geochemical surveys have been conducted in the Cordillera (e.g. Levson *et al.* 1994, 1997*a,b*; Plouffe 1995*a*; Plouffe & Jackson 1995; Weary *et al.* 1995; Sibbick *et al.* 1996; Bobrowsky & Sibbick 1996; Bobrowsky *et al.* 1997, 1998; Kerr & Levson 1997; Levson & Giles 1997; O'Brien *et al.* 1997; Plouffe & Williams 1998; Paulen *et al.* 2000). These studies provide overviews of till geochemistry and sampling techniques applicable in various parts of the Cordillera. A wealth of information is also contained in unpublished assessment reports filed with the B.C. Ministry of Energy and Mines (c.f. Kerr 1995). A number of till geochemistry case studies around mineral deposits and occurrences also have been published in recent years including work in the Tulameen Pt district (Cook & Fletcher 1993), at the Nickel Plate Au–Ag deposit (Sibbick & Fletcher 1993), Mount Milligan Cu–Au porphyry (Sibbick & Kerr 1995), Myra Falls Cu–Zn–Pb–Au–Ag mine (Hicock 1995), Grew Creek Au–Ag prospect, Yukon (Plouffe & Jackson 1995), the Bell Cu–Au porphyry deposit (Stumpf *et al.* 1997), at several properties in the southern Nechako region (Levson *et al.* 1997*c*; O'Brien *et al.* 1997), in the Babine porphyry belt (Stumpf *et al.* 1996, Levson *et al.* 1997*a*; Levson & Stumpf 1998) and in the Kootenay Terrane (Lett *et al.* 1998, 1999; Paulen *et al.* 1999).

Basal till as a geochemical sample medium

In this paper, basal till refers to unsorted or very poorly sorted sediment deposited at the base of a glacier with little or no reworking by water (c.f. Dreimanis 1990). Deposition usually occurs by lodgement or meltout processes but also by subglacial flow. 'Basal' in this usage does not imply a stratigraphic position at the base of a sequence and basal tills may occur at any depth between bedrock and the ground surface. Basal till is the preferred sampling medium used in till geochemistry programs in the Canadian Cordillera, for several reasons:

(1) basal till deposits are mainly a direct result of the last glacial event and, due in part to strong topographic control, the ice-flow histories of many parts of the Cordillera are dominated by one main direction throughout much of the last glacial period (Late Wisconsinan). This has resulted in generally linear (often valley-parallel), down-ice transport of material, thus making the tracing of basal till anomalies to source relatively easy compared to regions with complex ice-flow histories;
(2) basal tills are deposited in areas directly down-ice from their source and therefore mineralized material dispersed within the tills can be more readily traced to its origin than can anomalies in other sediment types. Processes of dispersal in supraglacial and englacial tills, glacigenic debris flows, glaciofluvial deposits, and glaciolacustrine sediments are more complex and they typically are transported farther than basal tills; and,
(3) due to the potential for the development of large dispersal trains, metal-rich material in basal tills may be readily detected in regional surveys.

Fig. 3. Massive sandy gravel (proximal glaciofluvial outwash) about 1 m thick, overlying 2 m of massive, silty diamicton (basal till). Note the sharp erosional contact (dashed line) between the glaciofluvial sediments and underlying till.

Distinguishing characteristics

Basal tills typically consist of dense, matrix-supported, silt and clay rich diamicton, contain many striated clasts, and display a strong unimodal fabric. They are generally over-consolidated and exhibit moderate to strong sub-horizontal fissility (Fig. 2). Sub-horizontal slickensided surfaces are sometimes present. Although not diagnostic, tills lying above the water table typically show vertical jointing and a blocky structure, commonly with Fe and Mn oxide staining along joint planes and partings. Clasts may range in size from granules to large boulders, usually with a mode in the medium to large pebble size range. Total clast content (>2 mm fraction) generally is between 10 and 30% but locally may be up to 50%. Subangular to subrounded clasts are most common and typically up to about 20% show evidence of glacial abrasion. Striated clasts may be bullet shaped or have faceted upper surfaces. The a-axes of elongate clasts are often aligned parallel to ice-flow direction. Lower contacts of basal till units are usually sharp and planar. All of these characteristics are consistent with a basal melt-out or lodgement till origin (Dreimanis 1990). During melt-out, debris banding in the basal ice may be preserved as crude, sub-horizontal bedding in the till. Sands and gravels deposited in subglacial tunnels may survive as plano-convex lenses within the till. Other varieties of basal till include subglacial flow tills and deformation tills. The former are similar in many respects to basal melt-out tills but individual flow deposits may be separated by sorted sediments. Injections of till into bedrock fractures locally indicate high pressure conditions at the base of the ice during deposition. These features, as well as the presence of sheared, folded and faulted bedrock or other sediments within the diamictons, indicate the local development of deformation tills (c.f. Bennett & Glasser 1996; Boulton 1996). A more complete discussion of the distinguishing characteristics of tills found in mountainous regions is provided by Levson & Rutter (1988).

Basal till deposits can be confused with other types of glacial or nonglacial sediments such as

Table 1. Distinguishing characteristics of basal till, supraglacial till, debris-flow deposits and colluvial diamicton[1]

Sediment type	Typical characteristics and sedimentary structures	Matrix texture/density	Clast provenance[2]/shape	Pebble fabric[3]	Lower contact	Associated landforms
Basal till[4]	massive, matrix-supported, diamicton; overconsolidated; fissile; oxidized blocky joint planes; shear structures; thrust faults	sand-silt-clay; very dense	mainly local; striated, faceted; subangular to subrounded	mod. to strong fabric, parallel to palaeoflow	sharp and planar (erosional)	level to rolling moraine, drumlins, flutings, crag-and-tails
Supraglacial till[5]	massive to crudely stratified, diamicton; matrix- or clast-supported; normally faulted or collapsed sand and gravel lenses	sandy; usually loose	far-travelled, from high elevations; large angular to subangular	chaotic or weak, random orientations	gradational, irregular	hummocky moraine, kames, kettles
Cohesive debris flows[6]	massive to crudely stratified, diamicton; matrix-supported, ungraded or coarse-tail inverse grading; thin discontinuous sandy or gravelly stratas	sand-silt-clay; loose to compact	local to distal; subangular to subrounded	weak fabric, parallel to slope or palaeoslope	clear to gradational; subhorizontal	streamlined or hummocky moraine; gentle to moderate slopes
Non-cohesive debris flows[7]	crudely stratified diamicton/poorly sorted gravel; clast-supported; ungraded or normal grading; often interbedded with trough-shaped sand and gravel lenses	sand-silt; usually loose	distal and local; well rounded to subangular	weak to moderate fabric; rare imbrication	gradational to sharp; often trough-shaped (scoured)	proximal glaciofluvial: sandurs, eskers, kame terraces, kame deltas, fans
Subaqueous flows[8]	massive to stratified, diamicton/gravelly mud; matrix-supported, silt and clay laminae; folded and convoluted strata	silt-clay-sand; compact	distal; subangular to subrounded; mainly pebbles	chaotic or weak *a*-axis fabric	sharp horizontal loaded	lacustrine basins in valley bottoms; often near large lakes
Colluvium	massive to crudely stratified, diamicton; clast-supported; strata parallel slope	sandy; often very loose	dominantly local; subangular to angular	weak to strong downslope dip	clear, parallel to slope	moderate to steep bedrock slopes

[1] based on data from Levson & Rutter (1988), Dreimanis (1990) and Levson & Giles (1997)
[2] local provenance indicates source rocks within a few kilometres; distal indicates source is more than a few kilometres
[3] fabric strength (S_1) is a normalized measure of clustering, varying from weak (S_1 < about 0.5) to strong (S_1 > about 0.7)
[4] mainly lodgement and melt-out till but also may locally include subglacial flow and deformation tills
[5] also commonly referred to as ablation till and deposited by a variety of mechanisms including glacigenic debris flow
[6] also referred to as mudflows; locally includes colluviated (remobilized) but not fluvially reworked till
[7] mainly gravelly and sandy debris flows; locally includes hyperconcentrated flood flows and fluvially reworked tills
[8] includes diamicton (gravelly mud) derived mainly from ice-rafted debris

Fig. 4. Valley glacier in the Coast Mountains, illustrating various depositional settings in the glacial environment and possible confusion of basal till with other sediment types: (1) basal till on a gentle slope with little reworking by water or gravity; (2) remobilized till on a steeper slope (just below sharp crest of lateral moraine); (3) supraglacial debris in a medial moraine that may eventually be deposited on top of basal till; (4) glaciofluvial outwash along the ice margin (outwash gravels are commonly interbedded with debris flow deposits); (5) glacial lake and glaciofluvial delta (proximal glaciolacustrine sediments may contain subaqueous debris flow deposits); and (6) colluvium along the valley sides and at the base of steep rocky slopes.

supraglacial tills, subaerial debris flow deposits, subaqueous debris flow deposits, debris-rich glaciolacustrine deposits, and poorly sorted colluvial deposits (Fig. 3). A summary of some characteristics useful for distinguishing basal tills from these other deposits is provided in Table 1. This distinction is critical as the dispersal characteristics of these different sediment types vary widely. Basal tills are first order derivative products whereas glacigenic debris flow deposits, for example, have undergone a second depositional phase, related either to the palaeo-ice surface or the present topography, and they are therefore more difficult to trace to their source.

The deposits most easily confused with basal tills are subaerial and subaqueous sediment gravity flows deposited in glacial (e.g. supraglacial), non-glacial (colluvial), or paraglacial environments. Supraglacial tills are most readily recognized by the presence of far-travelled, angular clasts and by association with hummocky topography (Table 1). Deposits that are particularly difficult to differentiate from basal tills are those sediments derived directly from basal debris within the glacier (e.g. ice-marginal debris flows) or from previously deposited basal tills that have been remobilized along slopes (i.e. colluviated tills; Fig. 4). These deposits are common throughout the Cordillera due to typically high relief and widespread ice-stagnation at the end of the last glaciation (Fulton 1991). Even in areas with relatively low slopes such as the British Columbia Interior Plateau (Fig. 1), it is rare to find undisturbed basal tills at the surface due to weathering and reworking of the tills by surface water. Surficial processes that alter the physical properties of tills and sometimes their chemical composition include: frost action, oxidation, wetting and drying, eluviation and other soil forming processes. Surface water percolation can result in the selective downward movement of fine-grained sediments, mobile elements and heavy minerals such as Au. This is especially true in wet-temperate mountain environments, characteristic of the Canadian Cordillera. Finally, and possibly most significantly, exposed tills on even the gentlest of slopes, may be washed by fluvial overland flows or they may become saturated and begin to move downslope as sediment gravity flows. These processes allow for fluvial and gravity sorting of the sediments and result in sandy surficial layers that are further depleted in fines, heavy minerals, and soluble elements. These latter processes were particularly important in the paraglacial environment that dominated at the end of the last glaciation, prior to the development of a vegetative cover (c.f. Fig. 4). All of these processes primarily effect till near the surface (upper 0.5 m) and for this reason samples collected below about 0.5 m depth, and deeper on slopes, are preferable.

Although differentiation of the origin of sediment gravity flows is often a complex exercise, they can be divided rheologically into cohesive and non-cohesive varieties with different sedimentological characteristics. Cohesive debris flow deposits are similar to basal tills as they are commonly massive, compact and clay rich. However, they may be distinguished from basal till by features such as the presence of sand and gravel interbeds, crude stratification, locally graded bedding, and weak fabric. Non-cohesive debris flows (where grain-support is largely due to clast collisions rather than cohesive matrix strength) are more distinctive, usually being dominated by loose granular materials and often exhibiting clast-support, stratification, normal grading, weak to moderate fabric and even imbrication (Table 1). The differentiation of subaerial and subaqueous debris flow deposits can often be accomplished by association with other sediment types. For example, basal tills reworked subaerially by gravity or water, generally contain lenses or interbeds of stratified silt, sand and gravel. Debris flow deposits associated with these sediments are typically loose, sandy, massive to crudely stratified diamicton, generally containing 20 to 50% gravel, although higher clast concentrations can occur locally. Subangular to subrounded clasts are most common, but angular clasts that reflect the incorporation of local bedrock during remobilization, also may be present. Subaqueous debris flows often are matrix-rich, have comparatively low clast contents and contain interstratified silt and clay laminae. All debris flow varieties may exhibit chaotic or weak pebble fabrics and in exposures they commonly are in gradational contact with underlying or overlying basal tills. Glaciolacustrine deposits containing abundant ice-rafted debris have similar characteristics to subaqueous debris flows and, texturally, they are often pebbly muds with minor sand. Colluvial diamictons derived mainly from local bedrock are differentiated from basal tills by their loose unconsolidated character, the dominance of coarse angular clasts, crude stratification and lenses of sorted sand and gravel (Table 1).

Dispersal characteristics of basal tills in the Cordillera

Many historical studies of mechanical dispersal in the Canadian Cordillera have been based on geochemical analysis of soils and on clast studies (e.g. Bradshaw *et al.* 1974; Boyle & Troup 1975; Mehrtens 1975; Levinson & Carter 1979; Fox *et al.* 1987; c.f. examples in Kerr & Levson 1995; Levson & Giles 1995). As a result, patterns of glacial dispersal and information on factors such as glacial transport distances are somewhat obscured by the masking effects of other sediment types and soil forming processes that vary from site to site. However, many of these soil geochemical surveys have been conducted in areas dominated by glacial tills and some generalizations regarding glacial dispersal can be made. In addition, a number of studies have been conducted in recent years in the Cordillera that characterize the size and shape of till geochemical anomalies, using basal tills as a specific sample media (e.g. Kerr *et al.* 1993;

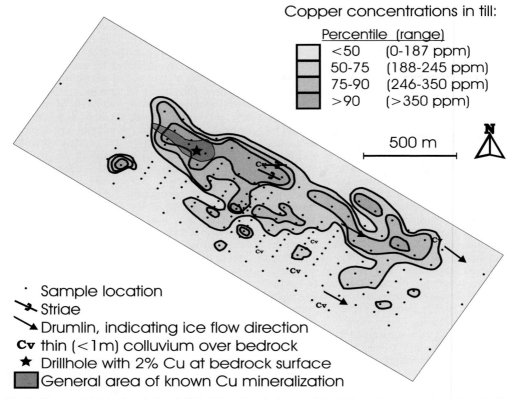

Fig. 5. Copper concentrations in basal till (−177 μm fraction) around the Galaxy Cu–Au porphyry deposit, 5 km SW of Kamloops (Lat. 50° 38′, Long. 120° 25′; Fig. 1), south-central British Columbia (modified from Kerr et al. 1993; drill hole data from Belik 1990). Note the general area of elevated Cu concentrations down-ice of mineralized bedrock and elongated parallel to the ice-flow direction. Isolated high Cu concentrations may reflect variable dispersal or other small areas of bedrock mineralization.

Plouffe & Jackson 1995; Levson & Giles 1997; O'Brien et al. 1997). These studies have shown that dispersal patterns of subglacial materials typically have a pronounced preferential elongation in one general direction (e.g. Fox et al. 1987; Kerr et al. 1993; Levson & Giles 1997; Stumpf et al. 1997). The dispersal trains generally are elongated parallel to the last dominant ice-flow direction and mineralized source rocks occur at or near their up-ice end. For example, Kerr et al. (1993) describe a Cu anomaly in till extending at least 1500 m down-ice of the Galaxy Cu–Au porphyry deposit (Fig. 5). Cu concentrations, averaging 136 ppm, occur in the most distal sample line from the deposit suggesting that a significant (>100 ppm) Cu anomaly extends farther down-ice. In general, till geochemical anomalies in the Cordillera are up to several kilometres long and >1 km wide but discontinuous occurrences of mineralized material, resulting from glacial dispersal of the same mineralized bedrock sources, may occur over much larger areas.

Dispersal trains are commonly narrow in comparison with their length and highly anomalous areas (e.g. >98th percentile) often have generally sharp lateral and vertical contacts with the surrounding till. Length to width ratios of 3:1 to 5:1 are typical. Progressive dilution of the mineralized material generally occurs in a down-ice direction until elevated geochemical concentrations can no longer be detected. Elongated anomaly patterns are particularly well developed along valleys in areas of high relief where dispersal trains extend parallel to the long-axes of the valleys. Hicock (1995), for example, found elevated Cu concentrations (300–500 ppm) in the <2 μm fraction of till at least 15 km down-ice from the Westmin Lynx Cu–Zn–Pb orebody. Zn and Pb concentrations, although erratic and generally decreasing down-valley, also were elevated (>200 ppm Zn and

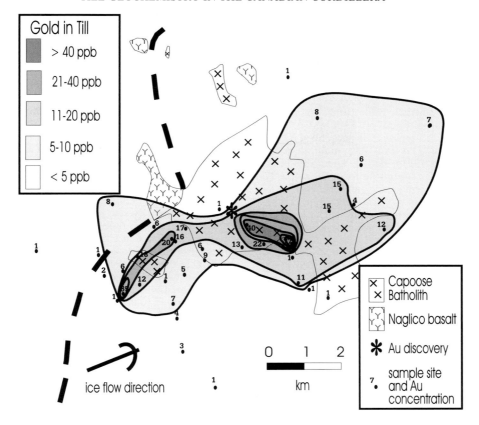

Fig. 6. Gold concentrations in basal till (−63 μm fraction) around the Capoose Batholith, northwest of Laidman Lake (Lat. 53° 10′, Long. 125° 14′; Fig. 1), central British Columbia. A regional till geochemical survey in the area (Levson et al. 1994) identified anomalous Au concentrations and subsequent exploration yielded a new gold discovery (indicated by a star on the figure) containing up to 19.6 g/t Au and 99.9 g/t Ag in quartz veins (Schimann 1995; Payne 1996). The dispersal pattern suggests that at least one other Au source in the area is yet to be discovered, up-ice (southwest) of the main till anomaly (north at the top of the figure).

> 40 ppm Pb) at several sites up to about 15 km from the orebody. A surprisingly similar glacial dispersal pattern also occurs in plateau areas within the Cordillera, even though topographic relief relative to former glacier thickness is low (e.g. O'Brien et al. 1997). In these areas, ice flow direction may have varied significantly during the course of the last glaciation but dispersal patterns tend to reflect the last dominant flow direction of subglacial ice and water as indicated by landforms such as drumlins, crag-and-tail ridges and flutes.

Till geochemical anomalies reflect up-ice bedrock sources and not the immediately underlying bedrock. Hicock (1995) found that the head of the dispersal train, where Cu, Pb and Zn concentrations in till were highest, was about 1 km down-glacier from the mineralized source.

Similarly, Levson & Giles (1995) reported that near-surface geochemical anomalies in till were displaced > 500 m down-ice from their bedrock source in areas where the till was several metres or more thick, and < 100 m in areas of shallow (< 1 m thick) till. Thus, subsurface exploration targets in drift covered areas should be located up-ice, rather than at the head, of the dispersal train and, in general, exploration activities should be conducted at greater up-ice distances as till thickness increases.

Few Cordilleran studies have investigated the three-dimensional geometry of till dispersal trains due to the difficulties of obtaining basal till samples at depth over a large area. Exploration drilling programs are common but cored overburden samples are rarely obtained. One exception to this was described by Levson et al.

(1997a) in the Babine Cu porphyry belt where diamond drill core of a thick till sequence overlying the Nak Cu porphyry prospect was obtained. Geochemical analyses of 58 intact till samples, recovered from 5 to 60 m depth, showed a median Cu concentration of 66 ppm. This value is elevated relative to the regional background (median value) of 41 ppm and the 75th percentile value of 53 ppm Cu in tills determined for regional surface till data from the adjoining Old Fort Mountain map sheet (Levson et al. 1997b; Fig. 1). The drill hole data show that Cu concentrations in till at the Nak property, are remarkably uniform over a substantial thickness (>60 m) and that compositional stratification in the till is not well developed. This vertical geochemical homogeneity reflects the relatively large size of the porphyry Cu prospect that underlies the till. Very high Cu concentrations (700-1000 ppm) in till were found only over one part of the deposit (at 2 of 19 holes sampled) and probably reflect locally enriched Cu mineralization within the porphyry.

Although till dispersal trains are usually long and narrow, variations in the dominant ice-flow direction in some areas have produced wider fan-shaped dispersal trains (c.f. Mehrtens 1975; Levson & Giles 1995). Complex ice flow patterns can be the result of topographic irregularities, shifting source areas, changing dynamics at the base of the ice, or multiple glaciations. Diffuse or irregularly-shaped dispersal trains (e.g. Fig. 6) that are difficult to trace to source also may reflect other controlling factors such as variable overburden thickness or multiple sources of bedrock mineralization. O'Brien et al. (1997) described the effects of topography and other factors on glacial dispersal at the CH, Blackwater-Davidson and Uduk Lake mineral properties in the Nechako River area. At each property, geochemical dispersal trains are typically elongated parallel to the last major ice-flow direction but the dimensions of the dispersal patterns are different. These differences are attributed to variations in glacier velocity and erosiveness and to variations in the size and type of mineralization of the bedrock source. Consequently, geochemical exploration programs in drift-covered regions must rely on an understanding of glacial processes and the glacial history of the area. In addition, more research is required in the Cordillera to evaluate relationships between regional glacial dispersal directions and both large-scale, streamlined landforms (e.g. drumlins) and smaller-scale subglacial ice and water erosion features (i.e. striae and sculpted bedrock forms).

Till geochemical surveys

Survey types

Till geochemical surveys conducted in the Cordillera can be grouped into three scales of surveys referred to here as reconnaissance, regional and property-scale surveys. Reconnaissance till geochemical surveys have been conducted in the Cordillera mainly by the Geological Survey of Canada (GSC) and are generally completed at 1:100,000 to 1:250,000 scales, with sample densities on the order of about 1 sample per 20–50 km^2 (e.g. Plouffe & Ballantyne 1993, 1994; Plouffe 1995a). Due to the large areas covered by reconnaissance till geochemical surveys, they are conducted mainly along roads and consequently sample density varies substantially across the survey areas due to variable road access. Regional surveys in the Cordillera have been conducted mainly by the British Columbia Geological Survey (BCGS), generally at 1:50,000 scale with an average sample density varying from 1 sample per 3 km^2 (e.g. Levson et al. 1997b) to 1 per 7 km^2 (e.g. Weary et al. 1995). A sample density of about 1 sample per 5 km^2 is probably sufficient to intersect narrow dispersal trains that are on the order of 10 km^2 in size, provided that the sample design is adequate (see below).

Till geochemical surveys are media specific (as opposed to soil geochemical surveys which are not) and can only be conducted in areas where basal tills are present at a density that can be reasonably sampled within the scope, scale and budget of the proposed survey. Extensive till cover occurs in most areas of the Cordillera especially in broad valley and plateau settings where typically about 50–90% of the region may be till covered. As a result, there are few areas in the Cordillera where reconnaissance till geochemical surveys could not be conducted. However, the use of regional and, particularly, property-scale surveys can be restricted by the absence or deep burial of tills (e.g. in areas where till is covered by glaciofluvial outwash or glaciolacustrine sediments). Property-scale exploration programs in these areas often have to rely on other sample media or other methods such as biogeochemical or geophysical surveys. Property-scale exploration geochemical studies have been completed around many mineralized areas in the Cordillera but most of these are soil geochemical sampling programs that do not discriminate between soils developed on basal tills from those developed on other surficial sediment types. A few typical examples of soil surveys from the Nechako region in central

British Columbia, including studies at the Wolf Au–Ag property (Delaney & Fletcher 1994; Fig.1); the Arrow Lake stibnite occurrence (Bohme 1988); and the CH Pb–Zn–Ag–Au–Cu property (Edwards & Campbell 1992), were reinterpreted in the context of glacial processes by Levson & Giles (1995) and O'Brien et al. (1995, 1997). The results of that work are discussed in a subsequent section comparing soil and till geochemical surveys.

Design considerations

Important factors to consider in a survey design include: purpose and type of survey; size of the study area; type and distribution of surficial materials; ice flow history; geomorphology and topography; and, size and type of exploration targets. Ideally, the sample density will be sufficient to provide complete coverage of the survey area and samples will be evenly distributed. Some economy of sampling can be achieved, however, if the size, shape and orientation of the target dispersal trains are known or can be inferred. For example, the probability of intersecting long, narrow, dispersal trains that are elongated parallel to the ice-flow direction is lower with a regularly spaced sample grid than with a grid where the sample locations on adjacent lines are offset (Figs. 7a & b). Sample lines with a close sample spacing along transects perpendicular to the ice-flow direction also can be relatively widely spaced (Fig. 7c). In addition, the orientation of sample lines can be adjusted to account for variable ice flow directions in the survey area (Fig. 7d).

Thus, an effective sample design clearly requires an understanding of the main direction of glacial dispersal as well as the size, shape and orientation of the buried exploration target. In the Babine porphyry belt, for example, Cu–Au porphyry deposits are generally large (about 0.5 to 1 km^2) and roughly equidimensional (Carter et al. 1995). The dispersal train from one such deposit at the Bell mine is 5–7 km in length and 1–2 km in width (Stumpf et al. 1997). Therefore, sample lines spaced <3 km apart and oriented perpendicular to ice flow should intersect a similar train twice. Likewise, with a sample spacing averaging about 1.5 km, at least one sample per line should intersect the train. Thus, following a similar design, a regional sample density of approximately 1 sample per 5 km^2 should be adequate to intersect such a dispersal train twice. With a wider line spacing of 5 km (sample density of 1 per 8 km^2), at least one line would cross the dispersal train.

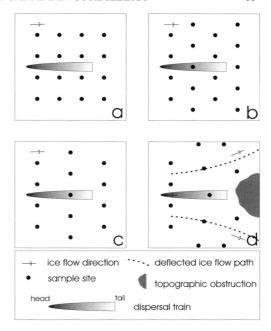

Fig. 7. Regular grid (**a**) and offset or staggered grid (**b**) sample designs for detection of narrow, till geochemical anomalies. The most effective design for anomaly detection utilizes offset sample lines, oriented perpendicular to ice flow direction (shown by arrows). (**c**) Offset grids oriented perpendicular to ice-flow direction allow for wider line spacing without significant loss of detection capability. (**d**) Lines orientation can be adjusted to reflect variations in ice flow, such as might occur around topographical obstructions (shaded area).

Field techniques

Sampling techniques used in till geochemistry programs in the Canadian Cordillera vary with the scale and purpose of the survey. The procedures summarized here are mainly those used during regional till geochemical sampling programs, particularly by the BCGS, but similar methods may be applied to both reconnaissance and property-scale surveys. A primary area of concern in the field is ensuring that sample media are correctly identified and that basal tills are not confused with other sediments. This requires the collection of sufficiently detailed and relevant data at each sample site. In particular, sedimentological data should be collected at all sample sites to aid in distinguishing basal till from sediments such as glacigenic debris flow deposits (possibly of supraglacial or englacial origin), colluvial sediments and glaciolacustrine diamicts. The different processes of transporta-

DATE (d, m, y)		SAMPLE #	
COLLECTOR		NTS	
TYPE	____ 0 - routine ____ 10 - first duplicate ____ 20 - second duplicate		
UTM EAST	UTM NORTH		ELEV (m)
MAP UNIT		SAMPLE MEDIUM	

DEPTH (cm):	
EXPOSURE: roadcut____ streamcut____ lakecut____ pit____ drill____ other____	
TOPO POSITION: ridge crest____ midslope____ lower slope____ level____ other____	
ASPECT	SLOPE (deg)
DRAINAGE: rapid____ well____ mod. well____ imperfect____ poor____ v. poor____	
VEGETATION:	
SOIL HORIZONS (cm): Dist.____ LFH____ Ah____ Ae____ Bm____ Bf____ other____	
FISSILITY: none____ weak____ moderate____ strong____	
DENSITY: very loose____ loose____ medium____ dense____ very dense____	
JOINTING: none____ weak____ moderate____ strong____	
OXIDATION: none____ weak____ moderate____ strong____	

MATRIX %	60 70 80 90 other:
COLOR	light grey____ dk grey____ light brown____ dk brown____ other____
TEXTURE	S____ ZS____ CS____ Z____ SZ____ CZ____ SZC____ other____
CLASTS	MODE: small pebble____ medium pebble____ large pebble____ MAX (cm):
modal SHAPE	angular____ subangular____ subrounded____ rounded____ well rounded____
% with STRIAE	none____ rare (<1%)____ common (1-10%)____ abundant (>10%)____

BEDROCK:
PHOTOS:
COMMENTS & SED STRUCTURES

Fig. 8. Example of field data form used in regional till geochemistry surveys in British Columbia (e.g. Levson *et al.* 1997*b*).

tion and deposition associated with these sediments must be recognized in order to understand mineral anomaly patterns. For example, local variations will be reflected in some sediments while regional trends may be observed in others. Analysis of these sediments will be useful only where their origin is understood. For these reasons, sedimentological data collected at each sample site should include descriptions of sample medium lithology (e.g. silty diamicton), primary and secondary structures (e.g. stratification, oxidation, jointing, and fissility, if present), matrix texture, density, descriptions of clasts (e.g. abundance, size, roundness, striae), and stratigraphic data. Information on soils (e.g. type and thickness of horizons), geomorphology (e.g. aspect, slope), topographic position (e.g. lower slope), terrain map unit (e.g. till plain), drainage, bedrock geology and vegetation should also be collected. An example of a form used to record field data is provided in Figure 8. In addition to the above sedimentological and geological data, field data forms should include fields for sample number, date of sampling, collector, map sheet number, sample type (routine or duplicate sample), location coordinates, elevation, sample depth, type of exposure (e.g. road cut), photo numbers and brief photo descriptions. Additional important information can be provided on the back of the data form such as pebble count data (clast lithology, abundance, shape etc.) and descriptions of mineralized clasts which are obviously critical for exploration purposes. A more complete discussion of field methods and collection of relevant data are provided by Plouffe (1995*b*) and Levson (1999).

Basal tills in high relief areas are commonly altered at surface by mass wasting, creep, slope wash, and other processes (Fig. 4). As a result, sample depths of usually at least 0.5 m and often about 1 m are required to obtain an undisturbed basal till sample. In some cases, such as areas where glaciofluvial outwash or other sediments occur at the surface, sample depths of several metres or more may be required. For this reason, regional sampling programs utilize pre-existing natural or man-made exposures (e.g. stream, river, lakeshore and road cuts) as much as possible. However, in many cases, till samples can only be obtained by hand digging pits or by augering. Caution must be exercised when using these latter methods as they restrict the amount of sedimentological data that can be obtained. In addition, basal tills are overconsolidated and difficult to penetrate with hand tools and, as a result, the sediment directly above the till often is sampled by mistake.

In most regional and reconnaissance till geochemistry surveys in the Cordillera, surficial geological mapping and studies of the ice-flow history and Quaternary geology are conducted in conjunction with the sampling program. These mapping activities include aerial photographic interpretation, field checking of surficial geology map units, stratigraphic and sedimentological investigations of Quaternary exposures and measurement of the orientation of crag-and-tail features, flutings, drumlins and striae. Care should be taken to distinguish between features formed by subglacial water erosion (e.g. Kor *et al*. 1991) and those formed by ice. Information on the ice flow history is required for adequate interpretation of the till geochemical data. Interdisciplinary programs, such as those that combine surficial and bedrock geological mapping, till geochemistry, lake sediment geochemistry, biogeochemistry and mineral deposit studies, have been particularly successful at locating new exploration targets (e.g. Cook *et al*. 1995).

Clast analysis

Till sampling programs should include an evaluation of clasts present in till to aid in geochemical interpretations and to identify evidence of mineralization. Analysis of glacially transported clasts also is useful for the following: relating till-clast lithology to bedrock lithology (for example, to aid in bedrock mapping), deciphering patterns of glacial dispersal, and/or to determine distances of glacial transport (e.g. using rates of clast abrasion and rounding). Since erratics trains (glacially transported clasts of a distinctive rock type or types) are often much longer and more easily detected than soil or till geochemical anomalies (Levson & Giles 1997), clast provenance investigations (e.g. boulder tracing) are often used as an initial phase in drift exploration programs. Clast studies typically involve the lithological identification of a representative sample of clasts (commonly 100 or more) from one or more size ranges (commonly medium to large pebbles). The number of clasts of each lithology identified is recorded as are clast sizes, angularity and abrasion characteristics. The use of general lithology and 'unidentifiable' categories is recommended when specific identifications are uncertain, as is often the case for small or weathered clasts. Depending on the purpose of the geochemical survey, the lithological identification of a large number of clasts can be an unjustifiably time-consuming process. However, it can be streamlined by grouping clasts into selectively useful and easily identifiable categories. For example, lithological analyses can focus specifically on the identification of mineralized clasts. At a minimum, general clast identification and observations of approximate clast abundance (frequency percent) can be made on site during the till sampling process.

Laboratory methods and quality control

Several authors (e.g. Kauranne *et al*. 1992; Lett 1995) have discussed laboratory methods and quality control procedures for till geochemistry and, consequently, only a brief outline of those commonly used in the Cordillera is provided here. Basal tills collected during geochemical surveys are typically air dried, split (with half of each sample archived) and sieved to −230 mesh ($< 62.5 \,\mu m$, silt and clay fraction). This fraction is commonly subjected to an aqua regia digestion and analysed by any of several methods such as inductively coupled plasma emission spectrometry (ICP-ES) and instrumental neutron activation (INAA). The −230 mesh fraction is used because it is dominated by phyllosilicates which are generally enriched in metals (Shilts 1993, 1995) and because it can be prepared for analysis rapidly and inexpensively. In some Cordilleran studies, such as those conducted by the GSC (e.g. Plouffe & Ballantyne 1993), the clay size ($<2\,\mu m$) fraction also is analysed, but due to the greater costs associated with clay separations, this is not standard industry or provincial survey practice in geochemical studies. Heavy mineral separations of the sand fraction (e.g. 0.063–0.250 mm) and accompanying geochemical and mineralogical analyses of the separates are sometimes conducted but, due to the additional costs, these analyses are often performed only on selected samples.

In order to discriminate geochemical trends related to geological factors from those that result from spurious sampling or analytical errors, a number of quality control measures should be implemented in both the field and laboratory components of till geochemistry programs. These include the use of field duplicate samples, blind analytical duplicate samples and standard reference materials which are randomly inserted into sample batches submitted for analysis. Field duplicates are taken from randomly selected field locations and are subjected to the identical laboratory preparation procedures as their replicate pairs. Analytical duplicates are sample splits taken after laboratory preparation procedures but prior to analysis. A more complete discussion of field and blind duplicates and control reference standards used in Cordilleran geochemical studies is provided by Lett (1995).

Evaluation of till geochemical anomalies

Threshold identification and background variations

One of the main problems in the interpretation of geochemical data is in the identification, from an exploration perspective, of 'significant' element concentrations. This identification is especially important for till geochemical data where differences between background and anomalous concentrations are small, compared to concentrations in colluvial or residual soils derived directly from nearby bedrock. For example, in soil survey results from many areas in the Cordillera, high metal concentrations often reflect thin soil developed on bedrock whereas relatively low metal concentrations occur in areas with thick till. Unfortunately, this relationship between metal concentrations and parent material is not always recognized. During data interpretation all sediment types are treated as a single population and anomalies are identified only in the shallow soil areas. As a result, the areas with the greatest potential for bedrock mineralization are not necessarily recognized. Instead, the distribution of residual or colluvial soils is simply highlighted, a process that could have been completed by surficial mapping alone without a geochemical survey. This example highlights the importance of distinguishing different types of parent material (sample media) when interpreting geochemical data. In such cases, the survey data should be separated for statistical analysis into populations reflecting different parent materials with distinctive geochemical backgrounds.

Identifying 'significant anomalies' is difficult, even in single-media basal till studies, because a large number of factors effect the ultimate concentration of any element in till. These include, but are not restricted to, the original concentration of the element in the source rocks, the erodability of the source rocks compared to the adjacent country rock, the processes of entrainment and deposition, position of transport and distance from source. These factors may lead to substantial natural geochemical variability and, since most of them are unknown, it is not always possible to predict element concentrations in till that are clearly indicative of significant mineralization in the underlying bedrock. One way to deal with this problem, in part, is to utilize data from case studies conducted around known mineral deposits to develop a local model of glacial dispersal (see discussion below). This model can then be used to explore for geologically similar targets within areas where similar glacial dynamics are expected to have been operating.

Although simple methods such as the use of probability plots, quantile plots, one-dimensional scatter plots and box plots (e.g. Kurzl 1988) can be helpful in defining distinctive populations in a till geochemical data set, there are no absolute values that can be considered to universally define thresholds between anomalous, background (often taken as the median in regional surveys; e.g. Cook *et al.* 1995) or intermediate populations for any specific deposit type. One approach for the presentation of till geochemical data uses specific percentile ranges (e.g. 50th, 70th, 90th, 95th) determined from ranked element concentrations (e.g. Fig. 9). This approach is a convenient way of presenting large volumes of data without assigning any specific 'significant anomaly' thresholds and, for this reason, it is often used for the presentation of

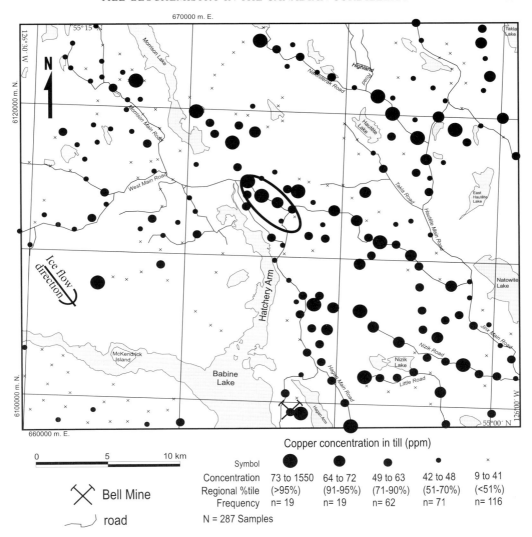

Fig. 9. Copper concentrations in basal till (−63 μm fraction) in the area covered by the Old Fort Mountain (north central British Columbia; Fig. 1) regional till geochemical survey (from Levson et al. 1997b). Copper concentrations in several areas with no known Cu mineralization (e.g. circled area at the north end of Hatchery Arm) are comparable to values (>95th percentile) found directly down-ice of known mineral deposits such as the Bell Cu–Au mine located on the southern edge of the map (Cu concentrations at sites in the circled area and down-ice of the Bell mine are shown on Fig. 10).

reconnaissance and regional till geochemical data sets. Interpretation of the significance of any particular element concentration is left to the user to evaluate for any specific area of interest. This method has the advantage that sites with relatively high (e.g. >99th percentile) concentrations of one or more elements can be easily identified and highlighted for further investigation. This approach is less suitable for data sets that are numerically or aerially small.

A disadvantage of the percentile method is that element concentrations that are elevated, but not highly anomalous (e.g. 90th percentile), may be overlooked even though they may be as significant or even more significant than higher values. One of the many possible reasons for the latter, is that the transport distance of the till from the mineralized source rocks will vary from

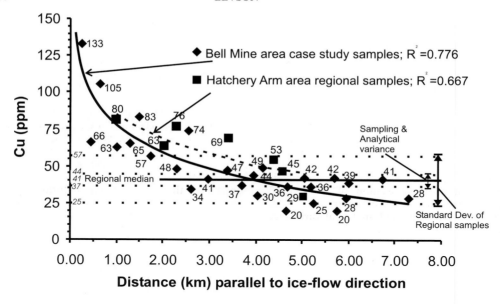

Fig. 10. Copper concentrations in basal till (−63 μm fraction) v. distance down-ice from the Bell Cu–Au mine (thick solid curve) compared with a selection of regional samples from the Hatchery Arm area (thick dashed curve; see Fig. 9 for site locations) where no mineral deposit has been identified (data from Levson et al. 1997b; Stumpf et al. 1997). Although only a few regional till samples were collected in the Hatchery Arm area, the similarity of the two curves suggests that the area may have potential for bedrock Cu concentrations similar to the Bell mine area. The regional median for the entire Babine Cu porphyry belt ($n = 937$) is shown as a horizontal line and one standard deviation above and below that line (dotted lines) is shown as an indication of the natural regional variability. The combined sampling and analytical variability (calculated as the average relative standard deviation of 57 field and analytical duplicates) is 7.5% and is shown in relation to the regional median. (R^2 is the square of the Pearson product moment correlation coefficient.) See text for discussion.

site to site. Thus, for example, a 90th percentile concentration at one site may reflect a more significant mineral source than a 99th percentile concentration at another site, if the former site is much further from the source mineralization. If the anomaly threshold is lowered to include elevated but not highly anomalous element concentrations, then the number of 'anomalies' becomes unmanageable from an exploration perspective. This shortcoming can be alleviated somewhat by evaluating more than one regional till geochemical site at a time. This is achieved by directly comparing element concentrations, at as many sites as possible down-ice of prospective areas, with glacial dispersal curves. These curves can be derived from local case study data obtained at known mineral occurrences or deposits (e.g. Fig. 10), that are similar to the exploration target(s). They show how rapidly element concentrations decrease with distance down-ice (i.e. the rate of decay). Using dispersal curves, an estimate can be made of the maximum distance that a mineral source target might occur up-ice of any particular regional till site (or a group of sites) with elevated element concentrations. This is accomplished by overlaying, on the known dispersal curve, a plot of transport distance versus element concentration for two, or preferably more, regional sites in the area of interest (Figs. 9 & 10).

If the curve from the elevated regional data shows a similar shape to the dispersal curve from a known deposit, with an exponential increase in the up-ice direction, then further work is warranted. If the curve is flat, then further work is less likely to be fruitful. In this way the significance of moderately elevated element concentrations that may reflect the center or tail of a dispersal train, can be evaluated. This is particularly useful as it is unlikely that the comparatively small area at the head of a dispersal train, where element concentrations are highest, will be encountered in a regional survey. In such an analysis, it is also important to remember that basal tills immediately down-ice from known mineral deposits can have relatively low element concentrations due to dilution by non-mineralized debris transported from locations up-ice of the mineral deposit; high values may not be encountered within the first several 100 m down-ice of the deposit.

One limitation of this method is the fact that elevated element concentrations down-ice of an ore body may be difficult to differentiate from background concentrations due to high levels of natural geochemical variability as well as field sampling and analytical variability. On Figure 10, the regional median for Cu at 937 till sites is shown as a horizontal line and is considered to be an approximation of background. One standard deviation above and below that line is also shown as an indication of the natural regional variability in Cu concentrations in till in the area. Sampling and analytical variability (represented by the average relative standard deviation of 57 field and analytical duplicates) are 4.8% and 2.7%, respectively. They are combined on Figure 10 and shown in relation to the regional median (7.5% of 41 ppm). Although 41 ppm is the regional median, it may be difficult to differentiate background from elevated Cu concentrations at levels below about 60 ppm when natural, sampling and analytical variability are taken into account.

Since background levels of various elements in tills are controlled in part by the background concentrations in their source rocks, the influence of bedrock geology must also be considered when interpreting regional till geochemical data. For example, to evaluate the effect of bedrock geology on the regional geochemical data set for the Fawnie Creek area, Levson *et al.* (1994) compared concentrations of Au, Ag, As, Sb, Pb, Zn, and Mo in till overlying different bedrock units. Background concentrations in the study, as defined by median values, are generally similar for tills underlain by mafic volcanic rocks and sedimentary rocks (Hazelton Group) but distinctive from tills underlain by rhyolitic rocks (Ootsa Lake Group) and Tertiary mafic volcanic rocks (Chilcotin Group). Ootsa Lake Group sites have the highest median Au concentrations whereas Chilcotin Group sites have the lowest Au, As and Sb concentrations. Sites underlain by these two groups also have low background concentrations of Pb, Zn and Cu relative to other rock types in the area.

Evaluating soil geochemical anomalies

The significance of soil geochemical anomalies, from an exploration perspective, often is not easily evaluated due to the highly variable and site dependent effects of pedogenic processes on contained element concentrations. As a result, it is difficult to differentiate the relative effects of geochemical processes acting within the soil from those that are reflecting the bedrock source. The heterogeneity of elemental concentrations in various soil horizons also is dependent on the overburden composition (e.g. Boyle & Troup 1975), as well as a number of other factors such as site drainage and vegetation. Since the development, thickness and properties of B-horizon soils generally are highly variable from site to site, the genesis of the actual material sampled can vary widely across a survey area. Consequently, element concentrations in B-horizon soils can be highly variable even though they are relatively uniform in the underlying bedrock. In contrast, samples collected from within the C-horizon are comparatively unaffected by the pedogenic processes operative in the A- and B-soil horizons (Agriculture Canada Expert Committee on Soil Survey 1987; Gleeson *et al.* 1989) and element concentrations commonly reflect mechanical rather than hydromorphic dispersal processes. The utility of sampling C-horizon soil developed on basal tills for outlining areas of mineralization has long been known (e.g. Shilts 1973*a, b*) but, until recently, exploration companies working in the Cordillera have generally favoured B-horizon soil sampling (Kerr 1995) partly because of ease of sampling, low costs and, probably, historical precedent. Although C-horizon soil samples can be effectively used to identify glacial dispersal trains (discussed further below), sampling the upper soil horizons also can provide important data, particularly when an understanding of the effects of local pedological and hydromorphic processes on elemental concentrations is required (Bradshaw *et al.* 1974; Gravel & Sibbick 1991; Sibbick & Fletcher 1993).

In areas of thick drift cover, traditional B-horizon soil geochemical surveys are considered less effective at identifying regional anomalies associated with mineralized bedrock than are basal till surveys. Recent investigations of glacial dispersal patterns using C-horizon soil developed on basal till, rather than B-horizon soil, indicate that dispersal trains in regions of thick drift can be detected over much larger areas in tills than in soils. In the Wolf area in central B.C. (Fig. 1), for example, a study of glacial dispersal, conducted in conjunction with the regional till sampling program (Levson *et al.* 1994), shows >5 km of down-ice dispersal from the known mineralization (Fig. 11). At these distances, Au concentrations in basal till are still elevated (>90th percentile compared to the regional data set). This contrasts with results of a detailed (50 m sampling interval) soil geochemistry sur-

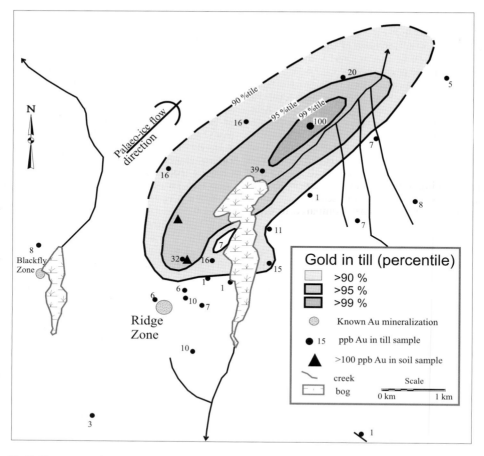

Fig. 11. Gold concentrations in basal till (–63 μm fraction, analyzed by INAA) down-ice of the Wolf Au–Ag prospect (Lat. 53°12′, Long. 125° 27′; Fig. 1) in the Interior Plateau of central British Columbia (see text for explanation). The dispersal train can be identified using relatively few basal till samples and that the head of the train occurs 500 m down-ice, not directly over top of mineralized bedrock. The highest Au concentrations in B-horizon soil samples (black triangles) also are displaced down-ice of the mineralization. The northern extent of the Au anomaly is uncertain (indicated by dashed line) due to the absence of data to the north.

vey conducted on the property (Dawson 1988). Concentrations of >15 ppb Au in soils were found at scattered sites located almost exclusively in thin colluvial soils around the Ridge Zone. Only a few sites with >15 ppb Au in soils were encountered >1 km down-ice from the deposit (see Levson & Giles 1995, fig. 6-4). The maximum down-ice extent of Au dispersal in till is not known but at least 5 km of transport is indicated (Fig. 11). The head of the Au dispersal train is about 500 m down-ice from the prospect and Au concentrations directly over the mineralized area are much lower. The same pattern is observed in the soil geochemistry data and highlights the importance of exploring up-ice, and not directly below, anomalous zones in areas with a thick till cover. The highest Au concentrations in soils (305 and 495 ppb) also occur at the head of the dispersal train (Fig. 11), up to 1.5 km down-ice from the mineralized area. The highest Au concentration in till is about 3.7 km down-ice from the prospect. These results also suggest that widely spaced, media specific basal till sampling would be more effective at detecting buried Au mineralization, at a regional scale, than would B-horizon soil sampling, even with a much higher sampling density.

Another reason for the typically larger size of till anomalies (up to several kilometres long; Levson & Giles 1997) compared to soil anomalies (usually less than a few kilometres in length) is due to indiscriminate sampling of inappropriate material in many soil surveys. Traditional soil surveys often sample soils developed on colluvial, glaciofluvial and glaciolacustrine sediments as well as englacial and supraglacial tills. Local variations are more poorly reflected in these sediments than in basal tills. In addition, the upper part of basal tills are often reworked by colluvial processes. This is particularly true on steep slopes such as are common in the Cordillera (Fig. 4). These processes can result in lower element concentrations in the reworked materials than in the undisturbed till due to gravity settling of heavy minerals (especially Au) and water suspension of fine sediments (which are generally metal-enriched compared to coarser sediments) and their resultant selective removal. Although useful information can be obtained from colluvially reworked glacial sediments, they should be interpreted separately from undisturbed basal tills. For this reason, a good understanding of the surficial geology in the area of a sampling program is required for both the design and interpretation phases of effective soil and till geochemistry surveys.

Effectiveness of regional till sampling programs

The effectiveness of regional till geochemistry as a tool for detecting buried mineralization can be evaluated by comparing regional till geochemical anomalies with the distribution of known mineral occurrences. Such a comparison was made by Cook et al. (1995) using regional till geochemical data from the Fawnie Creek area in central British Columbia (Levson et al. 1994). Known mineral occurrences in the survey area were confirmed by multi-element anomalies (in this case defined as >95th percentile concentrations of two or more of the following elements: Au, Cu, Pb, Zn, Mo, As, Sb), as well as mineralized bedrock not known to the samplers before the survey was conducted. In addition, nine new exploration targets with multi-element geochemical anomalies were identified (by the same criteria) from the regional till geochemical survey data. The till geochemical anomalies identified in the region show values comparable to those down-ice of advanced exploration prospects in the area. In addition, the significance of the geochemical anomalies was verified by the presence of anomalous (>95th percentile) metal concentrations in lake sediments at six of the nine new target areas; there were no nearby lakes at the remaining three sites.

The highest Au concentration encountered in the Fawnie Creek area occurred at a site contiguous with several other sites with anomalous Au, Ag, As and Sb. High Au concentrations in till were found along a zone about 1 km wide and several kilometres long, trending east, parallel to the local ice-flow direction (Fig. 6). Follow-up studies at the site led to the discovery of mineralized bedrock at the up-ice end of the anomaly which contains up to 2.5 g/t Au and 25.5 g/t Ag (Schimann 1995). Subsequent work revealed mineralized quartz veins in the initial discovery area containing up to 19.6 g/t Au and 99.9 g/t Ag (Payne 1996). In addition, the dispersal pattern in the area suggests that at least one other Au source in the area is yet to be discovered, southwest (up-ice) of the main till anomaly. Similarly, anomalous Au concentrations in till were found at three regional sites down-ice of an area identified by regional bedrock mapping (Diakow et al. 1997) that has since been extensively explored and drilled resulting in a number of new vein-hosted Au discoveries. These data strongly suggest that geochemical surveys, using basal tills as a sampling medium, are an effective tool for regional exploration, especially when integrated with other types of geological and geochemical surveys.

Several other examples from the British Columbia interior illustrate the effectiveness of regional till geochemical surveys in detecting buried Au mineralization. For example, results of detailed till geochemical sampling conducted at the Bell Cu–Au mine were compared with regional results from the Babine Cu porphyry belt (Fig. 9) by Levson et al. (1997b). Most known mineral properties in the map region were detected by the regional till geochemical data and tills in several areas with no recorded mineral occurrences were anomalous (>95th percentile of the regional data) in Cu or Au. Copper concentrations in tills in some of these areas with favourable geology are comparable to those found in tills down-ice of the Bell mine (Stumpf et al. 1997).

Summary

Basal till geochemical sampling programs are an effective tool for locating mineralized bedrock in drift-covered parts of the Canadian Cordillera, and the method has been increasingly used in recent years. These surveys are distinct from B-horizon soil geochemical surveys that have been standard industry practice in the Cordillera for many years. Till geochemical surveys are media-specific and can only be conducted in areas where basal tills can be sampled within the scope, scale and budget of the proposed survey. The proportion of till in any one area, in relation to other surficial sediment types, is highly variable but in most valley and plateau settings is sufficiently high (typically 70–90%) to allow for effective regional till sampling programs. To reflect mechanical dispersal processes, samples should be collected from within the C-horizon soil developed on basal till. Till geochemical surveys can be conducted at reconnaissance, regional or property scales, depending on the purpose of the survey, but the use of property-scale and sometimes even regional till geochemical surveys can be locally restricted due to the absence or deep burial of tills. Sampling programs should be designed to account for ice flow direction and the typically long and narrow shape of dispersal trains. In general, a sample density of at least 1 site per $5\,km^2$ is recommended for regional till geochemistry programs but significantly higher densities may be required if the exploration targets are small.

Till geochemical anomalies associated with glacial dispersal of mineralized bedrock in the Cordillera can be up to several kilometres long and >1 km wide, but more discontinuous, isolated anomalies associated with the dispersal trains may cover much larger areas. They show a pronounced elongation parallel to ice-flow direction, with mineralized source rocks occurring at or near the up-ice end of the trains and dispersal trains. Due to the strong topographic control in much of the Cordillera on ice flow patterns, basal till anomalies often can be traced to source along linear (often valley-parallel) transport paths. The till geochemistry reflects up-ice bedrock sources and not the immediately underlying bedrock. In areas of thick till, near-surface anomalies may be displaced by 500 m or more down-ice from their bedrock source. Subsurface exploration targets in these areas should be up-ice, rather than at the head, of the geochemical anomaly. In areas of thick drift, till geochemical anomalies are also more readily detected than soil anomalies. Conventional B-horizon soil geochemical studies are most useful for property-scale surveys where shallow residual or colluvial deposits dominate such as is in steep mountainous terrain. Interpretations of till geochemical data should consider background variations in element concentrations due to changes in bedrock geology and interpretations can be aided by comparison of regional and case study data.

Basal tills have distinctive characteristics but they can be confused with other facies of morainal sediments such as glacigenic debris flow deposits or with other poorly sorted sediments such as colluvial deposits. These sediments have different processes of transportation and deposition. Local variations will be reflected in some sediment types while regional trends may be evident in others. Analysis of these sediments will be useful only if their origin is understood. For this reason, sedimentological data should be collected at all sample sites by well trained staff. Rigorous quality control procedures should be applied to all surveys. Evaluations of clasts in till are considered an important component of till geochemistry programs because erratics trains often are much longer and more readily detected than till geochemical anomalies. Till geochemical programs should be conducted in conjunction with surficial geological mapping and other Quaternary studies because an understanding of ice-flow direction(s), glacial dispersal patterns, transport distances, Quaternary stratigraphy and the origin of different sampling media is essential for successful drift exploration programs. In addition, till geochemical surveys should be conducted in conjunction with bedrock geological mapping, other geochemical surveys (such as lake sediment geochemistry) and/or geophysical surveys where possible.

Much of the information presented in this paper was obtained during the course of till geochemical studies conducted by the British Columbia Geological Survey from 1993 to 1998 in the central Interior Plateau region of the Cordillera. The author extends thanks to numerous individuals who were involved in various ways in these studies including (in alphabetical order): C. Churchill, S. Cook, T. Giles, J. Hobday, D. Huntley, D. Kerr, D. Mate, D. Meldrum, E. O'Brien, S. Sibbick, A. Stewart, A. Stumpf and G. Weary. Access to mineral exploration properties was provided by several companies including Metall Mining, Placer Dome, Cogema, Noranda, and Hera Resources. Assistance with drafting of figures and editorial revisions was provided by K. Cleary and T. Ferbey. Drafts of this paper were reviewed by S. Cook, R. Lett and D. Mate. The manuscript was also much improved by the reviewers comments of R. Paulen and R. Stea.

References

AGRICULTURE CANADA EXPERT COMMITTEE ON SOIL SURVEY. 1987. *The Canadian System of Soil Classification*. Second Edition. Agriculture Canada Publication **1646**.

BELIK, G. 1990. *Percussion drilling report on the Venus 2, 4 and 11 Fraction*. British Columbia Ministry of Energy and Mines Assessment Report **20663**.

BENNETT, M. R. & GLASSER, N. F. 1996. *Glacial Geology: Ice Sheets and Landforms*. John Wiley & Sons Ltd, Chichester.

BOBROWSKY, P. T. & SIBBICK, S. J. 1996. *Till geochemistry of northern Vancouver Island area (NTS 92L/5, 6W, 11W, 12)*. British Columbia Geological Survey, Open File **1996-7**.

BOBROWSKY, P. T., LEBOE, E. R, DIXON-WARREN, A., LEDWON, A., MACDOUGALL, D. & SIBBICK, S. J. 1997. *Till geochemistry of the Adams Plateau - North Barriere Lake area (NTS 82M/4 & 5)*. British Columbia Geological Survey, Open File **1997-9**.

BOBROWSKY, P. T., PAULEN, R. C., LITTLE, E., PREBBLE, A., LEDWON, A. & LETT, R. E. 1998. *Till geochemistry of the Louis Creek - Chua Chua Creek area (NTS 92 P/1E and 92 P/8E)*. British Columbia Geological Survey, Open File **1998-6**.

BOHME, D. M. 1988. *Geological, geochemical and geophysical report on the White Claim group, Omineca Mining Division (NTS 93F/11E, 6E; Latitude 53° 30' N, Longitude 125° 05' W)*. British Columbia Ministry of Energy and Mines Assessment Report **18191**.

BOULTON, G. S. 1996. Theory of glacial erosion, transport and deposition as a consequence of subglacial sediment deformation. *Journal of Glaciology*, **42**, 43–62.

BOYLE, D. R. & TROUP, A. G. 1975. Copper-molybdenum porphyry mineralization in central British Columbia, Canada: an assessment of geochemical sampling media useful in areas of glaciated terrain. *In:* JONES, M. J. (ed.) *Prospecting in Areas of Glaciated Terrain*. Institution of Mining and Metallurgy, London, 6–15.

BRADSHAW, P. M. D., THOMSON, I., SMEE, B. W. & LARSSON, J. 1974. The application of different analytical extractions and soil profile sampling in exploration geochemistry. *Journal of Geochemical Exploration*, **3**, 209–225.

CARTER, N. C., DIROM, G. E. & OGRYZLO, P. L. 1995. Porphyry copper-gold deposits, Babine Lake area, west-central British Columbia. *In:* SCHROETER, T. G. (ed.) *Porphyry Deposits of the Northwestern Cordillera of North America*. Canadian Institute of Mining and Metallurgy and Petroleum, Special Volume, **46**, 247–255.

COOK, S. J. & FLETCHER, W. K. 1993. Distribution and behavior of platinum in soils, sediments and waters of the Tulameen ultramafic complex, southern British Columbia, Canada. *Journal of Geochemical Exploration*, **46**, 279–308.

COOK, S. J., LEVSON, V. M., GILES, T. R. & JACKAMAN, W. 1995. A comparison of regional lake sediment and till geochemistry surveys: a case study from the Fawnie Creek area, Central British Columbia. *Exploration and Mining Geology*, **4/2**, 93–110.

DAWSON, J. M. 1988. *Geological and geochemical report on the Wolf property, Omineca Mining Division*. British Columbia Ministry of Energy and Mines, Assessment Report **16995**.

DELANEY, T. A. & FLETCHER, W. K. 1994. *Quaternary geology/geochemical exploration studies in the Wolf-Capoose and AOK-Moondust areas of the central Interior Plateau of British Columbia*. University of British Columbia, Mineral Deposits Research Unit, Technical Report **MT-8**.

DIAKOW, L. J., WEBSTER, I. W. E., RICHARDS, T. A. & TIPPER, H. W. 1997. *Geology of the Fawnie and Nechako Ranges, southern Nechako Plateau, central British Columbia (93F/2, 3, 6, 7)*. British Columbia Geological Survey, Paper **1997-2**, 7–30.

DILABIO, R. N. W. 1981. *Glacial dispersal of rocks and minerals at the south end of Lac Mistassini, Quebec, with special reference to the Icon dispersal train*. Geological Survey of Canada, Bulletin **323**.

DILABIO, R. N. W. 1990. Glacial dispersal trains. *In:* KUJANSUU, R. & SAARNISTO, M. (eds) *Glacial Indicator Tracing*. A. A. Balkema, Rotterdam, 109–122.

DREIMANIS, A. 1960. Geochemical prospecting for Cu, Pb and Zn in glaciated areas, eastern Canada. 21st International Geological Congress, Norden, Pt. II, 7–19.

DREIMANIS, A. 1990. Formation, deposition and identification of subglacial and supraglacial tills. *In:* KUJANSUU. R. & SAARNISTO, M. (eds) *Glacial Indicator Tracing*. A. A. Balkema, Rotterdam, 35–59.

EDWARDS, K. & CAMPBELL, T. 1992. *Geological, geochemical, geophysical assessment report for the CH 10-16 mineral claims, Omineca Mining Division (NTS 93F/7E, 8W; Latitude 53° 31' N, Longitude 124°25' W)*. British Columbia Ministry of Energy and Mines Assessment Report **22027**.

FOX, P., CAMERON, R. & HOFFMAN, S. 1987. Geology and soil geochemistry of the Quesnel River gold deposit, British Columbia. *In:* ELLIOT, I. & SMEE, B. (eds) *Geoexpo '86*. Association of Exploration Geochemists, 61–67.

FULTON, R. J. 1991, A conceptual model for growth and decay of the Cordilleran Ice Sheet. *Géographie physique et Quaternaire*, **45**, 281-286.

GLEESON, C. F., RAMPTON, V. N., THOMAS, R. D. & PARADIS, S. 1989. Effective mineral exploration for gold using geology, Quaternary geology and exploration geochemistry in areas of shallow till. *In:* DILABIO, R. N. W. & COKER, W. B. (eds) *Drift Prospecting*. Geological Survey of Canada Paper **89-20**, 71–96.

GRAVEL, J. L. & SIBBICK, S. J. 1991. Geochemical dispersion in complex glacial drift at the Mount Milligan copper-gold porphyry deposit (93N/1E, 93O/4W). *In: Exploration in British Columbia 1990*. British Columbia Geological Survey, 117–134.

HICOCK, S. R. 1995. Glacial geology applied to drift prospecting in Buttle valley, Vancouver Island. *In:* BOBROWSKY, P. T., SIBBICK, S. J., NEWELL, J. M. & MATYSEK, P. F. (eds) *Drift Exploration in the Canadian Cordillera*. British Columbia Geological Survey, Paper **1995-2**, 33–41.

KAURANNE, L. K. 1959. Pedogeochemical prospecting in glaciated terrain. Geological Survey of Finland, Bulletin **184**, 1–10.

KAURANNE, K. SALMINEN, R. & ERIKSSON, K. 1992. *Regolith Exploration Geochemistry in Arctic and Temperate Terrains, Handbook of Exploration Geochemistry, Volume 5*. Elsevier Science Publishers, Amsterdam.

KERR, D. K. 1995. Drift exploration data from B. C. Ministry of Energy, Mines and Petroleum Resources assessment reports: NTS 93N (Manson River) south half, northern Quesnel Trough region, B.C.. *In:* BOBROWSKY, P. T., SIBBICK, S. J., NEWELL, J. M. & MATYSEK, P. F. (eds) *Drift Exploration in the Canadian Cordillera.* British Columbia Geological Survey, Paper **1995-2**, 265–276.

KERR, D. K. & LEVSON, V. M. 1995. Annotated bibliography of drift prospecting activities in British Columbia. *In:* BOBROWSKY, P. T., SIBBICK, S. J., NEWELL, J. M. & MATYSEK, P. F. (eds) *Drift Exploration in the Canadian Cordillera.* British Columbia Geological Survey, Paper **1995-2**, 277–303.

KERR, D. K. & LEVSON, V. M. 1997. Drift prospecting activities in British Columbia: an overview with emphasis on the Interior Plateau. *In:* DIAKOW, L. J. & NEWELL, J. M. (eds) *Interior Plateau Geoscience Project: Summary of Geological, Geochemical and Geophysical Studies.* Geological Survey of Canada, Open File **3448** and B.C. Geological Survey Branch, Paper **1997-2**, 159–172.

KERR, D. E., SIBBICK, S. J. & JACKAMAN, W. 1992. *Till geochemistry of the Quatsino map area (92L/12).* British Columbia Geological Survey, Open File **1992-21**.

KERR, D. E., SIBBICK, S. J. & BELIK, G. D. 1993. Preliminary results of glacial dispersion studies on the Galaxy property, Kamloops, B.C. (92I/9). *In:* GRANT, B. & NEWELL, J. M. (eds) *Geological Fieldwork 1992.* British Columbia Geological Survey, Paper **1993-1**, 439–443.

KLASSEN, R. A. 1997. Glacial history and ice flow dynamics applied to drift prospecting and geochemical exploration. *In:* GUBINS, A. G. (ed.) *Geophysics and Geochemistry at the Millennium.* Proceedings of Exploration 97: Fourth Decennial International Conference on Mineral Exploration, Toronto, 221–232.

KOR, P. S. G., SHAW, J. & SHARPE, D. R. 1991. Erosion of bedrock by subglacial meltwater, Georgian Bay, Ontario: a regional view. *Canadian Journal of Earth Sciences*, **28**, 623–642.

KURZL, H. 1998. Exploration data analysis: recent advances for the interpretation of geochemical data. *Journal of Geochemical Exploration*, **30**, 309–322.

LETT, R. E. 1995. Analytical methods for drift samples. *In:* BOBROWSKY, P. T., SIBBICK, S. J., NEWELL, J. M. & MATYSEK, P. F. (eds) *Drift Exploration in the Canadian Cordillera.* British Columbia Geological Survey, Paper **1995-2**, 215–228.

LETT, R. E., BOBROWSKY, P. T., CATHRO, M. & YEOW, A. 1998. Geochemical pathfinders for massive sulphides in the southern Kootenay Terrane. *In: Geological Fieldwork 1997.* British Columbia Geological Survey, Paper **1998-1**, 15–1 to 15–9.

LETT, R. E., JACKAMAN, W. & YEOW, A. 1999. Detailed geochemical exploration techniques for base and precious metals in the Kootenay Terrane (82 L/13, L/14, M/4, M/5, P/1). *In: Geological Fieldwork 1998.* British Columbia Geological Survey, Paper **1999-1**, 297–306.

LEVINSON, A. & CARTER, N. 1979. Glacial overburden profile sampling for porphyry copper exploration: Babine Lake area, British Columbia. *Western Miner*, **5**, 19–32.

LEVSON, V. M. 2001. Regional till geochemical surveys in the Canadian Cordillera: sample media, methods and anomaly evaluation. *In:* MCCLENAGHAN, M. B., BOBROWSKY, P. T., HALL, G. E. M. & COOK, S, J. (eds) *Drift Exploration in Glaciated Terrain.* Geological Society, London, Special Publications, **185**, 45–68.

LEVSON, V. M. & GILES, T. R. 1995. Glacial dispersal patterns of mineralized bedrock with examples from the Nechako Plateau, central British Columbia. *In:* BOBROWSKY, P. T., SIBBICK, S. J., NEWELL, J. M. & MATYSEK, P. F. (eds) *Drift Exploration in the Canadian Cordillera.* British Columbia Geological Survey, Paper **1995-2**, 67–76.

LEVSON, V. M. & GILES, T. R. 1997. Quaternary geology and till geochemistry studies in the Nechako and Fraser Plateaus, central British Columbia. *In:* DIAKOW, L. J. & NEWELL, J. M. (eds) *Interior Plateau Geoscience Project: Summary of Geological, Geochemical and Geophysical Studies.* Geological Survey of Canada Open File **3448** and B.C. Geological Survey Branch Paper **1997-2**, 121–145.

LEVSON, V. M. & RUTTER, N. W. 1988. A lithofacies analysis and interpretation of depositional environments of montane glacial diamictons, Jasper, Alberta, Canada. *In:* GOLDTHWAIT, R. P. & MATSCH, C. L. (eds) *Genetic Classifications of Glacigenic Deposits.* Balkema, Rotterdam, 117–140.

LEVSON, V. M. & STUMPF, A. J. 1998. Glacial controls on geochemical transport distance and direction in north-central Stikinia: implications for exploration. *In:* MUSTARD, P. & GAREAU, S. (eds) *Cordillera Revisited: Recent Developments in Cordilleran Geology, Tectonics and Mineral Deposits.* Geological Association of Canada, Cordilleran Section, Short Course Notes, 68–75.

LEVSON, V. M., GILES, T. R., COOK, S. J. & JACKAMAN, W. 1994. *Till geochemistry of the Fawnie Creek area (93F/03).* British Columbia Geological Survey, Open File **1994-18**.

LEVSON, V. M., MELDRUM, D. G., COOK, S. J., STUMPF, A. J., O'BRIEN, E. K., CHURCHILL, C., BROSTER, B. E. & CONEYS, A. M. 1997*a*. Till geochemical studies in the Babine porphyry belt: regional surveys and deposit-scale studies (NTS 93 L/16, M/1, M/8). *In:* LEFEBURE, D. V., MCMILLAN, W. J. & MCARTHUR, J. G. (eds) *Geological Fieldwork 1996.* British Columbia Geological Survey, Paper **1997-1**, 457–466.

LEVSON, V. M., COOK, S. J., HOBDAY, J., HUNTLEY, D. H., O'BRIEN, E. K., STUMPF, A. J. & WEARY, G. 1997*b*. *Till geochemistry of the Old Fort Mountain area, central British Columbia (NTS 93 M/1).* British Columbia Geological Survey, Open File **1997-10a**.

LEVSON, V. M., COOK, S. J. & GILES, T. R. 1997*c*. *Surficial geochemical signatures of buried mineral deposits in central British Columbia.* Canadian Institute of Mining, Paper **WPM1-G2** (CD-ROM).

MEHRTENS, M. B. 1975. Chutanli Mo deposit, British

Columbia; conceptual models in exploration geochemistry. *Journal of Geochemical Exploration*, **4/1**, 63–65.

MENZIES, J. (ed.) 1995. Modern Glacial Environments: Processes, Dynamics and Sediments. *Glacial Environments: Volume 1*. Butterworth-Heinemann Publishers, Oxford.

MENZIES, J. (ed.) 1996. Past Glacial Environments: Sediments, Forms and Techniques. *Glacial Environments: Volume 2*. Butterworth-Heinemann Publishers, Oxford.

O'BRIEN, E. K. BROSTER, B. E. GILES, T. R. & LEVSON, V. M. 1995. Till geochemical sampling: CH, Blackwater-Davidson, and Uduk Lake properties, British Columbia: report of activities. *In:* GRANT, B. & NEWELL, J. M. (eds) *Geological Fieldwork 1994*. British Columbia Geological Survey, Paper **1995-1**, 207–211.

O'BRIEN, E. K., LEVSON, V. M. & BROSTER, B. E. 1997. *Till geochemical dispersal in central British Columbia*. British Columbia Geological Survey, Open File **1997-12**.

PAULEN, R. C., BOBROWSKY, P. T., LETT, R. E., BICHLER, A. J. & WINGERTER, C. 1999. Till geochemistry in the Kootenay, Slide Mountain and Quesnel terranes. *In: Geological Fieldwork 1998*, British Columbia Geological Survey, Paper **1999-1**, 307–319.

PAULEN, R. C., BOBROWSKY, P. T., LETT, R. E., JACKAMAN, W., BICHLER, A. J. & WINGERTER, C. 2000. *Till geochemistry of the Chu Chua - Clearwater area, B.C. (parts of NTS 92P/8 and 92P/9)*. British Columbia Geological Survey, Open File **2000-17**.

PAYNE, C. W. 1996. *Geological and soil geochemical report on the Laidman property, Omineca Mining Division*. British Columbia Ministry of Energy and Mines Assessment Report **24234**.

PLOUFFE, A. 1995a. *Geochemistry, lithology, mineralogy and visible gold grain content of till in the Manson River and Fort Fraser map areas, central British Columbia (NTS 93K, N)*. Geological Survey of Canada, Open File **3194**.

PLOUFFE, A. 1995b. Drift prospecting sampling methods. *In:* BOBROWSKY, P. T., SIBBICK, S. J., NEWELL, J. M. & MATYSEK, P. F. (eds) *Drift Exploration in the Canadian Cordillera*. British Columbia Geological Survey, Paper **1995-2**, 43–52.

PLOUFFE, A. & BALLANTYNE, S. B. 1993. *Regional till geochemistry, Manson River and Fort Fraser area, British Columbia (93K, 93N), silt plus clay and clay size fractions*. Geological Survey of Canada, Open File **2593**.

PLOUFFE, A. & BALLANTYNE, S. B. 1994. *Regional till geochemistry, Mount Tatlow and Elkin Creek area, British Columbia (93O/5 and 93O/12)*. Geological Survey of Canada, Open File **2909**.

PLOUFFE, A. & JACKSON, L. E. J. 1995. Quaternary stratigraphy and till geochemistry in the Tintina Trench, near Faro and Ross River, Yukon Territory. *In:* BOBROWSKY, P. T., SIBBICK, S. J., NEWELL, J. M. & MATYSEK, P. F. (eds) *Drift exploration in the Canadian Cordillera*. British Columbia Geological Survey, Paper **1995-2**, 53–66.

PLOUFFE, A. & WILLIAMS, S. P. 1998. *Regional till geochemistry of the northern sector of Nechako River map sheet (NTS 93F)*. Geological Survey of Canada, Open File **3687**.

SALONEN, V. P. 1989. Application of glacial dynamics, genetic differentiation of glacigenic deposits and their landforms to indicator tracing in search for ore deposits. *In:* GOLDTHWAIT, R. P. & MATSCH, C. L. (eds) *Genetic Classifications of Glacigenic Deposits*. Balkema, Rotterdam, 183–190.

SCHIMANN, K. 1995. *Geological and geochemical survey, Laidman property, Omineca Mining Division*. British Columbia Ministry of Energy and Mines Assessment Report **23751**.

SHILTS, W. 1971. Till studies and their application to regional drift prospecting. *Canadian Mining Journal*, **92**, 45–50.

SHILTS, W. 1973a. *Glacial dispersion of rocks, minerals and trace elements in Wisconsin till, southeastern Quebec, Canada*. Geological Society of America, Memoir **136**, 189–219.

SHILTS, W. 1973b. *Till indicator train formed by glacial transport of nickel and other ultrabasic components: a model for drift prospecting*. Geological Survey of Canada, Paper **73-1**, 213–218.

SHILTS, W. 1976. Glacial till and mineral exploration. *In:* LEGGET, R. F. (ed) *Glacial Till, An Interdisciplinary Study*. Royal Society of Canada, Special Publication **12**, 205–224.

SHILTS, W. 1993. Geological Survey of Canada's contributions to understanding the composition of glacial sediments. *Canadian Journal of Earth Sciences*, **30**, 333–353.

SHILTS, W. 1995. Geochemical partitioning in till. *In:* BOBROWSKY, P. T., SIBBICK, S. J., NEWELL, J. M. & MATYSEK, P. F. (eds) *Drift Exploration in the Canadian Cordillera*. British Columbia Geological Survey, Paper **1995-2**, 149–164.

SIBBICK, S. J. & FLETCHER, W. K. 1993. Distribution and behavior of gold in soils and tills at the Nickel Plate mine, southern British Columbia, Canada. *Journal of Geochemical Exploration*, **47**, 183–200.

SIBBICK, S. J. & KERR, D. E. 1995. Till geochemistry of the Mount Milligan area, north-central British Columbia; Recommendations for drift-exploration for porphyry Cu-Au mineralization. *In:* BOBROWSKY, P. T., SIBBICK, S. J., NEWELL, J. M. & MATYSEK, P. F. (eds) *Drift exploration in the Canadian Cordillera*. British Columbia Geological Survey, Paper **1995-2**, 167–180.

SIBBICK, S. J., BALMA, R. G. & DUNN, C. E. 1996. *Till geochemistry of the Mount Milligan area (Parts of 93N/1 and 93O/4)*. British Columbia Geological Survey, Open File **1996-22**.

STUMPF, A., HUNTLEY, D. H., BROSTER, B. E. & LEVSON, V. M. 1996. Babine Porphyry Belt project: detailed drift exploration studies in the Old Fort Mountain (93M/01) and Fulton Lake (93L/16) map areas, British Columbia. *In:* GRANT, B. & NEWELL, J. M. (eds) *Geological Fieldwork 1995*. British Columbia Geological Survey, Paper **1996-1**, 37–44.

STUMPF, A. J., BROSTER, B. E. & LEVSON, V. M. 1997. Evaluating the use of till geochemistry to define buried mineral targets: a case study from the Bell mine property, (93 L/16, M/1) west-central British Columbia. *In:* LEFEBURE, D. V., MCMILLAN, W. J. & MCARTHUR, J. G. (eds) *Geological Fieldwork 1996*. British Columbia Geological Survey, Paper **1997-1**, 439–456.

WEARY, G. F., GILES, T. R., LEVSON, V. M. & BROSTER, B. E. 1995. Surficial geology and Quaternary stratigraphy of the Chedakuz Creek area (NTS 93F/7). British Columbia Geological Survey, Open File Map **1995-13** (1 : 50 000 scale).

WEARY G. F., LEVSON, V. M. & BROSTER, B. E. 1997. *Till geochemistry of the Chedakuz Creek map area (93F/7)*. British Columbia. British Columbia Geological Survey, Open File **1997-11**.

The application of heavy indicator mineralogy in mineral exploration with emphasis on base metal indicators in glaciated metamorphic and plutonic terrains

STUART A. AVERILL

Overburden Drilling Management Limited, 107-15 Capella Court, Nepean, Ontario K2E 7X1, Canada (e-mail: odm@storm.ca)

Abstract: Indicator mineralogy is used to explore for a wide variety of mineral commodities. The method utilizes minerals which are sufficiently heavy to be readily concentrated in the laboratory, often colourful and possess other useful physical and chemical properties. The minerals also must be source specific. Some indicator minerals are true resistate minerals. The others, although less resistant, are stable in oxidized glacial drift and many non-glacial sediments. A few, such as gold grains, are silt sized but most are coarse grained. Grain size has a major impact on indicator mineral dispersal patterns in glacial drift.

The coarse-grained indicator minerals are of two main types: (1) kimberlite indicator minerals (KIMs); and (2) metamorphosed or magmatic massive sulphide indicator minerals (MMSIMs). KIMs are enriched in Mg and Cr and most MMSIMs are enriched in Mg, Mn, Al or Cr. These indicator elements cannot be diagnosed geochemically in anomalous heavy mineral concentrates because the concentrates contain other, more plentiful non-indicator minerals containing the same elements in the same chemical form. Chalcopyrite is also a very useful MMSIM but the number of surviving grains in a dispersal train is too low for detection by selective geochemical analysis.

MMSIMs are derived from three main types of base metal deposits and their associated alteration or reaction zones: (1) volcanosedimentary massive sulphides (encompassing volcanogenic, Sedex and Mississippi Valley subtypes) in medium to high grade regional metamorphic terrains; (2) skarn and greisen deposits; and (3) magmatic Ni–Cu sulphides. The variety of MMSIMs associated with Ni–Cu deposits is astonishing, apparently reflecting mineral hybridization related to assimilation of sulphurous sedimentary rocks by ultramafic magmas. Cr–diopside is one of the best indicators of fertile Ni–Cu environments although not necessarily of the actual Ni–Cu deposits.

Heavy indicator mineralogy is much more sensitive than heavy mineral geochemical analysis and offers many exploration benefits in regional exploration programs including: (1) sampling efficiencies; (2) enlargement of both the bedrock target and dispersal train; (3) coverage of a wider range of mineral commodities; (4) undiminished sensitivity in areas of overabundant non-indicator heavy minerals; (5) visual evidence of points of origin of dispersal trains; and (6) indications of the economic potential of the source mineralization. It is most effective as a reconnaissance exploration tool and is particularly well suited for testing gneissic volcanosedimentary and plutonic terranes where base and precious metal deposits are highly modified and difficult to recognize by other prospecting methods.

Laboratory processing of drift samples for heavy minerals is often employed on geochemical exploration programs in glaciated terrains to enhance the geochemical signature of glacially dispersed metallic mineralization and thereby enlarge the geochemical target. Unweathered glacial drift obtained by drilling is an ideal heavy mineral sampling medium because base metals remain in primary sulphide minerals which are readily concentrated and analysed. However, weathered surface samples lacking sulphides can be used to search for selected metals such as Au, Pt, Sn and W that occur in chemically resistant minerals.

In addition to enhancing geochemical signatures, heavy mineral processing offers many potential benefits available through mineralogical examination of the concentrate. Two well

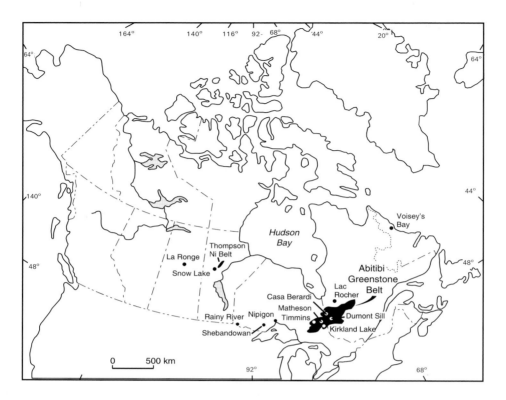

Fig. 1. Index to Canadian localities referenced in the text.

known examples are: (1) determining the sources of gold dispersal trains from the degree of wear of gold grains (Averill 1988; DiLabio 1990); and (2) identifying kimberlite pipes and their diamond potential from kimberlite indicator minerals (KIMs) (Gurney & Moore 1993; Fipke et al. 1995). In the case of gold grains, the anomaly signature in the concentrate is both geochemical (i.e. analytical) and mineralogical (i.e. visual). In the case of KIMs, it is strictly mineralogical because the number of indicator grains in the concentrate is small and these grains are enriched only in elements such as Mg and Cr which are overwhelmed by the higher Mg and Cr contributions of common non-indicator heavy minerals.

Heavy indicator mineralogy is dependent upon the minerals being source specific and having various useful properties that complement their high specific gravity. In the case of kimberlite indicators, the minerals are: (1) found in few if any rocks other than kimberlite; (2) coarse-grained; (3) visually distinctive (colourful and/or uniquely textured, altered or fractured); (4) sufficiently heavy (S.G. > 3.2) to be readily concentrated by gravity means; (5) amenable to further concentration by electromagnetic separation; and (6) relatively resistant to weathering, especially in immature glaciated terrains. Kimberlite pipes also tend to be large relative to metallic mineral deposits. Therefore even a pipe that is relatively KIM poor can be a significant KIM source. Moreover, KIMs are so source specific that only a few grains are needed in glacial sediments to recognize the presence of kimberlite. Consequently, KIM dispersal trains tend to be tens of kilometres long whereas the heavy mineral geochemical signatures of base metal dispersal trains in weathered, sulphide-depleted glacial drift are generally measured in hundreds of metres.

Recently it has been recognized that base metal indicator minerals with properties similar to KIMs are present in large alteration and reaction zones associated with certain types of deposits including: (1) volcanosedimentary massive sulphides (encompassing volcanogenic, Sedex and Mississippi Valley subtypes) in medium

Table 1. *Selected gold grain dispersal trains in glaciated terrains in Canada*

Fold belt	Deposit name	Train length* (m) Traced	Est. total	Gold grains per kg	Average gold grain diameter μm)
Abitibi	Belore	400	400	2	50–100
Abitibi	Cooke Mine†	800	1000	Encapsulated	–
Abitibi	Golden Pond West	400	‡400	3	50–100
Abitibi	Golden Pond	400	‡500	2	50–75
Abitibi	Golden Pond East	800	‡1000	6	25–75
Abitibi	Orenada	100	200	2	25–75
Abitibi	Kiena	100	300	3	10–75
Abitibi	Chimo	600	1000	4	50–75
La Ronge	EP** (Waddy Lake)	600	2000	10	10–100
La Ronge	Star Lake	300	800	2	10–50
La Ronge	Tower Lake	7000	‡7000	10	10–50
La Ronge	Bakos	2000	2000	20	25–50
Lynn Lake	Farley Lake	400	400	1	25–75
Humber	Devil's Cove	2000	2000	6	10–100
Rainy River	17 Zone	2000	††15 000	10	10–50

* Based on minimum ten gold grains of similar size and shape per standard 10 kg sample for free gold trains and coincident high gold and base metal assays in unweathered till for encapsulated gold trains.
† Encapsulated gold deposit.
‡ Train shortened and/or gapped by erosion in last ice advance.
** Deposit orientated parallel to glacial ice flow.
†† Train length enhanced by a 5 km^2 Au-bearing alteration zone surrounding Au deposit.

to high-grade (upper greenschist to granulite facies) regional metamorphic terrains; (2) skarn and greisen deposits; and (3) magmatic Ni–Cu sulphides. The author's company, Overburden Drilling Management Limited (ODM), has invoked the acronym MMSIMs® (for metamorphic or magmatic massive sulphide indicator minerals) for these mineral species and has tested several thousand samples from exploration projects worldwide for MMSIM anomalies. Many of these MMSIM projects included gold grain and KIM components, and many thousands of additional samples have been tested for gold grains alone or for KIMs plus gold grains. Most of the samples weighed 10 to 20 kg. Their <2 mm fraction was processed by gravity tabling followed by heavy liquid refining, typically at a specific gravity of 3.20 g cm^3. In some cases, samples of the source rocks supplying the indicator minerals were also processed. The indicator minerals were roughly sorted with an electromagnetic separator and visually identified by geologists using a binocular microscope, with resolution of difficult grains by energy dispersive x-ray spectrometer (EDS) analysis using a scanning electron microscope (SEM). Gold grains were micropanned from the table concentrates and counted, measured and classified by binocular microscope with further examination by SEM/EDS on special projects.

ODM's original 1970s morphological classification scheme for these gold grains (delicate/irregular/abraded; Averill & Zimmerman 1986; Averill 1988) was updated by DiLabio (1990) with the author's assistance using more generic terms (pristine/modified/reshaped) which allow for chemical as well as physical changes to gold grain morphology.

This paper describes the indicator minerals employed in the above surveys and the types of dispersal patterns observed with emphasis on MMSIMs in weathered glacial drift. The reader will appreciate that the MMSIM method is relatively new and most of the MMSIM survey results are confidential; therefore few explicit case histories are cited and emphasis has been placed on describing broad, repetitive dispersal patterns of general use to explorationists.

Gold grains

Numerous gold deposits in Canada have been discovered by identifying gold grain dispersal trains in glacial drift, primarily till. Examples (Fig. 1) include the Golden Pond East and West deposits at Casa-Berardi, Quebec (Sauerbrei *et al.* 1987), the Aquarius deposit at Timmins, Ontario (Gray 1983), the 17 Zone at Rainy River, Ontario (Averill 1998) and at least four

Gold Grain Morphology

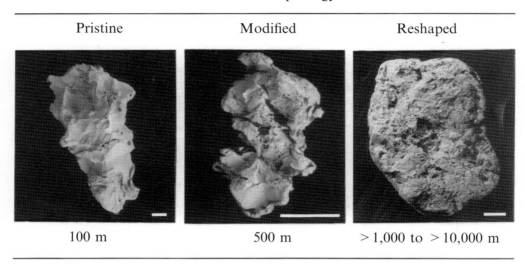

Fig. 2. Backscatter electron images of gold grains from till illustrating the relationship between grain wear and distance of glacial transport. The wear processes are compressional (infolding and compaction) and do not reduce the mass of the gold grain. Scale bars = 50 μm.

deposits in the La Ronge Belt, Saskatchewan (Averill & Zimmerman 1986; Sopuck et al. 1986; Chapman et al. 1990; Lehnert-Thiel 1998). Case studies have also been performed at several gold mines, primarily by reverse circulation drilling (Table 1) but in some cases by rotasonic drilling or backhoe sampling (McClenaghan 2001). Data of a more regional nature have been obtained from many government surveys (e.g. Thorleifson & Kristjansson 1990; Bajc 1991; Thorleifson & Matile 1993; McClenaghan 1994; Plouffe 1995; Morris et al. 1998). The principal gold grain dispersal patterns observed are:

(1) Gold grains are present in tills everywhere in Canada (Averill 1988). Their background abundance ranges from less than one grain per standard 10 kg exploration sample (0.1 grains/kg) in regions of thick Phanerozoic cover such as the central Prairies to more than 20 grains per sample (2 grains/kg) on the down-ice margins of large volcanosedimentary terrains such as the Abitibi Greenstone Belt (Fig. 1).
(2) Eighty to 90% of gold grains in till are silt sized (< 63 μ wide), mirroring the grain size of the parent bedrock mineralization.
(3) Most gold geochemical anomalies obtained from analysing the sieved fine sand and silt fractions of raw (unconcentrated) till or soil samples are due to these gold grains, not to gold chemically adsorbed on clay minerals or limonite. This relationship is easily demonstrated by micropanning anomalous samples that have previously been analysed by the non-consumptive instrumental neutron activation method.
(4) Being soft and malleable, gold grains are deformed rapidly during ice transport (Fig. 2), progressively transforming them from pristine to modified and reshaped forms (DiLabio 1990). This deformation occurs by infolding and compaction; the original mass of the grains does not change. The reshaping process is typically complete after 1 km of transport but required 5 to 10 km of transport at Rainy River (Fig. 1) where buoyant ice flowed rapidly through glacial Lake Agassiz (Averill 1998). That is, gold grain wear appears to be more closely related to transport time than to transport distance.
(5) Most gold grain dispersal trains related to significant mineralization are < 1 km long (Table 1); therefore their gold grains are primarily of the pristine and modified classes. The anomaly strength midway along the train is typically 2 to 4 grains/kg but in a

few trains, especially in the La Ronge Belt (Fig. 1), it reaches ten grains per kilogram.

(6) Esker sediments are depleted in gold grains relative to tills. This depletion occurs in part because esker sediments consist mainly of sorted medium to coarse sand grains whereas gold grains by nature are mostly silt sized. Although heavy, these small gold grains tend to be flushed from coarse esker sands and gravels into distal outwash silty sands due to the effective reduction in specific gravity that occurs with decreasing grain size (Stokes' law). Also the rapid rate of deposition of esker sediments does not permit the few available coarse gold grains to be concentrated into placer gold beds. Reported occurrences of placer gold such as in the Munro Esker in the Matheson district (Fig. 1) of the Abitibi Greenstone Belt (Ferguson & Freeman 1978) are generally hosted by glaciolacustrine beach deposits developed on eskers rather than in true esker sediments.

Kimberlite indicator minerals

KIMs are described in detail elsewhere in this volume (McClenaghan & Kjarsgaard 2001) and therefore will be mentioned only briefly here. Three garnet species are commonly used as KIMs (Dawson & Stephens 1975; Fipke et al. 1995): (1) peridotitic Cr–pyrope; (2) eclogitic pyrope-almandine; and (3) megacrystic Cr-poor pyrope. The other prime indicators are Mg–ilmenite, chromite and Cr–diopside. Mg-rich olivine (forsterite) and orthopyroxene (enstatite) are supplemental indicators; they are of limited use alone as they occur in peridotites and some metamorphic rocks in addition to kimberlite.

The chemistry of certain KIM species is a useful guide to the diamond potential of their source kimberlites (Gurney & Moore 1993; Fipke et al. 1995). Specifically, Ca-poor, harzburgitic 'G10' Cr–pyrope, Na-rich pyrope–almandine and ultra Cr-rich chromite indicate sampling of diamond-fertile layers in the mantle by the kimberlite magma and ultra Mg-rich ilmenite indicates reducing conditions favouring diamond preservation during the magma's ascent through the crust.

All KIM species are chemically stable in immature glacial drift; Cr–diopside and garnet are not selectively destroyed in mature nonglacial terrains (Mosig 1980). However certain KIMs, especially garnets, tend to occur as highly fractured grains in kimberlite as a consequence of hydration either during volcanism or by preglacial weathering, and their subsequent grain size and relative abundance in glacial drift are controlled in part by this preparatory fracturing (Averill & McClenaghan 1994; McClenaghan & Kjarsgaard 2001). Fracture-prone Cr–pyrope macrocrysts, for example, tend to break into large numbers of medium sand sized (0.25–0.5 mm) grains whereas fracture-resistant Mg–ilmenite generally remains at coarse sand size (0.5–1.0 mm) and therefore tends to be less abundant. In the dispersal train of the C14 pipe near Kirkland Lake, Ontario (Fig. 1), unfractured but cleavable Cr–diopside grains appear to have broken after fractured, uncleavable Cr–pyrope as the ratio of Cr–diopside to Cr–pyrope grains increases down-ice (Averill & McClenaghan 1994). Being naturally medium to coarse grained rather than silt sized like gold grains, KIMs are concentrated rather than depleted in esker sediments relative to tills, typically tenfold. Furthermore, medium sand sized Cr–pyrope tends to collect in sand beds and coarse sand sized Mg–ilmenite in gravel beds.

Metamorphic/magmatic massive sulphide indicator minerals

MMSIMs are heavy, coarse-grained, weathering-resistant minerals formed by any of the following processes: (1) recrystallization of volcanosedimentary massive sulphide deposits and their hydrothermal alteration halos by medium to high-grade regional metamorphism; (2) high-temperature magmatic metasomatism (skarns and greisens); or (3) reactions associated with the separation of Ni–Cu–Fe sulphides from ultramafic magmas and komatiites. Their resistance to weathering reflects enrichment in elements such as Mg, Mn, Al and Cr which are concentrated in acid-leached 'aluminous' hydrothermal alteration zones (Lydon 1989), introduced by magmatic metasomatism (Dawson & Kirkham 1996) or united through the assimilation of felsic, sulphurous sedimentary rocks by ultramafic magmas and lavas during the formation of Ni–Cu deposits (Naldrett 1989; Naldrett et al. 1996).

MMSIM dispersal trains tend to be large (sometimes as large as KIM trains) because the indicator minerals reflect both the deposit and its alteration envelope and some of the minerals are as unique as KIMs. However, some MMSIMs, such as kyanite, sillimanite, staurolite and orthopyroxene, are less useful than others because they are common throughout certain high-grade regional metamorphic terrains as well as in hydrothermal alteration zones. On the other hand, a few MMSIMs are enriched in ore metals instead of (or in addition to) alteration zone elements and thus are doubly useful indicators in the manner of G10 Cr–pyrope on KIM surveys. Common examples are gahnite ($ZnAl_2O_4$), zincian staurolite (($Fe,Mg,Zn)_2Al_9$($Si,Al)_4O_{22}(OH)_2$), willemite (Zn_2SiO_4) and franklinite (($Zn,Mn,Fe)(Fe,Mn)_2O_4$) in metamorphosed massive sulphide alteration zones, and scheelite ($CaWO_4$) and cassiterite (SnO_2) in skarns and greisens. If the mineral deposit contains accessory precious metals, native indicator minerals such as gold grains and PGE alloys may also be present in its dispersal train although the fine-grained fraction of the samples must be processed to utilize these minerals.

Arsenides such as sperrylite ($PtAs_2$), rammelsbergite ($NiAs_2$) and loellingite ($FeAs_2$) are also relatively stable, even in sediments in unglaciated terrains, and can be used as MMSIMs. Most sulphide minerals, in contrast, are very unstable. However, ODM has observed that chalcopyrite (and to a lesser degree sphalerite) is nominally stable in surficial sediments worldwide. In Cu-fertile regions it is not unusual to find 50 to 100 chalcopyrite grains in a till or alluvial sediment sample that does not contain pyrite, even if pyrite is ten times more plentiful than chalcopyrite in known Cu-deposits in the area. More research on mineral stabilities is needed to explain this useful phenomenon. However, the observed chalcopyrite grains are clearly survivors of a once-larger sulphide mineral population and the Cu anomalies are much stronger than the grain counts suggest.

Perhaps the earliest documented usage of chalcopyrite as a MMSIM was in 1994 at the Voisey's Bay Ni–Cu–Co deposit in Labrador (Fig. 1) as chronicled by McNish (1998). The first indications of the deposit were found the previous year by prospectors Al Chislett and Chris Verbiski when they examined a rusty outcrop while sampling stream sediments for KIMs on behalf of Diamond Field Resources Inc. The prospectors assayed their outcrop samples for Cu but not Ni and the full significance of their discovery was not recognized. Moreover, Diamond Fields was interested in diamonds, not copper. Consequently the project was refocused on kimberlite in 1994 and ODM was contracted to process the stored 1993 stream sediment samples. Much to the consternation of Diamond Fields, no KIMs were found. Three directors, Mike McMurrough, Richard Garnett and Rod Baker, placed a telephone conference call to the author on May 13, 1994, seeking an explanation. Following is McNish's (1998, p. 75) account of that conversation:

'The lab, Overburden Drilling of Ottawa, reported confusion over the test results. It was supposed to be testing for diamonds, but the results showed unusually high levels of sulphides, a chemical compound that typically indicates the presence of base metals. What did Diamond Fields want the lab to do about the sulphides? The three officials agreed that the lab should keep tracking the sulphide occurrences and send its reports to Garnett.

Garnett kept track of the extensive data by marking the occurrences in different colours, green highlights for copper showings and yellow for pyrite. After a few weeks, the area surrounding Voisey's Bay was so full of green and yellow spots that it looked like a crude pointillist painting. Surprised by the heavy concentration of sulphides, Garnett began to research the few geological reports about the region. He grew more intrigued after he read an eight-year-old report by Newfoundland government geologist Bruce Ryan about the strange orange gossan on a hill now claimed by the prospectors. There was something very unusual about Voisey's Bay.'

The discovery of MMSIMs in surficial sediments at Voisey's Bay while searching for KIMs illustrates the extended versatility of heavy indicator mineralogy compared to heavy mineral geochemical analysis. Moreover, the chalcopyrite grains were observed despite the fact that the heavy mineral concentrates were grossly oversized, weighing up to 2 kg rather than the usual 25 to 50 g, due to an overabundance of garnet and other heavy minerals eroded from the underlying Archean gneisses of the Nain Province. With so much heavy mineral dilution, the Cu in the chalcopyrite would represent only a few ppm Cu in the concentrates and thus would not be considered geochemically anomalous even if the analysis employed a sulphide-selective extraction.

A similar situation occurs in till over the Shebandowan Greenstone Belt (Figs 1 and 3) of northwestern Ontario (Bajc 1999, 2000) except that heavy mineral dilution here is due to pyroxene glacially transported from the laterally extensive, pyroxene-rich (40 to 50%) Nipigon Diabase intrusion 70 km to the northeast (Fig. 3). The local bedrock of the Shebandowan Greenstone Belt is well represented in the pebble fraction of the till but the rocks contain <0.5% heavy minerals and therefore are minor contributors to the overall heavy mineral fraction of the till. The relative paucity of locally derived

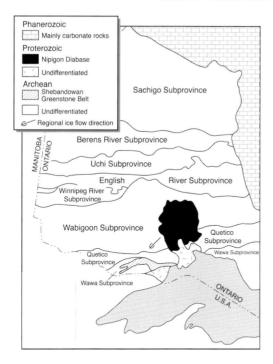

Fig. 3. Bedrock geology map of northwestern Ontario showing the location of the Shebandowan Greenstone Belt relative to Nipigon Diabase. Ice flow was towards the southwest across the diabase, polluting the till with pyroxene which suppresses heavy mineral geochemical responses of local mineralization in tills in the Shebandowan area.

heavy minerals has no bearing on the indicator mineral signature of local mineral deposits – indeed several significant gold grain and MMSIM dispersal trains are present – but the geochemical signatures of these trains are greatly suppressed by the exotic pyroxene. This type of dilution is seldom encountered in non-glacial terrains because the heavy mineral sources are geographically more restricted.

MMSIMs regularly observed by ODM geologists in dispersal trains from metamorphosed volcanosedimentary massive sulphide deposits, skarns/greisens and magmatic Ni–Cu sulphide deposits are described sequentially in the following sections. Only primary MMSIMs found in weathered immature glacial drift are included. Many of these primary minerals are also found in mature non-glacial sediments along with secondary base metal oxides, silicates, carbonates, sulphates and phosphates.

Metamorphosed volcanosedimentary massive sulphide indicator minerals

MMSIM species regularly observed in glacial dispersal trains associated with metamorphosed volcanosedimentary massive sulphide mineralization are listed in Table 2 together with the indicator elements in which the minerals are enriched. The minerals are diverse because they include species from all parts of the alteration zone as well as the actual sulphide deposit. In a Mn-bearing volcanogenic massive sulphide system, for example, metamorphism of the proximal, chalcopyrite-veined chloritic alteration pipe (Lydon 1989) would tend to increase the grain size of the chalcopyrite and produce anthophyllite ($(Mg,Fe)_7Si_8O_{22}(OH)_2$) + spessartine ($Mn_3Al_2Si_3O_{12}$) whereas metamorphism of the distal, semiconformable, epidote-quartz alteration zone would tend to produce red Mn–epidote ($Ca_2(Al,Fe,Mn)_3Si_3O_{12}OH$). The minerals observed in dispersal trains and their associated indicator elements include chalcopyrite (Cu, S), barite (Ba, S), gahnite (Zn, Al), spinel–staurolite–sapphirine (Mg, Al), kyanite–sillimanite (Al), anthophyllite–orthopyroxene (Mg), spessartine (Mn, Al), red epidote (Mn), red rutile (Cr) and loellingite (As). Note that most of the indicator elements are common to other, non-indicator heavy minerals and are too tightly bonded to respond to selective extraction in the analytical laboratory. Thus the anomalies, like KIM anomalies, are seldom discernible by geochemical analysis of the concentrates. Even Cu in chalcopyrite and Zn in gahnite are difficult to detect because anomalous concentrations of these minerals are in the order of just tens of grains per sample (50 chalcopyrite grains in 50 g of 0.25 to 0.5 mm concentrate approximates 15 ppm Cu). The only indicator minerals in Table 2 commonly reaching percentage levels in heavy mineral concentrates are kyanite, sillimanite, staurolite, spessartine, anthophyllite, orthopyroxene and barite.

As with KIM anomalies, the best MMSIM anomalies include two or more mineral species. Common mineral associations observed in the dispersal trains of metamorphosed volcanosedimentary massive sulphide deposits are tourmaline–barite, chalcopyrite–spessartine (or Mn–epidote), gahnite–chalcopyrite–staurolite (or anthophyllite), and red rutile–spinel–kyanite (or sillimanite).

The regional-scale abundance of kyanite, sillimanite or staurolite in metasedimentary terrains containing slightly aluminous pelitic horizons, and of orthopyroxene in granulite-

Table 2. *Common heavy indicator minerals of hydrothermal alteration zones associated with metamorphosed volcanosedimentary massive sulphide deposits in glaciated terrains*

Indicator mineral	Chemical composition	Indicator elements
sillimanite	Al_2SiO_5	Al
kyanite	Al_2SiO_5	Al
corundum	Al_2O_3	Al
anthophyllite	$(Mg,Fe)_7Si_8O_{22}(OH)_2$	Mg
orthopyroxene	$(Mg,Fe)_2Si_2O_6$	Mg
Mg–spinel	$MgAl_2O_4$	Mg, Al
sapphirine	$(Mg,Al)_8(Al,Si)_6O_{20}$	Mg, Al
staurolite	$(Fe,Mg,Zn)_2Al_9(Si,Al)_4O_{22}(OH)_2$	Mg ($\pm$Zn), Al
tourmaline	$(Na,Ca)(Mg,Fe)_3Al_6(BO_3)_3(Si_6O_{18})(OH)_4$	Al, B
dumortierite	$Al_7(BO_3)(SiO_4)_3O_3$	Al, B
Mn–epidote	$Ca_2(Al,Fe,Mn)_3Si_3O_{12}(OH)$	Mn
spessartine	$Mn_3Al_2Si_3O_{12}$	Mn, Al
gahnite	$ZnAl_2O_4$	Zn, Al
franklinite	$(Zn,Mn,Fe)(Fe,Mn)_2O_4$	Zn, Mn
willemite	Zn_2SiO_4	Zn
Cr–rutile	$(Ti,Cr)O_2$	Cr
barite	$BaSO_4$	Ba, S
chalcopyrite	$CuFeS_2$	Cu, S
cinnabar	HgS	Hg, S
loellingite	$FeAs_2$	As
native gold	Au	Au

Table 3. *Common heavy indicator minerals of skarn and greisen deposits in glaciated terrains*

Indicator mineral	Chemical composition	Indicator elements
SKARN:		
forsterite olivine	Mg_2SiO_4	Mg
knebelite olivine	$(Fe,Mn)_2SiO_4$	Mn
vesuvianite	$Ca_{10}Mg_2Al_4(Si_2O_7)_2(SiO_4)_5(OH)_4$	Mg
johannsenite	$CaMnSi_2O_6$	Mn
grossular	$Ca_3Al_2Si_3O_{12}$	Al$\pm$Cr
scheelite	$CaWO_4$	W
chalcopyrite	$CuFeS_2$	Cu, S
native gold	Au	Au
GREISEN:		
topaz	$Al_2SiO_4(F,OH)_2$	Al, F
tourmaline	$(Na,Ca)(Mg,Fe)_3Al_6(BO_3)_3(Si_6O_{18})(OH)_4$	Al, B
fluorite	CaF_2	F
cassiterite	SnO_2	Sn
wolframite	$(Fe,Mn)WO_4$	Mn, W
chalcopyrite	$CuFeS_2$	Cu, S

facies terrains, can restrict the role of these minerals to supplemental MMSIMs similar to forsterite and enstatite in diamond exploration. However, this restriction applies only to high-pressure, Barrovian-type metamorphic terrains; it is not a factor in high-temperature, Abukuma-type (Miayashiro 1973) terrains where such minerals are stable only in alteration zones. Moreover, the chemical compositions of hydrothermal alteration zones are often so specific that a particular metamorphic mineral will form only at these sites even in Barrovian terrains. For example, the alteration zones associated with the Snow Lake deposits in Manitoba (Fig. 1) are characterized by staurolite + kyanite (Walford & Franklin 1982) whereas the main aluminous mineral in the neighbouring unmineralized Kisseynew gneisses is sillimanite (Froese

Table 4. *Common heavy indicator minerals of magmatic Ni-Cu massive sulphide deposits in glaciated terrains*

Indicator mineral	Chemical composition	Indicator elements
hercynite	$FeAl_2O_4$	Al
olivine	$(Mg,Fe)SiO_4$	Mg
orthopyroxene	$(Mg,Fe)_2Si_2O_6$	Mg
low-Cr diopside	$Ca(MgCr)Si_2O_6$	Mg, Cr
chromite	$(Fe,Mg)(Cr,Al)_2O_4$	Cr, Mg, Al ($\pm$Zn)
uvarovite	$Ca_3Cr_2Si_3O_{12}$	Cr
Cr–rutile	$(Ti,Cr)O_2$	Cr
chalcopyrite	$CuFeS_2$	Cu, S
loellingite	$FeAs_2$	As
rammelsbergite	$NiAs_2$	Ni, As
sperrylite	$PtAs_2$	Pt, As
PGE alloys	PGE	PGE

& Moore 1980). In some cases, such as in the 17 Zone at Rainy River, Ontario (Fig. 1), hydrothermal alteration envelopes are so enriched in Mn or Al that minerals such as spessartine or kyanite will form even under lower greenschist-facies conditions of metamorphism where they would not normally be stable (Averill 1998). Such occurrences are primarily geological curiosities; most hydrothermal alteration zones in greenschist facies terrains are characterized by fine-grained, low-density minerals and therefore cannot be detected by the MMSIM method.

Skarn and greisen indicator minerals

MMSIM species regularly observed in glacial dispersal trains associated with skarn and greisen mineralization are listed in Table 3 together with the indicator elements in which the minerals are enriched. Skarn indicators include forsterite (Mg), knebelite (Mn), johannsenite (Mn), grossular (Al±Cr), chalcopyrite (Cu, S) and scheelite (W). The principal greisen indicators are topaz (Al, F), tourmaline (Al, B), fluorite (F), cassiterite (Sn) and wolframite (Mn, W). No glacial drift sampling has been done near Olympic Dam-type (Cu–Au–U) deposits but high-temperature indicator minerals analogous to those of skarns and greisens would be expected.

Ni–Cu massive sulphide indicator minerals

MMSIM species regularly observed in glacial dispersal trains associated with Ni–Cu massive sulphide mineralization hosted by komatiites and layered mafic/ultramafic intrusions are listed in Table 4 together with the indicator elements in which the minerals are enriched. These minerals include olivine (especially Mg-rich forsterite), orthopyroxene (especially Mg-rich bronzite/enstatite), Cr–diopside (Mg, Cr), chromite (Cr, Mg, Al), uvarovite (Cr), red rutile (Cr), hercynite (Al), chalcopyrite (Cu, S), loellingite (As), rammelsbergite (Ni, As), sperrylite (Pt, As) and native PGE alloys. Olivine, orthopyroxene and chromite occur in unmineralized as well as mineralized ultramafic rocks and therefore are useable only if their chemical compositions are unique or they are accompanied by other, more diagnostic indicator minerals.

The great variety of Ni–Cu indicator minerals contrasts with the simplicity of Ni–Cu sulphide deposits. This diversity appears to reflect the hybridization process (voluminous assimilation of sulphurous felsic sedimentary rocks by superheated ultramafic magmas) that separates a Ni–Cu–Fe–S liquid from ultramafic magma. Assimilation is known to generate distinctive hybrid Mg–Al rocks such as olivine and orthopyroxene gabbro (troctolite and norite); therefore it might also be expected to combine Mg, Fe or Cr with Al or Si in distinctive hybrid minerals such as chromite ($(Fe,Mg)(Cr,Al)_2O_4$), uvarovite ($Ca_3Cr_2Si_3O_{12}$) and hercynite ($FeAl_2O_4$). Many other combinations are possible. For example, if excess Cr were available following the crystallization of chromite or uvarovite it would tend to be accommodated in common, Cr-receptive minerals such as diopside ($CaMgSi_2O_6$) and rutile (TiO_2), transforming them into colourful indicator species ($Ca(Mg,Cr)Si_2O_6$ and $(Ti,Cr)O_2$, respectively).

Cr-bearing diopside is the principal clinopyroxene of pyroxenite and peridotite horizons associated with the Ni deposits at Outokumpu, Finland (Papunen *et al.* 1979) and in the Dumont sill (Duke 1986) and Lac Rocher intrusions (Averill 1999*a,b*, 2000), Quebec (Fig. 1). Being so plentiful, it is generally the most useful Ni–Cu indicator mineral in glacial drift. For example, a significant dispersal train

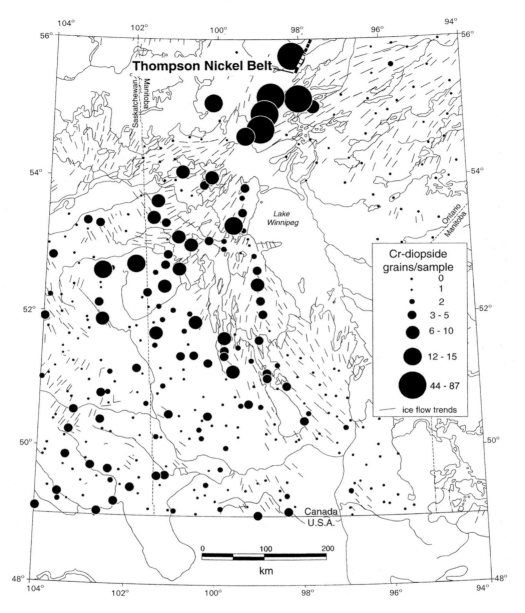

Fig. 4. Distribution of Cr–diopside in surface till in Manitoba, Saskatchewan and Alberta. Note the high concentrations extending 400 km SW (down-ice) from the Thompson Ni Belt (adapted from Thorleifson & Garrett 1993).

containing Cr–diopside but no other indicator minerals (Bajc 2000) is present down-ice from the Shebandowan Ni–Cu–PGE deposit in Ontario (Fig. 1). Also, a major Cr–diopside dispersal train extends 400 km southwest into Saskatchewan from the Thompson Nickel Belt in Manitoba (Fig. 4; Thorleifson & Garrett 1993; Matile & Thorleifson 1997). Closer to Thompson, the Cr–diopside in this train is accompanied by other Ni indicators such as chalcopyrite, hercynite and chromite which occur as minor accessory minerals in the Ni deposits (de Saboia 1978). Although the dispersal train is clearly related to the nickel belt, the actual source of the Cr–diopside within the belt is poorly constrained because little information is available on the chemistry of clinopyroxene in the peridotite and pyroxenite horizons. The Ni-bearing horizons are mostly harzburgites and orthopyroxenites but unmineralized lherzolites, wehrlites, websterites and clinopyroxenites are present (Peredery 1982). Most of the clinopyroxene in these phases has been altered to amphibole but three grains from the Pipe Mine 2 ultramafic body were analysed by de Saboia (1978) and all are excellent Cr–diopsides (0.47 to 1.09 wt.% Cr_2O_3). Orthopyroxenes from the same rocks also contain significant Cr_2O_3 (0.20 to 0.88 wt.%). At Lac Rocher, Quebec (Fig. 1), the main intrusive phase is clinopyroxenite and most of the clinopyroxene in both mineralized and unmineralized phases is Cr–diopside although it is largely hydrated to tremolite + Cr–magnetite (Averill 1999a, b, 2000). Orthopyroxene, where present, is also Cr–bearing but the Cr is not colour enhancing as it is in diopside. The Thompson Nickel Belt is much larger than the Lac Rocher intrusions and as little as 1% Cr–diopside in the ultramafic rocks would be sufficient to explain the observed Cr–diopside dispersal train in the till. As well, diopside-bearing carbonate horizons are common in the metasediments that enclose the pyroxenites and peridotites (R. Somerville, Inco Ltd., pers. comm. 2000) and Cr–metasomatism of these carbonate rocks by the ultramafic bodies during emplacement or metamorphism could also have produced considerable Cr–diopside as at Outokumpu (Papunen et al. 1979).

Cr–diopside that is used as a Ni–Cu indicator generally contains less Cr_2O_3 (<1.25 wt.%) than typical kimberlitic Cr–diopside derived from mantle lherzolites (>1.25 wt.%; Eggler et al. 1979) and tends to have a paler green colour. However, considerable overlap in both Cr_2O_3 content and colour between the types does occur. Kimberlites are also known to contain Cr–diopside of at least six parageneses having lower Cr contents than the bright green Cr–diopside of lherzolitic paragenesis that is used as a kimberlite indicator mineral (Mitchell 1986). Cr–diopside grains used as a Ni–Cu indicator are less easily recognized in rock specimens than in heavy mineral concentrates and have seldom, if ever, been recognized in drill core. Ideally any accompanying chromite grains in a Cr–diopside dispersal train are Zn–bearing, indicating that their source rocks have assimilated significant quantities of sedimentary sulphides (Lesher 1989). Thus, in Ni–Cu exploration as in diamond exploration, the chemistry of the chromite grains is as important as their association with other indicator minerals.

Conclusions

Heavy indicator mineralogy has many benefits not found in heavy mineral geochemistry. These benefits arise due to: (1) the high degree of chemical stability of indicator minerals relative to sulphide minerals upon which heavy mineral geochemistry is so dependent; (2) the presence in the indicator minerals of elements such as Mg, Mn, Al and Cr which favourably enhance colour, chemical stability and electromagnetic separability but cannot be detected by geochemical analysis of mineralogically anomalous concentrates; and (3) the opportunity to observe physical features of the mineralization. The benefits of heavy indicator mineralogy over heavy mineral geochemical analysis include:

(1) Major sampling efficiencies in base metal exploration through the substitution of weathered surface sediments for unweathered buried sediments, thereby reducing the need for deep sampling.
(2) Considerable enlargement of the bedrock target by adding the alteration zone to the actual mineralization. Since this enlargement occurs in three dimensions, it may also allow detection of otherwise blind mineralized targets.
(3) Manyfold enlargement of the dispersal train by effectively lowering the detection limit of both alteration elements such as Mg, Mn, Al and Cr and base metals such as Zn, Cu and Ni to the single mineral level (e.g. Mg and Al in staurolite, Mn in spessartine, Cr in diopside, Zn in gahnite, Cu in chalcopyrite, Ni in rammelsbergite).
(4) Undiminished sensitivity in terrains with overabundant non-indicator heavy minerals.
(5) Coverage of a wider range of mineral commodities in a single survey.
(6) Visible evidence of the dispersal distance including the degree of reshaping of gold and platinum grains and the degree of separation of KIMs from their alteration rinds.
(7) Evidence of the worth of the source mineralization including the type and number of indicator mineral species present in the dispersal train and the chemistry of KIMs such as Cr–pyrope or MMSIMs such as gahnite, staurolite and chromite.

With the targets being so large and so many mineral commodities being identifiable from a single set of samples, heavy indicator mineralogy is most effective as a reconnaissance exploration tool. The reconnaissance capabilities are well established for KIMs and are probably even greater for MMSIMs. Extensive, highly metamorphosed, often gneissic Archean and Proterozoic volcanosedimentary and plutonic terrains have undergone negligible base metal exploration due to the difficulty of recognizing fertile protoliths and recrystallized hydrothermal alteration zones beneath their metamorphic masks. Heavy indicator mineralogy offers much promise for the future development of these terrains.

Most of the heavy mineral relationships described in this paper were identified over a 25-year period from tens of thousands of bulk samples submitted to ODM by more than a hundred individual prospectors and corporate and government clients. Many of these samples were from pioneering projects requiring considerable faith in heavy mineral sampling as a viable exploration tool. BHP World Explorations Inc., WMC International Ltd., the Geological Survey of Canada and the Ontario Geological Survey have been particularly helpful in demonstrating the utility of the MMSIM method. R. Somerville and P. Golightly of Inco Ltd. supplied critical information on mineral chemistry in the Thompson Nickel Belt. Nuinsco Resources Ltd. granted permission to use data from the author's consulting reports on the Lac Rocher intrusions. The paper benefited from insightful reviews by J. M. Franklin, L. Hulbert, P. Taufen, and L. H. Thorleifson and thoughtful editing by M. B. McClenaghan.

References

AVERILL, S. A. 1988. Regional variations in the gold content of till in Canada. *In*: MACDONALD, D. R. & MILLS, K. A. (eds) *Prospecting in Areas of Glaciated Terrain – 1988*. Canadian Institute of Mining and Metallurgy, Geological Division, Halifax, 271–284.

AVERILL, S. A. 1998. *Reverse circulation drilling discoveries on the Rainy River property of Nuinsco Resources Limited, 1994 to 1998*. Unpublished oral presentation. Minnesota Department of Natural Resources, Birchdale-Indus Field Trip.

AVERILL, S. A. 1999a. *SEM/EDS analysis of drill core, holes NLR-99-26, 28 and 31, Lac Rocher, Quebec*. Unpublished report prepared for Nuinsco Resources Limited.

AVERILL, S. A. 1999b. *SEM/EDS analysis of drill core from holes NLR-99-41 and 44 on the Anomaly A intrusion, Lac Rocher, Quebec*. Unpublished report prepared for Nuinsco Resources Limited.

AVERILL, S. A. 2000. *Intrusive phases and Ni-in-olivine content of the ultramafic Salamandre intrusion, Lac Rocher West area, Quebec*. Unpublished report prepared for Nuinsco Resources Limited.

AVERILL, S. A. & MCCLENAGHAN, M. B. 1994. *Distribution and character of kimberlite indicator minerals in glacial sediments, C14 and Diamond Lake kimberlite pipes, Kirkland Lake, Ontario*. Geological Survey of Canada Open File **2819**.

AVERILL, S. A. & ZIMMERMAN, J. R. 1986. The riddle resolved: the discovery of the Partridge gold zone using sonic drilling in glacial overburden at Waddy Lake, Saskatchewan. *Canadian Geology Journal of the Canadian Institute of Mining and Metallurgy*, **1**, 14–20.

BAJC, A. F. 1991. *Till sampling survey, Fort Frances area, results and interpretation*. Ontario Geological Survey Study **56**.

BAJC, A. F. 1999. *Results of regional humus and till sampling in the eastern part of the Shebandowan Greenstone Belt, Northwestern Ontario*. Ontario Geological Survey Open File **5993**.

BAJC, A. F. 2000. *Results of regional till sampling in the western part of the Shebandowan Greenstone Belt*. Ontario Geological Survey Open File **6012**.

CHAPMAN, R., CURRY, G. & SOPUCK, V. 1990. The Bakos deposit discovery - a case history. *In*: BECK, L. S. & HARPER, C. T. (eds) *Modern Exploration Techniques*. Saskatchewan Geological Society, Special Publications, **10**, 195–212.

DAWSON, K. M. & KIRKHAM, R. V. 1996. Skarn copper. *In*: ECKSTRAND, O. R., SINCLAIR, W. D. & THORPE, R. I. (eds) *Geology of Canadian Mineral Deposit Types*. Geological Survey of Canada, **8**, 460–476.

DAWSON, J. B. & STEPHENS, W. E. 1975. Statistical classification of garnets from kimberlite and associated xenoliths. *Journal of Geology*, **83**, 589–607.

DILABIO, R. N. W. 1990. Classification and interpretation of the shapes and surface textures of gold grains from till on the Canadian Shield. *Current Research, Part C*. Geological Survey of Canada Paper **88-1C**, 61–65.

DUKE, J. M. 1986. *Petrology and economic geology of the Dumont Sill: an Archean intrusion of komatiitic affinity in northwestern Quebec*. Geological Survey of Canada Economic Geology Report **35**.

EGGLER, D. H., MCCALLUM, M. E. & SMITH, C. B. 1979. Megacryst assemblages in kimberlite from northern Colorado and southern Wyoming: Petrology, geothermometry-barometry and areal distribution. *Second International Kimberlite Conference* **2**, 213–226.

FERGUSON, S. A. & FREEMAN, E. B. 1978. *Ontario occurrences of float, placer gold, and other heavy minerals*. Ontario Geological Survey, Mining Department, Circular **17**.

FIPKE, C. E., GURNEY, J. J. & MOORE, R. O. 1995. *Diamond exploration techniques emphasizing indicator mineral geochemistry and Canadian examples*. Geological Survey of Canada, Bulletin **423**.

FROESE, E. & MOORE, J. M. 1980. *Metamorphism in the Snow Lake area, Manitoba*. Geological Survey of Canada Paper **78-27**.

GRAY, R. S. 1983. *Overburden drilling as a tool for gold*

several case studies (Table 1) were initiated to document indicator mineral and till geochemical dispersal patterns in glacial sediments around known kimberlites (e.g. Averill & McClenaghan 1994; McClenaghan *et al.* 1996, 1998, 1999*a, b, c*; Thorleifson & Garrett 2000). This paper provides a summary of published information on diamond exploration techniques in the glaciated terrain of Canada, focusing on indicator minerals and till geochemistry.

Kimberlite

Kimberlites are CO_2- and H_2O-rich ultrabasic rocks of magmatic origin, derived from limited partial melting of the mantle at very high pressure (10 GPa or 300 km depth). The rocks have a distinctive inequigranular texture, and often appear hybrid in nature as they may contain mantle xenoliths, mantle xenocrysts, macrocrysts (a nongenetic term for large crystals 1 to 20 cm in size, i.e. megacrysts and xenocrysts), crustal xenoliths and euhedral to subhedral phenocrysts set in a groundmass matrix crystallized from the kimberlite magma. Macrocrysts include minerals derived from disaggregated mantle xenoliths, plus the kimberlite megacryst suite of minerals (Mg–ilmenite, Cr-poor Ti–pyrope garnet, Cr–diopside, phlogopite, enstatite, zircon and olivine). Primary minerals include euhedral olivine and phlogopite phenocrysts and microphenocrysts in a fine-grained matrix of one or more of the following minerals: spinel, ilmenite, perovskite, monticellite, apatite, phlogopite–kinoshitalite mica, carbonate, and serpentine.

The diverse mineralogy and associated mineral chemistry of kimberlites reflect the unusual major and trace element composition of these ultrabasic rocks. Kimberlites have a characteristic geochemical signature, being rich in the 'incompatible' elements Sr, Ba, LREE (La, Ce, Sm, Nd), Nb, Ta, Hf, Zr, P, and Ti, as well as having high concentrations of first order transition elements Mg, Ni, Cr, Co, which are of ultramafic affinity. The presence of a distinctive suite of resistant and abundant minerals (pyrope, Mg–ilmenite Cr–diopside, Cr–spinel) known as kimberlite indicator minerals, coupled with a characteristic geochemical signature form the basis for drift prospecting methods for diamonds in Canada.

Kimberlites in Canada

Distribution

Locations of known kimberlites in Canada are shown in Figure 1. Exploration activity for kimberlites during the past decade in the Slave Province of northern Canada has resulted in the discovery of 250 kimberlites (Armstrong 1998, 1999). The Lac de Gras area of the central Slave Province hosts the producing Ekati diamond mine and also the Diavik prospect, which is scheduled to go into production in 2003. Greater than 80% of known kimberlites are found in a corridor from Kennady Lake in the southeast through Lac de Gras and Rocking Horse Lake to Kikkerk Lake in the northwest. A few additional kimberlites are also known from the south-central (e.g. Snap Lake), west-central (e.g. Cross), and southern parts of the Slave Province (e.g. Dry Bones). Seven kimberlites have been discovered recently on Victoria Island, Nunavut, which is possibly underlain by a northern extension of the Slave Province. Many of the kimberlites in the Slave Province are diamond-bearing, some richly so. In contrast, kimberlites in the Churchill Province of northern Canada appear to be poor to void of diamonds, including kimberlites on Somerset Island (e.g. Mitchell 1975; Kjarsgaard 1996*a*), the Rankin Inlet area (Miller *et al.* 1998; Heaman & Kjarsgaard 2000), and the most recently discovered kimberlites on north Baffin Island (Northern Miner 1999*a*).

Two major kimberlite fields have been discovered on the Canadian Prairies, both of which contain some moderately diamondiferous kimberlite pipes. The Fort à la Corne field in central Saskatchewan is comprised of at least 70 kimberlites (Kjarsgaard 1995, 1996*b*; Leckie *et al.* 1997*a*), and the associated Candle Lake cluster, 70 km north, contains two additional pipes. The recently discovered Buffalo Hills field of north-central Alberta (Carlson *et al.* 1998; Carlson *et al.* 1999*b*) contains 25 kimberlites. A few additional widely scattered bodies are also known from the Prairie region. Unfortunately, relatively little is known about the kimberlite dykes reported in the Wekusko Lake area of Manitoba. The Cross diatreme in the southeastern Canadian Rocky Mountains is described by Hall *et al.* (1989) as a small multi-phase body. The Mountain Lake ultrabasic volcanics in the Peace River area of Alberta consist of one (twinned) pipe (Leckie *et al.* 1997*b*), although it is not kimberlite.

Sage (1996, 1998, 2000) provides the most recent summary on the geographic distribution, and the geology and mineralogy of kimberlites in the northeastern Ontario portion of the Archean Superior Province. Additional information about kimberlites in the Kirkland Lake and Lake Timiskaming fields can be found in

Fig. 1. Kimberlites and selected ultrabasic locales discussed in the text, shown with a backdrop of the major geological terrains of Canada. Localities as follows: (1) North Baffin Island cluster; (2) Brodeur Penninsula pipe; (3) Somerset Island field; (4) Victoria Island cluster; (5) Kikkerk Lake cluster; (6) Rocking Horse/Contwoyto Lakes field; (7) Ranch Lake pipe; (8) Lac de Gras field; (9) Snare/Carp Lakes clusters; (10) Snap Lake dyke; (11) Kennady Lake cluster; (12) Dry Bones Bay pipe; (13) Birch Mountains cluster; (14) Buffalo Hills field; (15) Mountain Lake ultrabasic pipes; (16) Cross pipe; (17) Fort à la Corne field; (18) Candle Lake cluster; (19) Snow Lake-Wekusko dyke; (20) Rankin Inlet cluster; (21) Kyle Lake cluster; (22) Attawapiskat field; (23) Whitefish Lake; (24) Kirkland Lake field; (25) Lake Timiskaming field; (26) Batchelor Lake; (27) Indicator Lake; and, (28) Torngat Mountains ultrabasic dykes.

McClenaghan et al. (1996, 1998, 1999a, b, c), Schulze (1996), Heaman & Kjarsgaard (2000) and references therein. Kimberlite pipes have also been discovered along the Attawapiskat River in the James Bay Lowland of northern Ontario. At least 18 kimberlites are known, 15 being diamond-bearing (Kong et al. 1999). A recent discovery is the Kyle Lake field of at least five Proterozoic age kimberlites, 75 km W of the Attawapiskat field (Janse et al. 1995). A few kimberlites are described in the literature from the Superior Province in central Quebec, including Batchelor Lake (Watson 1955) and Indicator Lake. No confirmed occurrences of kimberlite have been reported from the Archean Nain craton of Labrador and northern Quebec. There are, however, many known localities of aillikite (an ultrabasic lamprophyre). Recent reports indicate that some of these rocks in northern Quebec are diamond-bearing (Northern Miner 1999b).

Shape and size

Kimberlites occur as small (2 km) but variably sized intrusive and/or extrusive bodies usually grouped together to form small clusters. A number of clusters within an area form a kimberlite field (e.g. the Lac de Gras field, NWT). Kimberlites in Canada occur in two different forms: (1) 'pipes' of extremely variable size and geometry (50 m to 1500 m diameter) which are comprised of diatreme facies and/or volcaniclastic facies kimberlite; and (2) thin dykes and sills, and small, regular to irregular shaped blows (200 m in size) of hypabyssal facies kimberlite.

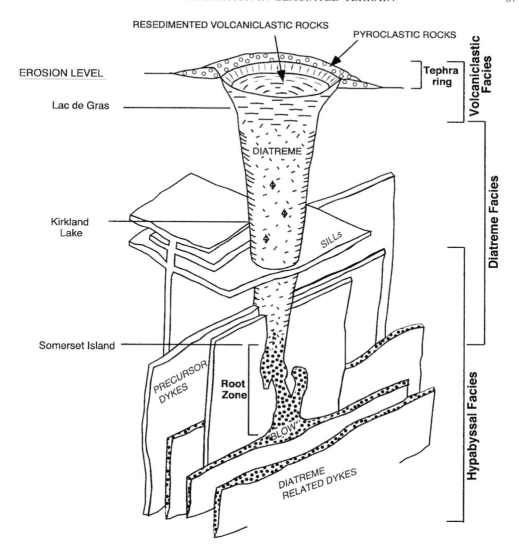

Fig. 2. Generalized model of the classic South African type of kimberlite magmatic system, showing volcaniclastic, diatreme, and hypabyssal facies rocks. Crater facies rocks consist of pyroclastic (tephra ring) and resedimented volcaniclastic rocks (crater in-fill); diatreme facies rocks consist of tuffisitic kimberlite and of tuffisitic kimberlite breccias; and hypabyssal facies rocks are found in the root zone of the diatreme and consist of 'blows' (enlarged dykes), dykes and sills. Also shown are the present erosion levels of some representative economic and Canadian kimberlites. Model after Mitchell (1986), with revised terminology for 'crater facies' rocks.

Highly explosive near-surface volcanism is consistent with the formation of kimberlite pipes. The classical South African type kimberlite pipe (Clement 1975) forms a carrot-shaped diatreme (Fig. 2), with steeply dipping (75 to 85°) country rock contacts. Examples of this type of pipe include a number of kimberlites in the Kirkland Lake field and the Jericho pipe, Nunavut. The surface area and shape of diatreme pipes in Canada range from 5 to 30 hectares with long axis dimensions of 150 m to 600 m. Rarely, volcaniclastic facies rocks (described below) are preserved at the top of the diatreme (Fig. 2). Diatreme facies rocks consist mainly of tuffisitic kimberlite breccias, which often have entrained large amounts of crustal material. With increasing depth, kimberlite diatremes grade into root zones (Fig. 2), consisting of kimberlite dykes, blows, and sills. Several distinct intrusive phases of hypabyssal kimberlite can occur in the root zone of a kimberlite pipe.

Pipes, comprised exclusively of volcaniclastic facies kimberlite with no diatreme facies kimberlite, have only recently been recognized (Kjarsgaard 1996b, 1996c; Kirkley et al. 1998). They consist of kimberlite volcaniclastic rocks (i.e. pyroclastic rocks plus resedimented kimberlite volcaniclastic-dominated rocks) (Fig. 3). Volcaniclastic facies kimberlite pipes have variable size, geometry and morphology. In the Fort à la Corne field they consist of stacked, intercalated, pyroclastic tephra cones and their fluvial and marine reworked equivalents (Leckie et al. 1997a; Kjarsgaard 1998a). These kimberlites range from 150 to 1500 m in diameter, with the larger examples consisting of multiple coalesced eruptive centres. Kimberlites from the Buffalo Hills field, Alberta (Carlson et al. 1998, 1999a) outcrop locally and subcrop in positive relief compared to the less resistant Cretaceous sedimentary host rocks. The volcaniclastic facies kimberlites from Buffalo Hills appear to have some morphological similarities to those at Fort à la Corne. In contrast, Lac de Gras volcaniclastic facies kimberlite pipes tend to be small (50 to 250 m in diameter), steep sided (75 to 90°) bodies (Kjarsgaard 1996d). The pipe infill consists of volcaniclastic-dominated material that was resedimented into the previously excavated pipe, with potential additional input from pyroclastic volcanism (Kirkley et al. 1998). Examples of this type of volcaniclastic facies kimberlite include the Misery and Diavik A-154N kimberlites.

Hypabyssal facies kimberlite (Fig. 3) can form simple sills or dykes, typically 1 to 10 m in width (e.g. Snap Lake, NWT) which may be laterally extensive. Larger (to 200 m), irregularly shaped, multiple lobed root zone complexes consisting of hypabyssal kimberlite are also found (e.g. 5034 kimberlite at Kennady Lake, NWT; Batty Complex, Somerset Island, Nunavut), as are more simple plug-like bodies of hypabyssal kimberlite (e.g. Peddie kimberlite, Lake Timiskaming field, Ontario).

Weathering

At the time of emplacement, kimberlite macrocrysts and primary groundmass minerals are variably replaced at low temperature by serpentine and calcite. Further, the highly porous nature of crater and diatreme facies kimberlite

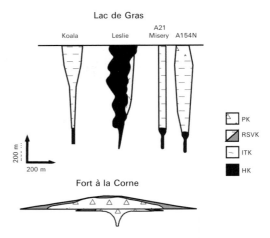

Fig. 3. Simplified, schematic cross-sections of volcaniclastic facies kimberlite pipes from Canada. Note the difference between kimberlites from Lac de Gras and those at Fort à la Corne. Thin hypabyssal dykes (not always observed) form feeder conduits for the volcanics. PK, pyroclastic kimberlite; RSVK, resedimented volcaniclastic kimberlite; ITK, intrusive tuffisitic kimberlite (diatreme facies); HK, hypabyssal kimberlite.

makes these rocks susceptible to alteration by weathering processes after emplacement. Over time, this alteration leads to the development of clay-rich 'blue' and 'yellow' ground. Weathered kimberlite is more susceptible to glacial erosion than the surrounding bedrock, thus many kimberlites in Canada are found in low or swampy ground, under small lakes (Fig. 4), or covered by thick glacial sediments (Fig. 5). Glacial erosion removes varying amounts of the preglacially weathered soft kimberlite, and in some cases the hard competent kimberlite below. For example, glacial erosion of the A4 kimberlite in Kirkland Lake removed all preglacially weathered kimberlite, leaving a fresh, hard subcropping surface beneath 30 m of glacial sediments (McClenaghan et al. 1999b). In contrast, glacial erosion of the B30 kimberlite (Fig. 5) 10 km to the west was minimal, leaving a 10 m thick cap of soft and highly weathered dark green clay-rich kimberlite (McClenaghan et al. 1996). Where glacial erosion has completely removed weathered kimberlite, the underlying unweathered kimberlite may be left as outcrop, as a fresh striated subcrop surface beneath glacial sediments (Fig. 6), or as eroded fragments in glacial sediments.

The kimberlite indicator mineral suite

The unique mineralogical signature of kimber-

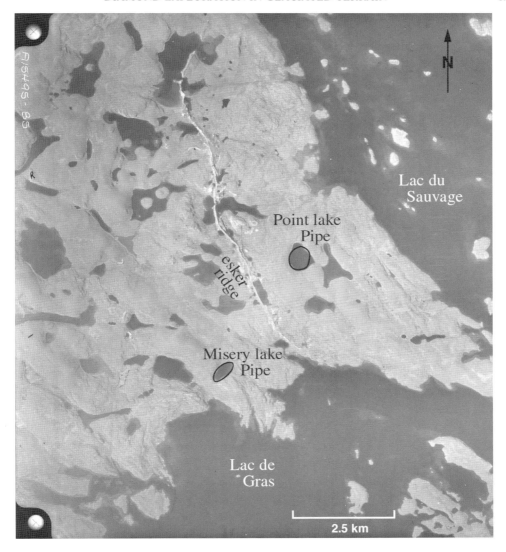

Fig. 4. Aerial photograph showing the glaciated landscape of the thin to moderately drift covered bedrock in the Lac de Gras region and the location of the Point lake and Misery lake kimberlites underlying deeply scoured lakes.

lites enables the application of indicator mineral sampling exploration techniques. The identification of resistant minerals that can indicate the potential presence of a kimberlite has been widely and successfully applied as an exploration technique in South Africa, Yakutia, Australia (Gurney 1984; Atkinson 1989) and Canada (Blusson 1998). Several minerals found in glacial sediment down-ice of kimberlites are useful indicators of the presence of kimberlite and, to a certain extent, in evaluation of their diamond potential. These minerals are: far more abundant in kimberlite than diamonds; visually and chemically distinct; sand-sized (0.25 to 2.0 mm); and, sufficiently dense to be concentrated by gravity methods. They have survived preglacial weathering and subsequent glacial transport (Dummett *et al.* 1987; Averill 2001). Kimberlite indicator minerals include: xenocrysts derived from disaggregated peridotite and eclogite mantle xenoliths (olivine, enstatite, Cr–diopside, Cr–pyrope garnet, Cr–spinel, py-

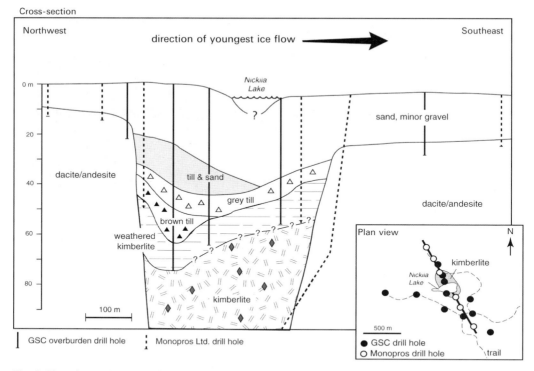

Fig. 5. Plan view and cross-section of the B30 kimberlite, Kirkland Lake, showing the deep level of kimberlite erosion below the surrounding bedrock surface, and a thick cover of glacial sediment masking the depression (from McClenaghan *et al.* 1996).

rope–almandine garnet, omphacitic pyroxene, and diamond); the associated megacryst suite of minerals (low-Cr Ti–pyrope, Mg–ilmenite, Cr–diopside, phlogopite, zircon and olivine); and, kimberlite-derived olivine, spinel and ilmenite.

It is important to note that these minerals (including diamond) are not conclusive of kimberlite magmatism as they can also, in part, be found in other ultrabasic rock types of deep-seated origin (e.g. alnoites; aillikites and other 'ultrabasic lamprophyres'; olivine lamproites). Nevertheless, in the glaciated terrain of Canada, application of drift prospecting for the most common kimberlite indicator minerals (Cr–pyrope, Mg–ilmenite, low-Cr Ti–pyrope garnet, pyrope–almandine garnet, Cr–diopside, Cr–spinel and olivine) has proved highly successful. The physical characteristics of the commonly used kimberlite indicator minerals in glaciated terrain are summarized in Table 2.

Indicator minerals recovered from various types of sampling medium (see following section) can be identified by electron microprobe analysis by determining concentrations of major oxides (e.g. Sobolev 1977; Fipke *et al.* 1989).

Although this is a relatively straightforward process, there can be problems in attempting to differentiate indicator minerals from kimberlite v. those from other bedrock sources. Identification of kimberlitic ilmenite, for example, utilizes a minimum threshold of 4 wt. % MgO combined with elevated (0.5 to 19 wt. %) Cr_2O_3 concentrations (Fig. 7). In the early 1990s in Canada, a number of sampling studies reported the occurrence of high Fe 'eclogitic G5' (nomenclature of Dawson & Stephens 1975) garnets. Subsequently, Schulze (1995) suggested that many of these garnets are in fact of crustal origin. More recently, Fipke *et al.* (1995*a*) and Schulze (1999) have published a threshold of 22 wt. % FeO to separate mantle-derived garnets from crustal garnets. Pyrope garnet derived from the megacryst suite and from peridotite xenoliths can be differentiated from crustal garnet on the basis of elevated Cr_2O_3 and MgO contents. In this regard, Garrett & Thorleifson (1996) and McClenaghan *et al.* (1999*a, b, c*) have developed semi-empirical sets of thresholds to discriminate mantle-derived kimberlitic minerals from those from crustal sources.

Fig. 6. Fresh, striated kimberlite surface on the Peddie kimberlite, Lake Timiskaming, overlain by 2 m of till and exposed in a backhoe trench (from McClenaghan *et al.* 1999c). Arrow indicates ice flow direction towards the southwest.

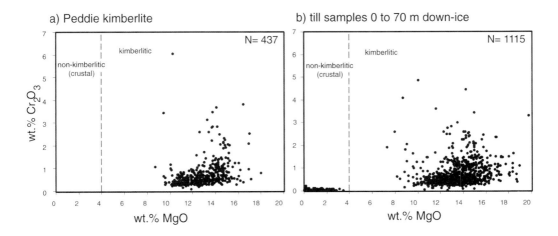

Fig. 7. Cr_2O_3 v. MgO bivariate plot of ilmenite from: (**a**) the Peddie kimberlite (Lake Timiskaming); and, (**b**) till samples collected down-ice (from McClenaghan *et al.* 1999c). Note ilmenites from kimberlite contain >4 wt.% MgO and elevated Cr_2O_3.

Table 2. *Summary of physical features of kimberlite indicator minerals (modified from Muggeridge 1995)*

Mineral	Composition[a]	Crystal system	Colour	Maximum grain size (cm) normal (to rare)	Response to hand magnet	Specific gravity	Hardness (Moh)	Visible diagnostic features[b]	Main source rocks
Pyrope garnet	Mg Al silicate, *Fe, Ca, Cr, Ti*	Isometric	purple, crimson red, mauve, orange	1 (to 15)	none to very weak	3.51	7.5	anhedral, kelyphite rim, characteristic colours, orange peel surface	Peridotite, kimberlite, lamproite, lamprophyre *(certain basic volcanic rocks)*
Mg–ilmenite (picroilmenite)	Mg Fe Ti oxide *Cr, Mn, Al*	Trigonal	black	2 (to 10)	medium to strong *(ferromagnetic)*	4.5 to 5.0	5 to 6	anhedral, rounded or blocky, white porcellaneous leucoxene coating, perovskite overgrowth, glassy lustre on broken surfaces, conchoidal fracture	Kimberlite *(certain basic volcanic) rocks*
Cr-diopside	Ca Mg silicate, *Fe, Cr, Al, Na*	Monoclinic	emerald green	2 (to 15)	none to very weak	3.2 to 3.6	5 to 6	anhedral, blocky, cleavage in 2 directions, characteristic colour	Kimberlite, lamproite *(garnet-) peridotite, certain ultrabasic rocks*
Cr-spinel	Mg Fe Cr Al oxide, *Mn, Ti*	Isometric	black, reddish brown	0.2 (to 0.8)	weak to medium *(paramagnetic)*	4.3 to 4.57	5.5	octahedral to irregular shape, reddish brown on edges of broken grains, glassy surface on broken grains	Lamproite, kimberlite, carbonatite, various ultramafic and basic rocks
Forsteritic olivine	Mg silicate, *Fe, Ni, Mn*	Orthorhombic	pale yellow to green	2 (to 15)	none to very weak	3.2 to 3.33	6 to 7	characteristic colour, irregular crystal apices, vermiform etching	peridotite, carbonatite, kimberlite, lamproite
Diamond	C native *(N, B)*	Isometric	colourless, pale colours (especially yellow and brown)	1 (to 3)	none	3.52	10	adamantine lustre, crystal form, resorption features, step layering	Kimberlite, lamproite *(certain lamprophyres and high grade metamorphic rocks)*

[a] Minor and trace elements listed in italics
[b] by naked eye, hand lens, or binocular microscope

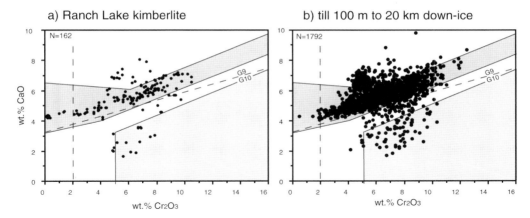

Fig. 8. CaO v. Cr_2O_3 bivariate plot of garnet from: (**a**) Ranch Lake kimberlite (Lac de Gras); and, (**b**) till 100 m to 20 km down-ice ice (from McClenaghan *et al.* 2000). Dashed diagonal line between G9 and G10 garnets is from Gurney (1984). Dashed vertical line at 2 wt. % Cr_2O_3 is from Fipke *et al.* (1995a). Upper shaded field for lherzolite garnets and lower shaded field for subcalcic garnets from the diamond stability field are from Sobolev (1977, 1993).

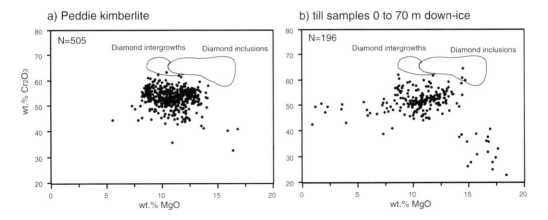

Fig 9. Cr_2O_3 v. MgO bivariate plot of Cr-spinel from: (**a**) the Peddie kimberlite (Lake Timiskaming); and (**b**) till samples down-ice (from McClenaghan *et al.* 1999c). Diamond inclusion and intergrowth fields are from Fipke *et al.* (1995a).

More problematic minerals to discriminate as being of kimberlitic origin include olivine, spinel and diopside. These minerals form phenocrysts in a wide variety of ultrabasic and basic rocks, and may also have primitive mineral chemistry (i.e. elevated Mg/Mg + Fe and Cr/Cr + Al ratios, high MgO and Cr_2O_3 concentrations). Kimberlitic olivine is MgO-rich (forsteritic; Mg/ Mg + Fe > 84) and contains variable, but high trace Ni (Mitchell 1986; Fipke *et al.* 1995a), as do a variety of other rocks. Kimberlites and mantle xenoliths, however, are the only rocks which contain very Cr-rich diopside with > 1.5 wt.% Cr_2O_3 (Deer *et al.* 1982; Fipke *et al.* 1989, 1995a). However, these rocks also may contain diopside with < 1.5% wt.% Cr_2O_3. Hence the identification of Cr–diopside with > 1.5 wt.% Cr_2O_3 in a surficial sample survey needs further discrimination regarding its source. Morris *et al.* (1999) recently described a preliminary Na–Al–Cr ternary discriminant diagram to address this problem. Griffin *et al.* (1997), Grütter & Apter (1998) and Kjarsgaard (1998b) have described revised discrimination diagrams for distinguish-

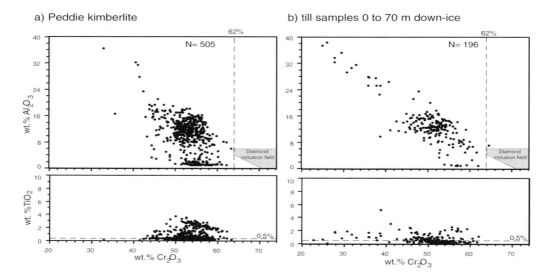

Fig. 10. TiO$_2$ and Al$_2$O$_3$ v. Cr$_2$O$_3$ bivariate plots of Cr–spinel from: (**a**) the Peddie kimberlite (Lake Timiskaming); and (**b**) till samples down-ice (from McClenaghan et al. 1999c). Diamond inclusion field and dashed lines are from Sobolev (1977).

ing spinel from different source rocks.

Because diamond is a rare mineral in kimberlite, usually ≪1.4 ppm, a subset of the kimberlite indicator minerals, termed 'diamond indicators', is used to indicate the *potential* presence of diamond in these rocks. This is based on studies of silicate and oxide inclusions in diamond and minerals from diamond-bearing mantle xenoliths (e.g. Sobolev 1971; Gurney & Switzer 1973; Sobolev et al. 1973; Gurney 1989; Fipke et al. 1995a). Specific diamond indicator minerals include: subcalcic Cr–pyrope, commonly referred to as G10 pyrope (garnet-bearing harzburgite/dunite source rock); Cr–pyrope, commonly referred to as G9 pyrope (garnet-bearing lherzolite source rock); high Cr–Mg chromite (chromite-bearing harzburgite/dunite source rock); and high Na–Ti pyrope–almandine garnet (eclogite source rock). Many diamond inclusion peridotitic garnets have low Ca, high Cr composition (Fig. 8) and thus these garnets are sought in diamond exploration (Gurney & Switzer 1973; Sobolev et al. 1973; Fipke et al. 1989, 1995a). In eclogitic garnet, elevated Ti and Na contents are viewed as being diamond favourable (Fipke et al. 1989, 1995a; Schulze 1999); however Grütter & Quadling (1999) have recently re-examined the validity of utilizing Na$_2$O contents as a discriminant. Cr–spinel, with >60% wt.% Cr$_2$O$_3$ and >12% wt.% MgO, are judged to have a diamond inclusion composition (Fipke et al. 1989, 1995a), although kimberlites contain spinels with a broader range of Cr$_2$O$_3$ and MgO (Fig. 9). Sobolev (1971, 1977) provides further refinement of spinel discrimination by examination of Al$_2$O$_3$ and TiO$_2$ concentrations in conjunction with Cr$_2$O$_3$ and MgO contents (Fig. 10). As shown in Figures 7 to 10, the chemistry of indicator minerals recovered from glacial sediments are also examined as they indicate the presence of kimberlite up-ice and its diamond potential.

Glacial sediment sampling

Methods and equipment used to collect glacial sediments in permafrost and non-permafrost terrains are reviewed in McMartin & McClenaghan (2001). The main types of glacial sediment available for sampling for indicator mineral and geochemical surveys are outlined below.

Till is a freshly crushed, first-cycle sediment transported by glaciers, blended with reworked sediments and plastered onto the bed or released by melting at its base or surface (Dreimanis 1990; Kauranne et al. 1992; Klassen 2001). The bulldozing and homogenizing action of the glacier produces an unsorted mixture of rock and mineral fragments, from boulder to clay sized. Till is useful at both regional and more detailed scales for detecting the presence of kimberlite and can be sampled to recover indicator minerals and for geochemical analysis.

Kimberlite boulders, produced by glacial erosion of exposed outcrop, can be deposited down-ice within, and on top of till (e.g. Dummett *et al.* 1987; Stewart *et al.* 1988; Papunen 1995; Melnyk 1997; Carlson *et al.* 1999*a*). Kimberlite boulders are also found within eskers (Fig. 11; e.g. Baker 1982; Brummer *et al.* 1992*a*; McClenaghan *et al.* 1998) and modern fluvial sediments (Kong *et al.* 1999). Preservation of kimberlite fragments in till and esker sediments is variable. Fresh, dense, hypabyssal facies kimberlite boulders and cobbles often survive to the present day. In contrast, highly porous and fragmental crater and diatreme facies kimberlite are highly susceptible to rapid breakdown in the postglacial (< 9000 years) surface weathering environment. As a result, glacially transported crater and diatreme facies kimberlite clasts rarely survive near the surface. Boulder distribution should be mapped systematically and sampled for indicator minerals.

Glaciofluvial sediments result from recycling of till by glacial meltwater and are sampled from eskers, braided subaerial outwash, proximal subaqueous outwash, kames, and moraines, or beaches formed on these deposits (Lilliesköld 1990). The principal criterion for sampling glaciofluvial sediments, in most cases, is the presence of medium to very coarse sand and gravel to collect sediment suitable for the recovery of indicator minerals. These surficial materials are useful at a regional-reconnaissance scale for detecting the presence of kimberlite (e.g. Lee 1965; Golubev 1995; McClenaghan *et al.* 1998) and are sampled to recover indicator minerals.

Sandy glaciolacustrine sediments are generally not ideal sample media for diamond exploration because they are often of limited areal extent, have undergone several cycles of erosion and transport, and their mineralogical composition is derived from a broad range of sediments and bedrock within a palaeolake basin (Morris & Kaszycki 1997). However, indicator minerals recovered from beach deposits developed on glacial sediments can provide an indication of the presence of kimberlite in a region, as they did in the initial discovery of the Lac de Gras kimberlites (Fipke *et al.* 1995*b*; Blusson 1998). Beach sediments in the Arkhangelsk region of Russia contain the same kimberlite indicator mineral assemblage as do the local tills from which they were derived. Golubev (1995) classified this as a 'detached halo', where indicator minerals cannot be traced directly back to their

Fig. 11. Weathered diatreme facies kimberlite boulders and cobble found in the poorly sorted gravel core of the Munro Esker, near Kirkland Lake, *c.* 3 km down-ice from the closest known kimberlite. Kimberlite boulders and cobble are indicated by white dashed outline and arrows. Geological hammer for scale.

source. Fine-grained (silt and clay) glaciolacustrine deposits are not suitable for diamond exploration.

Stream sediments or alluvium, derived from erosion of bedrock and recycling of glacial sediments in modern day streams, can be useful reconnaissance sampling media in glaciated terrains having sufficient topographic relief (e.g. Dummett *et al.* 1987; Golubev 1995; Morris *et al.* 1997, 1998*a*; Steenfelt *et al.* 1999) or where surface sediments, such as peat or glaciolacustrine sediments, mask the underlying till (e.g. Brown *et al.* 1967; Wolfe *et al.* 1975; Morris *et al.* 1998*b*; Kong *et al.* 1999). Caution should be used when sampling fluvial sediments in Shield terrain, as the sediments have undergone several cycles of transport from their bedrock source and are more difficult to trace up-stream than till which has a simpler transport history and more uniform composition. Stream sediments are suited for regional-reconnaissance scale studies and are sampled to recover indicator minerals.

Sample size

For indicator mineral recovery and examination, at least 10 to 20 kg (c. 5 to 10 litres) of glacial sediment, or more where the till is silt- or clay-rich, are required (Averill 1990, 2001; Thorleifson 1999). If samples are screened in the field to remove larger clasts (>4 mm) fraction, a separate sample of pebble-sized (2 to 6 cm) clasts should be collected for lithological examination. For those samples not pre-screened in the field, the pebble fraction is removed during sample processing (Fig. 12) and put aside for lithological studies. A 1 kg (c. 0.5 litre) of till may also be collected for geochemical analysis of the matrix and for archiving.

Sampling strategies

Diamond exploration is generally carried out in a sequential series of phases (e.g. Atkinson 1989; Coopersmith 1993; Mckinlay et al. 1997), commencing with area selection. Following selection of a region favourable for diamond exploration in Canada, one needs to carefully examine both the surficial and bedrock geology maps for the selected area, at an appropriate scale. Depending on the type of surficial cover, one or more specific techniques are selected to optimize the exploration program. Once a region is selected, regional-scale surveys are conducted to identify specific areas for more focused exploration, followed up by a study of specific targets. If present, fluvial, glaciofluvial or beach sediments are sampled initially instead of till (e.g. Fipke et al. 1995b; Mckinlay et al. 1997; Blusson 1998), because they can represent the bedrock composition of a large area and sample collection is simple. These samples are examined for the presence of indicator minerals. Systematic follow-up study is then completed using regional indicator mineral sampling of till (e.g. Fipke et al. 1995b; Cookenboo et al. 1998). Further sampling in areas of interest is then conducted at a higher till sampling density, combined with airborne geophysical surveys (e.g. Brummer et al. 1992a, b; Jennings 1995; Mckinlay et al. 1997; Carlson et al. 1999a; Graham et al. 1999; Kong et al. 1999). In some cases however, till sampling may indicate the presence of kimberlites with weakly contrasting geophysical signatures. In areas with extensive cover of glaciolacustrine sediments and thick organic deposits overlying till (e.g. areas of the Hudson Bay–James Bay Lowlands), stream sediments may be an adequate sampling medium (Kong et al. 1999). However, where relief is minimal and streams are sluggish, regional airborne magnetic surveys still represent a viable method for finding kimberlites (e.g. Kyle Lake field; Janse et al. 1995).

In areas where till is thin and exposed at the surface, samples are collected down-ice of targets or geophysical anomalies along several lines perpendicular to ice flow to test indicator mineral abundances and chemistry (some dispersal trains from indicator-poor kimberlites may have only a till geochemical anomaly). Sound knowledge of regional and local ice-flow history is required at this stage to ensure optimal orientation of sample lines, detection of dispersal trains and interpretation of data. If not already known, ice flow patterns can be determined from striations on bedrock, the orientation of elongate clasts in till (fabric), and glacial landforms (Hirvas & Nenonen 1990; Kujansuu 1990; Lundqvist 1990). In general, dispersal trains vary in shape depending on ice flow directions, ice sheet dynamics and bedrock topography (DiLabio 1990). For example, dispersal trains from individual kimberlite pipes within the Lac de Gras field are typically narrow (e.g. hundred of metres to 2 km wide), sharp-edged linear ribbons that extend several kilometres to tens of kilometres down-ice (e.g. Cookenboo 1996; McClenaghan et al. 2000). The Strange Lake Zr–Nb–Y–Be–REE dispersal train (McConnell & Batterson 1987; Batterson 1989), although from a larger point source, is a good analogue for the ribbon-shaped dispersal trains from kimberlites in the Lac de Gras region. In the Kirkland Lake area, dispersal trains from individual kimberlites are diffuse fans that extend a few tens of metres to tens of kilometres down-ice.

Biogeochemical surveys directly over geophysical and/or at the head of or up-ice from geochemical/indicator mineral anomalies in till can also assist in target evaluation. Favourable geophysical, indicator mineral, till geochemical and biogeochemical targets are then drilled and sampled for kimberlite. Further characterization including indicator mineral studies, diamond recovery and bulk sampling may be warranted.

Sample preparation

Heavy mineral concentrates are prepared from large till samples to concentrate indicator minerals and to remove components such as quartz, feldspar, most ferromagnesian minerals,

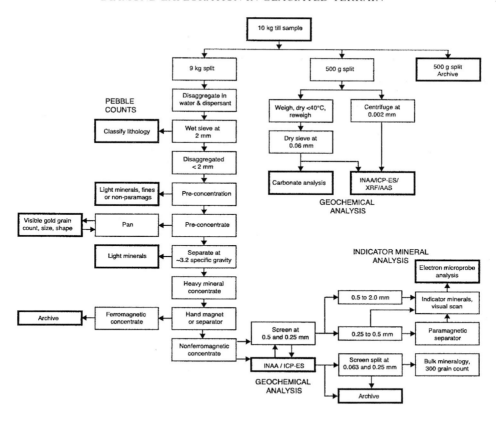

Fig. 12. Generalized sample processing flow sheet for recovery of indicator minerals from, and geochemical analysis of, till samples collected by the Geological Survey of Canada (GSC Ottawa).

carbonates and rock fragments that may mask or dilute the indicator mineral signature (Shilts 1975). The combinations of processing techniques used by exploration companies or government agencies for recovering indicator minerals vary (e.g. Gregory & White 1989; Towie & Seet 1995; Davison 1993; McClenaghan et al. 1996). The generalized flow sheet outlined in Figure 12 is an example of that used by the Geological Survey of Canada (GSC), Ottawa.

Prior to any indicator mineral recovery, small 500 g (c. 0.25 litre moist) subsamples can be removed from the samples (Fig. 12) for preparation of the fine fraction for geochemistry and for archiving. The remaining material is disaggregated, typically by agitation in a dispersant, and the gravel fraction (>2 mm) is removed for lithological analysis (pebble counts). Because of the large weight (10 to 25 kg) of >2 mm material to be processed, laboratories use pre-concentration methods by density, grain size or magnetic susceptibility before final density separation is made. Density pre-concentration, which may include jig, table, spiral, dense media separator, or pan, may be combined with the use of heavy liquid separation methods (e.g. using tetrabromoethane), prior to final concentration. Tabling has the advantages of speed, relatively low cost (Gregory & White 1989) and provides the opportunity to capture indicators of other commodities such as gold, base metals, and REE. Screening to recover the medium to very coarse sand-sized fraction, or rejection of non-paramagnetic minerals, principally quartz and feldspar, may be used instead of density pre-concentration if recovery of gold grains is not a priority. Final density concentration is completed using heavy liquids such as methylene iodide diluted with acetone or a Magstream separator using a specific gravity threshold of 3.1

to 3.2 g/cm^3 to ensure more complete recovery of Cr–diopside and diopside (e.g. McClenaghan et al. 1999a, b, c), the least dense of the kimberlite indicator minerals (Table 2). The ferromagnetic fraction, composed almost exclusively of magnetite, is removed using a hand magnet or roll separator, weighed and stored. This magnetic fraction is sometimes examined for Mg–ilmenite.

A significant proportion of a diamond exploration budget may be spent on sample collection, especially in remote areas accessible only by air. Typically, sample collection involves some field processing as a means of pre-concentration. Ideally this should be carried out in such a manner to concentrate kimberlite indicator minerals. Concentrates may also contain indicators of other commodities, such as gold, sulphides, REE minerals, scheelite, or cassiterite, and these minerals may be lost if certain grain sizes or magnetic susceptibility classes are discarded during processing designed exclusively for diamonds (Thorleifson 1999).

Depending upon the sampling medium chosen for the survey, specific preparation methods prior to geochemical analysis are utilized. McMartin & McClenaghan (2001) describe general methods for till sample preparation, Cook & McConnell (2001) describe lake sediment sample preparation, Dunn (2001) describes geobotanical sample preparation, and Björklund et al. (1994) describe stream sediment preparation. Regardless of the sample medium, it is important to choose an analytical method(s) which will produce reliable data for the elements of interest (Atkinson 1989). Various applicable analytical methods include XRF, INAA, and total or partial acid dissolutions followed by detection using AAS, ICP-ES or ICP-MS.

Indicator mineral methods

Kimberlite *in situ* contains tens of thousands of indicator mineral grains per 10 kg sample. The non-ferromagnetic heavy mineral concentrate is examined to recover and analyse indicator minerals in kimberlite and glacial sediments. Although they occur in the fine sand size (0.06 to 0.25 mm) fraction, indicator minerals in this fraction are expensive and time consuming to recover. Instead, indicator minerals typically are picked from: (1) medium sand (0.25 to 0.5 mm); (2) coarse sand (0.5 to 1.0 mm); and (3) very coarse sand (1.0 to 2.0 mm) sized fractions. Of these three size ranges, indicator minerals typically are most abundant in the 0.25 to 0.5 mm fraction (e.g. Thorleifson & Garrett 1993; Dredge et al. 1997; McClenaghan et al. 1999a) and, therefore, this size fraction is most often selected. Indicator minerals are picked from concentrates during a visual scan under a stereoscopic microscope. The 0.25 to 0.50 mm fraction may require further sorting based on magnetic susceptibility (e.g. McCallum & Vos 1993), depending on regional bedrock geology and mineralogy, in order to reduce the volume of material to be scanned for picking (Thorleifson 1999). Picking of a concentrate may take 0.1 to 2 hours, and a few tens to several hundred grains may be obtained. The occurrence of a single grain in a regional survey can be highly significant, and hence, the ability of a laboratory to produce high quality concentrates which are then examined by experienced mineral pickers is exceptionally important. Physical features summarized in Table 2, primarily colour (Fig. 13), are used to visually identify potential kimberlite indicator minerals.

Analysis

Visually picked grains are mounted on glass slides or formed into circular epoxy mounts. Mineral grains can be further screened at this stage by rapid quantitative analysis using an automated routine on a Scanning Electron Microscope (SEM) equipped with an EDS detector. Subsequently, mineral grains may undergo qualitative electron microprobe (EM) analysis to determine the major (and trace) element composition. The identity of mineral grains as being kimberlitic is then determined using the chemical criteria described previously.

Pyrope, Cr–diopside and olivine are the easiest minerals to identify visually and therefore the number of grains picked usually is similar to the number confirmed by microprobe analysis. In contrast, Mg–ilmenite is visually similar to ilmenite from crustal rocks as well as strongly resorbed Cr–spinel that has lost its octahedral form (Fig. 13e). SEM or EM analysis is the only means of distinguishing between them. As a result, the number of grains picked can be much greater than the number confirmed by microprobe analysis.

Surface features and shape

Surface features and textures on garnet may provide clues as to their derivation from kimberlite as well as distance and nature of the transport medium e.g. glacial v. fluvial (Mosig 1980; Afanase'ev et al. 1984; McCandless 1990; Sobolev et al. 1993; Averill & McClenaghan

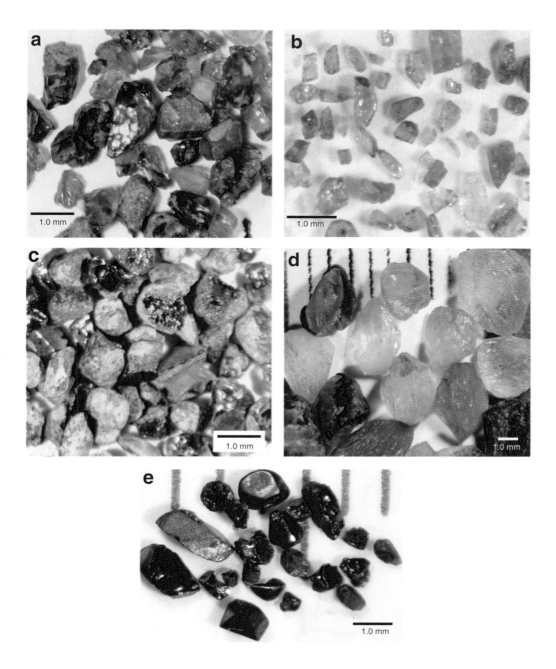

Fig. 13. Examples of typical colours and surface features of kimberlite indicator minerals: (**a**) purple and mauve Cr–pyrope, some retaining kelyphite (dark); (**b**) green Cr–diopside; (**c**) Mg–ilmenite, both single crystals and polycrystalline morphologies are shown; (**d**) yellow to colourless olivine; and, (**e**) opaque Cr–spinel, some exhibiting slightly resorbed octahedral form.

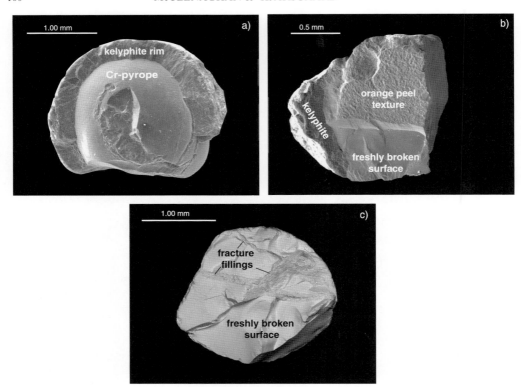

Fig. 14. Secondary electron images showing surface features on Cr–pyrope from kimberlite: (**a**) kelyphite rim on round garnet grain; (**b**) kelyphite rim and orange peel texture on angular grain; and (**c**) broken round garnet with angular sub-grains. Note the fracture fillings between angular sub-grains.

1994; Golubev 1995; Dredge et al. 1996a). Garnet may be partially covered with up to 2.0 mm of thick greenish–black kelyphite reaction rims (Figs. 13a & 14a) composed mainly of pyroxene with spinel, phlogopite or serpentine. The garnet rims form during a decompression reaction with the high-temperature kimberlite magma during ascent (Reid & Dawson 1972; Garvie & Robinson 1984). Kelyphite layers break off readily during abrasion (McCandless 1990) and therefore, the presence/absence and the thickness of kelyphite on individual pyrope grains may give some indication of the relative distance of glacial transport (Averill & McClenaghan 1994; Dredge et al. 1996a; McClenaghan et al. 2000).

A pitted or 'orange peel' textured surface pyrope underlies the kelyphite (Fig. 14b). This texture, formed on the surface of pyrope, consists of minute triangular or rhombic pits which are impressions of radially arranged pyroxene crystals (Garvie & Robinson 1984).

Similar to kelyphite, the presence or absence of an orange peel texture on pyrope may indicate the distance of glacial transport. As kelyphite is eroded from pyrope, the amount of orange peel surface texture exposed on grains increases (Mosig 1980; Averill & McClenaghan 1994; Dredge et al. 1996a; McClenaghan et al. 2000).

Pyrope in kimberlite occurs as spherical nodules (Fig. 14b) to angular fragments (Figs. 13a, 14b, c) and therefore roundness cannot be used to estimate distance of glacial transport. Very little wear, rounding or breakdown of indicator mineral grains occurs in till or glaciofluvial sediments down-ice of kimberlites in the Kirkland Lake area. Similar indicator mineral shape preservation has been reported during fluvial transport in Siberia (Afanase'ev et al. 1984). In contrast, indicator minerals transported by sheet wash and soil creep in Australia show pronounced wear and degradation over short transport distances (Mosig 1980; Atkinson 1989).

Size range

Indicator mineral size in glacial sediments is controlled by the processes that form the minerals in the mantle rocks or in the kimberlite magma. Cr–pyrope in kimberlite is commonly 0.1 to 1.0 cm in diameter (Smirnov 1959; Mitchell 1986) and is highly susceptible to fracturing (Fig. 15) during decompression and cooling (Garvie & Robinson 1984; McCandless 1990). As a glacier entrains and transports kimberlite, Cr–pyrope preferentially breaks along the pre-existing fractures into angular grains that are dominantly sand-sized (<2.0 mm) particles. These grains appear to remain at this size regardless of glacial transport distance (Averill & McClenaghan 1994). Olivine is the most abundant mineral grain in the 0.25 to 0.5 mm fraction of kimberlite and till. Mg–ilmenite tends to occur among coarser sand-sized fractions in till (e.g. Garrett & Thorleifson 1993; Averill 2001). Cr–diopside and Cr–spinel tend to occur in the smaller size fraction (0.5 mm) in kimberlite (e.g. Sage 1995) and glacial sediments.

Relative abundance

Several studies have focussed on prediction of fluvial transport distance and resistance to abrasion based on abundances of indicator minerals. In this regard, indicator mineral concentrations should always be reported with respect to the weight of material sampled. To compare indicator mineral counts between sampling programs, survey areas or kimberlite fields, indicator counts for till samples should be standardized to a constant sample weight (Jennings 1990). For example, McClenaghan et al. (1999a,b,c) standardized indicator mineral counts to a 10 kg sample weight before comparing results from till samples down-ice of kimberlites in the Kirkland Lake area.

Indicator mineral abundance in till is not affected by degradation or physical breakdown of grains during glacial transport. All kimberlite indicator minerals survive long distance glacial transport. The relative abundance of indicator minerals in individual kimberlites varies significantly (e.g. Smirnov 1959; Mitchell 1986; Schulze 1993a,b) and it is these variations in the kimberlite mineralogy that control the relative amounts of indicator minerals in glacial sediments down-ice (e.g. Kong et al. 1999). The Diamond Lake kimberlite near Kirkland Lake, for example, contains c. 6800 Mg–ilmenite grains in a 10 kg sample (McClenaghan et al. 1998). In contrast, the C14 kimberlite, 20 km to the northwest is Mg–ilmenite poor and contains only three grains in a 10 kg sample (McClenaghan et al. 1999a). The large difference in Mg–ilmenite content in glacial sediments down-ice of the two kimberlites reflects the significantly different Mg–ilmenite content of each kimberlite source. Decreases in concentrations down-ice are the result of dilution, not degradation.

Geochemical methods

The use of till geochemistry is gaining popularity in kimberlite exploration because of its versatility (regional and local surveys) and initial low cost as compared to indicator mineral processing and analysis. Similarly, lake sediment geochemistry appears to have good potential for regional exploration surveys. In contrast, biogeochemistry and soil geochemistry have been little used and only for local, follow-up studies and target evaluation. The latter technique is widely used in tropical or arid terrains where in-situ soils develop on weathered kimberlite (e.g. Webb 1956; Jedwab 1958; Aicard 1959; Gregory & Tooms 1969; Mathur & Alexander 1983). However, in Canada conventional (aqua regia) soil geochemistry has not been used for kimberlite exploration often because soils are developed on glacial sediments that have been clastically transported several metres to hundreds of kilometres. Instead, new selective leach geochemical methods applied to soils are being evaluated (e.g. Eccles 1998a).

The discrimination afforded by these geochemical methods is based on the unusual major and trace element composition of kimberlites. A characteristic suite of kimberlite pathfinder elements is utilized, including the first order transition elements Mg, Ni, Cr, Co, V, Mn and Fe and a suite of incompatible elements including La, Ce, Nd, Sm, Nb, Ti, P, Ba, Ta, Hf, Zr and Sr (Atkinson 1989; Kjarsgaard 1996e). These elements are useful pathfinders in various combinations which depend in part on kimberlite composition, but more importantly on knowledge of the regional bedrock geochemistry. Geochemical anomalies are generated by the contrast between kimberlite and host rocks. Hence a set of pathfinder elements may be specific to one region, or location. For example, Cr, Ni, Ta, Sr, Ba, Co, Mg, Ca, Fe, K, and LREE are geochemical pathfinder elements in the Kirkland Lake field (McClenaghan et al. 1999b) whereas in the Lac de Gras area, the occurrence of a suite of incompatible element-rich granitoid rocks reduces the list of pathfinder elements to Ni, Cr, Ba, Co, Sr, K, Mg and Mn (Wilkinson et al. 1999).

Fig. 15. Fractured Cr–pyrope grain (1 cm across) in a mantle xenolith from the C14 kimberlite, Kirkland Lake. Cr–pyrope grains break along these pre-existing fractures during glacial transport resulting in many more smaller (<2.0 mm) Cr–pyrope grains in till.

In the Kirkland Lake area, partial Sr determined by aqua regia/ICP-ES is a better indicator of the presence of kimberlitic debris in till as compared to total Sr determined by XRF. The aqua regia digestion dissolves carbonate minerals in till, which in the Kirkland Lake area are derived from two sources: Sr-rich carbonates from kimberlite and Sr-poor carbonates from Palaeozoic sedimentary rocks from the Hudson Bay Lowland. Therefore, till containing significant amounts of kimberlite will contain much greater Sr (determined by aqua regia) than tills derived from the local Archean metavolcanic and Palaeozoic sedimentary rocks.

Preliminary studies of the influence of size range on geochemical detection of kimberlite in till has been conducted around kimberlites in the Kirkland Lake and Lake Timiskaming fields (McClenaghan et al. 1999c), for the following fractions: silt + clay (<0.063 mm); very fine to fine sand (0.063 to 0.25 mm); medium sand (0.25 to 0.5 mm); and coarse sand (0.5 to 2.0 mm). Results indicate that the <0.063 mm (silt + - clay) and the 0.5 to 2.0 mm (coarse sand) fractions provide the strongest contrast between background and anomalous trace element concentrations related to kimberlite. Less often, non-ferromagnetic heavy mineral concentrates are analysed for kimberlite pathfinder elements. Studies around the Kirkland Lake kimberlites indicate that anomalously high concentrations of Ni, Cr, Ti, Ta and LREE, and depletions of Ca and heavy REE in till heavy mineral concentrates are the best indicators of the presence of kimberlite debris in the till. Anomalously high concentrations of Cr in the heavy mineral fraction are likely to reflect the high abundance of Cr–pyrope and Cr–spinel, while higher levels of Ta reflect the presence of Mg–ilmenite and perovskite. The lower concentrations of heavy REE and Ca in the heavy mineral fraction of till reflect the lower concentrations of these elements in the Kirkland Lake kimberlites compared to the surrounding country rocks. These results have important implications for regional and local scale till sampling programs

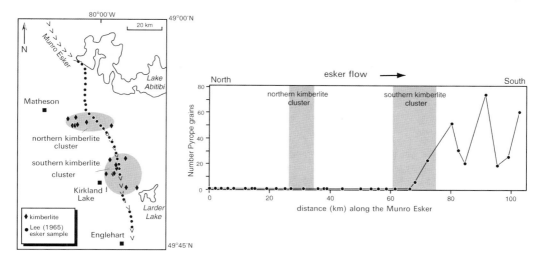

Fig. 16. Abundance of pyrope in the 0.5 to 1.23 mm fraction of 1.3 ft^3 sand samples collected along a 100 km segment of the Munro Esker near Kirkland Lake by Lee (1965). The location of kimberlites subsequently discovered is shown by the grey shaded boxes (modified from Lee 1965). Note pyropes appear in esker sediments just south of the southern kimberlite cluster.

conducted in the past 20 years because INAA analysis (for kimberlite pathfinder elements) of archived heavy mineral concentrates prepared for gold and base metal exploration programs may provide new insights into the diamond potential of an area.

Regional surveys

In the past 40 years, diamond exploration companies have completed regional indicator mineral surveys across large parts of Canada. Unfortunately, most data from these surveys remain proprietary. In recent years, government agencies have also conducted regional indicator mineral and geochemical surveys in support of diamond exploration (Table 1). Results from some of these surveys are highlighted below.

Kirkland Lake, Ontario

While studying the Au content of esker sediments along a 100 km segment of the Munro Esker near Kirkland Lake, Lee (1965) noted the presence of pyrope in the southern 40 km of the transect (Fig. 16). Many years later, this pyrope discovery was recognized as one of the earliest indications of the presence of kimberlite in the Kirkland Lake region. The location of the northern-most pyrope grains in the Munro esker corresponds to the area where several kimberlite pipes were subsequently discovered 20 years later (Fig. 16). Between 1979 and 1982, and prior to the discovery of kimberlite pipes in the region, the Ontario Geological Survey (OGS) sampled till and glaciofluvial sand in 171 reverse circulation overburden holes and 200 surface pits in a 2500 km^2 area around Kirkland Lake, directed mainly at Au exploration (Fortescue et al. 1984). Pyrope, along with Cr–diopside grains, were found in some of the OGS samples (Fig. 17). The results were the first published report (Fortescue et al. 1984) of the presence and regional distribution of kimberlite indicator minerals in Canada and aided in the discovery of the southern Kirkland Lake kimberlite cluster in the early 1980s (Pegg et al. 1990; Brummer et al. 1992a, b). Pyrope (Fig. 17) and Cr–diopside concentrations are highest over a 25 km wide area centered on the Munro Esker and extending southeast for 35 km. The highest pyrope concentrations are displaced 5 km southeast (down-ice) of the southern kimberlite cluster. Cr–diopside occurs in ultramafic rocks as well as kimberlite in the Kirkland Lake region (Fortescue et al. 1984). However, the similarity of pyrope and Cr–diopside distribution patterns in the OGS samples and the presence of high concentrations of both minerals down-ice of the kimberlites indicate that most grains in the OGS samples are kimberlite-derived.

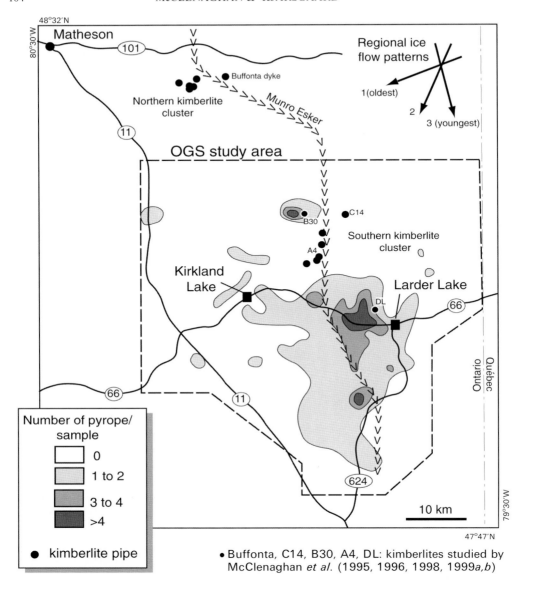

Fig. 17. Distribution of pyrope in 5 to 8 kg till and glaciofluvial sediments collected by the Ontario Geological Survey as part of the KLIP regional geochemical survey (1979–1984) and location of the subsequently discovered kimberlites (modified from Brummer *et al.* 1992a; kimberlite locations from Zalnieriunas & Sage 1995).

Lac de Gras, NWT

Between 1969 and 1971 the GSC conducted a regional lake sediment survey in the NWT, including the Lac de Gras area, for Au and base metal exploration (Allan *et al.* 1972). The survey analysed the silt + clay (<0.063 mm) fraction of near shore clastic lake sediments (till samples in the near shore zone) for a variety of elements. Interestingly, the marginal notes for the Ni distribution map for the Lac de Gras sheet (Allan & Cameron 1973) describe elevated Ni contents for which the source was unknown based on bedrock maps of that time (i.e.

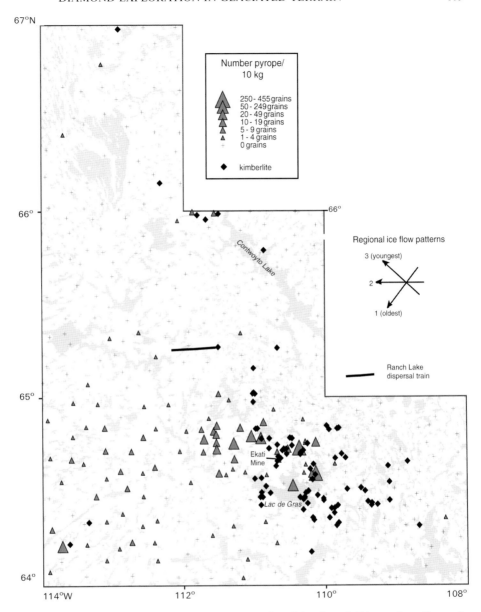

Fig. 18. Abundance of Cr–pyrope in the 0.25 to 0.5 mm heavy mineral fraction of 10 kg surface till samples collected by the Geological Survey of Canada across the Lac de Gras region, NWT (modified from Dredge et al. 1997 and Kerr 2000). Cr–pyrope grains have been dispersed by three phases of ice to the southwest, then to the west and finally to the northeast. The highest concentrations are overlying and just down-ice from the main kimberlite cluster centered around Lac de Gras.

undiscovered basic or ultrabasic rocks). The BHP/DiaMet 'Point lake' discovery kimberlite coincidentally is located within this Ni anomaly. More recently, some of the original lake sediment samples from the Lac de Gras area were reanalysed by INAA for additional kimberlite pathfinder elements (e.g. Nb, Ta, REE) as these were not determined in the original study (Kjarsgaard et al. 1992). This new study confirmed the earlier results of Allan et al.

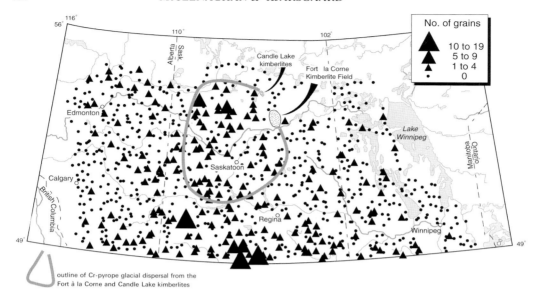

Fig. 19. Abundance of Cr–pyrope in the 0.25 to 2.0 mm heavy mineral fraction of 25 kg surface till samples collected by the Geological Survey of Canada across the Canadian Prairie. The high concentrations of grains NE of Saskatoon are derived from the Fort à la Corne kimberlite cluster and high concentrations along southern border are eroded from underlying Tertiary gravels.

(1972) and Allan & Cameron (1973) and demonstrated that Nb, Ta and some REE are also useful pathfinder elements for kimberlite.

In 1991, the GSC initiated a regional indicator mineral survey over an area of 50 000 km^2 around Lac de Gras at an average sample spacing of one sample per 15 to 20 km. Surface till samples, weighing c. 10 kg, were collected from hand-dug pits excavated in mudboils at 0.5 to 1.0 m depths. Kimberlite indicator mineral abundances in the regional till samples clearly indicate the presence of a significant kimberlite field in the region. At the regional scale, Cr–pyrope is the most abundant indicator mineral in till followed by, in decreasing order of abundance, Cr–diopside, Cr–spinel and Mg–ilmenite (Dredge et al. 1997). Background concentrations of Cr–pyrope in 10 kg till samples vary across the area (Fig. 18), and depend on location relative to the central part of the kimberlite field. North and southeast of the central kimberlite cluster, background counts are 0 grains per sample. In contrast, southwest of the central cluster background counts are 9 grains per sample, whereas within the central part of the field background counts are as high as 50 grains per sample. Regional Cr–pyrope and Cr–diopside distribution patterns are the net result of three phases of ice flow, initially to the southwest, then west and finally northwest of the Lac de Gras field (Ward et al. 1995, 1996; Dredge et al. 1995, 1997). The lack of indicator minerals in till over the eastern part of the kimberlite field is an enigma. Possible explanations include the presence of allocthonous till over this area, or thick till deposits which may include multiple till sheets that mask the underlying dispersal trains. The GSC regional survey provides the context in which regional- and property-scale indicator mineral surveys conducted by exploration companies in the Lac de Gras region can be interpreted. Exploration companies have outlined indicator mineral dispersal trains from several kimberlites (e.g. Cookenboo 1996; Armstrong 1999; Carlson et al. 1999a; Doyle et al. 1999; Graham et al. 1999) using these methods.

Canadian Prairie

Indicator minerals picked from regional till and glaciofluvial sediment samples collected over large parts of the Canadian Prairie (Alberta, Saskatchewan, Manitoba) document indicator mineral distribution patterns and provide insights into the diamond potential of the region. Reconnaissance-scale GSC surface till samples cover a 840 000 km^2 area (Fig. 19) at an average spacing of 80 km between samples (Garrett & Thorleifson 1993, 1995, 1996; Thorleifson & Garrett 1993; Thorleifson et al. 1994). GSC data complement indicator mineral data from regional-scale surveys (Table 1) conducted in Alberta (e.g. Fenton Pawlowicz 1993; Fenton et al. 1994), Saskatchewan (e.g. Simpson 1993; Swanson & Gent, 1993) and Manitoba (e.g. Matile et al. 1996; Fedikow et al. 1997; Matile & Thorleifson 1997). Published indicator mineral counts for the Prairie region reveal broad distribution patterns. Pyrope, Cr–diopside and Mg–ilmenite are widespread across the Prairie with a conspicuous absence of Cr–spinel. Elevated Cr–pyrope concentrations in till form a broad, 300 km wide plume in western and central Saskatchewan which is derived from the Fort à la Corne and Candle Lake kimberlites in central Saskatchewan (Garrett & Thorleifson 1996). In contrast, elevated Cr–pyrope concentrations in southern Saskatchewan are derived from local Tertiary gravels which contain reworked material from Montana (Simpson 1993; Kjarsgaard 1995). The large Cr–diopside dispersal train extending 400 km SW across Manitoba and into Saskatchewan (Averill 2001, fig. 4) is not kimberlite-derived. Instead, it is thought to originate from the Thompson area of Manitoba (Matile & Thorleifson 1997; Averill 2001).

Local surveys (within kimberlite fields)

Local-scale orientation studies (e.g. Levinson 1980) of mineralogical and geochemical signatures of individual kimberlites have been conducted in the Lac de Gras, Fort à la Corne and Kirkland Lake kimberlite fields and at Mountain Lake, Alberta. Although many studies have been completed by exploration companies, most of the published studies have been conducted by government agencies (Table 1). Highlights of their work are summarized below.

Kirkland Lake and Lake Timiskaming, Ontario

McClenaghan et al. (1996, 1998, 1999a,b,c) have tested the effectiveness of a variety of sample media over the Kirkland Lake and Lake Timiskaming fields for diamond exploration in glaciated terrain. Most of their research has focussed on indicator minerals in glacial sediments. In the Kirkland Lake area, anomalous till samples down-ice contain tens to thousands

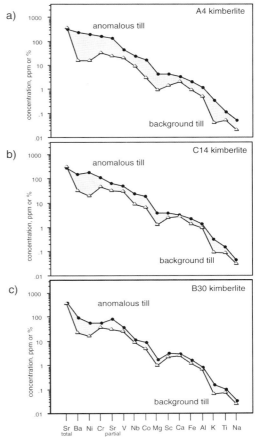

Fig. 20. Comparison of major and trace element concentrations in the <0.063 mm fraction (aqua regia/ICP-ES) of background and anomalous till (till enriched in kimberlite) down-ice of three kimberlites in the Kirkland Lake area: (a) A4; (b) C14; and (c) B30. Note both Sr_{total} (XRF) and $Sr_{partial}$ (aqua regia-ICP-ES) are reported.

of indicator minerals per 10 kg sample. Background concentrations of each indicator mineral increase from north to south across the Kirkland Lake field. For example, north of the B30 kimberlite (Fig. 17) background counts for pyrope are c. 0 to five grains in a 10 kg sample while one kilometre north of the A4 kimberlite, background is 22 pyrope grains in a 10 kg sample.

Relative abundance of indicator minerals is different in individual kimberlites. The C14 kimberlite contains very few (<10 grains per 10 kg sample) Mg–ilmenite but Mg–ilmenite is the most abundant indicator mineral (>5000

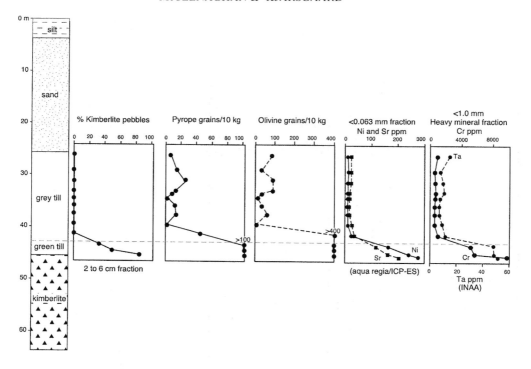

Fig. 21. Comparison of kimberlite pebble and indicator mineral abundances and till geochemistry for a thick till sequence overlying the A4 kimberlite, Kirkland Lake. The top of the till sequence represents background concentrations similar to the regional till. In contrast, the 'green till' at the base of the till profile shows a strong local (kimberlite) signature in all size fractions (from McClenaghan *et al.* 1999*b*).

grains per 10 kg) in the Diamond Lake (DL) kimberlite. Olivine is the most abundant indicator mineral in the A4 and Peddie kimberlites (>10 000 grains per 10 kg) but is much less abundant (<400 grains per 10 kg) in the B30 kimberlite. The dramatic differences in relative abundance are shown in the glacial sediments down-ice of each kimberlite. Chemical characteristics of indicator minerals differ between kimberlites and these differences are also reflected in the glacial sediments down-ice. For example, Mg–ilmenite in the Kirkland Lake kimberlites displays distinct MgO v. Cr_2O_3 compositions that can be used to identify individual kimberlite sources of Mg–ilmenite in glacial sediment and kimberlite boulders down-ice (e.g. McClenaghan *et al.* 1998).

Utilizing the <0.063 mm size fraction of till for geochemical detection of the presence of kimberlite was evaluated by comparing the average concentration of elements in indicator mineral-rich or 'anomalous' till (solid circles) to background till (open triangles) (Fig. 20). The farther apart the two symbols plot, the greater the usefulness of an element as a kimberlite pathfinder element. The largest differences are displayed by Ba, Ni, Cr, V, Co, Mg, K and partial Sr. Till overlying kimberlites in the Kirkland Lake area displays a range in intensity of kimberlite geochemical signature from strong (A4 kimberlite), to intermediate (C14 kimberlite), to weak (B30) (Fig. 20a, b, c). Glacial erosion and incorporation of kimberlite was strongest at A4, as indicated by the dark green (instead of grey) colour of till and abundant kimberlite fragments and indicator minerals in the till (Fig. 21). As a result, till over the A4 kimberlite displays the greatest contrast between background and anomalous till in the <0.063 mm fraction (Fig. 20a). Glacial erosion and incorporation of the B30 kimberlite, in contrast, was minimal, as indicated by the intact capping of soft preglacially weathered kimberlite, the paucity of indicator minerals in till and marginal contrast between background and anomalous trace element concentrations in the <0.063 mm fraction (Fig. 20c).

However, in the Kirkland Lake–Lake Timiskaming region, the <0.063 mm size fraction of till may not always be optimal for detecting a kimberlitic geochemical response. Comparison of four different size fractions (<0.063 mm, 0.63 to 0.25 mm, 0.25 to 0.5 mm, 0.5 to 2.0 mm) of 'anomalous' till down-ice of kimberlites shows that the optimal size fraction varies for two different kimberlites. Transition and incompatible element abundances are highest in the 0.5 to 2.0 mm (coarse to very coarse sand) fraction of till from the Peddie pipe (Fig. 22a), an olivine-rich hypabyssal kimberlite. The coarse fraction geochemistry best reflects the incorporation of kimberlite debris because of the high concentrations of kimberlitic olivine in the coarse to very coarse sand fraction (McClenaghan *et al.* 1999*c*). In contrast, for the A4 kimberlite the 0.5 to 2.0 mm fraction is optimal for Cr and La, and the <0.063 mm fraction is optimal for Ba and Th (Fig. 22b).

Non-destructive INAA analysis of the <1.0 mm non-ferromagnetic heavy mineral fraction of till prior to indicator mineral picking may also provide useful information (McClenaghan *et al.* 1996, 1998, 1999*a, b, c*). Geochemical patterns in indicator mineral-rich till overlying the A4 kimberlite show significant enrichments as compared to the surrounding local till. Anomalous, kimberlite-rich till is enriched in Cr, Ta (Fig. 21), La, Ce, Eu and Nd and depleted in Ca, Fe, Hf, Tb, Yb and Lu. Overall, the heavy mineral fraction of indicator mineral-rich till down-ice of kimberlites in the Kirkland Lake area display strong enrichments in Cr, Ta, and light REE and depletions in Ca and heavy REE.

Biogeochemical transects across the C14 and Diamond Lake kimberlites (both of which are covered by thick glacial sediments) and the Buffonta kimberlite dyke (with a thin surficial cover) reveal geochemical responses in black spruce (*Picea mariana*) bark and twigs, balsam fir (*Abies balsamea*) twigs, and pin cherry (*Prunus pennsylvanica*) twigs (McClenaghan & Dunn 1995; Dunn & McClenaghan 1996). Most tree tissue from sites over and adjacent to the kimberlites are enriched in Sr and Rb (Fig. 23) and depleted in Mn. However, no geobotanical response is noticeable.

Lac de Gras, NWT

The GSC collected closely spaced till samples down-ice of the Ranch Lake kimberlite and used proprietary indicator mineral data supplied by the property owner to document glacial dispersal patterns from a single kimberlite (McClenaghan *et al.* 2000). A well formed indicator mineral dispersal train trends west from Ranch Lake for at least 70 km (Cookenboo 1996). At its head, the train is 500 m wide and gradually increases to 1.5 km in width 16 km down-ice and 2 km in width at 30 km down-ice (Fig. 24). Lateral edges of the train are sharp, being defined by the total absence of indicator minerals in till. The narrow pencil shape and sharp edges of the dispersal train indicate that it was formed by a single and sustained ice flow event towards the west.

The Ranch Lake kimberlite contains thousands of grains in a 10 kg till sample with Cr–diopside being the most abundant, followed in decreasing order of abundance by pyrope, Cr–spinel and Mg–ilmenite. Down-ice, samples contain hundreds to thousands of indicator minerals with the same relative abundance of indicator mineral species. Concentrations of indicator minerals in the 0.25 to 0.5 mm till fraction reach a maximum 8 km down-ice, containing 570 pyrope, 969 Cr–diopside, 3 Mg–ilmenite and 17 Cr-spinel. Background concentrations of indicator minerals in till immediately up-ice of the Ranch Lake kimberlite and north and south of the dispersal train are zero (Ward *et al.* 1997). Of the two size fractions of heavy minerals examined (0.25 to 0.5 mm and 0.5 to 0.84 mm), indicator minerals are most abundant in the 0.25 to 0.5 mm fraction of both kimberlite and till. Figure 18 shows the extent of the Ranch Lake dispersal train in relation to GSC regional till sample sites across the Lac de Gras region. The absence of indicator minerals in regional samples down-ice of the kimberlite is not unexpected as the train is extremely narrow (0.5 to 2 km wide) and regional samples were collected 15 to 20 km apart.

Fort à la Corne, Saskatchewan

Thorleifson & Garrett (2000) examined the lithology of a 100 m thick Pleistocene glacial sequence overlying interbedded Cretaceous shale and volcaniclastic facies kimberlite within the Fort à la Corne kimberlite field near Smeaton. They report the till intersection from 52 to 68 m depth (over 45 m above the No. 169 kimberlite body) contains elevated abundances of indicator minerals, and suggest this represents a dispersal train from a kimberlite source up-ice. Indicator mineral abundance increases upward within the train from 2 Cr–pyrope and 5 Mg–ilmenite to 12 Cr–pyrope and 15 Mg–ilmenite in a 25 l sample. In contrast, Cr–diopside occurs throughout the 100 m thick sequence of till and sand and shows no correlation to the Cr–pyrope and Mg–

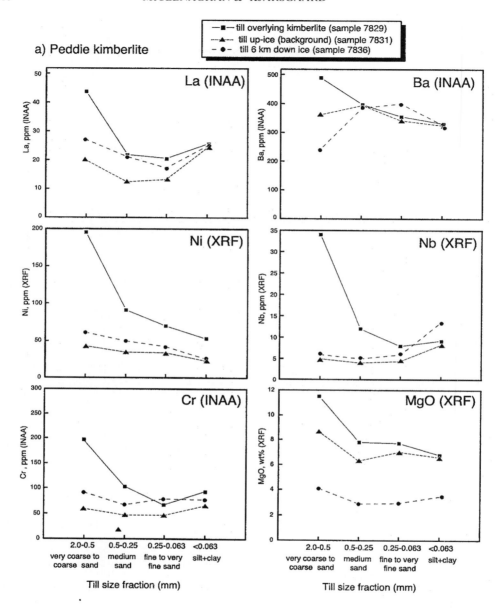

Fig. 22. Comparison of element concentrations in four size fractions of till samples collected: (**a**) up-ice, overlying and down-ice of the Peddie kimberlite, Lake Timiskaming.

ilmenite abundance. No geochemical signature of the kimberlite was recognized in the till.

Dunn (1993) sampled tree tissue over a 200 m by 125 m glacially transported megablock of crater-facies kimberlite sitting on top of Cretaceous shale (Kjarsgaard 1995), referred to as the Sturgeon Lake 01 kimberlite block. The kimberlite block is overlain by up to 10 m of till and, at the time of sampling, the block was thought to be in place. Trembling aspen (*Populus tremuloides*), red-osier dogwood (*Cornus stolonifera*) and beaked hazelnut (*Corylus conuta*) over the kimberlite are enriched in Ni, Rb, Sr, Cr, Nb, Mg and P and depleted in Mn and Ba as

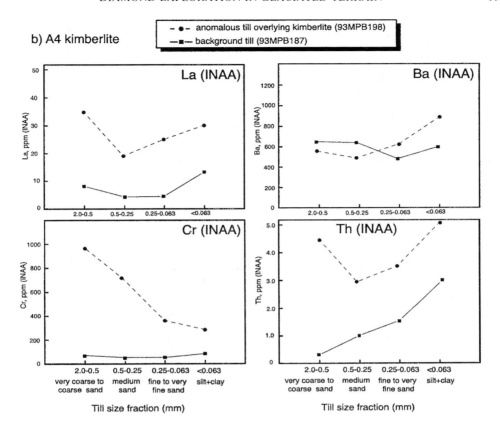

Fig. 22. Comparison of element concentrations in four size fractions of till samples collected: (**b**) overlying the A4 kimberlite, Kirkland Lake.

compared to the surrounding area, indicating the kimberlite block was of sufficient size to generate a biogeochemical response.

Mountain Lake ultrabasics, Alberta

The Mountain Lake ultrabasic volcaniclastic occurrence, although not a kimberlite, is noted here because its geochemical signature is very similar to kimberlite. In 1998, Eccles (1998b) sampled trembling aspen (*Populus tremuloides*) bark, twigs and stems over the Mountain Lake occurrence in northwestern Alberta. Glacial sediment thickness is 1 to 2 m over the ultrabasic. Bark, twig and stem samples displayed significant geochemical signatures for Co, Cr, Mg, Ni, La, Rb, Sm, Au, Fe, Na, V and W over or adjacent to the Mountain Lake ultrabasic. Stems, followed by bark and then twigs provided the best overall geochemical correlation with soil and whole rock geochemical patterns, especially for kimberlite pathfinder elements. Eccles (1998b) also detected a geobotanical response over the Mountain Lake body, consisting of a 'ring-like' vegetation pattern of healthy and larger trees growing at the edge, with smaller trees growing over the middle. Eccles (1998a) also used a selective leach for amorphous Mn-oxide (Enzyme Leach) to detect a geochemical signature in B-horizon soil over the ultrabasic rocks for numerous elements, including Cr, Ni, Co, Nb, Co, P_2O_5 and LREE.

Discussion

Kimberlite is a mineralogically and chemically distinct point source which forms discrete dispersal trains in glacial sediments. Its variable size (50 to 1500 m) and competence will influence the size and extent of the dispersal train(s), and whether they consist of kimberlite boulders, indicator minerals or trace elements in

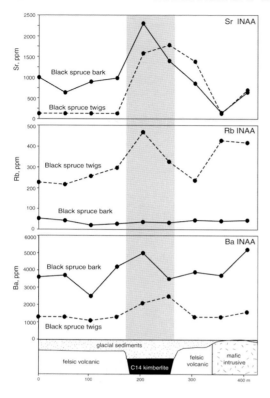

Fig. 23. Distribution of Sr, Rb and Ba, determined by INAA, in the ash of black spruce bark and twigs along a N–S transect across the C14 kimberlite, Kirkland Lake (from Dunn & McClenaghan 1996).

till. Understanding ice flow history, recognizing the genetic history of different glacial sediments (till, glaciofluvial sediments) sampled, and identifying multiple till sheets in thick drift areas, such as the Prairies, Kirkland Lake and parts of the Lac de Gras region, are essential to successful sampling, interpretation and follow-up of dispersal trains. Local-scale orientation studies around known kimberlites are crucial sources of information on their geochemical, mineralogical and lithological signatures in till and glaciofluvial sediments, on appropriate sampling methods and size fractions to examine or analyse, and on local glacial stratigraphy and ice flow patterns (Jennings 1990). As a significant proportion of an exploration budget is spent on the initial sample collection, especially in remote areas accessible only by air, it is cost-effective for exploration programs to include commodities other than diamonds. This can be accommodated if the sample processing is carried out such that indicator minerals of other commodities such as precious or base metals are also recovered (Thorleifson 1999).

Indicator minerals

Kimberlites contain different proportions of indicator minerals (Smirnov 1959; Mosig 1980; McClenaghan et al. 1999a,b,c). In unglaciated terrains, chemical and mechanical degradation of indicator minerals changes these proportions in soils, wind blown sediments, sheet wash or stream sediments and they will vary with proximity to the kimberlite source (Mosig 1980; Jennings 1990). Under Australian conditions, for example, Cr–spinel is most resistant to weathering followed by, in decreasing order of resistance, Mg–ilmenite, Cr–pyrope, Cr–diopside and olivine (Atkinson 1989). In glaciated terrain, however, the relative abundance of indicator minerals down-ice of a kimberlite is a function of their varying proportions in the kimberlite and not from mechanical degradation during glacial transport.

Use of Cr–pyrope, Cr–diopside, Mg–ilmenite and Cr–spinel in regional and local-scale till sampling programs is well established. To date, few examples of olivine glacial dispersal patterns have been published; however, preliminary results indicate that it easily survives glacial transport and is visually distinct from olivine derived from other rocks (e.g. Afanase'ev et al. 1984; McClenaghan et al. 1999c). Regional indicator mineral distribution patterns define the net effect of glacial dispersal, often in different ice flow directions. For example, Cr–pyrope distribution in till across the Lac de Gras region (Fig. 20) is the result of three main phases of ice flow, to the southwest, west and finally northwest (Dredge et al. 1997).

Exploration companies have identified numerous indicator mineral dispersal trains down-ice of known kimberlites as well as from as yet undiscovered kimberlites. In the Lac de Gras region, the trains often extend for tens of kilometres in the direction of ice flow (e.g. Cookenboo 1996; Armstrong 1999; Carlson et al. 1999a; Doyle et al. 1999; Graham et al. 1999) and are sharp edged linear ribbons (McClenaghan et al. 2000). In the James Bay Lowland and Kirkland Lake area, the trains may be broader (e.g. Kong et al. 1999).

Till and soil geochemistry

Till geochemistry is gaining popularity in diamond exploration (e.g. Thomas & Gleeson

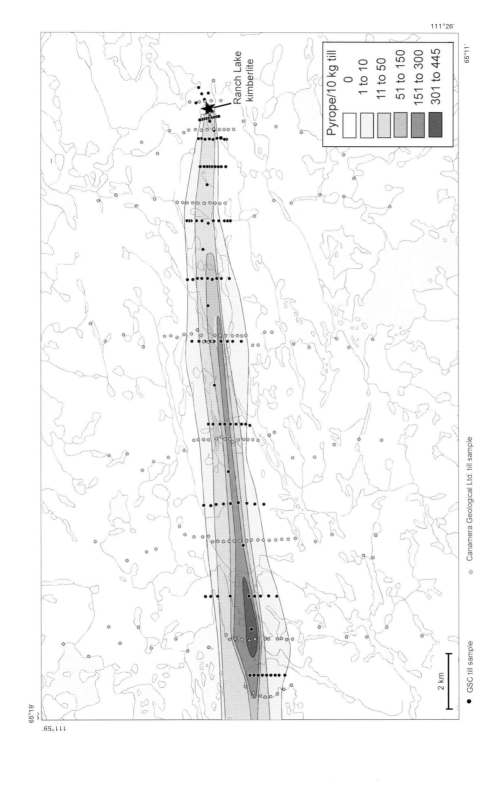

Fig. 24. Indicator mineral dispersal train (shaded grey) trending westward from the Ranch Lake kimberlite, Lac de Gras as defined by Cr–pyrope concentrations in 10 kg till samples (McClenaghan et al. 2000).

1999) because of its significantly lower cost as compared to indicator mineral analysis and rapid analysis time. Important kimberlite pathfinder elements that provide good contrast in till geochemical surveys include Ni, Cr, Ba, Co, Sr, Rb, Nb, Mg, Ta, Ca, Fe, K, Ti and LREE, the combination of which will depend on kimberlite composition as well as that of the surrounding bedrock. Geochemical signatures in till cannot be detected as far down-ice as a few tens of sand-sized indicator mineral grains, and therefore, sample spacing for regional or local-scale till geochemical surveys must be smaller than for indicator mineral surveys. Till geochemistry may prove to be useful as a pre-screening tool for selecting samples for further, more costly indicator mineral analysis. Observations from limited partitioning studies of different till size fractions are preliminary and more orientation studies need to be completed. However, specific size ranges of till such as the medium + coarse sand (0.5 to 2.0 mm) and silt + clay (<0.063 mm) fractions will likely prove to be the optimal size fractions for geochemical analysis because the medium + coarse sand fraction includes kimberlite indicator minerals and the silt + clay fraction includes clays-sized minerals derived from the fine-grained kimberlite matrix. Analysing the −80 mesh (<0.177 mm) fraction of till is not recommended because it includes abundant sand-sized quartz and feldspar which dilute and mask the geochemical signature of the kimberlite. Analytical methods most useful for geochemically detecting kimberlite dispersal include INAA, XRF and ICP-MS.

Recently, interest has been growing in surface geochemical exploration techniques capable of detecting sulphide mineralization (e.g. Clark 1993; Bajc 1998) and, more recently kimberlites, buried by thick glacial sediments. It is believed that during oxidation, elements from the ore zone or kimberlite intrusion are mobilized at depth and migrate vertically to the surface where they are deposited on mineral grains in near-surface soil. Transport mechanisms invoked to account for soil geochemical anomalies overlying mineralization include some combination of diffusion, advective groundwater, gaseous or electrochemical transport (Hamilton 1998, 2000). The labile ions in the soil are loosely bound and readily detectable by weak, selective chemical digestions/ICP-MS methods. Eccles (1998b) demonstrated that a selective leach for amorphous Mn-oxide (Enzyme Leach) was able to detect a geochemical signature in B-horizon soil over the thin drift covered Mountain Lake ultrabasic. Future selective leach studies over more deeply (>20 m) buried kimberlites should provide further insight as to the effectiveness of selective leach-soil sampling methods for kimberlite exploration in glaciated terrain.

Biogeochemistry

Geobotanical and biogeochemical techniques for diamond exploration in the glaciated terrain of Canada are not well tested or documented. Several site-specific studies of vegetation growing over known kimberlites indicate the potential for developing these techniques. Kimberlite may have strong differences in water content compared to the surrounding bedrock. Altered or weathered kimberlite usually has a high clay-mineral content, which promotes water retention (Fipke et al. 1995a). In contrast, fine-grained, unweathered kimberlite may have very low porosity and thus lower water contents than the surrounding rocks. These differences in water content combined with the K- and P- rich nature of kimberlites may enhance or deter growth of certain vegetation types or cause seasonal stresses to the vegetation during dry or rainy seasons that are readily visible. These differences can be used to help identify potential kimberlites for further exploration (Mannard 1968). Several studies have been conducted over kimberlites in different terrains around the world to determine the distribution pattern among trees (geobotany) and the distinctive chemistry of tree tissue (biogeochemistry) that reflect the underlying kimberlite (e.g. Cole 1980; Mathur & Alexander 1983; Alexander & Shrivastava 1984).

The few site-specific studies that have been conducted over the Siberian kimberlites in the glaciated terrain of northern Russia indicate the potential of biogeochemistry for diamond exploration (e.g. Kobets & Komarov 1958; Litinsky 1964; Buks 1965; Komogorova et al. 1986). More recently, biogeochemical studies over kimberlites in Canada support the Russian observations of geochemical signatures in vegetation growing on glacial sediments over kimberlite. Future studies will provide insight into, and evaluation of, the method as an exploration tool in glaciated terrain.

Conclusions

- Application of indicator mineral methods to diamond exploration in the glaciated terrain of Canada has intensified in the past ten years. Diamond exploration in Canada uses established glacial sediment sample collection and processing methodologies and is based on an

advanced knowledge of indicator mineral chemistry.
- Rapid and inexpensive multi-element geochemical techniques combined with improved and more readily available indicator mineral methods now allow the diamond potential and other commodities, such as precious and base metals, of a region to be evaluated using glacial sediment sampling.
- Case studies around known kimberlites are crucial sources of information on their mineralogical and geochemical signatures, appropriate sample media and sample size, suitable processing and analytical methods, as well as glacial dispersal patterns.
- During the last decade, regional indicator mineral surveys conducted in Canada have dramatically increased our knowledge of the distribution, composition and provenance of till. This knowledge has had a significant impact on survey design and sampling and analytical methods used in exploration programs for all commodities.
- Continued research into geochemical methods for kimberlite exploration in glaciated terrain may encourage the use of less expensive geochemical methods for regional exploration and target evaluation.
- Surface till sampling methods are effective for detecting kimberlite dispersal trains in most parts of the Lac de Gras kimberlite field. In other kimberlite fields covered by thick glacial sediments, such as Fort à la Corne, Kirkland Lake-Lake Timiskaming and the James By Lowland, subsurface till sampling may also be needed to detect glacial dispersal from kimberlite.
- Although the methods and examples described here are from Canada, they can be utilized for diamond exploration in glaciated terrain around the world.

The authors thank reviewers A. J. A. Janse, A. A. Levinson and R. N. W. DiLabio for their thorough and thoughtful comments, and G. E. M. Hall for her editorial review. M. Beaumier, J. Campbell, L. Dredge, R. Eccles, D. Kerr, T. Morris, V. Rampton, R. Sage and H. Thorleifson provided important insights and information on regional kimberlite indicator mineral surveys and kimberlite occurrences across Canada. GSC Contribution No. 1999220.

References

AFANASE'EV, V. P., VARLAMOV, V. A. & GARANIN, V. K. 1984. The abrasion of minerals in kimberlites in relation to the conditions and distance of their transportation. *Geologiya i Geofizika*, **25/10**, 119–125.

AICARD, P. 1959. Application of geochemical methods for chromium and nickel to prospecting for kimberlite pipes – results obtained on a known pipe at Kenieba, French Sudan. *Annales des Mines*, **Fevrier**, 103–110 (in French).

ALEXANDER, P. O. & SHRIVASTAVA, V. K. 1984. Geobotanical expression of a blind kimberlite pipe, central India. *In:* KORNPROBST, J. (ed.) *Proceedings of the Third International Kimberlite Conference, Kimberlites, I: Kimberlites and Related Rocks.* Elsevier, Amsterdam, 33–40.

ALLAN, R. J. & CAMERON, E. M. 1973. *U, Zn, Pb, Mn, Fe, Cu, Ni and K content in sediments: Bear-Slave operation.* Geological Survey of Canada Maps **9-1972** to **15-1972.**

ALLAN, R. J., CAMERON, E. M. & DURHAM, C. C. 1972. *Reconnaissance geochemistry using lake sediments of a 36,000 square mile area of the northwestern Canadian Shield.* Geological Survey of Canada Paper **72-50**.

ARMSTRONG, J. 1998. *Slave province kimberlite location digital database. Indian and Northern Affairs Canada, Yellowknife.* Economic Geology Series Open file **1999-03**.

ARMSTRONG, J. 1999. *KIDD database. Indian and Northern Affairs Canada, Yellowknife.* Economic Geology Series Open file **1999-03**.

ATKINSON, W. J. 1989. Diamond exploration philosophy, practice, and promises: a review. *In:* MEYER, H. O. A. LEONARDOS, O. H. (eds) *Proceedings of the Fifth International Kimberlite Conference, Brazil, v. 2, Diamonds: Characterization, Genesis and Exploration.* CPRM Special Publication, **1B** 1075–1107.

AVERILL, S. A. 1990. Drilling and sample processing methods for deep till geochemistry surveys: making the right choices. *In:* AVERILL, S., BOLDUC, A., COKER, W. B., DILABIO, R. N. W., PARENT, M. & VEILLETTE, J. (eds) *Application de la géologie du Quaternaire à l'exploration minérale.* Association professionelle des géologues et des géophysiciens du Québec, 139–173.

AVERILL, S. A. 2001. The application of heavy indicator mineralogy in mineral exploration with emphasis on base metal indicators in glaciated terrains. *In:* MCCLENAGHAN, M. B. BOBROWSKY, P. T., HALL, G. E. M. & COOK, S. J. (eds) *Drift Exploration in Glaciated Terrain.* Geological Society, London, Special Publications, **184**, 69–81.

AVERILL, S. A. & MCCLENAGHAN, M. B. 1994. *Distribution and character of kimberlite indicator minerals in glacial sediments, C14 and Diamond Lake kimberlite pipes, Kirkland Lake, Ontario.* Geological Survey of Canada Open File **2819**.

BAJC, A. F. 1998. A comparative analysis of enzyme leach and mobile metal ion selective extractions: case studies from glaciated terrain, northern Ontario. *Journal of Geochemical Exploration*, **61**, 113–148.

BAKER, C. L. 1982. Report on the sedimentology and provenance of sediments in eskers in the Kirkland Lake area and on finding of kimberlite float in Gauthier Township. *In:* WOOD, J., WHITE, O. L., BARLOW, R. B. & COLVINE, A. C. (eds) *Summary of*

Field Work-1982. Ontario Geological Survey, Miscellaneous Publication **106**, 125–127.

BATTERSON, M. J. 1989. Glacial dispersal from the Strange Lake alkalic complex, northern Labrador. *In:* DILABIO, R. N. W. COKER, W. B. (eds) *Drift Prospecting*. Geological Survey of Canada, Paper **89-20**, 31–40.

BEAUMIER, M., DION, D. J., LASALLE, P. & MOORHEAD, J. 1993a. *Exploration du diamant au Témiscamingue*. Ministère Ressources Naturelle Pro **93-08**.

BEAUMIER, M., LASALLE, P., LASALLE, Y. & WARREN, B. 1993b. *Minéraux indicateurs de kimberlite dans les eskers du Nord-Ouest Québécois*. Ministère Ressources Naturelle, **MB93-60**.

BEAUMIER, M., LEFÈBVRES, D. & RIVARD, P. 1994. *Contribution de la géochimie et de la géophysique à la recherche de diamants le long du rift du Lac Témiscamingue*. Ministère Ressources Naturelle, **MB94-63**.

BEDNARSKI, J., LECKIE, D. & DE PAOLI, G. 1998. *Gold recovery and kimberlite/diamond indicators from Del Bonita Upland, Alberta*. Geological Survey of Canada Open File **3601**.

BJÖRKLUND, A., LEHMUSPELTO, P., OTTESEN, R. T. & STEENFELT, A. 1994. Drainage geochemistry in glaciated terrain. *In:* HALE, M. PLANT, J. A. (eds) *Drainage Geochemistry: Handbook of Exploration Geochemistry, Volume 6*. Elsevier, Amsterdam, 307–340.

BLUSSON, S. L. 1998. Key steps to discovery of the Slave Craton diamond field, N.W.T., Canada. *In:* WALTON, G. & JAMBOR, J. (eds) *Pathways '98 Abstract Volume*. Cordilleran Roundup, Vancouver, 1998, British Columbia Yukon Chamber of Mines, 25.

BROWN, D. D., BENNETT, G. & GEORGE, P. T. 1967. *The source of alluvial kimberlite indicator minerals in the James Bay Lowland*. Ontario Department of Mines, Miscellaneous Paper **MP 7**.

BRUMMER, J. J., MACFADYEN, D. A. & PEGG, C. C. 1992a. Discovery of kimberlites in the Kirkland Lake area, northern Ontario, Canada. Part I: early surveys and the surficial geology. *Exploration and Mining Geology*, **1**, 339–350.

BRUMMER, J. J., MACFADYEN, D. A. & PEGG, C. C. 1992b. Discovery of kimberlites in the Kirkland Lake area, northern Ontario, Canada. Part II: kimberlite discoveries, sampling, diamond content, ages and emplacement. *Exploration and Mining Geology*, **1**, 351–370.

BUKS, I. I. 1965. The use of geobotanical method in the search for kimberlite tubes in the Yakutian polar region. *In:* CHIKISHEV, A. G. (ed.) *Plant Indicators of Soils, Rocks and Subsurface Waters*. Consultants Bureau, New York, 173–175.

CARLSON, S. M., HILLIER, W. D., HOOD, C. T., PRYDE, R. P. & SKELTON, D. 1998. The Buffalo Hills kimberlite province, north-central Alberta, Canada. *In: The 7th International Kimberlite Conference*, extended abstracts, Cape Town, 138–140.

CARLSON, J. A., KIRKLEY, M. B., THOMAS, E. M. & HILLIER, W. D. 1999a. Recent Canadian Kimberlite Discoveries. *In:* GURNEY, J. J., GURNEY, J. L., PASCOE, M. D. RICHARDSON, S. H. (eds) *The J.B. Dawson Volume, Proceedings of the VIIth International Kimberlite*, Cape Town, 81–89.

CARLSON, J. A., HILLIER, W. D., HOOD, C. T., PRYDE, R. P. & SKELTON, D. N. 1999b. The Buffalo Hills kimberlites: a newly discovered diamondiferous kimberlite province in north-central Alberta, Canada. *In:* GURNEY, J. J., GURNEY, J. L., PASCOE, M. D. & RICHARDSON, S. H. (eds) *The J.B. Dawson Volume, Proceedings of the VIIth International Kimberlite*, Cape Town, 109–116.

CLARK, J. R. 1993. Enzyme-induced leaching of B horizon soils for mineral exploration in areas of glacial overburden. *Transactions of the Institute of Mining and Metallurgy*, **B102**, 19–29.

CLEMENT, C. R. 1975. The emplacement of some diatreme-facies kimberlites. *Physics and Chemistry of the Earth*, **9**, 51–59.

COLE, M. M. 1980. Geobotanical expression of orebodies. *Transactions of the Institute of Mining and Metallurgy*, **B89**, 73–91.

COOK, S. & MCCONNELL, J. 2001. Lake sediment geochemical methods in the Shield and Cordillera. *In:* MCCLENAGHAN, M. B. BOBROWSKY, P. T., HALL, G. E. M. & COOK, S. J. (eds) *Drift Exploration in Glaciated Terrain*. Geological Society, London, Special Publications, **184**, 125–149.

COOKENBOO, H. 1996. Ranch Lake kimberlite in the central Slave craton; the mantle sample. *The Gangue*. GAC Mineral Deposits Division, Report **52**, 12–13.

COOKENBOO, H. (ed.) 1999. Diamond Exploration Methods and Case Studies. *In: Association of Exploration Geochemists Short Course Notes, 19th International Geochemical Exploration Symposium*, Vancouver, April, 1999.

COOKENBOO, H. O., DUPUIS, J. & FOULKES, J. 1998. Discovery and evaluation of the Jericho and 5034 kimberlite fields in the Slave Craton, northern Canada. *In:* WALTON, G. & JAMBOR, J. (eds) *Pathways '98 Abstract Volume, Cordilleran Roundup 1998*. British Columbia Yukon Chamber of Mines, 166–169.

COOPERSMITH, H. G. 1993. Diamond mine discovery. *In:* SHEAHAN, P. & CHATER, A. (chairmen) *Diamond: Exploration, Sampling and Evaluation*. Prospectors and Developers Association of Canada, Short Course Proceedings, Toronto, Ontario, March 27, 1993, 73–108.

DAVISON, J. G. 1993. Diamond exploration samples: laboratory processing. *In:* SHEAHAN, P. & CHATER, A. (chairmen) *Diamond: Exploration, Sampling and Evaluation*. Prospectors and Developers Association of Canada, Short Course Proceedings, Toronto, March 27, 1993, 315–341.

DAWSON, J. B. & STEPHENS, W. E. 1975. Statistical classification of garnets from kimberlite and associated xenoliths. *Journal of Geology*, **83**, 589–607.

DEER, W. A., HOWIE, R. A. & ZUSSMAN, J. 1982: *Rock Forming Minerals, Volume 1A, Orthosilicates*, Second Edition. Longman Group Ltd., New York, USA.

DILABIO, R. N. W. 1990. Glacial dispersal trains. *In:* KUJANSUU, R. & SAARNISTO, M. (eds) *Glacial*

Indicator Tracing. A. A. Balkema, Rotterdam, 109–122.

DiLabio, R. N. W. & Knight, R. D. 1997. *Kimberlite indicator minerals in GSC's archived till samples: Victoria Island and Hay River - Fort Smith area, Northwest Territories.* Geological Survey of Canada Open File **3505**.

DiLabio, R. N. W. & Knight, R. D. 1998. *Kimberlite indicator minerals in GSC's archived till samples: results of analysis of samples from Baker Lake area and northern Baffin Island, Northwest Territories.* Geological Survey of Canada Open File **3643**.

Doyle, B. J., Kivi, K. & Scott Smith, B. H. 1999. The Tli Kwi Cho (DO27 and DO18) diamondiferous kimberlite complex, Northwest Territories, Canada. *In:* Gurney, J. J., Gurney, J. L., Pascoe, M. D. & Richardson, S. H. (eds) *The J.B. Dawson Volume, Proceedings of the VIIth International Kimberlite,* Cape Town, 194–204.

Dredge, L., Kjarsgaard, I., Ward, B., Kerr, D. & Stirling, J. 1995. *Distribution and chemistry of kimberlite indicator minerals, Aylmer Lake map area, NWT (NTS 76C).* Geological Survey of Canada Open File **3080**.

Dredge, L. A., Ward, B. C. & Kerr, D. E. 1996a. Morphology and kelyphite preservation on glacially transported pyrope grains. *In:* LeCheminant, A. N., Richardson, D. G., DiLabio, R. N. W. & Richardson, K. A. (eds) *Searching For Diamonds in Canada.* Geological Survey of Canada Open File **3228**, 197–204.

Dredge, L., Kjarsgaard, I., Ward, B., Kerr, D. & Stirling, J. 1996b. *Distribution and chemistry of kimberlite indicator minerals, Point Lake map area, NWT (NTS 86H).* Geological Survey of Canada Open File **3341**.

Dredge, L. A., Kerr, D. E., Kjarsgaard, I. M., Knight, R. D. & Ward, B. C. 1997. *Kimberlite indicator minerals in till, central Slave Province, Northwest Territories.* Geological Survey of Canada, Open File **3426**.

Dreimanis, A. 1990. Formation, deposition and identification of subglacial and supraglacial tills. *In:* Kujansuu, R. & Saarnisto, M. (eds) *Glacial Indicator Tracing.* A. A. Balkema, Rotterdam, 35–60.

Dufresne, M. B., Olson, R. A., Schmitt, D. R., McKinstry, B., Eccles, D. R., Fenton, M. M., Pawlowicz, J. G., Edwards, W. A. D. & Richardson, R. J. H. 1994. *The diamond potential of Alberta: a regional synthesis of structural and stratigraphic setting and other preliminary indications of diamond potential.* Alberta Research Council Open File Report **1994-10**.

Dufresne, M. B., Eccles, D. R., McKinstry, B., Schmitt, D. R., Fenton, M. M., Pawlowicz, J. G. & Edwards, W. A. D. 1996. *The diamond potential of Alberta.* Alberta Research Council Bulletin No. **63**.

Dummett, H. T., Fipke, C. E. & Blusson, S. L. 1987. Diamond exploration geochemistry in the North American Cordillera. *In:* Elliot, I. L. Smee, B. E. (eds) *Geoexpo/86 Exploration in the North American Cordillera,* Association of Exploration Geochemists, 168–176.

Dunn, C. E. 1993. Diamondiferous kimberlite in Saskatchewan, Canada – a biogeochemical study. *Journal of Geochemical Exploration,* **47**, 131–141.

Dunn, C. E. 2001. Biogeochemical exploration methods in the Canadian Shield and Cordillera. *In:* McClenaghan, M. B., Bobrowsky, P. T., Hall, G. E. M. & Cook, S. J. (eds) *Drift Exploration in Glaciated Terrain.* Geological Society, London, Special Publications, **184**, 159–164.

Dunn, C. E. McClenaghan, M. B. 1996. Biogeochemical studies of kimberlites. *In:* Le Cheminant, A. N., Richardson, D. G., DiLabio, R. N. W. & Richardson, K. A. (eds) *Searching For Diamonds in Canada.* Geological Survey of Canada Open File **3228**, 219–223.

Eccles, D. R. 1998a. *Enzyme leach-based soil geochemistry of the Mountain Lake Diatreme, Alberta.* Alberta Geological Survey Open File Report **1998-01**.

Eccles, D. R. 1998b. *Biogeochemical orientation survey of the Mountain Lake diatreme, Alberta.* Alberta Geological Survey Open File Report **1998-06**.

Fipke, C. E., Gurney, J. J., Moore, R. O. & Nassichuk, W. W. 1989. *The development of advanced technology to distinguish between diamondiferous and barren diatremes.* Geological Survey of Canada Open File **2124**.

Fipke, C. E., Gurney, J. J., & Moore, R. O. 1995a. *Diamond exploration techniques emphasizing indicator mineral geochemistry and Canadian examples.* Geological Survey of Canada Bulletin **423**.

Fipke, C. E., Dummett, H. T., Moore, R. O., Carlson, J. A., Ashley, R. M., Gurney, J. J. & Kirkley, M. B. 1995b. History of the discovery of diamondiferous kimberlite in the Northwest Territories, Canada. *In: Extended Abstract Volume, Sixth International Kimberlite Conference,* Novosibirsk, Siberia, 158–160.

Fedikow, M. A. F., Nielsen, E., Conley, G. G. & Matile, G. L. D. 1997. *Operation Superior: multimedia geochemical survey results from the Echimamish River, Carrot River and Munro Lake greenstone belts, Northern Superior province, Manitoba (NTS53L and 63I) Part 6: kimberlite indicator minerals.* Manitoba Energy and Mines Open File **97-2**.

Fenton, M. M. & Pawlowicz, J. G. 1993. *Reconnaissance mineral and geochemical survey with emphasis on northern Alberta; report for the end of fiscal year, 1992–93, MDA Project number M92-04-006.* Alberta Research Council Open File Report **1993-16**.

Fenton, M. M., Pawlowicz, J. G. & Dufresne, M. B. 1994. *Reconnaissance mineral and geochemical survey with emphasis on northern Alberta; report for the end of fiscal year, 1993–94, MDA Project number M92-04-006.* Alberta Research Council Open File Report **1994-21**.

Fortescue, J. A. C., Lourim, J., Gleeson, C. F., Jensen, L. & Baker, C. L. 1984. *A synthesis and interpretation of basal till geochemical and mineralogical data obtained from the Kirkland Lake (KLIP) area (1979-1982), Part 1.* Ontario Geological Survey Open File Report **5506**.

GARRETT, R. G. & THORLEIFSON, L. H. 1993. *Prairie kimberlite study - soil and till geochemistry and mineralogy, low density orientation survey traverses, Winnipeg-Calgary-Edmonton-Winnipeg, 1991.* Geological Survey of Canada Open File **2685**.

GARRETT, R. G. & THORLEIFSON, L. H. 1995. Kimberlite indicator mineral and till geochemical reconnaissance, southern Saskatchewan. *In: Investigations Completed by the Saskatchewan Geological Survey and the Geological Survey of Canada Under the Geoscience Program of the Canada-Saskatchewan Partnership Agreement on Mineral Development (1990–1995)*. Geological Survey of Canada Open File **3119**, 227–253.

GARRETT, R. G. & THORLEIFSON, L. H. 1996. Kimberlite indicator mineral and soil geochemical reconnaissance of the Canadian Prairie region. *In:* LECHEMINANT, A. N., RICHARDSON, D. G., DILABIO, R. N. W. & RICHARDSON, K. A. (eds) *Searching for Diamonds in Canada*. Geological Survey of Canada Open File **3228**, 205–211.

GARVIE, O. G. & ROBINSON, D. N. 1984. The formation of kelyphite and associated subkelyphitic and sculpted surfaces on pyrope from kimberlite. *In:* KORNPROBST, J. (ed.) *Proceedings of the Third International Kimberlite Conference, Kimberlites I, Kimberlites and Related Rocks*. Elsevier, Amsterdam, 371–382.

GENT, M. R. 1992. *Diamonds and precious gems of the Phanerozoic Basin, Saskatchewan: preliminary investigations*. Saskatchewan Energy and Mines, Saskatchewan Geological Survey Open File Report **92-2**.

GOLUBEV, Y. K. 1995. Diamond exploration in glaciated terrain: a Russian perspective. *Journal of Geochemical Exploration*, **53**, 265–275.

GRAHAM, I., BURGESS, J. L., BRYAN, D., RAVENSCROFT, P. J., THOMAS, E., DOYLE, B. J., HOPKINS, R. & ARMSTRONG, K. A. 1999. Exploration history and geology of the Diavik kimberlites, Lac de Gras, Northwest Territories, Canada. *In:* GURNEY, J. J., GURNEY, J. L., PASCOE, M. D. & RICHARDSON, S. H. (eds) *The J.B. Dawson Volume, Proceedings of the VIIth International Kimberlite*, Cape Town, 262–279.

GREGORY, P. & TOOMS, J. S. 1969. Geochemical prospecting for kimberlites. *Quarterly of the Colorado School of Mines*, **64**, 265–306.

GREGORY, G. P. & WHITE, D. R. 1989. Collection and treatment of diamond exploration samples. *In:* Ross, J. (ed.) *Kimberlites and Related Rocks, Volume 2, Their Crust/Mantle Setting, Diamonds and Diamond Exploration*. Geological Society of Australia, Special Publication **14**, Blackwell Scientific Publications, Oxford, 1123–1134.

GRIFFIN, W. L., FISHER, N. I., FRIEDMAN, J. H. & RYAN, C. G. 1997. Statistical techniques for the classification of chromites in diamond exploration. *Journal of Geochemical Exploration*, **59**, 233–249.

GRÜTTER, H. S. & APTER, D. B. 1998. Kimberlite- and lamproite-borne chromite phenocrysts with 'diamond-inclusion'type chemistries. *In: 7th International Kimberlite Conference, extended abstracts*, Cape Town, 280–281.

GRÜTTER, H. S. & QUADLING, K. E. 1999. Can Sodium in Garnet be used to monitor Eclogitic Diamond Potential? *In:* GURNEY, J. J., GURNEY, J. L., PASCOE, M. D. & RICHARDSON, S. H. (eds) *The J.B. Dawson Volume, Proceedings of the VIIth International Kimberlite)*, Cape Town, 314–320.

GURNEY, J. J. 1984. A correlation between garnets and diamonds in kimberlites. *In: Kimberlite Occurrence and Origin: a basis for conceptual models in exploration*. Geology Department and University Extension, University of Western Australia, Publication No. **8**, 143–166.

GURNEY, J. J. 1989. Diamonds. *In:* Ross, J. (ed.) *Proceedings of the Fourth International Kimberlite Conference, volume 1. Kimberlites and Related Rocks: Their Composition, Occurrence, Origin and Emplacement*. Geological Society of Australia, Special Publication **14**, Blackwell Scientific Publications, Oxford, 935–65.

GURNEY, J. J. & SWITZER, G. 1973. The discovery of garnets closely related to diamonds in the Finsch Pipe, South Africa. *Contributions to Mineralogy and Petrology*, **39**, 103–116.

HALL, D. C., HELMSTAEDT, H. H. & SCHULZE, D. J. 1989. The Cross diatreme, British Columbia, Canada: a kimberlite in a young orogenic belt. *In:* Ross, J. (ed.) *Proceedings of the Fourth International Kimberlite Conference, volume 1, Kimberlites and Related Rocks: Their Composition, Occurrence, Origin and Emplacement*. Geological Society of Australia, Special Publication **14**, Blackwell Scientific Publications, Oxford, 97–108.

HAMILTON, S. M. 1998. Electrochemical mass-transport in overburden: a new model to account for the formation of selective leach geochemical anomalies in glacial terrain. *Journal of Geochemical Exploration*, **63**, 155–172.

HAMILTON, S. M. 2000. Spontaneous potentials and electrochemical cells. *In:* HALE, M. (ed.) *Handbook of Exploration Geochemistry, Volume 7. Geochemical Remote Sensing of the Subsurface*. Elsevier, Amsterdam, 81–119.

HEAMAN, L. M. & KJARSGAARD, B. A. 2000. Timing of eastern North American kimberlite magmatism; continental extension of the great meteor hotspot track. *Earth and Planetary Science Letters*.

HELMSTAEDT, H., SCHULZE, D. J. & KAMINSKY, F. 1995. *Diamonds-theory and exploration, Short Course 20*. Geological Association of Canada, February 6, 1995, Vancouver, B.C.

HENDERSON, P. J. 1996. *Kimberlite indicator mineral data from the Bissett – English Brook – Wallace Lake area (NTS 62P/1, 52 M/3, 52 M/4, 52 L/14), Rice Lake Greenstone Belt, southeastern Manitoba*. Geological Survey of Canada Open File **3367**.

HIRVAS, H. & NENONEN, K. 1990. Field methods for glacial indicator tracing, Chapter 12. *In:* KUJANSUU, R. & SAARNISTO, M. (eds) *Glacial Indicator Tracing*. A. A. Balkema, Rotterdam, 217–248.

JANSE, A. J. A., NOVAK, N. A. & MACFAYDEN, D. A. 1995. Discovery of a new type of highly diamondiferous kimberlite rock in the James Bay Lowland, northern Ontario, Canada. *In: Extended Abstracts Volume, Sixth International kimberlite conference*.

Novosibirsk. United Institute of Geology, Geophysics and Mineralogy, Siberian Branch of the Institute of the Russian Academy of Sciences, 260–262.

JEDWAB, J. 1958. Looking for kimberlitic diamond deposits by geochemical prospection. *Mineral Wealth of Madhya Pradesh*, **1**(3), 68–71.

JENNINGS, C. M. H. 1990. Exploration for diamondiferous kimberlites and lamproites. *In:* BECK, L. S. & HARPER, C. T. (eds) *Modern Exploration Techniques*. Saskatchewan Geological Society London, Special Publications **10**, 139–148.

JENNINGS, C. M. H. 1995. The exploration context of diamonds. *Journal of Geochemical Exploration*, **53**, 113–124.

KAURANNE, K., SALMINEN, R. & ERIKSSON, K. (eds) 1992. *Handbook of Exploration Geochemistry, Volume 5. Regolith Exploration Geochemistry in Arctic and Temperate Terrains*. Elsevier, Amsterdam.

KERR, D. 2000. Kimberlite indicator minerals in till, Lac de Gras region. Poster presentation at the Prospectors Developers Convention, Toronto, Ontario, March 5–10, 2000.

KERR, .D., KJARSGAARD, I., WARD, B., DREDGE, L. & STIRLING, J. 1995. *Distribution and chemistry of kimberlite indicator minerals, Winter Lake map area, NWT (NTS 86A)*. Geological Survey of Canada Open File **3081**.

KERR, .D., KJARSGAARD, I., WARD, B., DREDGE, L. & STIRLING, J. 1996. *Distribution and chemistry of kimberlite indicator minerals, Napaktulik Lake map area, NWT (NTS 86I)*. Geological Survey of Canada Open File **3355**.

KERR, D., KJARSGAARD, I., KNIGHT, R. D., DREDGE, L. & STIRLING, J. 1999. *Distribution and chemistry of kimberlite indicator minerals, Contwoyto Lake map area, NWT (NTS 76E, north half)*. Geological Survey of Canada Open File **3768**.

KERR, D. E., SMITH, D. & WILSON, P. 2000. *Anomalous kimberlite indicator mineral and gold grain abundances, Dry Bones Bay and Yellowknife area, Northwest Territories*. Geological Survey of Canada Open File **D3861**.

KETTLES, I. M. 1994. *Till geochemistry in the Manitouwadge-Hornepayne region, Ontario*. Geological Survey of Canada Open File **2933**.

KIRKLEY, M. B., KOLEBABA, M. R., CARLSON, J. A., GONZALES, A. M., DYCK, D. R. & DIERKER, C. 1998. Kimberlite emplacement processes interpreted from Lac de Gras examples. *In: 7th International Kimberlite Conference, Extended Abstracts*, Cape Town, 429–431.

KJARSGAARD, B. A. 1995. Research on kimberlites and applications of diamond exploration techniques in Saskatchewan. *In:* RICHARDSON, D. (ed.) *Investigations Completed by the Saskatchewan Geological Survey and the Geological Survey of Canada Under the Geoscience Program of the Canada-Saskatchewan Partnership Agreement on Mineral Development (1990-1995)*. Geological Survey of Canada Open File **3119**, 167–181.

KJARSGAARD, B. A. 1996a. Somerset Island kimberlite field, District of Franklin, Northwest Territories. *In:* LE CHEMINANT, A. N., RICHARDSON, D. G., DILABIO, R. N. W. & RICHARDSON, K. A. (eds) *Searching For Diamonds in Canada*. Geological Survey of Canada Open File **3228**, 61–66.

KJARSGAARD, B. A. 1996b. Prairie kimberlites. *In:* LE CHEMINANT, A. N., RICHARDSON, D. G., DILABIO, R. N. W. & RICHARDSON, K. A. (eds) *Searching For Diamonds in Canada*. Geological Survey of Canada Open File **3228**, 67–72.

KJARSGAARD, B. A. 1996c. Occurrence, distribution, age and economic potential of kimberlites in western Canada. Geological Survey of Canada Minerals Colloquium, Program with Abstracts, Ottawa, 15.

KJARSGAARD, B. A. 1996d. Kimberlites. *In:* LE CHEMINANT, A. N., RICHARDSON, D. G., DILABIO, R. N. W. & RICHARDSON, K. A. (eds) *Searching For Diamonds in Canada*. Geological Survey of Canada Open File **3228**, 29–37.

KJARSGAARD, B. A. 1996e. Kimberlite-hosted diamond. *In:* ECKSTRAND, O. R., SINCLAIR, W. D. & THORPE, R. I. (eds.) *Geology of Canadian Mineral Deposit Types*. Geological Survey of Canada, Geology of Canada, No. **8**, 557–566.

KJARSGAARD, B. A. 1998a. *Kimberlites in the Western Canada Sedimentary Basin*. Manitoba Mining and Minerals Convention '98, Winnipeg, Manitoba, November 15–16, 22.

KJARSGAARD, B. A. 1998b. Compositional trends of spinel and mica in alkali minettes, southern Alberta, Canada. *In: The 7th International Kimberlite Conference, extended abstracts*, Cape Town, 435–437.

KJARSGAARD, B. A., FRISKE, P. W. B., MCCURDY, M. W., LYNCH, J. J., DAY, S. J. & DURHAM, C. C. 1992. *Re-analysis of selected lake sediment samples from the Bear-Slave Operation, Northwest Territories (Parts of 76B and 76D)*. Geological Survey of Canada Open File **2578**.

KLASSEN, R. A. 2001. A Quaternary geological perspective for geochemical exploration in glaciated terrain. *In:* MCCLENAGHAN, M. B. BOBROVSKY, P. T., HALL, G. E. M. & COOK, S. J. (eds) *Drift Exploration in Glaciated Terrain*. Geological Society, London, Special Publications, **184**, 1–17.

KOBETS, N. V. & KOMAROV, B. V. 1958. Some problems of methodology in prospecting for primary diamond deposits by aero methods. *Bulletin of the Academie of Sciences of the USSR*, **2**, 80–86. (A.G.I. translation from Russian).

KOMOGOROVA, L. G., STADNIK, YE. V. & FEDEROV, V. I. 1986. Phytogeochemical surveys within kimberlite bodies. Translated from: Fitogeokhimicheskiye issledovaniya v konturakh kimberlitovykh. *Doklady Akademii Nauk SSSR, 1987*, **29**, 468–470.

KONG, J. M., BOUCHER, D. R. & SCOTT-SMITH, B. H. 1999. Exploration and geology of the Attawapiskat kimberlites, James Bay Lowland, northern Ontario, Canada. *In:* GURNEY, J. J., GURNEY, J. L., PASCOE, M. D. & RICHARDSON, S. H. (eds) *The J.B. Dawson Volume, Proceedings of the VIIth International Kimberlite*, Cape Town, 446–448.

KUJANSUU, R. 1990. Glacial flow indicators in air photographs. *In:* KUJANSUU, R. & SAARNISTO, M. (eds) *Glacial Indicator Tracing*. A. A. Balkema,

Rotterdam, 71–86.

LeCheminant, A. N., Richardson, D. G., DiLabio, R. N. W. & Richardson, K. A. (eds) 1996. *Searching For Diamonds in Canada*. Geological Survey of Canada Open File **3228**.

Leckie, D. A., Kjarsgaard, B. A., Bloch, J., McIntyre, D., McNeil, D., & Stasiuk, L. 1997a. Emplacement and re-working of diamond-bearing, crater-facies kimberlite in Albian sediments of central Saskatchewan. *Geological Society of America, Bulletin*, **109**, 1000–1024.

Leckie, D. A., Kjarsgaard, B. A. Collins, M., Dufresne, M. D., Eccles, D. R., Heaman, L. M., Pierce, J., Stasiuk, L. & Sweet, R. 1997b. *Geology of a Late Cretaceous possible kimberlite at Mountain Lake, Alberta - chemistry, petrology, indicator minerals, aeromagnetic signature, age, stratigraphic position and setting*. Geological Survey of Canada Open File **3554**.

Leckie, D. A., Bednarski, J, & De Paoli, G. 1998. *Gold recovery and kimberlite/ diamond indicators from Cripple Creek, Rocky Mountain Foothills, Alberta*. Geological Survey of Canada Open File **3602**.

Lee, H. A. 1965. *Investigation of eskers for mineral exploration*. Geological Survey of Canada Paper **65-14**.

Letendre, J. 1992. *Report on the examination of samples from Manitoba Energy and Mines*. Manitoba Energy and Mines, Miscellaneous Publication **92-1**.

Letendre, J. 1994. *Report on the examination of samples from Manitoba Energy and Mines*. Manitoba Energy and Mines, Miscellaneous Publication **94-1**.

Levinson, A. A. 1980. *Introduction to Exploration Geochemistry*. Applied Publishing Ltd., Wilmette, Illinois.

Lillieskõld, M. 1990. Lithology and transport distance of glaciofluvial material. *In:* Kujansuu, R. & Saarnisto, M. (eds) *Glacial Indicator Tracing*. A. A. Balkema, Rotterdam, 151–164.

Litinsky, V. A. 1964. Application of metallometry and kappametry in prospecting for kimberlite bodies. *International Geology Review*, **6**, 2027–2035.

Lundqvist, J. 1990. Glacial morphology as an indicator of the direction of glacial transport. *In:* Kujansuu, R. & Saarnisto, M. (eds) *Glacial Indicator Tracing*. A. A. Balkema, Rotterdam, 61–70.

Mannard, G. W. 1968. The surface expression of kimberlite pipes. Geological Association of Canada, Proceedings Volume **19**, 15–21.

Mathur, S. M. & Alexander, P. O. 1983. Preliminary pedogeochemical and biogeochemical studies on the Hinota kimberlite, Panna District, MP. *Indian Academy of Sciences (Earth and Planetary Science) Proceedings*, **92/1**, 81–88.

Matile, G. L. D. & Thorleifson, L. H. 1997. *Till geochemical and indicator mineral reconnaissance of northeastern Manitoba*. Manitoba Energy and Mines Open File **97-3**.

Matile, G. L. D., Nielsen, E., Thorleifson, L. H. & Garrett, R. G. 1996. *Kimberlite indicator mineral analysis from the Westlake Plain: follow-up to the GSC Prairie kimberlite study*. Manitoba Energy and Mines Open File **96-2**.

McCallum, M. E. & Vos, W. P. 1993. Ilmenite signature: utilization of paramagnetic and chemical properties in kimberlite exploration. *In:* Sheahan, P. & Chater, A. (eds) *Diamonds: Exploration, Sampling and Evaluation*. Prospectors and Developers Association of Canada, 109–146.

McCandless, T. E. 1990. Kimberlite xenocryst wear in high-energy fluvial systems: experimental studies. *Journal of Geochemical Exploration*, **37**, 323–331.

McClenaghan, M. B. & Dunn, C. E. 1995. *Biogeochemical survey over kimberlites in the Kirkland Lake area, northeastern Ontario*. Geological Survey of Canada Open File **3005**.

McClenaghan, M. B., Kjarsgaard, I. M., Stirling, J. A. R., Pringle, G. & Crabtree, D. 1993. *Chemistry of kimberlitic indicator minerals in drift from the Kirkland Lake area, northeastern*. Geological Survey of Canada Open File **2761**.

McClenaghan, M. B., Kjarsgaard, I. M., Crabtree, D. & DiLabio, R. N. W. 1995. *Mineralogy and geochemistry of till and soil overlying the Buffonta kimberlite dyke, Kirkland Lake, Ontario*. Geological Survey of Canada Open File **3007**.

McClenaghan, M. B., Kjarsgaard, I. M., Schulze, D. J., Stirling, J. A. R., Pringle, G. & Berger, B. R. 1996. *Mineralogy and geochemistry of the B30 kimberlite and overlying glacial sediments, Kirkland Lake, Ontario*. Geological Survey of Canada Open File **3295**.

McClenaghan, M. B., Kjarsgaard, I. M., Schulze, D. J., Berger, B. R., Stirling, J. A. R. & Pringle, G. 1998. *Mineralogy and geochemistry of the Diamond Lake kimberlite and associated esker sediments, Kirkland Lake, Ontario*. Geological Survey of Canada Open File **3576**.

McClenaghan, M. B., Kjarsgaard, I. M., Stirling, J. A. R., Pringle, G., Kjarsgaard, B. A. & Berger, B. R. 1999a. *Mineralogy and geochemistry of the C14 kimberlite and associated glacial sediments, Kirkland Lake, Ontario*. Geological Survey of Canada Open File **3719**.

McClenaghan, M. B., Kjarsgaard, I. M., Kjarsgaard, B. A., Stirling, J. A. R., Pringle, G. & Berger, B. R. 1999b. *Mineralogy and geochemistry of the A4 kimberlite and associated glacial sediments, Kirkland Lake, Ontario*. Geological Survey of Canada Open File **3769**.

McClenaghan, M. B., Kjarsgaard, B. A., Kjarsgaard, I. M., Paulen, R. C. & Stirling, J. A. R. 1999c. *Mineralogy and geochemistry of the Peddie kimberlite and associated glacial sediments, Lake Timiskaming, Ontario*. Geological Survey of Canada Open File **3775**.

McClenaghan, M. B., Ward, B. C., Kjarsgaard, B. A., Kjarsgaard, I. M., Stirling, J. A. R., Dredge, L. & Kerr, D. 2000. *Mineralogy and geochemistry of the Ranch Lake kimberlite and associated glacial sediments, Lac de Gras region, NWT*. Geological Survey of Canada Open File **D3924**.

McConnell, J. W. & Batterson, M. J. 1987. The Strange Lake Zr-Y-Nb-Be-REE deposit, Labrador, a geochemical profile in till, lake and stream sediment, and water. *Journal of Geochemical Exploration*, **29**, 105–127.

McKinlay, F. T., Williams, A. C., Kong, J. & Scott-Smith, B. H. 1997. An integrated exploration case history for diamonds, Hardy Lake Project, N.W.T. *In:* Gubins, A. G. (ed.) *Exploration '97, Proceedings of the Fourth Decennial International Conference on Mineral Exploration*, 1029–1038.

McMartin, I. & McClenaghan, M. B. 2001. Till geochemistry and sampling techniques in glaciated Shield terrain. *In:* McClenaghan, M. B. Bobrowsky, P. T., Hall, G. E. M. & Cook, S. J. (eds) *Drift Exploration in Glaciated Terrain*. Geological Society, London, Special Publications, **184**, 19–44.

McMartin, I. & Pringle, G. 1994. *Regional kimberlite indicator mineral data and till geochemistry from the Wekusko Lake area (63J), north-central Manitoba*. Geological Survey of Canada Open File Report **2844**.

Melnyk, W. 1997. Snap Lake, NWT. Cordilleran Roundup Abstract Volume, Fourteenth Annual Cordilleran Geology and Exploration Roundup, Vancouver, January, 1997, 38.

Millard, M. J. 1994. *Kimberlite indicator minerals in the Meadow Lake Area, Saskatchewan*. Saskatchewan Research Council Publication No. **R-1210-1-E94**.

Miller, A. R, Seller, M. H. Armitage, A. E., Davis, W. J. & Barnett, R. L., 1998. Late Triassic kimberlitic magmatism, western Churchill structural province, Canada. *In: 7th International Kimberlite Conference, Extended Abstracts*, Cape Town, 591–593.

Mitchell, R. H. 1975. Geology, magnetic expression and structural control of the central Somerset Island kimberlites. *Canadian Journal of Earth Sciences*, **12**, 757–764.

Mitchell, R. H 1986. *Kimberlites, Mineralogy, Geochemistry and Petrology*. Plenum Press, New York.

Morris, T. F. 1999. *Geochemical, heavy mineral and pebble lithology data, surficial sediment sampling program, Wawa region, Northeastern Ontario*. Ontario Geological Survey Open File Report **5981**.

Morris, T. F. & Kaszycki, C. A. 1997. *Prospector's guide to drift prospecting for diamonds, northern Ontario*. Ontario Geological Survey, Miscellaneous Paper **167**.

Morris, T. F., Murray, C. & Crabtree, D. C. 1994. *Results of overburden sampling for kimberlite heavy mineral indicators and gold grains, Michipicoten River- Wawa area, northern Ontario*. Ontario Geological Survey Open File Report **5908**.

Morris, T. F., Bajc, A. F., Bernier, M. A., Kaszycki, C. A., Kelly, R. I., Murray, C. & Stone, D. 1995. *Kimberlite indicator heavy mineral indicator release*. Ontario Geological Survey Open File Report **5934**.

Morris, T. F., Crabtree, D. C. & Pianosi, S. 1997. *Results of modern alluvium sampling for kimberlite indicator minerals, Kinniwabi Lake area, northeastern Ontario*. Ontario Geological Survey Open File Report **5956**.

Morris, T. F., Crabtree, D. C., Sage, R. P. & Averill, S. A. 1998a. Types, abundances and distribution of kimberlite indicator minerals in alluvial sediments, Wawa-Kinniwabi area, northeastern Ontario: implications for the presence of diamond-bearing kimberlite. *Journal of Geochemical Exploration*, **63**, 217–235.

Morris, T. F., Crabtree, D. C. & Averill, S. A. 1998b. *Kimberlite, base metal and gold exploration targets based upon heavy mineral data derived from surface material, Kapuskasing, northeastern Ontario*. Ontario Geological Survey Open File Report **5967**.

Morris, T. F., Sage, R. P. & Ayer, J. A. 1999. Cr-diopside as an indicator of kimberlite: applications to overburden studies. GAC-MAC Joint Annual Meeting, Sudbury 1999, Abstract Volume **24**, 86.

Morris, T. F., Sage, R. P., Crabtree, D. C. & Pitre, S. A. 2000. *Kimberlite, base metal, gold and carbonatite exploration targets derived from overburden heavy mineral data, Killala Lake area, northwestern Ontario*. Ontario Geological Survey Open File Report **6013**.

Mosig, R. W. 1980. Morphology of indicator minerals, a guide to proximity of source. *In:* Glover, J. E. & Grovers, D. E. (eds) *Kimberlites and Diamonds*. Geology Department and Extension Service, University of Western Australia, Publication **No. 5**, 81–87.

Muggeridge, M. T. 1995. Pathfinder sampling techniques for locating primary sources of diamond; recovery of indicator minerals, diamonds and geochemical signatures. *Journal of Geochemical Exploration*, **53**, 183–204.

Northern Miner. 1999a. 'Partners find kimberlite on Baffin Island in Canada's North', September 27, 1999, B5.

Northern Miner. 1999b. Twin Gold recovers more diamonds, December 13, 1999, 6.

Papunen, H. 1995. Diamonds of northern Europe – a review. *In:* Pasava, J., Kríbek, B. & Zák, K. (eds) *Mineral Deposits: From their Origin to their Environmental Impacts*. A. A. Balkema, Rotterdam, 617–619.

Pegg, C. C, Brummer, J. J. & MacFadyen, D. A. 1990. Discovery of kimberlite diatremes in the Kirkland Lake area, Ontario. *Canadian Institute of Mining and Metallurgy Bulletin*, **83**, 90.

Reid, A. & Dawson, J. 1972. Olivine-garnet reaction in peridotites from Tanzania. *Lithos*, **5**, 115–124.

Sage, R. P. 1995. Kimberlites of Ontario. *In:* Summary of Field Work and Other Activities-1995. Ontario Geological Survey, Miscellaneous Publication **MP 164**, 111–112.

Sage, R. P. 1996. *Kimberlites of the Lake Timiskaming structural zone*. Ontario Geological Survey Open File Report **5937**.

Sage, R. P. 1998. Structural patterns and kimberlite emplacement. *In: Summary of Field Work and Other Activities-1998*. Ontario Geological Survey, Miscellaneous Publication **MP 169**, 224–229.

Sage, R. P. 2000. *Kimberlites of the Lake Timiskaming structural zone: Supplement*. Ontario Geological Survey Open File Report **6018**.

SCHULZE, D. J. 1993a. An introduction to the recognition and significance of kimberlite indicator minerals. *In: Techniques in Exploration For Diamonds*, Short Course Notes, Canadian Institute of Mining, Metallurgy and Petroleum Geology, May 13–14, 1993, Calgary, Alberta.

SCHULZE, D. J. 1993b. Garnet xenocryst populations in North American kimberlites. *In:* SHEAHAN, P. & CHATER, A. (chairmen) *Diamond: Exploration, Sampling and Evaluation*. Prospectors and Developers Association of Canada, Short Course Proceedings, Toronto, Ontario, March 27, 1993, 359–377.

SCHULZE, D. J. 1995. A guide to the recognition and significance of kimberlite indicator minerals. *In: Diamonds-Theory and Exploration*. Geological Association of Canada, Short Course 20, 1–39.

SCHULZE, D. J. 1996. Kimberlites in the vicinity of Kirkland Lake and Lake Timiskaming, Ontario and Quebec. *In:* LECHEMINANT, A. N., RICHARDSON, D. G., DILABIO, R. N. W. & RICHARDSON, K. A. (eds) *Searching for Diamonds in Canada*. Geological Survey of Canada Open File, **3228**, 73–78.

SCHULZE, D. J. 1999. The significance of eclogite and Cr-poor megacryst garnets in diamond exploration. *Exploration and Mining Geology*, **6**, 349–366.

SHEAHAN, P. & CHATER, A. (chairmen) 1993. *Diamonds: Exploration, Sampling and Evaluation*. Short course Proceedings, Prospectors and Developers Association of Canada, Toronto, Ontario, March 27, 1993.

SHILTS, W. W. 1975. Principles of geochemical exploration for sulphide deposits using shallow samples of glacial drift. *Canadian Institute of Mining and Metallurgy Bulletin*, **68**, 73–80.

SIMPSON, M. A. 1993. *Kimberlite indicator minerals in southwestern Saskatchewan*. Saskatchewan Research Council, Publication No. **R-1210-8-E91**.

SMIRNOV, G. I. 1959. Mineralogy of Siberian kimberlites. *International Geology Review*, **1**, 21–39.

SOBOLEV, N. V. 1971. On mineralogical criteria of a diamond potential of kimberlites; *Geologiya i Geofizika*. **12 (3)**, 70–78 (in Russian).

SOBOLEV, N. V. 1977. *Deep seated inclusions in kimberlites and the problem of the composition of the upper mantle*. American Geophysical Union, Washington.

SOBOLEV, N. V. 1993. Kimberlites of the Siberian Platform: their geological and mineralogical features. *In:* SHEAHAN, P. & CHATER, A. (chairmen) *Diamonds: Exploration, Sampling and Evaluation*. Short Course Proceedings, Prospectors and Developers Association of Canada, Toronto, Ontario, March 27, 1993, 343–357.

SOBOLEV, N. V., LAVRENT'EV, Y. G, POSPELOVA, L. N. & SOBOLEV, E. V. 1973 Chrome-rich garnets from the kimberlites of Yakutia and their paragenesis. *Contributions to Mineralogy and Petrology*, **40**, 39–52.

SOBOLEV, N. V., POKHILENKO, N. P. & AFANAS'EV, V. P. 1993. Kimberlitic pyrope and chromite morphology and chemistry, as indicators of diamond grade in Yakutian and Arkhangelsk provinces. *In:* DUNNE, K. P. E. & GRANT, B. (eds) *Mid-continent Diamonds*. Geological Association of Canada, Symposium Volume, 63–69.

STEENFELT, A., JENSEN, S. M., LARSEN, L. M. & STENDAL, H. 1999. Diamond exploration in southern West Greenland, *In:* COOKENBOO, H. (ed.) *Diamond Exploration Methods and Case Studies*. Association of Exploration Geochemists Short Course Notes, 19th International Geochemical Exploration Symposium, Vancouver, B.C. April, 1999, 76–84.

STEPHENSON, D. M., MORRIS, T. F. & CRABTREE, D. C. 1999. *Kimberlite, base metal and gold exploration targets based upon heavy mineral data derived from surface materials, Opasatika Lake area, northeastern Ontario*. Ontario Geological Survey Open File Report **5982**.

STEWART, R. A., MAYBERRY, S. W. & PICKERILL, M. J. 1988. Composition of till in the vicinity of the Lake Ellen kimberlite and implications for the source of diamonds in glacial sediments of eastern Wisconsin. *In:* MACDONALD, D. R & MILLS, K. A (eds) *Prospecting in Areas of Glaciated Terrain – 1988*. Canadian Institute of Mining and Metallurgy, 103–120.

STONE, D., MORRIS, T. F. & CRABTREE, D. C. 1999. *Heavy mineral indicator database derived from overburden for kimberlite, metamorphosed magmatic sulphide indicators and gold, Stull Lake area, Northwestern Ontario*. Ontario Geological Survey, Miscellaneous Release-**Data 45**.

SWANSON, F. J. & GENT, M. R. 1993. *Results of reconnaissance sampling for kimberlite and lamproite indicator minerals, Saskatchewan*. Saskatchewan Geological Survey Open File Report **93-4**.

THOMAS, R. D. & GLEESON, C. F. 1999. Use of regional till geochemistry to outline the area of occurrence of diamondiferous lamprophyre dykes in the Wawa area. Geological Association of Canada-Mineralogical Association of Canada Annual Meeting, Sudbury 1999, Abstract Volume **24**, 128.

THORLEIFSON, L. H. 1999. Kimberlite indicator minerals in glacial sediments. *In:* MCCLENAGHAN, M. B., BOBROWSKY, P. T. & COOK, S. (eds) *Drift Exploration in Glaciated Terrain*, Association of Exploration Geochemists Short Course Notes, 19th International Geochemical Exploration Symposium, Vancouver, April, 1999.

THORLEIFSON, L. H. & GARRETT. R. G. 1993. *Prairie kimberlite study - till matrix geochemistry and preliminary indicator mineral data*. Geological Survey of Canada Open File **2745**.

THORLEIFSON, L. H. & GARRETT. R. G. 2000. *Lithology, mineralogy, and geochemistry of glacial sediments overlying kimberlite at Smeaton, Saskatchewan, Canada*. Geological Survey of Canada Bulletin **551**.

THORLEIFSON, L. H. & MATILE. G. 1993. *Till geochemical and indicator mineral reconnaissance of southeastern Manitoba*. Geological Survey of Canada Open File **2750**.

THORLEIFSON, L. H., GARRETT, R. G. & MATILE. G. L. D. 1994. *Prairie kimberlite study – indicator mineral geochemistry*. Geological Survey of Canada Open File **2875**.

TOWIE, N. J. & SEET, L. H. 1995. Diamond laboratory techniques. *Journal of Geochemical Exploration*, **53**, 205–212.

WARD, B. C., KJARSGAARD, I. M., DREDGE, L. A., KERR, D. E. & STIRLING, J. A. R. 1995. *Distribution and chemistry of kimberlite indicator minerals, Lac de Gras map area, NWT (NTS 76D)*. Geological Survey of Canada Open File **3079**.

WARD, B. C., DREDGE, L. A., KERR, D. E. & KJARSGAARD, I. M. 1996. Kimberlite indicator minerals in glacial deposits, Lac de Gras area, N.W.T. *In:* LECHEMINANT, A. N., RICHARDSON, D. G., DILABIO, R. N. W. & RICHARDSON, K. A. (eds) *Searching for Diamonds in Canada*. Geological Survey of Canada Open File **3228**, 191–195.

WARD, B. C., KJARSGAARD, I. M., DREDGE, L. A., KERR, D. E. & STIRLING, J. 1997. *Distribution and chemistry of kimberlite indicator minerals, Contwoyto Lake map area, NWT (NTS 76E, south half)*. Geological Survey of Canada Open File **3386**.

WATSON, K. D. 1955. Kimberlite at Batchelor Lake, Quebec. *American Mineralogist*, **40**, 656-679.

WEBB, J. S. 1956. Observations on geochemical exploration in tropical terrains. *In:* Twentieth International Geological Congress, Mexico, *Symposium de Exploration Geoquimica*. Tomo **I**, 163–164.

WILKINSON, L., HARRIS, J. R. & KJARSGAARD, B. A. 1999. Searching for kimberlite: Weights of evidence modeling of the Lac de Gras area, NWT using GIS technology. *In: The 13th International Conference on Applied Geologic Remote Sensing, Vancouver, Canada, March 1–3, 1999, Extended Abstracts*, 63–72.

WOLFE, W. J., LEE, H. A. & HICKS, W. D. 1975. *Heavy mineral indicators in alluvial and esker gravels of the Moose River Basin, James Bay Lowland*. Ontario Geological Survey Geoscience Report **126**.

ZALNIERIUNAS, R. V. & SAGE, R. P. 1995. *Known kimberlites of eastern Ontario*. Ontario Geological Survey Preliminary Map **P. 3321**.

Lake sediment geochemical methods in the Canadian Shield, Cordillera and Appalachia

STEPHEN J. COOK[1] & JOHN W. McCONNELL[2]

[1] *British Columbia Geological Survey, 5-1810 Blanshard Street, Victoria, British Columbia V8V 1X4, Canada (e-mail: stephencook@angloamerican.ca)*
Current address: Hudson Bay Exploration and Development Co. Limited, 800-700 West Pender St., Vancouver, British Columbia V6C 1G8, Canada
[2] *Geological Survey of Newfoundland and Labrador, P.O. Box 8700, St. John's, Newfoundland A1B 4J6, Canada*

Abstract: Lake sediment geochemistry has been used in Canada since the 1970s for mineral exploration and resource evaluation in glaciated regions of Appalachia, the Canadian Shield and the western Cordillera which are of low to moderate relief. Geochemical signatures of bedrock and mineralization within a lake's catchment basin are commonly reflected in the chemical constituents of the organic-rich Holocene sediment which has been transported from source by a combination of mechanical and hydromorphic processes. Lake sediment geochemical surveys have been used successfully to discover base metal, Au, Mo, W, Sb, Sn, U, and REE mineralization. The scale of such surveys determines the size and density of lake sediment sampling. In regional surveys, large lithological targets such as greenstone belts or chemically distinctive intrusives can be identified by low density (c. 1 per 10–15 km^2) sampling of relatively larger lakes. Smaller targets such as mineralized areas require higher density (c. 1 per 4–5 km^2) sampling using a greater number of smaller lakes. Regional surveys are typically helicopter-borne, and employ a tubular grab sampler which permits rapid reconnaissance-scale coverage of large areas. Centre-basin profundal lake sediment, or gyttja, is an ideal sample medium because of its homogeneity and abundance of fine-grained organic matter, which complexes with many trace elements. Multi-element analytical techniques in common use include inductively coupled plasma emission spectrometry (ICP-ES), ICP mass spectrometry (ICP-MS) and instrumental neutron activation analysis (INAA). Reliablility of data is assessed by analysis of known standards, site duplicates and sample splits. PC statistical software packages facilitate the interpretation of geochemical data, and desktop GIS packages aid in further interpretation and generation of multi-layer maps.

Lake sediment geochemical surveys of northern glaciated terrains have been conducted primarily in Canada and Fennoscandia. In Canada, use of lake sediment geochemistry for mineral exploration and resource assessment purposes has achieved wide acceptance in glaciated regions of both the Canadian Shield and the western Cordillera. This is due in part to the ubiquitous distribution of lake sediment in both areas, and its ability to accumulate metals derived from potentially economic mineralization. Lake sediments are present in the numerous lakes comprising disorganized drainage in Shield areas, and in the extensive central Plateau areas of British Columbia, where a lack of active sediment limits the use of stream sediment geochemistry. Unlike many other commonly used geochemical media such as soils and stream sediments, in which the element distributions can be profoundly affected by numerous environmental variables, deep profundal lake sediments typically provide a more homogenous sample medium over wide areas. Regional surveys have been conducted by the mineral exploration industry, and by federal and provincial geological surveys. Most publicly funded regional surveys have been conducted in Shield and Appalachian regions of central and Atlantic Canada, comprising c. 181 000 sites which cover an area of about 2.6×10^6 km^2 (Davenport *et al.* 1997). Undertaken for over two decades to the standards of the Geological Survey of Canada's

(GSC) National Geochemical Reconnaissance (NGR) program, these surveys provide a wealth of high quality geochemical data for mineral exploration, geochemical mapping and environmental assessment purposes. They have contributed to the discovery of several mineral deposits and prospects, such as the Strange Lake Y–Zr–Be deposit in Labrador (McConnell & Batterson 1987), the Percy Lake Zn–Pb–Cu–Ag prospect in northern Ontario (Hamilton et al. 1995) and the Eastmain Au–As showing in northern Quebec (Davenport et al. 1997).

This paper is not intended as an exhaustive review of the application of lake sediments to mineral exploration, resource assessment or environmental studies. Rather, its purpose is to complement the earlier reviews of Coker et al. (1979) and Friske (1991) by: (1) reviewing the theoretical basis of lake sediment geochemical exploration; (2) documenting the application of lake sediment geochemical mapping to regional and property-scale mineral exploration, with recommendations for their effective use; (3) discussing the practical aspects of sample collection, preparation and analytical methodologies; (4) considering methods of data interpretation and presentation and finally; and (5) presenting some case study results of the successful use of lake sediment geochemistry in regional-scale mineral exploration. The paper focuses on the glaciated regions of the western Cordillera, the Canadian Shield and Appalachia, with particular reference to central British Columbia (BC) and Labrador.

Lake sediments and their use in mineral exploration

Organic-rich Holocene sediments in modern lake basins have accumulated since the Wisconsin deglaciation approximately 10 000–12 000 BP. Sediment accumulation rate varies from one part of the country to another, depending on factors such as organic production, which is greater in forested, relative to tundra, regions. In general, lake sediment geochemical methods in mineral exploration are based on the detection of elements which are liberated from rock, till or other material during weathering, and subsequently transported to and fixed in bottom sediments. These methods can be an effective tool to delineate regional geochemical patterns and anomalous metal concentrations related to mineral deposits (Hoffman 1976; Coker et al. 1979) in central BC, Shield regions of Canada and Appalachia. In much of the Canadian Cordillera and northern Labrador, stream sediments are the standard sampling medium for reconnaissance-scale geochemical surveys, as high relief and a scarcity of lakes preclude the effective use of lake sediment geochemistry. In BC, Regional Geochemical Surveys (RGS) conducted over the past 20 years comprise c. 45 000 sites covering about 65% of the province at a site density of approximately one per 10–13 km^2. Organic-rich profundal lake sediments are, however, a more appropriate geochemical medium in the Nechako Plateau, a region of subdued topography with an abundance of lakes located in the central part of the Cordillera within BC. Use of stream sediment geochemical methods are limited here by a paucity of active sediment in drainage channels. Mineral exploration has in the past been impeded in this area by extensive drift and forest cover, poor bedrock exposure and a barren Tertiary volcanic cover.

Factors affecting the composition of lake sediments

Geochemical and physical composition of organic-rich Holocene lake sediments is influenced by such factors as bedrock geology (Hoffman 1976; Gintautas 1984), presence of nearby mineral deposits or prospects, surficial geology, climate, soils, vegetation, limnological factors and, in some areas, human activity. There are three general varieties of such sediment, depending upon their location within the lake basin: organic gel, organic sediment and inorganic sediment (Jonasson 1976). Organic gels, or 'gyttja', are homogenous mixtures of particulate organic matter, inorganic precipitates and mineral matter (Wetzel 1983). Green–grey to black, these sediments are characteristic of deepwater, or profundal, lake basins. Organic sediments are immature mixtures of organic gels, organic debris and mineral matter occurring in shallow water and near drainage inflows (Jonasson 1976). Inorganic sediments are composed of mineral particles with relatively little organic matter. Of the three, organic gels are most suitable as a geochemical exploration medium because of (a) their homogeneity, and (b) abundance of very fine-grained organic matter which complex with many trace elements. The centres of profundal basins where these sediments accumulate are considered to be ideal sites for regional geochemical sampling (Friske 1991). It is important to differentiate between these modern Holocene organic-rich lake sediments and the silty glaciolacustrine sediments found in some low-lying areas of central BC. The latter, products of short-lived sedimentation in ice-dammed lakes during deglaciation, are not locally derived.

Processes of geochemical transport and accumulation of metals

In glaciated regions, geochemical dispersion of base and precious metals into lake basins occurs by hydromorphic and mechanical processes which involve groundwater, stream water, stream sediments or combinations thereof. Means of hydromorphic transport is by dissolved ions and species in ground and stream waters. Means of mechanical transport include both elements complexed with suspended particulate organic matter in stream water, and those associated with suspended mineral grains in stream sediment.

Within the lake basins, subsequent accumulation and fixation of elements occurs by several means:

- adsorption by organic matter of elements from solution;
- settling of organic suspensates;
- precipitation and settling of dissolved metal-organic complexes;
- co-precipitation of metals from solution with hydrous oxides of Fe, Al and Mn;
- local deposition of silt/clay particles at stream inflows.

Timperley & Allan (1973) provide further information on trace element transport and accumulation in lake basins of the southern Shield region.

Limnological factors

Limnological factors that affect sediment composition have their origin in the temperature (T) and dissolved oxygen (DO) stratification of lake waters in northern temperate regions during the warm summer months, which subsequently overturn with seasonal temperature changes in the spring and fall. Eutrophic lakes are small, nutrient-rich lakes with high organic production and almost complete oxygen depletion with increasing depth. Conversely, oligotrophic lakes are deep, large, nutrient-poor lakes with low organic production and a much more constant oxygen content with depth. Polymictic or unstratified lakes are relatively shallow and are not thermally stratified. There are distinct geochemical differences between the sediments of eutrophic and oligotrophic lakes in central BC, particularly with respect to the abundance of organic matter and hydrous oxides of Fe and Mn (Hoffman & Fletcher 1981; Earle 1993). In general, high organic matter content is characteristic of eutrophic lakes, whereas sediment precipitates of hydrous Fe and Mn oxides are products of the more oxygen-rich conditions within typically larger oligotrophic lakes. Friske (1995) documented the effects of different limnological environments on trace element distributions within sediment cores of a single lake in central BC (Tatin Lake) and concluded that they are an important control on the concentrations of several elements.

Lake sediment exploration for different deposit types

The potential of lake sediment geochemistry in the search for mineral deposits was recognized over 30 years ago by the mineral exploration industry, which first employed them in the search for base metals in the Cordillera and uranium in the Shield. They have since been widely applied across Canada in the search for U, base metals, Au and other elements such as rare earth elements (REEs) and platinum-group elements (PGEs). Results of these surveys have led, in whole or in part, to the discovery of many mineral deposits and prospects, including the Percy Lake Zn–Pb–Cu–Ag prospect in northern Ontario (Hamilton *et al.* 1995), the Eastmain Au–As showing in northern Quebec (Davenport *et al.* 1997) and the Mac Mo deposit in BC (Cope & Spence 1995).

Over the past decade, Au and other elements have been determined on archived lake sediment survey samples from various regions, and the new data released to augment existing trace element data. Similarly, further analysis of lake sediments from the region of the East Kemptville Sn deposit in Nova Scotia, obtained prior to its discovery in 1978, was described by Rogers & Garrett (1987). In this case, the samples were re-analysed for the lithophile elements Sn, Rb, F and Cl, which were associated with hydrothermal mineralization in the area.

Uranium

Regional lake sediment surveys were carried out by the exploration industry in the U exploration boom during the 1970s. Uranium was a target of the first publicly funded federal survey, such as the Uranium Reconnaissance Program (URP), precursor to the NGR program, which began during this period. Coker *et al.* (1979) provides a good review of the application of regional surveys to exploration for U and other metals during this period, which led to the discovery of the Key Lake U deposit.

Base metals

Lake sediment geochemical surveys have been widely used in base metal deposit exploration, in particular for porphyry-style Cu or Cu–Mo deposits in the Cordillera and for volcanogenic massive sulphide (VMS) deposits in the Shield. Many Cordilleran surveys have focused on porphyry-style targets, which are widespread in the Stikine and Quesnel Terrane rocks which underlay much of the Plateau region of central BC. The associated mineralization and alteration systems of porphyry deposits are often areally more extensive than those of stratabound deposits, increasing the likelihood of their detection. These surveys have also been conducted for sedimentary-exhalative style (SEDEX) Pb–Zn deposits in the Kechika Trough of northern BC and the Selwyn Basin in the Yukon.

In eastern Canada, the pioneering studies of Schmidt (1956) were conducted in lakes adjacent to base metal mineralization in northern New Brunswick. Since then, reconnaissance-scale and orientation surveys for base metal deposits have been conducted, for example, in the Shield areas of Ontario (Coker & Nichol 1975; Hamilton *et al.* 1995), the NWT (Jackson & Nichol 1974) and Labrador (McConnell 1984, 1999*b*). Target deposit types have included VMS Zn–Pb–Cu–Ag and Cu–Zn–Ag–Au deposits and magmatic Ni–Cu deposits.

Gold

The application of lake sediment geochemistry to Au exploration became more widespread during the 1980s with the advent of instrumental neutron activation analysis (INAA) and the availability of more reliable Au determinations at the low concentration levels typical of lake sediments. In the Cordillera, regional lake sediment surveys have proven useful in the discovery of epithermal Au deposits. Follow-up exploration of an industry Ag–Zn–As–Mo lake sediment anomaly (Dawson 1988) led to the discovery of the Wolf Au deposit in the southern Nechako Plateau. A later detailed study of this site showed high sediment Au concentrations of up to 56 ppb (Cook 1995) relative to a regional median of only about 1 ppb. The release of regional lake sediment geochemical results also played a subordinate role in the discovery of the nearby Tsacha Au prospect.

In Appalachia in 1984, a pilot study applying lake sediment geochemistry to Au exploration identified what was later to become the Hope Brook orebody in Newfoundland, discovered independently by industry (McConnell 1985). Subsequently, re-analysis of over 17 000 archived reconnaissance lake sediment samples from the island of Newfoundland delineated several Au belts and led directly to major staking 'rushes' and the discovery of many new Au occurrences (Davenport & Nolan 1989). In the Shield, the Bakos Au deposit was discovered upon follow-up of a 30 ppb Au anomaly in lake sediment in the La Ronge Au belt of northern Saskatchewan (Chapman *et al.* 1990). More recently, a high density lake sediment survey in the Nain Province of Labrador identified several known Au occurrences and targeted new anomalies (McConnell 1999*b*).

Other targets

The advent of inexpensive multi-element geochemical packages reporting, for example, many of the REEs, has led to the discovery of deposit types other than those originally sought. In the case of the REEs, the best example is the discovery of the Strange Lake Y–Zr–Be deposit in Labrador (McConnell & Batterson 1987). Conversely, surveys for elements such as the PGEs in lake sediments require the use of much more element-specific analytical methods. Several small regional surveys were conducted by Friske *et al.* (1990) to determine the distribution of PGEs in lakes bordering ultramafic rocks in northwestern Ontario. Platinum concentrations of up to 15 ppb (Hornbrook 1989) were found in sediments of lakes adjacent to the Lac des Iles PGE deposit north of Thunder Bay.

Focus and scale of lake sediment surveys

Consideration of scale is essential at the design stage of a survey. Sampling strategy is dependent on the scale of the target sought. Regional or reconnaissance-scale surveys are designed to identify large or even crustal features such a greenstone belts or metalliferous intrusions, areas of metal enrichment in which to focus further attention. In such surveys, samples are collected from relatively large lakes (*c.* 0.25–1 km^2) with large catchment areas at a low site density of about one per 10–15 km^2. In higher density surveys, the focus narrows and the targets become the delineation of anomalies within belts or areas with the expectation of generating targets for ground follow-up or geophysical surveys. Typically such surveys have

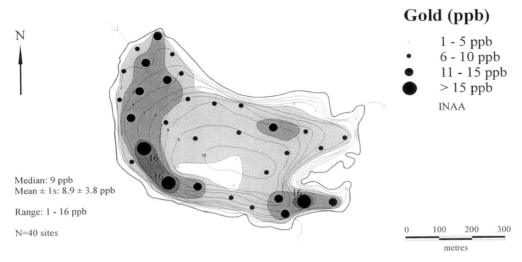

Fig. 4. Distribution of Au (ppb) in sediments of Clisbako Lake, in the Fraser Plateau area of central BC. Creeks entering the lake from the south and the northwest drain epithermal mineralization and/or alteration zones of the Clisbako property (Dawson 1991).

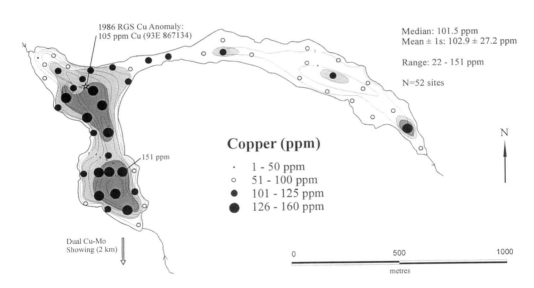

Fig. 5. Distribution of Cu (ppm) in sediments of Hill-Tout Lake, in the western part of the Interior Plateau. Creeks entering the lake from the south and the northwest drain the area of the Dual Cu–Mo showing. Bathymetry after Webber (1984).

(Fig. 4; Cook 1995, 1997a, b) occur in sediments from Clisbako Lake, which is located about 100 km west of Quesnel in the Fraser Plateau. Two of these, located at relatively shallow water stream inflows, indicate the presence of updrainage argillic alteration and/or epithermal Au zones of the Clisbako Au–Ag prospect (Dawson 1991). Regional background, as defined by the median Au concentration reported from nearby surveys, is 1 ppb (Cook & Jackaman 1994). Results illustrate that for Au, centre-lake sediments do not necessarily contain the highest Au

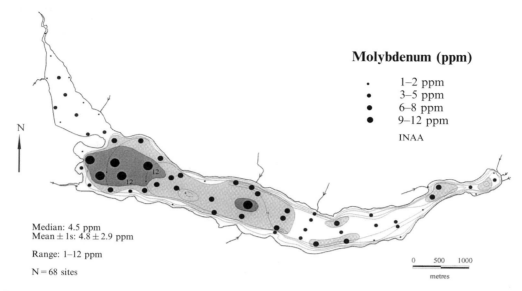

Fig. 6. Distribution of Mo (ppm) in sediments of Kuyakuz Lake, southern Nechako Plateau. Bathymetry after Burns (1978).

concentrations in lakes adjacent to epithermal Au deposits (Cook 1995, 1997a, b). In Shield regions of central Canada, Coker et al. (1982) and Fox et al. (1987) also noted that organic-rich sediments with the highest Au concentrations may be near-shore sediments as well as those of the profundal basin. The presence of elevated Au values in lake-margin sediments and adjacent to stream inflows gives rise to lateral metal zoning patterns in the lake sediment, which may be mapped out to provide a vector towards a potential source.

Hill-Tout Lake, located in the Nadina Lake area south of Houston, is situated about 2 km down-drainage of the Dual Cu–Mo showing, a pyrite–chalcopyrite–molybdenite occurrence within a quartz monzonite porphyry. A single lake sediment site from the 1986 RGS survey of the Whitesail Lake area (NTS 93E) had a value of 105 ppm Cu (Fig. 5). A subsequent detailed lake sediment study defined two discrete zones of elevated Cu concentrations (up to 151 ppm by aqua regia digestion/ICP-ES) in the two westernmost sub-basins of the lake. The two basins are adjacent to stream inflows which drain the Dual showing. Even the shallower and more distal eastern basin of the lake contains sediment Cu values which are considerably greater than the 34 ppm median value for the combined Whitesail and Smithers map areas (Johnson et al. 1987a, b). Kuyakuz Lake, a large unstratified lake in the Nechako Range south of Vanderhoof, is not located near any known mineralization. However, Mo (median: 4.5 ppm; Fig. 6) and As (median: 9.2 ppm) sediment zoning patterns near a stream inflow at the western end of the lake suggest a possible buried source in the heavily forested region to the west, where Diakow & Levson (1997) reported a large buried intrusive body near Tsacha Mountain. Many of the small lakes to the west of Kuyakuz Lake have low pH values, contain elevated Al concentrations in water (up to 490 ppb; Cook et al. 1999), and receive upslope and upstream drainage from gossanous and clay-altered felsic volcanic flows. Sediment Mo (INAA) and As concentrations in the western part of Kuyakuz Lake reach 12 ppm and 20 ppm, respectively, and elevated Au and Zn concentrations also occur at some sites.

Seepage lakes. Seepage lakes and ponds receive groundwater or spring input predominantly from below the lake surface, and lack significant stream outflow. They are generally smaller than drainage lakes, and centre-basin geochemical results are more likely to be representative of the entire lake and watershed than sites within the more heterogeneous drainage lakes (Cook 1997a). As a result of this more uniform sediment distribution of metals, detailed sampling of seepage lakes may be a less useful tool for property-scale follow-up surveys of regional geochemical anomalies. Delineation of wa-

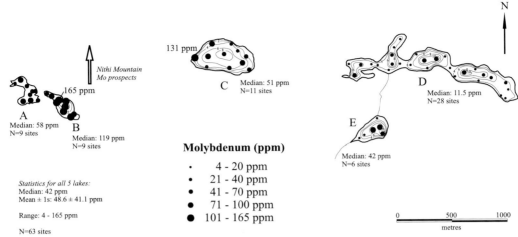

Fig. 7. Distribution of Mo (ppm) in sediments of five of the Counts lakes, located on the south side of Nithi Mountain in the Fraser Lake area, Nechako Plateau. The three westernmost lakes are downslope from numerous porphyry Mo showings on the south side of Nithi Mountain. Contours indicate sediment sample depth (metres).

tershed boundaries, followed by surface prospecting and till geochemical surveys, may be a more useful procedure in locating potential metal sources. Some examples of small seepage lakes in the Francois Lake region of central BC are shown in Figure 7. The five mostly eutrophic small lakes belong to the Counts lakes, an E–W trending chain of interconnected lakes and ponds occupying Counts lakes valley on the south side of Nithi Mountain near Fraser Lake. They are notable for their high sediment Mo concentrations (up to 165 ppm, ICP-ES), in a region where background Mo values in lake sediments are typically in the range of 1–5 ppm. The overall median Mo value for the five lakes is 42 ppm; median values for individual lakes are in the range 12–119 ppm. The highest sediment Mo concentrations are in the three westernmost lakes. These are directly downslope of numerous porphyry Mo prospects exposed on Nithi Mountain, where molybdenite occurs as disseminations and in narrow quartz veins.

Sample collection, preparation and analysis

Sampling methods

Critical to the success of any survey is the acquisition of high quality data, and central to this is proper sampling. Factors to consider include sample site density, collection procedures, sample size, suitability of sites within a lake, site duplicates and field observations. Although the discussion of lake water geochemistry is beyond the scope of this paper, water samples are usually obtained at all sediment sites during publicly funded surveys. These samples are collected from c. 15 cm below the lake surface using 250 ml polyethylene bottles. Water samples should be collected prior to the sediment to avoid inclusion of particulate matter and the risk of contamination. Collection of water samples concurrently with the sediment survey is recommended as data from these provide additional information on lake geochemistry at little additional expense (McConnell 1999b).

Sample site density: reconnaissance v. detailed. The question of sample site density should be addressed in the planning stages of the survey. The selected density is a function of both the expected size of the target anomaly and the number of samples desired to fall within it. A limitation to sampling density may be imposed by the natural distribution and density of the lakes themselves. In many areas of Labrador, for example, a practical lower limit is about one sample per 1–4 km^2, whereas in central BC it is about one sample per 8–10 km^2.

Mode of transport. Depending on the location, and the scope and scale of the survey, various modes of transport may be employed to conduct the sampling. In most cases, a float-equipped helicopter with a two-person sampling crew is the best solution. A Bell 206-B turbine helicopter with a fuel range extender and the rear door removed is perhaps the most satisfactory machine. Its low profile on the water and large rear passenger area make for rapid and easy manoeuvring for the person sampling. It has good visibility between the navigator in the front and the sampler in the back, which is essential for communication between the two. In mountainous terrain, however, it is prone to buffeting by winds and suffers from poor climbing ability relative to more powerful helicopters such as the Hughes 500.

For small, detailed-scale surveys, a wintertime survey using snowmobiles may be an economical alternative to a helicopter. One disadvantage is the inability to spot 'holes' or basins within lakes visible only from the air. In rare instances, sampling from a boat may be possible. In general, however, lack of road access and the time required for moving a boat between lakes make this method unattractive for regional surveys. Boats or canoes are, however, suitable for detailed-scale surveys of individual lakes (Cook 1995, 1997a).

Sampling devices: grab or core retrieval. The use of 'grab' rather than 'core' samplers is generally preferable as the time required to sample is shorter and the equipment less costly. Only where a detailed stratigraphic study of a lake's sediment is being made or where there is concern about otherwise unavoidable anthropogenic contamination (Johnson *et al.* 1986; Fortescue & Vida 1990) is the use of a core sampler required. For routine exploration or regional survey purposes, a sediment sample is collected at each site with a 'Hornbrook-type' sampler and placed in Kraft paper bag. The Hornbrook sampler, described by Coker *et al.* (1979), is a tubular, torpedo-type sampler of a hollow-pipe configuration. It is designed specifically for the collection of organic-rich lake sediments, is inexpensive and can be made by any well equipped machine shop. Standard sampling procedures have been discussed by Friske (1991) and Friske & Hornbrook (1991). The sampler, with an attached rope, is dropped from the float of the helicopter and allowed to penetrate into the sediment until it comes to a stop. A stainless-steel butterfly valve at the base of the sampler secures the contained sediment in place as the sampler is pulled to the surface. The 30–35 cm long slurry of wet sediment (volume *c.* 1100 ml) is then removed from the sampler using a plastic scoop, and placed in a large Kraft paper bag.

The Hornbrook sampler is designed to bypass the upper 10 cm of the sediment column (Friske 1991), with which modern anthropogenic contaminants and diagenetic processes are usually associated. In lakes with thick, soft sediment, several centimetres of the surface sediment will be ejected through the top of the sampler during impact, effectively removing recently deposited (and possibly contaminated) sediment from the sample data. In most cases, a sample of pre-industrial sediment is obtained from at least 30 cm below the sediment-water interface.

Another type of grab sampler is the 'jaw-type' or 'clamshell-type', in which a pair of hinged jaws snap together trapping the sediment. This has the disadvantage of also retaining the sediment from the surface of the lake bottom (i.e. the potentially contaminated layer). Advantages of the Hornbrook sampler relative to jaw-type samplers are: (1) speed of operation, with a sampling rate of at least 12–14 sites per hour being easily obtained; (2) avoidance of the upper part of the sediment column; and (3) robust and inexpensive construction.

Site location and sample size. The preferable sample site is often in the centre and usually the deepest part of the lake. Commonly the profundal basin can be seen from the helicopter as the lake is approached. In large or irregularly shaped lakes with isolated bays, the bays may be considered as separate catchment basins if they receive significant stream inflow. In contrast to centre-lake samples, near-shore samples may be composed largely of adjacent bank material that is unrepresentative of the catchment basin as a whole. These sediments are best avoided during regional scale surveys. Similarly, samples should not be collected from the centres of very large and deep lakes ($>10\,km^2$, $>40\,m$ deep), artificial reservoirs, or from swamps and bogs. Much of the source of the metals in lake sediment in glaciated terrain are the glacial deposits themselves. If these deposits are not derived locally, the lake sediment geochemistry at that site is unlikely to give a strong reflection of local bedrock. In some areas, extensive glaciogenic sand and clay plains derived from distal sources may mask the local bedrock signal and render lake sediment surveys ineffective. In coastal regions, extensive areas of glacial marine overlap may exist, complicating the glacial stratigraphy. In Labrador for example, sea level in the recent past was in excess of 100 m above present (McConnell 1999b). As a result, valleys far inland may be underlain by marine clays. An understanding of the history of glacial deposition is critical for successful application of lake sediment geochemistry. In many areas surficial geological maps are available, often in digital format, to identify potential problem areas.

For most elements, only a few grams or less of material are required for analysis. For some elements, however, notably Au, at least a 10 g sample for analysis is recommended. Ideally, and particularly if Au is the target, an orientation survey will be conducted over a known deposit in the area to determine the character of the associated anomaly, the applicability of standard-sized sediment samples, and sampling variability. Typically, fine-grained Au deposits give a more consistent pattern in lake sediment than do coarse-grained deposits where the nugget effect becomes a factor. In central BC, for example, results of paired standard and larger sized sediment samples from a lake near the Clisbako epithermal Au prospect showed there to be no significant difference in Au concentrations of the two field sample sizes (Cook 1997b). In this case, there was no evidence to suggest that use of larger, more costly, sediment samples would be more effective indicators of fine-grained epithermal Au mineralization than standard sized samples.

Site duplicates. An important component of assessing data quality and representivity is the inclusion of pairs of site, or field, duplicate samples as a routine component of a survey. Usually one randomly selected site in 20 is sampled in duplicate (Fig. 8), in accordance with established NGR and provincial protocols. Collection procedure of duplicates involves collecting two samples from the same lake, perhaps 50–100 m apart. Suitable coding parameters on the field data sheet should allow for ready identification and comparison of data from duplicate pairs at the interpretation stage. Data from such site duplicates are used to provide an assessment of total variability of the sampling and analytical phases of the survey.

Field observations. Several parameters pertaining to sample medium, site and local terrain should be recorded at the time of sampling to facilitate interpretation. These parameters include site coordinates, water depth, sediment colour and composition, general relief and potential sources of contamination (Garrett 1974). These are best recorded in numeric formats easily processed by computer. If sampling by helicopter, the navigator member of the sampling team records field data while the sampler is collecting from the float. It is critical that the sampler be in headphone communication with the navigator and pilot to pass on his observations or instruct the pilot to move to a more suitable spot. Of singular importance are the site coordinates, the most commonly used of which are those of the Universal Transverse Mercator (UTM) system or latitude and longitude. The actual site

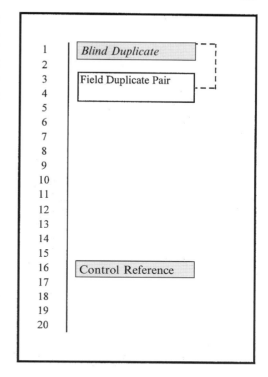

Fig. 8. Typical block of 20 sample collection scheme used in regional lake sediment geochemical surveys conducted to NGR and RGS standards. The sampling block incorporates 17 routine samples, a site duplicate, a blind duplicate and a control reference standard.

coordinates can be determined from a Geographic Positioning System (GPS), now almost standard equipment in helicopters, or from a 1:50 000 scale topographic map. In the latter case, field site locations are transferred to master basemaps and then digitized to obtain UTM site coordinates.

Water or sample depth provides useful information and often correlates with trace element content. It is easily determined by marking the sampler retrieval rope at 1 m intervals with a permanent felt-tip marker. Distinctive coloured markings every 5 m facilitate observation. A float-mounted depth-sounder may also be used to record water depth, which may vary slightly from sample depth. Sample colour is easily noted and may be a useful parameter. Similarly, sample texture/composition is useful, particularly to distinguish unusually peat-rich or clastic-rich samples from the more ideal 'ooze' or gyttja compositions. Lake area can be measured by overlaying, on the topographic map, a transparent plastic grid divided into 100 m cells. The cells underlain by the lake are counted and recorded. In some surveys, elevation may be important to record. This can be easily noted from the helicopter's altimeter. Any suspected contamination of the site or lake should be recorded. Observations might also include abnormal sample locations (e.g. near-shore because a mid-lake sample could not be obtained), the presence of an esker complex, or a new gossan observed in the catchment basin. Variables such as lake area and site geology, which reflect the dominant bedrock geological unit of each lake catchment, are coded after sample collection.

Sample drying and storage. Sediment is conveniently stored in water-resistant Kraft paper sample bags. At the end of each sampling run, the samples should be laid out on a chemically 'clean' area to drain, then strung together and left to air dry for a few days before transporting to a laboratory. A large ventilated drying tent with mesh racks and a source of heat may employed for drying samples in cool or wet weather. If moved when too wet, the bags may split apart or the contents may squeeze out causing cross-contamination. When dry, profundal sediment samples typically form small, hard, dense brick-like aggregates, weighing 5–10% of the wet sample weight.

Preparation and analysis

Sample preparation. Regardless of the analytical method(s) used, the recommended sample preparation procedures prior to analysis are similar. Samples from the field are oven-dried at 40°C (max) or thoroughly air-dried, and then are either ball-milled or disaggregated in their entirety (max: *c.* 250 g) using a ceramic ring mill. The resulting powder is screened with a 177 μm (80 mesh) sieve to remove any remaining aggregate material, and the fines retained for analysis.

Quality Control. The ability to discriminate genuine geochemical trends from those resulting from sampling and analytical variation is of considerable importance in the interpretation of geochemical data. Variations in element concentrations due to regional geological and geochemical variations may be the result of different bedrock lithologies and surficial materials, the absence or presence of mineralization, or limnological variations. Good analytical precision, readily obtainable by modern instrumental analytical methods, is of limited importance if the sample collection and preparation error is indistinguishable from the regional geochemical variation. Accordingly, a rigorous quality control program should be maintained to monitor the precision and accuracy of regional survey analytical results. In addition to the field duplicates described above, it is recommended that every batch of 20 analyses (Fig. 8) include a pair of randomly selected laboratory duplicates to measure analytical precision. A duplicate is prepared by splitting a sample pulp from the ball-milled and sieved material. Each block of 20 should also include a control standard of known composition to assess analytical accuracy. The best are standards prepared from bulk samples of real lake sediment material: CANMET (Natural Resources Canada) in Ottawa markets the series LKSD 1–4 established by the GSC (Lynch 1990, 1999). Sample blocks are routinely re-analysed if control and blind duplicate sample results exceed preset tolerances. In BC, repeat INAA analyses are conducted on, and reported for, all samples with Au concentrations in the top ten percentile of the data.

Analytical methods. Analysis of routine lake sediment and water samples are conducted by in-house or contract laboratories in accordance with established NGR and, in BC, RGS methods and procedures. These methods are carefully monitored to ensure consistent and reliable results from survey to survey regardless of the region, year or laboratory. In these regional surveys, two analytical splits are taken for geochemical analysis: one subsample (*c.* 20–30 g) for determination of Au and a suite of additional precious metal pathfinder and REEs by INAA, and a second, smaller subsample for determination of Zn, Cu, Pb, Ni, Ag and several other trace elements by either atomic absorption spectroscopy (AAS) or ICP-ES, following an aqua regia digestion. ICP-ES and ICP-MS are the standard analytical tools in use today. They provide low cost, multi-element data over a wide concentration range. INAA provides a relatively low cost and non-destructive method of multi-element analysis with sufficiently low detection limits for Au (1–2 ppb) as well as associated pathfinder elements such as As, Sb and W. It also handles up to 30 g sample weights, allowing for a more representative sample.

In most cases, the choice of analytical method is less critical with lake sediment than with rock, soil or stream sediment as most elements of interest are more loosely bound than in other media. In lake sediment, it is thought that most metals are transported to the lake in solution, as weak complexes with organic matter or as very fine clastic material; thus a wet chemical attack such as aqua regia can be considered as 'total' for most metals of interest in exploration.

Another variable which is useful when evaluating lake sediment data is the organic carbon content, commonly reported as the surrogate loss-on-ignition (LOI). Many elements have a strong positive or negative correlation with organic matter and their signatures can be levelled by regression techniques. Other elements may be added for specialized surveys. For example, F is useful for granitoid-related exploration and is not included in all multi-element packages. Complete details of analytical procedures and associated quality control methods for any given provincial or federal survey are provided in the relevant open file reports (e.g. Wagenbauer *et al.* 1983; Cook *et al.* 1997, 1998; Finch 1998).

Data interpretation and presentation

The goal of a lake sediment survey is the successful identification of geochemical patterns related to mineralization or mineralizing processes. Before this final step is undertaken, the data must be assembled into a coherent database (in which all the parameters recorded for a given sample are associated with a single record) that can be queried, manipulated, sorted, massaged and finally plotted, usually in conjunction with other geo-referenced databases.

Statistical analysis

Univariate statistics. Although univariate statistics can be computed by calculator or even manually, the availability of microcomputers has greatly simplified their calculation. Statistics software packages can vary from high powered versions (SPSS, SAS, MINITAB and SYSTAT) to ones that provide adequate calculations and are available as freeware or shareware. In addition to cost, factors to consider in choosing a package include ease of use, availability of support and compatibility with other software you may be using such as spreadsheets and GIS programs. Univariate statistics provide measures of the distribution characteristics of a given element or variable, such as the mean, median, range and standard deviation. The mean and median are measures of central tendency of a population. The median is the central value and is a more robust statistic than the mean, the value of which can be strongly influenced by the presence of a few extreme values. The standard deviation measures how closely the population clusters around the mean and is sometimes used as a rough method of separating 'background' samples from 'anomalous'; other techniques discussed below may be preferable. The median, for example, is a good approximation of 'background' values for sites draining any given region or geological unit. The distributions of most trace elements in lake sediment more closely approximate log distributions than arithmetic. In these cases, the calculation of the geometric mean and log standard deviation is preferable.

Multivariate statistics. Although univariate statistics provide a simple and useful approach to characterizing the distribution of an element, they treat the element in isolation. However, the distribution of an element of economic interest may be influenced by the distribution of other elements or may be associated with non-economic elements that give a stronger pattern. A variety of multivariate statistical tools are available to measure the interrelationship of elements.

Correlation analysis gives a measure of the tendency for pairs of elements or variables to vary sympathetically or inversely. For example, a common feature in lake sediment geochemistry is that elements such as Zn, Co, Ni, As and others correlate strongly with Fe and/or Mn and/or organic matter. Correlation analysis provides a method to identify 'false' anomalies due to scavenging effects. Correlation coefficients vary from +1 (perfect correlation) to −1 (perfect inverse correlation). In general, and not to be applied rigorously, correlation coefficients of < |0.6| suggest that the effects of oxide scavenging are too weak to disguise the primary geochemical signal (i.e. plots of the raw data should provide a satisfactory guide to element distribution in the landscape). Values > |0.6| may call for remedial measures such as the calculation of residuals, using regression analysis. Various types of correlation coefficients can be calculated. Although they are broadly similar, calculations that employ ranking procedures, such as the Spearman correlation, make no assumption about the distribution nature of the population (e.g. log v. arithmetic) and are recommended. Aside from highlighting oxide scavenging effects, mineralization-related element associations may be identified. Two examples include the frequent association of Sb or As with Au mineralization. In some areas, the Au pathfinder provides a broader or stronger dispersion pattern around mineralization than does Au. Such correlation may also provide information about the nature of the targeted mineral deposit itself.

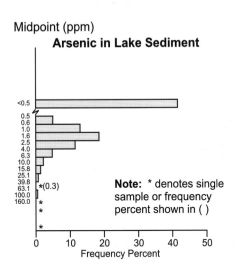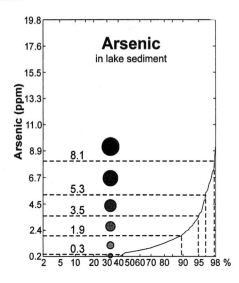

Fig. 9. Histogram and cumulative frequency plot of As in lake sediment, $n = 578$ (after McConnell 1999b).

If correlation analysis indicates that the primary signal of an element is being obscured by non-mineralization factors such as oxide scavenging, regression analysis may remove much of the 'environmental noise' and permit the primary signal to be plotted. In essence, regression analysis is a procedure in which the equation of a best-fit line is calculated between the target variable and one or more 'noise-inducing' variables such as Fe, Mn, LOI, and depth. The equation can then be used to calculate residual values of the target variables (i.e. the value remaining after the component caused by element association has been removed). Most statistical packages include regression calculation procedures. Ideally, they also calculate residuals and provide a means of exporting them in a format readable by a plotting package.

Factor analysis provides another approach to group data by element association. Factor analysis is a statistical approach that can be applied to identify relationships among several variables and to explain these variables in terms of their common underlying dimensions (factors). This statistical approach consists of reducing the information contained in a number of original variables into a smaller set of dimensions (factors) with minimal loss of information. In a geochemical survey, the statistical program could be asked to produce a range of factor models, perhaps containing from 3–7 factors, possibly reducing 40–80 variables to a manageable number. One would then apply geological knowledge of the area to interpret the resulting element groupings in the factors in terms of bedrock type, mineralization, surficial processes, etc. For example, a factor that is dominated by loadings on As and Sb with minor loadings on Au might emerge from survey data collected over a greenstone belt, suggesting a Au mineralization factor. A plot of this factor might give a more useful pattern than any single element alone. Somewhere else, a Ni–Cr factor might emerge that would map out ultramafic bedrock in an area with poor exposure.

Graphical aids. Histograms provide a rapid view of the shape of a distribution; however, depending on the intervals used to group the data, population breaks or sub-populations can be easily missed. More informative are cumulative frequency plots (CFPs). Using specialized plotting paper or a statistical program such as UNISTAT (Nolan 1990), data from a normal population plot as a straight line. Sub-populations will cause the line to bend and these inflection points can be used to recognize population breaks and are useful for assigning intervals on contour or symbol plots. Often, however, these inflections are too subtle to be seen in log plots and can be more readily recognized in arithmetic plots despite the overall curve of the data. Examples of a histogram and CFP are shown in Figure 9 where data from Labrador for As in lake sediments are plotted.

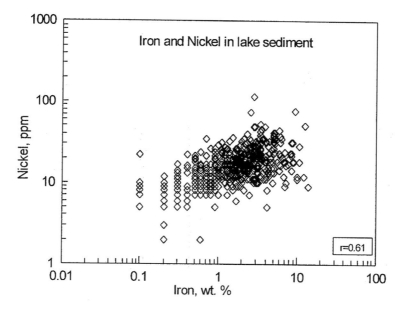

Fig. 10. Scatterplot of Ni and Fe in Labrador lake sediment, $n = 578$ (McConnell 1999b).

The histogram shows a population skewed to the right even after log transformation; the only apparent population break occurs at about 30 ppm. The CFP shows more detail of the distribution and indicates other breaks at 5.3 ppm and 8.1 ppm. When these breaks are used to define intervals for a 'dot plot' map, the top two groups correlate with lithology: 11 of the 12 samples with As > 8.1 ppm, and 6 of 8 samples with As between > 5.3 and 8.1 ppm lie in areas underlain by greenstone belts.

Relationships between elements known to be correlated can be shown graphically in scatterplots or X–Y plots, in which one variable is plotted along the X-axis and the other along the Y. An example is shown, from a high-density Labrador lake sediment survey, of a plot of Ni against Fe with a correlation coefficient of 0.61 (Fig. 10). The graph indicates a moderately strong relationship between Fe and Ni suggesting the presence of oxide scavenging.

Maps and data presentation

The final product of most lake sediment surveys is the presentation of the data as some form of map. Increasingly, explorationists recognize the benefit of integrating geochemical data with other geo-referenced data such as drainage, topography, bedrock geology, surficial geology, geophysics and mineral occurrences. Several commercial PC-based geographical information systems (GIS) are available to manipulate and display the various data sets. The geochemical data can be presented as various types of layers: point, line or interpolated surface depending on the information sought (Davenport et al. 1994; Bonham-Carter 1994). Figure 11 is an example of a symbol plot of As from the high-density lake sediment survey in Labrador discussed previously is shown together with 1:250 000 drainage, geology, mineral occurrences, coastline, NTS boundaries and an area exempt from staking (McConnell 1999b). In the case of Newfoundland's high-density surveys and BC's RGS, page-size element distribution maps are created using GIS technology. Results are released to the public in booklet form, with survey documentation, interpretation, data listings, summary statistics, element distribution maps, sample location maps, and digital data on disk or CD-ROM (e.g. Cook et al. 1997, 1998; McConnell 1999a, 2000). The regional survey data in Newfoundland and in Labrador are available as interactive CD-ROMs where the user can exercise control over the plotting of various layers of geochemical, geological, geophysical and topographical data with a simple GIS viewer provided (Davenport et al. 1999a,b).

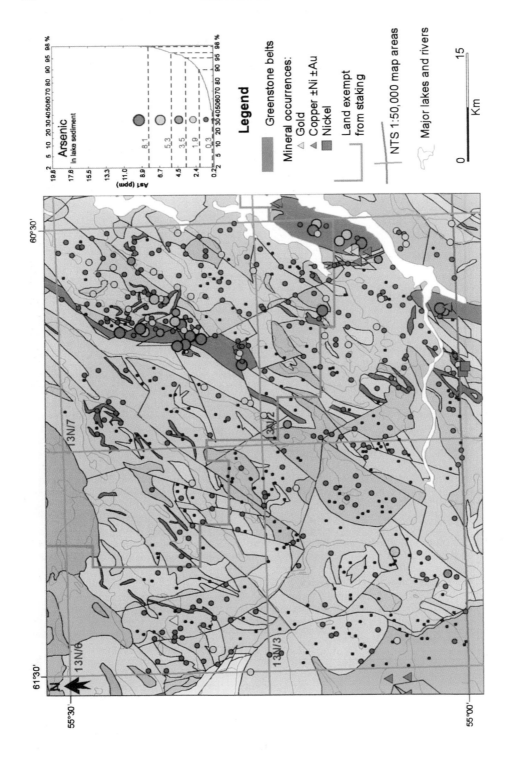

Fig. 11. Arsenic in lake sediment in high-density Labrador survey (after McConnell 1999b).

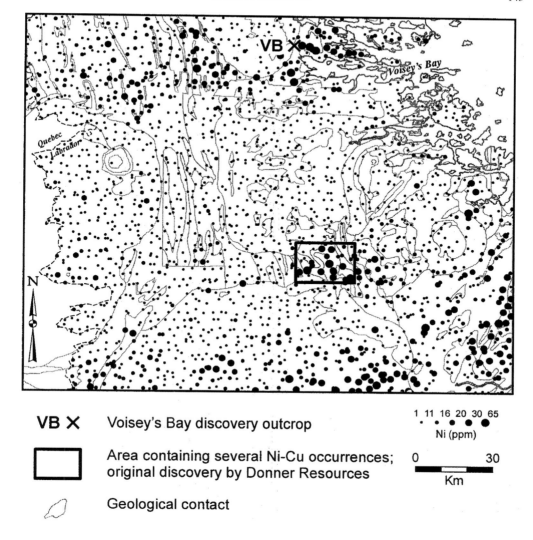

Fig. 12. Nickel in regional Labrador lake sediment survey, and locations of Voisey's Bay and Donner discoveries.

Examples from the Shield and Cordillera

Shield deposits

The examples discussed here are drawn from the Churchill, Nain and Superior Provinces in Labrador because these are areas with which the authors are most familiar. However, numerous instances of mineralization having associated lake sediment anomalies can be taken from Shield areas across Canada, including: Key Lake (U; Parslow 1981) and the Au deposits of the La Ronge belt (Coker *et al.* 1982; Chapman *et al.* 1990) in Saskatchewan; the Hemlo Au camp (Friske 1991) and Percy Lake Zn–Pb–Cu–Ag prospect (Hamilton *et al.* 1995) in Ontario; the Eastmain Au–As showing in northern Quebec (Davenport *et al.* 1997); the East Kemptville Sn deposit in Nova Scotia (Rogers & Garrett 1987); and the Lupin Au deposit (Friske 1991) in the NWT.

Ni–Cu deposits. Reconnaissance lake sediment and water surveys in Labrador during the 1970s and 1980s provide data from over 18 000 sites. The area in which the Voisey's Bay Ni–Cu–Co deposit was discovered was surveyed in 1985 and the data released in 1986 (Friske *et al.* 1993).

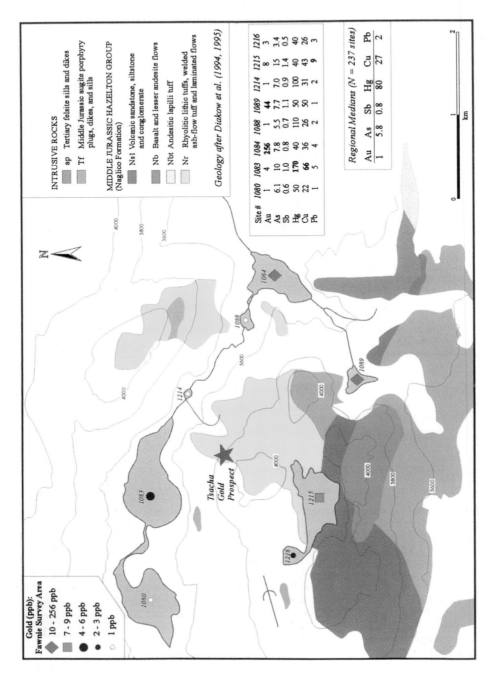

Fig. 13. Locations of regional lake sediment survey sites with elevated Au concentrations near the Tsacha epithermal Au prospect, southern Nechako Plateau, BC. Regional geochemical data from Cook & Jackaman (1994); elements in bold type exceed 95th percentile for that survey (Au, Hg in ppb; others in ppm). Geology after Diakow et al. (1994, 1995).

elements and exploration using centre-lake bottom sediments from the East Kemptville area, southern Nova Scotia, Canada. *Journal of Geochemical Exploration*, **28**, 467–478.

SCHMIDT, R. C. 1956. *Adsorption of Copper, Lead and Zinc on some Common Rock Forming Minerals and its Effect on Lake Sediments*. PhD Thesis, McGill University.

SPILSBURY, W. & FLETCHER, W. K. 1974. Application of regression analysis to interpretation of geochemical data from lake sediments in central British Columbia. *Canadian Journal of Earth Scientists*, **11**, 345–348.

TIMPERLEY, M. H. & ALLAN, R. J. 1973. The formation and detection of metal dispersion halos in organic lake sediments. *Journal of Geochemical Exploration*, **3**, 167–190.

WAGENBAUER, H. A., RILEY, C. A. & DAWE, G. 1983. Geochemical laboratory. In: *Current Research 1983*. Newfoundland Department of Mines and Energy, Mineral Development Division Report **83-1**, 133–137.

WEBBER, T. N. 1984. *A Reconnaissance Survey of Hill-Tout Lake*. British Columbia Ministry of Environment, Lands and Parks, unpublished Fisheries Branch report.

WETZEL, R. G. 1983. *Limnology*. Second edition. Saunders, Philadelphia.

Biogeochemical exploration methods in the Canadian Shield and Cordillera

COLIN E. DUNN

Geological Survey of Canada (Emeritus), 9860 West Saanich Road, Sidney, British Columbia V8L 4B2, Canada
(e-mail cdunn@nrcan.gc.ca or colindunn@home.com)

Abstract: This review article focuses on field methods in biogeochemical exploration and is based largely on the author's experience[1]. Consideration is given to reasons for applying biogeochemical methods as alternatives or supplements to other surficial sampling media that can be used in the exploration for mineral deposits in glaciated terrain. Extensive root systems can absorb metals from the substrate and integrate the geochemical signature of large volumes of sediment, groundwater and sometimes bedrock, thereby providing a more representative reflection of the chemical environment than that obtained from some other media. Sampling methods and precautions that should be taken are outlined. Variables that govern plant chemistry include the heterogeneity of composition among plant species and plant tissues, and the modifying effects of the seasons and contamination from external sources. Studies indicate that biogeochemical methods can provide a more proximal indication of concealed mineralization than the distal indications typical of till geochemistry programmes. Consequently, comparisons of till and biogeochemical data can help to define vectors toward mineralized sources such that the two methods are complementary.

The underlying rationale for applying biogeochemical methods to mineral exploration is that trees and shrubs absorb metals present in the ground and transfer these metals via their root systems to the growing plant. Metals are absorbed from soil, from groundwater, and locally from bedrock in those environments where roots penetrate faults, joints and along cleavage planes. In geochemical exploration, the significant advantage of applying biogeochemical methods is that the root system of a large tree may penetrate through many cubic metres of the substrate. Consequently, the root system can integrate the geochemical signature of a large volume of all soil horizons, the contained groundwater, and bedrock where it is covered by only a few metres of overburden. Conifers have fairly shallow root systems compared to those of deciduous species. However, depth of root penetration is not critical for a biogeochemical response, because root systems can access elements that migrate upward from considerable depth in solution, by diffusion, by capillary action, from microbubbles or nanoparticles, and in electrochemical cells. In general, the more arid the environment, the deeper that roots of trees and shrubs can penetrate into the substrate, while continually drawing nutrients and other elements. The biogeochemical signature is an integrated signature of the substrate.

Those who decide to test the application of biogeochemical methods to exploration have commonly had some experience in other geochemical methods. Usually, this experience has included soil or till sampling, and therefore some preconceived ideas may have developed as to interpretation of results. However, just as two types of geophysical surveys may provide different types of information, so will two types of geochemical surveys, especially in terrain with transported overburden.

Many texts suggest that for biogeochemical

---

[1] Much of this account and the Tables are abstracted from Dunn *et al.* 1993 and Dunn 1995. Case histories can be found in these references and other texts such as Kovalevsky (1987). The present text focuses on methods, building upon the extensive pioneering studies of the late Professor Harry Warren and his colleagues at the University of British Columbia (e.g. Warren & Delavault 1950; Warren *et al.* 1968).

From: McCLENAGHAN, M. B., BOBROWSKY, P. T., HALL, G. E. M. & COOK, S. J. (eds) 2001. *Drift Exploration in Glaciated Terrain*. Geological Society, London, Special Publications, **185**, 151–164.
0305-8719/01/$15.00 © Geological Society of London 2001.

exploration to be successful there should be a high correlation between the metal content of the soil and that of the plant. This is a valid concept for some parts of the world where there are residual soils. However, a good positive correlation between plant and soil chemistry may not always occur, especially where transported overburden such as wind-blown loess or glacially derived deposits have been deposited on top of mineralized bedrock. Glacial dispersal of elements in the ore minerals may be modified further by element migration from an ore source through groundwater movements. In such environments, plant roots may absorb elements from zones beneath the development of soil, or extract elements dissolved in groundwater. In the latter situation, the physicochemical environment of the soil may not be conducive to element adsorption from groundwater, whereas plant roots can absorb elements in solution and concentrate them in their tissues. A further complication is that many plants establish barriers to metal uptake so that the metal content of a plant may not be proportional over a wide range of concentrations to the metal content of the soil (Kovalevsky 1987, 1995).

Whereas plant:soil coefficients can be established in laboratory experiments, the real world is rarely that simple. In attempting to determine the relationship between the chemistry of the soil and that of a tree, the usual procedure is to collect a bag of soil and a bag of tree tissue. However, there arise some fundamental questions:

- which soil horizon should be collected?
- which size fraction of the soil should be analysed?
- which type of tree tissue (and from which part of a tree, top or bottom, north or south) should be collected for comparison with the underlying soil?

Typically, each soil horizon has a different metal content, as does each size fraction of that soil. Similarly, each vegetation tissue type has a different ability to collect and store metals, and concentrations of living tissue change with the seasons. The problem is compounded by the fact that a soil sample is usually no more than a handful of a single horizon, and as such represents a minuscule sample compared to the volume of material sampled by the root system of a large tree.

Table 1 shows correlation coefficients (r) between Au and As concentrations in bark ash and in different soil horizons. The data show that for both tree species the correlation between the bark and C-horizon soil is stronger than between the bark and other soil horizons. The implication is that the root systems of these trees have absorbed more Au and As from the C-horizon than the other soil horizons and transferred these metals to the trees' extremities.

First considerations

Since all plant species have different chemical composition, some prior knowledge of metal accumulation characteristics of these different species is required before a successful biogeochemical survey can be undertaken. Interpretation of results also requires some understanding of the chemical requirements of the many plant species. For example, Zn is an element essential for plant metabolism and therefore high concentrations do not necessarily indicate the presence of mineralization. Conversely, Au is not required for plant growth so the presence of high Au concentrations in plant tissue may indicate enrichment in the substrate,.

Many factors are involved in distributing metals among the diverse components of soils and trees. Some comprehensive accounts are given by Brooks (1983); Kovalevsky (1987); and

Table 1. *Correlation coefficients (r) between Au and As in bark ash and underlying soil horizons near the Nickel Plate gold mine, Hedley, southern British Columbia*

Soil Horizon	Douglas-fir Bark $n = 12$		Engelmann Spruce Bark $n = 13$	
	Au	As	Au	As
Forest Litter	0.13 (ns)	0.10 (ns)	0.48 (ns)	0.58 (s)
A - Horizon	0.63 (s)	0.63 (s)	0.65 (s*)	0.65 (s*)
B - Horizon	0.60 (s)	0.55 (s)	0.79 (s**)	0.80 (s**)
C - Horizon	0.76 (s*)	0.64 (s)	0.90 (s**)	0.88 (s**)

ns, not significant (P > 0.05); s, significant (0.05 > P > 0.01); s*, highly significant (0.01 P > 0.001); s**, highly significant (P < 0.001)

Table 2. *Element concentrations in the ash of tree tissues from a single location near Au mineralization at Doctor's Point, Harrison Lake, southern British Columbia*

Species	Tissue	Au ppb*	As ppm**	Mo ppm	Sb ppm
Douglas-fir	Twig	35	1600	<1	1
Douglas-fir	Needle	23	130	<1	2
Douglas-fir	Bark	53	250	<1	8
Western hemlock	Twig	200	710	<1	8
Western red cedar	Twig	7	11	4	1
Western red cedar	Needle	5	6	<1	1
Western red cedar	Bark (all)	8	12	<1	1
Western red cedar	Bark (outer)	31	46	<1	11
Red alder	Twig	14	4	57	0.5
Red alder	Bark	<5	4	4	0.3
Douglas maple	Twig	12	6	4	1

*ppb, ng/g (μg/kg); **ppm, μg/g (mg/kg). For this text, all concentrations are shown as ppb or ppm.

Table 3. *Variation of Au concentrations within single jack pine and black spruce trees; Rottenstone PGE-gold mine, northern Saskatchewan*

	Au (ppb) in:	
	Dry Tissue	Ash
Jack Pine		
Outer bark	2.10	140
Inner bark	0.61	32
Needles	0.36	15
Young twigs	0.36	24
Old twigs	0.15	17
Outer trunk wood	0.08	32
Inner trunk wood	0.04	14
Black Spruce		
Outer bark	0.90	50
Twigs	0.62	28
Trunk wood	0.09	19
Needles	<0.16	<5

Brooks *et al.* (1995). In many situations the information supplied by the tree chemistry is different from that derived from the soils or glacial deposits; each provides its own 'layer' of geochemical information in the same way that different geophysical measurements provide different information on the physics of the Earth.

There are great differences in the uptake of metals by different species of plant. Table 2 shows, from single analyses, the variations that occur in trees rooted in the thin drift cover that overlies Au mineralization at Doctor's Point, on the west side of Harrison Lake, British Columbia. Note in particular the wide range in concentrations of As.

Within a single tree, there are substantial differences in the element content of its various components. An example of this is given in Table 3, which shows significant differences in the Au content of two tree species, jack pine and black spruce, common to the northern Cordillera and the boreal forests of Canada. These examples, from single trees close to the abandoned precious metal (PGE and Au) Rottenstone mine in northern Saskatchewan, show that for these species, the highest gold enrichment is in the outer bark. The outer bark is an appropriate sample medium for many species and many types of mineralization.

Another example of element variation within a single tree is taken from a site close to the Sullivan Pb–Zn mine at Kimberley, southern British Columbia. Table 4 shows the distribution of metals in a single lodgepole pine rooted in outcrop of tourmalinite associated with base metals on North Star Hill. Remarkably high concentrations of Pb, Zn, Cd, Ag, and Mn occur in the roots. However, Ni, Cu, B, and Cs are at their highest concentrations in the treetop. This substantial variation within a single plant should be appreciated before commencing a sampling program – it emphasizes the importance of being consistent in collecting a similar amount of the same tissue type at each sample station.

Unless dead tissue (such as bark) is to be sampled, seasonal changes in plant chemistry must be considered. Table 5 shows the substantial changes of Au concentrations in alder that occur during the year. All samples were from mature shrubs, from which the most recent three years of growth was collected on each sampling occasion. At each of the 19 sample sites, the changes in values were consistent, i.e. in early August all of the shrubs recorded lower Au concentrations than in early June, and the pattern remained the same for the other sampling occasions.

Table 4. Concentrations of elements in the ash of tissues from a single lodgepole pine above mineralized tourmalinite, North Star Hill, near the Sullivan Pb–Zn mine, Kimberley, British Columbia

	Top Stem	Lower Twigs	Outer Bark	Roots
Ag ppm	1	3	13	77
As ppm	9	9	52	190
Au ppb	< 5	< 5	20	19
B ppm	1150	400	260	580
Ba ppm	48	310	1000	500
Cd ppm	52	95	143	135
Cr ppm	6	18	18	10
Cs ppm	110	9	5	38
Mn ppm	13000	27000	4230	63000
Ni ppm	180	22	14	24
Pb ppm	150	2950	4900	16400
Sb ppm	2	3	13	5
Zn ppm	6100	7350	5700	12800

Table 5. Average concentrations of Au in the ash of alder twigs collected from the same 19 shrubs at different times of the year – La Ronge Belt, Saskatchewan

Season	Gold (ppb) in ash
Early Summer	28
Late Summer	10
Early Fall	17
Spring	69

Each plant species exhibits its own variations in different elements throughout the year. Consequently, a survey using live tissues should be conducted in as short a time as possible (e.g. within a 2 to 3 week period), because metal concentrations in a tree sampled in the spring will be different from those in the same tree during the summer.

Selection of plant species for sampling

Orientation

The first step in preparing to conduct a biogeochemical survey is to look for the most widespread species within the area of interest, then determine from published information if is likely to be informative (i.e. do the easy-to-reach parts accumulate the elements of interest, and are they known to correlate with the presence of ore?). If published information is not available, an orientation survey should be conducted (Rose et al. 1979). The survey area should always be reconnoitred prior to sample collection to assess which species are dominant. If time permits, an orientation survey should be undertaken to collect a few samples of common species and submit them to a laboratory for multi-element analysis. This will provide preliminary insight to the levels of metals present and the species that might be of greatest value for a survey.

Sometimes one species is not sufficiently widespread to sample at every survey station. It is not uncommon for a proposed survey area to extend through several vegetation zones. In the temperate forests of western North America there may be, for example, Douglas-fir at low elevations, with lodgepole pine on surrounding slopes, and sub-alpine fir at higher elevations. In this situation, it is necessary to establish overlap zones where two species grow, so that samples of both species can be collected at a few sites. By establishing the relative element concentrations of pairs of species it is possible to normalize the data to a common benchmark. However, because of the difference in element requirements and tolerances among species, this normalization commonly proves to be a rather crude process.

Temperate coniferous forest

The diversity of flora in areas of temperate forest is, at first sight, quite daunting. The dominant north–south trend of the mountain ranges and valleys in western North America gives rise to different climatic zones within which there are considerable differences in rainfall and temperature. The result is a series of N–S trending zones of vegetation that have been defined as biogeoclimatic zones, i.e. an ecological classification based on vegetation, soils and climate (Pojar et al. 1987). Each zone has its characteristic flora, and locally, within a distance of only a few hundred metres, because of change in elevation or aspect there may be a different assemblage of trees.

Throughout the temperate forests, the conifers are in general the most useful and chemically informative plant types, especially: lodgepole pine (*Pinus contorta*); Pacific silver fir (*Abies amabilis*); sub-alpine fir (*Abies lasiocarpa*); western hemlock (*Tsuga heterophylla*); mountain hemlock (*Tsuga mertensiana*); Engelmann spruce (*Picea engelmannii*); Douglas-fir (*Pseudotsuga menziesii*); western redcedar (*Thuja plicata*); western larch (*Larix occidentalis*).

In different climatic settings there occur other species of pine, spruce, fir, larch, yew and cedar. Of the many deciduous species that may occur, alder, birch, maple, willow and poplar are the most common. Temperature and moisture regime and the elevation guide the choice of sample medium. In coastal regions at moderate to low elevations Douglas-fir, western red cedar, and western hemlock are three of the most common tree species. Farther inland and at higher elevations, lodgepole pine (one of the most useful trees), Pacific silver fir, mountain hemlock, sub-alpine fir, and Engelmann spruce are most common. In the drier interior, there is Ponderosa pine, characterised by its soft, thick orange to grey bark that falls off in large scales.

Temperate deciduous forest

Biogeochemical surveys in deciduous forests are commonly more difficult to conduct than those in the coniferous forests. Typically, the deciduous forest has a diverse flora and inconsistent distributions of species such that the desired species may be absent from many of the proposed sample stations. Commonly, tree bark is smooth, branches are out of reach, and growth increments are irregular and more difficult to discern than those of the conifers. Furthermore, many tree species (poplar, maple, oak, elm, beech, red alder and willow) take up relatively low concentrations of the elements that are of interest for locating mineralized bedrock. In general, deciduous trees will generate biogeochemical anomalies that are of low magnitude, and therefore quite difficult to detect. Leaves may be more informative than the twigs, but surveys should be conducted within a short time frame because of the rapid changes in leaf chemistry that take place with time.

Growing amongst the trees, there is often an understorey of shrubs and bushes that may be both more informative than the trees and certainly more practical to collect. Shrubs that have been used with some success include: mountain (or green) alder (*Alnus crispa*); beaked hazelnut (*Corylus cornuta*); and, red-osier dogwood (*Cornus stolonifera*). Locally, there may be many other possible choices of shrub that could be used. Since comprehensive databases on many plant species are not available, when confronted with a field situation where data are lacking for a widespread species, an orientation survey to sample that species (or several species) should be conducted. Provided appropriate precautions are taken to collect twigs of similar age of growth from the same species at each test site, the sensitivity of a plant to reflect mineralization or different substrates and environmental conditions can be determined. Although there are some plant species that are better than others for biogeochemical surveys, there are few that provide no information of value when chemically analysed.

Boreal (northern) forest

Large tracts of the northern latitudes are covered with boreal forest. Fortunately, there are only a few common tree species that are likely to occur in sufficient abundance to be of use for a biogeochemical survey. This makes the boreal forest one of the most favourable regimes for biogeochemical prospecting. The most common conifers are spruce, pine, fir, and tamarack. Birch and aspen are the dominant deciduous trees and the most common shrubs are alder, willow and Labrador tea. A short list of the most favourable species for biogeochemical prospecting would include black spruce (*Picea mariana*), jack pine (*Pinus banksiana*), balsam fir (*Abies balsamea*), mountain alder (*Alnus crispa*), and Labrador tea (*Ledum groenlandicum*). There are few regions of boreal forest that do not contain one or more of these species.

Sampling

Rings or metal jewellery should not be worn whilst collecting or handling biogeochemical samples, because they may contaminate the samples and generate false anomalies. Sampling procedures are mostly simple, but before conducting a survey a number of precautions need to be taken. The basic rule is to 'be consistent', collect the same type of plant tissue and the same amount of growth, all from the same species, and try to collect from trees of similar appearance and state of health.

Field accessories

The only additions to the usual field equipment

of the geologist are:

- a pair of anvil-type pruning snips, preferably Teflon-coated;
- a paint scraper or hunting knife for scraping bark, and either a dust-pan or paper bag for collecting the flakes of bark (a hatchet is useful for surveys involving collection of thick bark, such as that of Douglas-fir);
- 'kraft' brown paper soil bags for bark samples; for twigs use fairly large bags (about 20×30 cm) made either of heavy duty coarse brown paper if conditions are dry (e.g. 16 lb (7 kg) hardware bags), or light-weight cloth if conditions are wet; the slightly smaller plasticized aerated bags with drawstrings are tough, light, and convenient, but samples should not be left in these bags for several weeks or they will grow mould. If mould should develop, the samples are less pleasant to handle and some remobilization of elements from the plant tissues to the mould takes place. Therefore, should this situation arise, it would be best to process the entire sample, rather than attempt any separation of tissue types (i.e. leaves from twigs). If the preferred sample bags are not available, any clean cloth or paper bag will suffice. Bags that have been treated with fungicide (commonly Sb-based) should be avoided because of the potential for sample contamination;
- a roll of masking tape (preferred) or a stapler to keep paper bags shut;
- a large back pack, because if twigs are the chosen sample medium the volume of material collected soon becomes quite large (but not heavy). For large surveys, heavy-duty plastic garbage bags are useful, and they can be left full of samples at the ends of cut lines to be picked up at the end of the day;
- a $\times 10$ hand lens to help in species identification, and in counting growth rings on twigs.

Selection of tissue type

As a basic premise, the same type of plant tissue should be collected from the same species of tree or shrub, unless there is prior knowledge that no significant chemical differences occur between two or more species. Most heavy metals, and especially the precious metals, concentrate in the extremities of plants – the outer bark, the twig ends, and tree tops. Laboratory studies by Girling & Peterson (1978), using radioactive Au, have shown that Au also accumulates in the tips of leaves. Many metals collect in the roots but these are impractical to collect. An effective biogeochemical survey needs to be simple and practical.

Bark

Throughout the forests of North America, Europe and Russia the sample medium that is commonly the most informative (and the easiest to obtain and process) is the scaly outer bark of conifers. Some conifers (mostly the firs) have smooth barks that are impractical to collect, and are not particularly informative. Of importance to a biogeochemical survey is the knowledge that inner bark has a chemical composition that is substantially different from outer bark. Table 6 shows that in an area of Au mineralization in eastern Canada, the concentrations of Au, As, Sb, Cr, Fe, and REE (with La as an example) were significantly higher in the outer than in the inner bark of red spruce. However, the reverse is true for Ba, Zn, and Ca, all of which concentrate in woody tissue.

Table 6. *Concentrations (in ash) of elements in inner and outer bark from two red spruce trees, central Nova Scotia*

	Tree A		Tree B	
	Inner Bark	Outer Bark	Inner Bark	Outer Bark
Au ppb	<5	51	9	126
As ppm	2	56	93	300
Sb ppm	0.1	10	0.7	3.5
Cr ppm	1	41	7	18
Fe ppm	500	16000	2200	16000
La ppm	0.5	16	3	18
Ba ppm	3600	1500	5100	2500
Zn ppm	3300	1600	9200	3900
Ca %	30	18	32	28

Table 7. *Comparison of element distribution along branches of a single western hemlock tree at the Ladner Creek Au deposit, near Hope, southern British Columbia*

	Thick (oldest) (> 10 mm dia.)	Medium (5–10 mm dia.)	Thin (youngest) (< 5 mm dia.)
Au ppb	530	650	1590
As ppm	22	31	82
Cr ppm	32	26	84
Co ppm	11	12	21
Ca %	29	24	14
Fe %	0.8	1.1	2.3
Na %	0.4	0.4	1.1
La ppm	2	3	6
Br ppm	19	18	18
Cs ppm	2	2	2
Sr ppm	430	480	450
Zn ppm	1500	1400	1900

Concentrations in ash.

Twigs

For surveys using twigs, each sample should comprise a similar number of years of growth, because the chemistry of a twig varies along its length. The highest concentrations of most heavy and toxic metals occur toward the tips, probably because the ratio of twig bark (containing most of the metals) to twig wood increases as the twig diameter decreases. In the boreal forest, ten years growth is commonly 25 to 30 cm in length, with a maximum twig diameter of 3 to 5 mm. This is a practical amount to collect, and 7 to 10 twigs of this size are usually sufficient for most analytical programmes. If twigs of three years of growth are collected at one site and twigs of ten years growth from another, a comparison of the element concentrations would give misleading results because of the differences in composition of twigs of differing ages.

Table 7 is another example of chemical variation along a twig/branch, in this case a western hemlock from close to Au mineralization at the Ladner Creek gold deposit in British Columbia. The differences in Au, As, and Cr distributions are particularly striking, with each being most concentrated toward the twig ends. As noted in the case of inner and outer bark, not all elements follow the same trend. Calcium is more enriched in the thick part of the branch, whereas Sr, Zn, Br, and Cs are homogeneously distributed.

Leaves and needles

Metals accumulate in the leaves of many deciduous species and, on a dry weight basis, the concentrations are commonly similar to, or greater than the concentrations in twigs. However, if analysis is to be performed on the ash of the tissues, then the lower ash yield of the twigs results in higher concentrations of elements in the twig ash. Furthermore, seasonal changes in the chemistry of the leaves are more rapid than for twigs because in hot weather there is appreciable evaporation and nucleation of salts that are readily washed from leaves during rains or blown away with spalling cuticle. Consequently, a biogeochemical survey that uses leaves should be conducted within a short time frame (a week or two), and any heavy rains, hot days or high winds should be noted.

Trunk wood

Concentrations of most elements of exploration significance are substantially lower in trunk wood of conifers that in outer bark and twigs. An exception is Ag, which is more highly concentrated in the trunk wood than elsewhere within coniferous trees. Whereas the ash of trunk wood commonly contains 10 to 20 ppm Ag as background concentrations, the ash of twigs has <2 ppm Ag. Traces of metals do, however, concentrate in the wood, and if there is uncertainty as to whether twigs or bark might be contaminated with air-borne dust, an analysis of the wood will help to resolve this problem. In general, concentrations of many elements are so low that they cannot be detected from conventional analysis (i.e. ICP-ES, INAA) of dry material. However, the ash yield of conifer trunk wood is very low (<0.5%; Table 8) and by reducing the wood to ash the elements are concentrated 200 to 400 fold. Some biogeo-

Table 8. Percentage ash yield commonly obtained from tissues of various plants from temperate and northern forests

	Plant Organ	% Ash Yield compared to original weight of samples (ashed at 470°C)
Coniferous Trees:	Twigs	2–3
	Needles	3–5
	Bark (outer)	1–3
	Bark (inner)	2–4
	Trunk wood	0.2–0.5
	Cones	0.5–1
Deciduous trees and shrubs	Twigs	3–4
	Leaves	5–8
	Bark (all)	4–6
	Trunk wood	0.4–0.8

Table 9. Concentrations of selected elements in ash of tissues (top, twig, bark) from black spruce, Brewery Creek, Yukon

	Site 1			Site 2			Site 3		Site 4	
	Top	Twig	Bark	Top	Twig	Bark	Top	Twig	Top	Twig
Au ppb	57	85	22	120	335	134	7	12	<5	10
As ppm	70	130	29	150	420	210	3.9	11	2.1	6.3
Ba ppm	13000	24000	17000	12000	14000	11000	480	2100	1400	3000
Cr ppm	33	37	17	50	49	30	24	24	26	24
Cs ppm	11	6.9	3.1	7	4.9	2.9	26	6.6	1.7	1.9
Cu ppm	155	144	117	133	117	88	162	209	137	162
Ni ppm	132	103	28	91	56	33	5	11	21	16
Sb ppm	130	200	47	300	710	290	1.9	5.0	1.8	3.0

Sites 1 and 2: close to Au mineralization (Canadian Zone): Sites 3 and 4: loess covered northwest-facing slopes with permafrost (no known mineralization)

chemical surveys using trunk wood have been undertaken (e.g. the Red Mountain Stockwork, Idaho; Erdman *et al.* 1985).

Tree tops

Some elements concentrate in the tops of plants. Consequently, if tree tops can be sampled, biogeochemical information can be obtained that may be of use to mineral exploration. In an area where tree growth is stunted, such as the northern part of the boreal forest, top stems can be reached by bending the trees. Table 9 is an example from permafrost terrain in the Yukon, showing the distribution of several elements in trees growing above Au mineralized (with As, Sb and Ba) and unmineralized bedrock. These data indicate that significant variations in the chemistry of the underlying terrain can be recognized by sampling treetops, instead of bark or twigs, even though in this environment the highest concentrations occur in the outer bark. In the temperate forests of central Canada, the tops of trembling aspen have been found of use in delineating kimberlite pipes (Dunn 1993; Eccles 1998).

Treetop sampling is particularly appropriate in heavily forested or rugged terrain where ground access is difficult, and a helicopter is available from which to collect samples by hovering at the top of the tree. This method provides a means of quickly covering a large area to delineate regional geochemical trends, and to obtain a focus for more detailed ground surveys (Dunn & Scagel 1989; Fedikow *et al.* 1997, 1998, 1999). In northern climates, it has the added advantage over collection of surface materials in that it can be conducted year-round, even when the ground is frozen or covered by thick snow.

Table 10. *Basic rules to be applied at each sampling station when conducting a biogeochemical survey*

Basic Rules	Reason
1. Collect same species.	Every species has a different chemical composition, and trace element requirements and tolerances (Table 2).
2. Collect same plant organ.	Each plant organ has a different capacity to store trace elements (Tables 3 and 4).
3. Collect same amount (i.e. age) of growth from same area of tree (e.g. chest height), preferably from all sides.	There are chemical variations along a twig (Table 7). Heterogeneity in bark scales can be minimized by scraping from around the tree.
4. Try to collect samples from plants of similar age and appearance.	This is the basic inter-site consistency that is required for any geochemical sample medium.
5. If living tissue is the selected medium, collect at same time of year (i.e. conduct survey in 2–3 week period).	For some elements there are significant seasonal changes in plant chemistry (Table 5).
Dead tissue (e.g. outer bark) can be collected at any time	No appreciable seasonal change
6. Do not return to a previously sampled tree and expect to obtain exactly the same analyses.	This is unrealistic in view of the heterogeneity of element distributions and seasonal variations in composition (and to a lesser extent annual variations). Be satisfied if an anomaly is the same order of magnitude.

Cones

The tops of many coniferous trees are laden with cones that can be collected with stems during helicopter-supported tree top sampling surveys. Alternatively, they can be collected when they eventually fall to the ground. Cones yield very little ash (0.2–0.5%), therefore when reduced to ash the metal content is high, and a strong background/anomaly ratio results. Old cones yield higher concentrations of metals than new green cones and therefore the two should not be mixed when preparing samples for analysis.

Summary

The procedures and precautions to be taken when collecting tree tissue samples can be summarized as a list of basic rules (Table 10).

Sample preparation

Contamination

When conducting any geochemical survey close attention should be paid to the possibilities of sample contamination. The 'gold ring effect' of sample contamination from a ring worn by the sampler has given rise to many false Au anomalies and short-lived enthusiasm on the results of surveys involving many different sample media. The potential for contamination of biogeochemical samples is particularly acute, because samples may require more handling than either rocks or soils. Concentrations are low compared to other media, and there are many possible local sources of contamination (e.g. metal waste). Samplers should make careful note of the environment (e.g. tree types, soil types, topography, soil moisture) surrounding each biogeochemical sample. Any signs of disturbance should be noted to help in later interpretation of biogeochemical data. Clearly, if elevated concentrations of a metal are the result of environmental disturbance, then the results are unlikely to be of value in an exploration programme. Follow-up tests can be undertaken to determine if anomalous results are attributable entirely to contamination or to natural uptake by the plant. For example, Table 11 shows that Au and As enrichment in lodgepole pine are not only on the surface of the tree (i.e. outer bark), but also in the inner bark and trunkwood. Typical background levels are <10 ppb Au and <5 ppm As.

Some contamination may not be evident at the time of sample collection, but may show up as unusual metal enrichments in certain samples, and require some thoughtful interpretation. The biogeochemist must remain alert to potential sources of contamination, and inexperienced personnel who have not been schooled into carefully observing their environment and identifying plant species should not conduct surveys.

Table 11. Concentrations of Au and As in lodgepole pine near the Nickel Plate gold mine, Hedley, southern British Columbia

	Gold (ppb) in ash			Arsenic (ppm) in ash		
	Outer Bark	Inner bark	Trunk wood	Outer bark	Inner bark	Trunk wood
Pine #1	420	114	128	220	25	59
Pine #2	308	28	56	160	22	41
Pine #3	238	32	36	150	20	33

Table 12. Effects of thorough washing in distilled water on the chemical composition of different plant tissues (1 hour in ultrasonic bath) from the Nickel Plate gold mine, Hedley, British Columbia

	Sagebrush Twig		Sagebrush Leaf		Lodgepole Pine Bark	
	Unwashed	Washed	Unwashed	Washed	Unwashed	Washed
Au (ppb)	270	294	279	267	293	298
As (ppm)	100	95	50	64	150	160
Ba (ppm)	330	300	140	150	590	590
Co (ppm)	4	4	2	2	11	0
Fe (ppm)	6300	5500	2500	2800	17600	17200
K (%)	26.3	24.3	17.4	13.2	3.2	1.5
Mo (ppm)	11	10	9	11	2	3
Sb (ppm)	1.7	1.5	0.7	1.1	4.2	4.3
Zn (ppm)	570	550	530	610	1300	1400

Washing

Samples from dusty areas should be washed. Rinsing samples in a stream or lake or using tap water is usually sufficient, although more thorough washing in a laboratory may be needed if samples are very dusty, and particles are lodged in the plant tissues. Samples from many areas of the northern and temperate forests need not be washed because they are regularly rinsed by rain. In most cases, the dust comprises mostly silicates that are unlikely to be enriched in precious and base metals. Table 12 shows element concentrations for washed, and unwashed portions of three samples collected near the Nickel Plate gold mine at Hedley, and confirms that there is insignificant loss of elements (except K) even after the most rigorous washing with distilled water.

Drying

Samples should be spread out to dry while still in their sample bags, in a clean environment, if possible on the day of collection. If they are left in paper bags in a backpack, box or pail, moisture released from the vegetation will soon cause the bags to disintegrate. If samples are stored in plastic bags, they will soon grow mould and begin to rot, making sample handling very unpleasant. Furthermore, redistribution of chemical elements among tissue types may occur. Mould can grow, too, on cloth bags. The samples do not need to be removed from the bags in which they were collected, provided the bags are moderately porous so that the air can get to the samples to dry them out. If plastic bags are used, samples should be removed on the day of collection. This is not required for plasticized aerated bags, but they should not be left closed up in a damp place for several weeks or mould will grow.

It takes several weeks for samples to dry in a warm, dry atmosphere. Faster methods include drying samples in an oven for 24 hours at about 100°C, or placing them in a microwave oven for 10 to 40 minutes, depending upon how wet they are. No metal objects (such as a staples) should be put in the microwave oven with the bags or they will catch fire. Microwaving should be carefully monitored. If Hg is to be determined, a microwave should not be used, and the temperature in a drying oven should be kept at < 40°C to prevent volatilization.

Separation

Once the moisture has been removed it is a simple process (for most species) to remove the foliage from the twigs by pummelling the bag, then removing the brittle leaves. Hands should be thoroughly washed and rings removed before doing this. This separation procedure is always advisable, because as noted above, the chemistry of the different tissue types is not the same, and the density of foliage may vary from one sample to the next (therefore the twig : foliage ratio will vary, providing the potential for false anomalies).

Bark needs no further separation, since any separation of inner from outer bark should have been performed in the field when it is much easier to do so because the samples are moist. Drying bakes the layers together. The dried and separated material is then ready for either maceration (chopping) for non-destructive direct analysis by instrumental neutron activation analysis (INAA) or, for a more limited range of elements and generally higher detection levels, analysis can be by X-Ray Fluorescence (XRF). Alternatively, samples can be reduced to ash to pre-concentrate the metals prior to analysis by non-destructive and/or wet chemical methods. Ashing is particularly useful if concentrations for elements not readily determined by INAA are required (e.g. Pb, Ni, Cu, Cd, V, Sn, Li, B, Bi, Se, Te, Ga, Tl, P, Mg, Mn, Al, and low levels of Ag).

Maceration

If analysis of dry tissue is the proposed method, material should be homogenized by macerating it in an appropriate blender or mill. It can then be pressed into pellets for trace element analysis by INAA or XRF. Pellets for INAA are prepared by pressing 8 g, 15 g or 30 g aliquots of material at 35 tons in an XRF press. Commonly the 15 g sample size is adequate. For XRF, 5 g is a more common sample weight. Wet chemical analysis can be performed on the dry powder, but most procedures are tedious and detection limits are commonly inadequate

Reduction to ash

Concentration of the vegetation by controlled ignition brings many metals to levels that are easily detectable by inductively coupled plasma emission spectrometry (ICP-ES), inductively coupled plasma mass spectrometry (ICP-MS), atomic absorption spectrometry (AAS), or simple colorimetry. Maceration is not usually necessary. Typically, between 50 and 100 g of dry material is placed in an aluminium tray and, after bringing the temperature slowly up to 470°C, it is held at that temperature for 12–24 h until all charcoal has disappeared. It is important that the material should just *smoulder*; if it ignites some elements will volatilize.

The ash is then ready for analysis by whatever chemical method is available and appropriate. Tests performed on the analysis of dry tissues and those reduced to ash indicate that a few elements (e.g. Br, Hg) volatilize during the controlled reduction to ash. There may be loss of a small portion of other elements, but data indicate that loss is a fairly consistent percentage.

There are some species (not commonly used in biogeochemical surveys), especially those belonging to the rose family, that contain cyanogenic glycosides. These compounds combine with Au in the plant causing volatilization during reduction to ash. Therefore, the ash yields little or no Au. Conversely, Pd forms a very stable monoxide during ashing to 470°C, requiring that the temperature be raised to 870°C to fully break this bond prior to wet chemical analysis.

Analysis

Over the past 20 years, the two methods most commonly employed in the analysis of plant material for mineral exploration are INAA and ICP-ES, both of which have been discussed briefly in the previous section. Recently, commercial laboratories have introduced ICP-MS for the analysis of plant ash.

INAA measures the total content of elements in the sample, regardless of how they are bound with other elements. It is particularly appropriate for measuring small traces of elements in dry vegetation or ash. Drawbacks to the method are that it cannot determine certain elements (e.g. Pb, Bi, Tl), and it has either high detection limits or requires a separate irradiation for some other elements (e.g. Ag, Cd, Cu, Ni, Mg, Mn, V etc.). If the analytical program requires mainly Au, As, Sb, Co, Cr and any of the other 30 elements available in commercial packages, INAA is the best method.

ICP-ES following an aqua regia digestion of ash provides 'total' concentrations for most elements, although some elements may be bound with others such that the aqua regia digestion does not release them all into solution, or there may be spectral interference among high levels of some elements. Data for some elements, especially Ba and Sr, are only partial, and detection limits are usually too high to be of use for Au, U and a few other elements. The determination of As, Sb, Se, Te, Bi, and Ge by hydride generation can be obtained for an additional cost, providing useful data on these 'pathfinder' elements. ICP-ES on an acid digestion of dry vegetation provides data of only limited value because of the low concentrations of elements present.

ICP-MS following either a nitric acid or aqua regia digestion is a method that is becoming increasingly important, because it can detect very low levels of more than 60 elements, some of which are not part of the usual suite of elements obtained by INAA or ICP-ES. The method is, therefore, providing new insight into the distribution of low levels of many elements.

Standards

Whichever analytical method is employed, it is *essential* that for adequate quality control at least one standard sample of known composition (and similar matrix) and one duplicate pair are inserted within every batch of 20 'regular' field samples. These samples provide control on the accuracy (using standards) and precision (determined from duplicates) of the data.

Comparisons of till and biogeochemical data

Geochemical study of a till sample typically involves analysis of one size fraction (e.g. <0.063 mm) from a bulk sample weighing approximately 7 to 10 kg dug from a single small pit (typically 1 m deep x 0.75 m wide). The tree roots, however, may extend through several cubic metres of soil (all horizons) and till, on occasion reaching and penetrating joints and fractures in bedrock. As noted earlier, a tree extracts elements from a large volume of material of diverse composition, including groundwater. Some elements dissolved in groundwater and readily extracted by the tree roots may not precipitate or be adsorbed on till and soil particles.

A second factor of importance is the barrier mechanism established at the root/sediment interface by some plants for some elements. Similarly, barriers may exist among different tissues of an individual plant. Because each species of plant has a different requirement for, and tolerance to, a range of chemical elements, some partitioning of elements takes place and there is selective absorption and translocation into the plants. For biogeochemical surveys, conifers are a good sample medium because they are primitive plants that have a wide tolerance to many trace elements. The outer bark may, by analogy, be equated with biotite in rocks, in that it is something of a repository for many elements that do not fit elsewhere or are not required for the metabolic function of the tree.

A third factor is that slight enrichments of metals in till samples are unlikely to be reflected in the vegetation as weak biogeochemical anomalies. This is especially true in the ppb (Au) and ppm ranges of concentration common in the fine fraction (<0.063 mm) of till. Some of these metals may not be present in a chemical form *available* for uptake (e.g. Cr structurally bound in chromite). Some may be excluded from uptake at the roots or only partially absorbed, and some may be taken up but dispersed among tree tissues to the extent that inter-site variations are so small that they cannot be detected.

The net result of these factors is that the geochemical information supplied by the vegetation is different from that of the till. To reiterate a statement in the 'Introduction', just as two methods of geophysical survey will provide totally different information, so will two different media used for geochemical surveys. A high correlation between distribution patterns of two geochemical sample media is the exception rather than the rule. In geological environments where there is sufficient concentration of metals to form a mineral deposit, such a 'critical mass' of elements may be sufficient to generate biogeochemical anomalies above (by upward diffusion), or around (by movement in electrochemical cells) the mineral source. Where there is sufficient topography, hydromorphic dispersion may displace biogeochemical anomalies down-slope, although there may remain an anomaly directly above the mineralized source (Dunn & Ray 1995). Tills, however, usually have geochemical anomalies displaced down-ice from the mineralized source (DiLabio 1990; Coker & Shilts 1991; McClenaghan et al. 1997; Levson 2001). Comparisons of till and biogeochemical anomalies are given in Coker et al. (1991); Rogers & Dunn (1993); MacDonald & Boner (1993); Dunn et al. (1996); Sibbick et al. (1996); and McClenaghan & Kjarsgaard (2001). Such factors need to be taken into consideration when interpreting geochemical results. Valuable exploration information can be obtained from the analysis of till samples. When this information is coupled with analysis of vegetation samples, a powerful combination is provided for assisting in the exploration for mineral deposits in glaciated terrain.

References

BROOKS, R. R. 1983. *Biological Methods of Prospecting for Minerals Second Edition*. John Wiley and Sons, New York, Toronto.

BROOKS, R. R., DUNN, C. E. & HALL, G. E. M. 1995. *Biological Systems in Mineral Exploration and Processing*. Ellis Horwood, Hemel Hempstead (UK), Toronto, New York.

COKER, W. B. & SHILTS, W. W. 1991. Geochemical exploration for gold in glaciated terrain. *In*: FOSTER, T. P. (ed.) *Gold Metallogeny and Exploration*. Blackie, New York, 336–359.

COKER, W. B., DUNN, C. E., HALL, G. E. M., RENCZ, A. N., DILABIO, R. N. W., SPIRITO, W. A. & CAMPBELL, J. E. 1991. The behaviour of platinum group elements in the surficial environment at Ferguson Lake, N.W.T., Rottenstone Lake, Sask., and Sudbury, Ont., Canada. *Journal of Geochemical Exploration*, **40**, 165–192.

DILABIO, R. N. W. 1990. Glacial dispersal trains. In KUJANSUU, R. & SAARNISTO, M. (eds) *Glacial Indicator Tracing*. A. A. Balkema, Rotterdam, 109–122.

DUNN, C. E. 1993. Diamondiferous kimberlite in Saskatchewan, Canada – a biogeochemical study. *Journal of Geochemical Exploration*, **47**, 131–141.

DUNN, C. E. 1995. Biogeochemical prospecting for metals. Chapters 19 and 20. *In*: BROOKS, R. R., DUNN, C. E. & HALL, G. E. M. (eds) *Biological Systems in Mineral Exploration and Processing*. Ellis Horwood, Hemel Hempstead (UK), Toronto, New York.

DUNN, C. E. & RAY, G. E. 1995. A comparison of lithogeochemical and biogeochemical patterns associated with gold mineralization in mountainous terrain of southern British Columbia. *Economic Geology*, **90**, 2232–2243.

DUNN, C. E. & SCAGEL, R. K. 1989. Tree-top sampling from a helicopter – a new approach to gold exploration. *Journal of Geochemical Exploration*, **43**, 255–270.

DUNN, C. E., HALL, G. E. M. & SCAGEL, R. K. 1993. *Applied Biogeochemical Prospecting in Forested Terrain*. Short Course Notes, Association of Exploration Geochemists, Nepean, ON.

DUNN, C. E., BALMA, R. G. & SIBBICK, S. J. 1996. *Biogeochemical survey using lodgepole pine bark: Mount Milligan, Central British Columbia (parts of 93N/1 and 93O/4)*. Geological Survey of Canada Open File **3290**.

ECCLES, D. R. 1998. *Biogeochemical orientation survey over the Mountain Lake diatreme, Alberta*. Alberta Geological Survey Open File Report **1998-06**.

ERDMAN, J. A., LEONARD, B. F. & MCKOWN, D. M. 1985. *A case for plants in exploration gold in Douglas-fir at the Ted Mountain stockwork, Yellow Pine district, Idaho*. U.S. Geological Survey Bulletin **1658A-S**.

FEDIKOW, M. A. F., NIELSEN, E., CONLEY, G. C. & MATILE, G. L. D. 1997. *Operation Superior: Multimedia geochemical survey results from the Echimamish River, Carrot River and Munro Lake greenstone belts, northern Superior Province, Manitoba (NTS 53L and 63I)*. Manitoba Energy and Mines Open File **OF97-2**.

FEDIKOW, M. A. F., NIELSEN, E., CONLEY, G. C. & MATILE, G. L. D. 1998. *Operation Superior: Multimedia geochemical survey results from the Edmund Lake and Sharpe Lake greenstone belts, northern Superior Province, Manitoba (NTS 53K)*. Manitoba Energy and Mines Open File **OF98-5**.

FEDIKOW, M. A. F., NIELSEN, E., CONLEY, G. C. & MATILE, G. L. D. 1999. *Operation Superior: Multimedia geochemical survey results from the Webber Lake, Knife Lake and Echimamish River greenstone belts, northern Superior Province, Manitoba (NTS 53L and 53K)*. Manitoba Energy and Mines Open File **OF99-8**.

GIRLING, C. A. & PETERSON, P. J. 1978. Uptake, transport and localization of gold in plants. *Trace Substances in Environmental Health*, **12**, 105–108

KOVALEVSKY, A. L. 1987. *Biogeochemical Exploration for Mineral Deposits. Second Edition*. VNU Science Press, Utrecht, Netherlands.

KOVALEVSKY, A. L. 1995. Chapter 16: The biogeochemical parameters of mineral prospecting. *In*: BROOKS, R. R., DUNN, C. E. & HALL, G. E. M. (eds) *Biological Systems in Mineral Exploration and Processing*. Ellis Horwood, Hemel Hempstead (UK), Toronto, New York.

LEVSON, V. M. 2001. Regional till geochemical surveys in the Canadian Cordillera, sample media, methods and anomaly evaluation. *In*. MCCLENAGHAN, M. B., BOBROWSKY, P. T., HALL, G. E. M. & COOK, S. J. (eds) *Drift Exploration in Glaciated Terrain*, Geological Society, London, Special Publications, **185**, 45–68.

MACDONALD, M. A. & BONER, F. J. 1993. Multi-media geochemistry and surficial geology of the Yava Pb deposit, southeastern Cape Breton Island, Nova Scotia, Canada. *Journal of Geochemical Exploration*, **48**, 39–69.

MCCLENAGHAN, M. B., THORLEIFSON, L. H. & DILABIO, R. N. W. 1997. Till geochemical and indicator mineral methods in mineral exploration. *In*: GUBINS, A. G. (ed.) *Proceedings of Exploration 1997: Fourth Decennial International Conference on Mineral Exploration*, 233–248.

MCCLENAGHAN, M. B. & KJARSGAARD, B. A. 2001. Geochemical and indicator mineral for diamond exploration in glaciated terrain of Canada *In*: MCCLENAGHAN, M. B., BOBROWSKY, P. T., HALL, G. E. M. & COOK, S. J. (eds) *Drift Exploration in Glaciated Terrain*, Geological Society, London, Special Publications, **185**, 83–123.

POJAR, L., KLINKA, K. & MEIDINGER, D. V. 1987. Biogeoclimatic ecosystem classification in British Columbia. *Forest Ecological Management*, **22**, 119–154.

ROGERS, P. J. & DUNN, C. E. 1993. Trace element chemistry of vegetation applied to mineral exploration in eastern Nova Scotia, Canada. *Journal of Geochemical Exploration*, **48**, 71–95.

ROSE, A. W., HAWKES, H. E. & WEBB, J. S. 1979. *Geochemistry in mineral exploration* Second edition. Academic Press, London.

SIBBICK, S. J., BALMA, R. & DUNN, C. E. 1996. *Till*

geochemistry survey: Mount Milligan, central British Columbia (Parts of NTS 93N/1 and 93O/4). Joint publication of British Columbia Geological Survey Open File **1996-22**, and Geological Survey Canada Open File **3291**.

WARREN, H. V. & DELAVAULT, R. E. 1950. Gold and silver content of some trees and horsetails in British Columbia. *Bulletin of the Geological Society of America*, **61**, 123–128.

WARREN, H. V., DELAVAULT, R. E. & BARAKSO, J. 1968. The arsenic content of Douglas fir as a guide to some gold, silver, and base metal deposits. *Bulletin of the Canadian Institute of Mining and Metallurgy*, **61**, 860–867.

Analysis of geochemical data for mineral exploration using a GIS — A case study from the Swayze greenstone belt, northern Ontario, Canada

J. R. HARRIS[1], L. WILKINSON[1] & M. BERNIER[2]

[1] *Geological Survey of Canada, 615 Booth Street, Ottawa, Ontario K1A 0E9, Canada*
(e-mail:harris@gis.nrcan.gc.ca)
[2] *Asquith Resources, Toronto, Canada*

Abstract: Geographic Information Systems (GIS) provide the geologist with a powerful tool, when used in concert with statistical and geostatistical analysis, for archiving, manipulating, analysing and visualizing geochemical data. This paper uses geochemical (Zn, Cu) data obtained from various media (rock, lake sediments, till, soil and humus) over the Swayze greenstone belt in northern Ontario, to explore methods for analysing and visualizing geochemical data with a focus to mineral exploration applications.

The behaviour of Zn and Cu in both bedrock and the surficial environment is studied using statistical and geostatistical techniques. Interpretation and uses of traditional statistics and dot plots are contrasted with interpolated geochemical maps as well as red–green–blue (RGB) ternary maps. Techniques for multimedia comparison and geochemical anomaly detection and screening are presented. The processing methods presented in this paper can be utilized and adapted by other geologists for exploring their own geochemical data. Many of the algorithms presented here are available within standard GIS software packages, or can be written easily using a GIS macro language.

The traditional approach to the analysis of geochemical data involves the use of spreadsheets and/or statistical software packages. Often the results are plotted using proportional symbol plots, where the size of each dot is proportional to the concentration of a given element, usually ranked by percentiles. A Geographic Information System (GIS) can offer the geologist significant advantages over traditional approaches to archiving, visualizing, manipulating, analysing, integrating and presenting geochemical data. GIS is now a firmly established technology that many mining companies are adding to their arsenal of exploration tools. These systems can and should be used by the exploration geochemical community to facilitate the storage and display of geochemical data. Furthermore, the GIS facilitates the spatial analysis and comparison of large volumes of geochemical data.

As with a *Computer Aided Design* System (CAD), a GIS employs graphic primitives (points, lines and polygons) which are the building blocks for constructing digital maps. These form the basis of the vector model which the GIS uses to store and view spatial data. The GIS also uses the raster model in which spatial information is divided into grid cells (*pixels*) containing a number (referred to as a *digital number* or *DN*) that has some user or data-defined meaning. Unlike a CAD system, the GIS utilizes a database which contains descriptive information (*attributes*) for each point, line or polygon. Thus, a geochemical sample point has a specific location, as well as a number of attributes that describe the sample, stored in an internal or external database. The data can be displayed and queried, by any attribute or combination of attributes in the database, using a database query language (e.g. SQL - Structured Query Language) or the proprietary query language of that particular GIS software. Although the GIS can be used as a cartographic tool for producing traditional geochemical maps (proportional symbols), it also provides spatial analysis and visualization tools that can be used to analyse the statistical and spatial characteristics of geochemical data. At the present stage of GIS development, other software packages (statistical and geostatistical) are still required by the geologist in addition to the GIS, to thoroughly analyse geochemical data, both statistically and spatially. However, this is beginning to change as many GIS vendors are

From: MCCLENAGHAN, M. B., BOBROWSKY, P. T., HALL, G. E. M. & COOK, S. J. (eds) 2001. *Drift Exploration in Glaciated Terrain.* Geological Society, London, Special Publications, **185**, 165–200. 0305-8719/01/$15.00 © The Geological Society of London 2001.

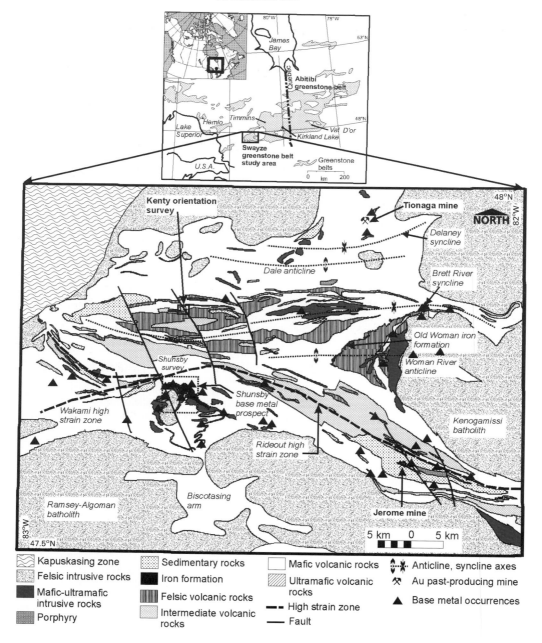

Fig. 1. Study area – generalized bedrock geology (Heather & Shore 1999).

now integrating more geostatistical and statistical tools into the GIS environment.

In this paper the GIS, in concert with statistical and geostatistical analysis software, is used to demonstrate techniques for:

(a) separating geochemical concentrations into anomalous and background populations;
(b) visualizing geochemical data;
(c) identifying lithologic signatures in surficial media (till, soil, humus);

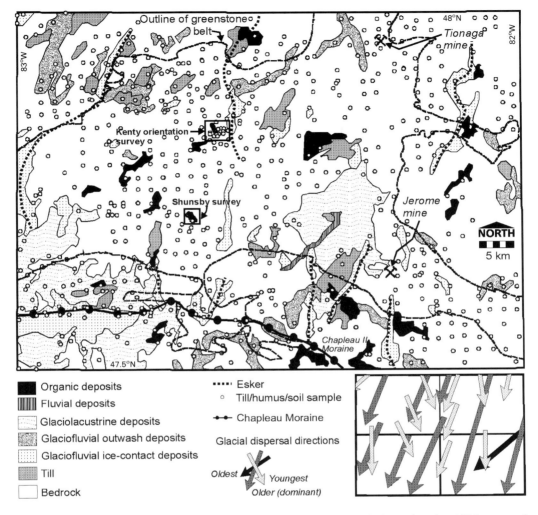

Fig. 2. Study area – generalized surficial geology (Bernier 1994) and geochemical sample points (till, humus and soil).

(d) comparing geochemical responses in different geochemical sampling media; and
(e) screening surficial geochemical data (soil) for false anomalies (e.g. anomalies not due to alteration or mineralization).

The geochemical data used in this study include humus, till and soil samples collected by the Ontario Geological Survey (OGS) (Bernier & Goff 1993; Bernier 1994; Bernier & Kaszycki 1995), lake sediment data collected by the Geological Survey of Canada (GSC) and lithogeochemical data obtained from the OGS, GSC and Falconbridge Ltd., all collected over the Swayze greenstone belt in Ontario (the lithogeochemical data is now published in Harris et al. 1999). Airborne magnetic data (OGS 1997) are also used to assist in the interpretation of geochemical patterns.

Study area

The Swayze greenstone belt (SGB) (Fig. 1) is the westernmost extension of the mineral-rich Abitibi greenstone belt (AGB), and has recently been re-mapped by both the OGS (Ayer 1995 a, b), and the GSC (Heather & Shore 1999). Like the AGB, the SGB contains a number of fold-repeated 2730–2680 Ma mafic-felsic metavolcanic packages, which are unconformably overlain by Timiskaming-type metasedimentary rocks and cut by high strain zones thought to be extensions of the major *breaks* (Destor-Porcupine and Cadillac-Larder Lake faults) found in the AGB (Ayer & Theriault 1993; Heather 1993; Heather & van Breeman 1994; Heather et al. 1995, 1996). These high strain zones are the hosts of several large Au deposits in the Timmins and Kirkland Lake areas.

The SGB shares many features in common with the mineral-rich AGB to the east, but lacks any significant mineral production. A detailed mineral occurrence database has been completed for the belt (Fumerton & Houle 1993; Fumerton et al. 1993). The two principal commodity types found within the current map area are base metals and Au.

The surficial geology of the study area (Fig. 2) is dominated by thin, sandy till mapped as bedrock–drift complex (Bernier & Goff 1993; Bernier 1994; Bernier et al. 1996). Glaciofluvial outwash and other glacial and post-glacial sediments occur throughout the study area, as well as several large esker complexes orientated approximately N–S to NNE–SSW. Relief is generally low except in the southern portion where rolling topography reflects underlying granitoid rocks. The EW-trending Chapleau Moraine forms a dominant topographic feature.

Geochemical data

A total of 845 geochemical samples were collected concurrently from humus, B-horizon soil and C-horizon till over the SGB, by the OGS from 1993 to 1995 (Bernier & Goff 1993; Bernier 1994; Bernier & Kaszycki 1995; Bernier et al. 1996). Table 1 is a summary of the survey methodology and Table 2 presents a summary of the analytical methods. Figure 2 shows the location of each humus, soil and till geochemical sample site within the study area. The Shunsby and Kenty mine areas (Figs 1 and 2) were sampled in more detail to characterize the behaviour of metals in the locally sampled media with respect to glacial dispersion (i.e. orientation surveys).

The lake sediment geochemical data (Fig. 3), consisting of over 350 samples, were collected as part of the National Geochemical Reconnaissance (NGR) program undertaken by the GSC (GSC 1987, 1988).

Approximately 4500 whole rock samples over the SGB were acquired from four principal sources: GSC, OGS, Falconbridge Ltd. and provincial assessment files (Table 3). Wilkinson et al. (1999) provide a summary on the compilation and levelling of this dataset. Harris et al. (1999a) have divided the database into altered and unaltered samples, using a variety of statistical methods.

The trace elements, Cu and Zn, were chosen as the focus of this paper due to their relevance in regional base metal exploration, their chemical similarity (transition elements), and their differences with respect to mobility in the surficial environment. Thus, different responses with respect to various analysis techniques applied to the data, and behaviour in different sampling media, would be expected.

The surficial geochemical data (humus, soil, till) were evaluated for quality using split and field duplicate analyses. Approximately 1 in 15 samples were split for duplicate analyses and 1 field duplicate was collected for every 20 samples. Zinc and Cu results were evaluated for quality control using Thompson-Howarth precision control plots (Thompson & Howarth

Table 1. *Survey methodology, Swayze greenstone belt NODA geochemical program*

Scale	Density	Media	Objectives

Regional	1 sample/4–7 km^2	C-horizon till
		B-horizon soil
		Humus
			Identify variations related to:
			  Changes in lithology
			  Mineralization
			  Changes in drift composition
			Evaluate mineral potential and
			  outline new exploration targets

Local (Orientation)	50 m spacing	C-horizon till B-horizon soil Humus	Characterize signatures of various types of mineralization Identify pathfinder components Map scale and form of dispersal

Site specific
(Vertical profiling)
	25–50 cm intervals
	(Systematic)
		A_0-A_2 and B-
		horizon soils
		C-horizon tills
			Identify surface weathering trends,
			  in-situ variability, and variations in
			  composition related to stratigraphy

Table 2. *Analytical methods, Swayze greenstone belt NODA geochemical program*

Medium	Size fraction	Analysis method*	Elements
Humus	-80 mesh (<177µm)	INAA	Au, As, Ba, Sb, Bi, Cr, REEs, Th, U, Rb, Sc
		ICP-ES (AR)	Base metals, Ag, As, Co Ba, Mo, V, Mn
B-horizon soil	-230 mesh (<63µm)	ICP-ES (HF)	Majors, base metals, Cr, Li, Ba, Co, Nb, Mo, V
C-horizon till	-230 mesh (<63µm) (silt+clay)	CV-ASS	Hg
		Carbonate analysis	% calcite, % dolomite, % total carbonate
C-horizon till	-10 mesh (<2 mm) (sand+silt+clay)	Heavy mineral separation (>3.2 specific gravity), mineralogical studies	Gold grains; kimberlite indicator minerals; total ferromagnetic and non-ferromagnetic heavy mineral abundance

* INAA, Instrumental neutron activation analysis; ICP-ES, Inductively coupled plasma emission spectrometry; AR, aqua regia; HF, Hydrofluoric acid; CV-AAS, Cold vapour atomic absorption spectrometry.

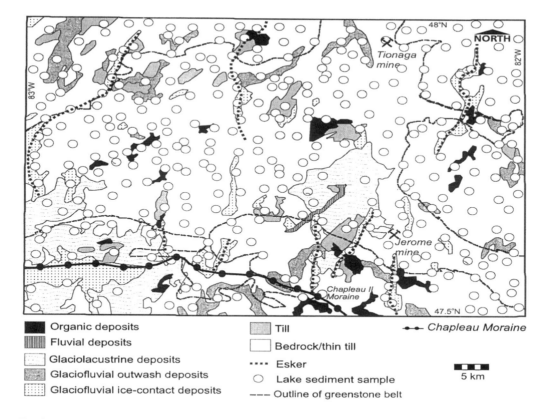

Fig. 3. Location of lake sediment samples.

Table 3. *Lithogeochemistry data source*

Dataset	No. of samples	Years collected	Analytical methods*	Source†
J. Ayer (JA)	136	1991–1993	91-XRF/ICP-MS/ICP-ES/ CV-AAS 92-XRF/ICP-ES/ICP-MS	OGS
K. Heather (KH)	157	1993–1995	XRF/ICP-MS/CV-AAS/ DIONEX	GSC
PETROCH (PT)	64	1976–1993		OGS
S. Fumerton (SF)	1304	?	XRF/ICP-MS/CV-AAS/FA	Assessment files
Falconbridge (FA)	1425	1978–1979	XRF	Falconbridge Ltd.
Texas Gulf (TG)	943	?	XRF/ICP-ES	Falconbridge Ltd.

*XRF, X-ray fluorescence; ICP-MS, Inductively coupled plasma mass spectrometry; ICP-ES, Inductively coupled plasma emission spectrometry; CV-AAS, Cold vapour atomic absorption spectrometry; FA, fire assay.
† OGS, Ontario Geological Survey; GSC, Geological Survey of Canada.

1976, 1978): both elements in all media were found to be within 10% precision levels. The lake sediment geochemical data were subjected to strict quality control measures under the NGR program (see Friske et al. 1988). Wilkinson et al. (1999) review the quality control procedures applied to the lithogeochemical data used in this paper.

Data analysis

The following section describes GIS techniques used in concert with statistical and geostatistical methods for visualizing, manipulating and analyzing Zn and Cu sampled from various media.

Characterizing geochemical distributions

Before applying map visualization procedures, geochemical data should be evaluated to characterize statistical distributions. Howarth (1983), Garrett (1991) and Grunsky (1986, 1997) provide thorough reviews of techniques used to characterize frequency distributions as well as exploratory data analysis (EDA) techniques for geochemical data. These techniques form the basis for further analysis, which includes methods for separating anomalous geochemical concentrations from background and methods for visualizing the spatial characteristics of the data.

Descriptive statistics such as histograms, normal probability plots and box and whisker plots provide insights into the statistical distribution of geochemical data. GIS complements statistical analysis by providing many methods with which to visualize the spatial characteristics (distributions) of the data. Table 4 is a summary of descriptive statistics for Zn and Cu from the various media sampled for this study. Log-transformation of the geochemical data produces distributions that are close to normal allowing for parametric statistical testing. Box and whisker plots (Tukey 1977) are useful for characterizing and comparing geochemical distributions and for assessing the shape of distributions (e.g. positively or negatively skewed). Figure 4 present box and whisker plots of log-transformed Zn and Cu concentrations in all media sampled for the entire study area without regard to lithological or surficial divisions. A box and whisker plot encloses the middle 50% of a data distribution by a box where the median value is represented by a line and the notch represents the 95% confidence interval around the median value. The lines (whiskers) are drawn from the lower and upper quartiles to the smallest and largest point within 1.5 interquartile ranges, respectively. Outliers beyond these ranges are displayed as points. The mean value is plotted as a point within the box. Distributions in which the notches overlap are not significantly different.

In this study the distributions for Zn and Cu are log normal or close to log normal. Copper concentration is highest in rock and lowest in soil whereas Zn concentration is highest in lake sediments and also lowest in soil. Copper concentrations amongst all media are significantly different as notches in the box and whisker plots do not overlap. The median value for Zn in lake sediments is significantly different from the median in the other media, although there is not a significant difference between soil and till nor between rock and humus. These

Table 4. *Descriptive statistics for Cu and Zn populations in all sampling media*

Element	No of samples	Mean (ppm)	Median (ppm)	Min (ppm)	Max (ppm)
		Cu			
Rock	4195	75.3	60.0	0	4500
Humus	734	114.0	66.0	9	4616
Soil	734	26.0	19.0	9	351
Till	734	38.2	20.0	7	2258
Lake sediment	357	100.8	87.0	10	500
		Zn			
Rock	2733	116.0	55.0	2	25100
Humus	734	16.7	12.5	2	562
Soil	734	9.3	6.0	1	132
Till	734	33.5	19.0	3	1704
Lake sediment	357	30.4	25.0	3	140

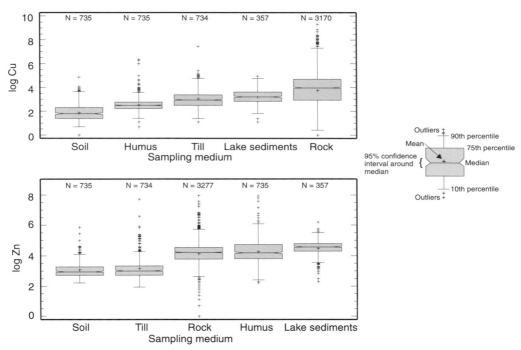

Fig. 4. Box and whisker plots of Zn and Cu concentration by media (rock, till, soil, humus, lake sediments).

differences between media may reflect differences in the behaviour of Cu and Zn in the surficial environment as well as variations due to bedrock and surficial geology. These variations will be investigated with the aid of statistical and GIS analysis techniques in the following sections.

An important goal for mineral exploration, with respect to the statistical and spatial analysis of geochemical data, is the detection of zones of elevated concentrations of major or trace elements that may reflect the presence of alteration and/or mineralization. A geochemical anomaly represents an area of elevated concentrations of a single element or group of elements that exceed the threshold expected for regional background (Howarth 1983; Garrett 1991). Many statistical and geostatistical techniques have been used to highlight or enhance trends or

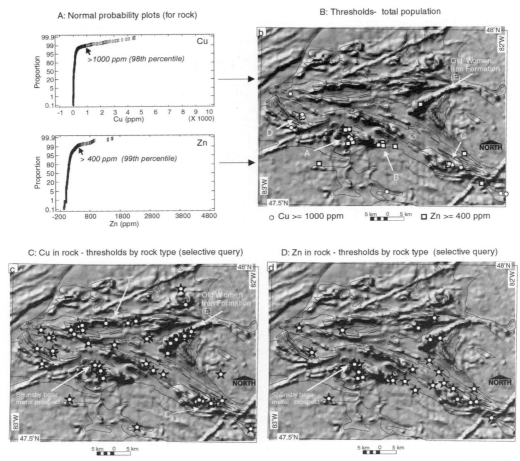

Fig. 5. Cu and Zn anomalies in rock, (**a**) normal probability plots of Cu and Zn in rock, (**b**) map of Cu and Zn anomalies irrespective of lithology, (**c**) Cu anomalies determined from selective query, (**d**) Zn anomalies determined from selective query.

patterns in geochemical data. These anomalous populations, particularly in lithogeochemical data, may reflect specific geological processes, such as fractionation, regional alteration, hydrothermal alteration and metasomatism. Grunsky (1997), Grunsky et al. (1992) and Harris et al. (1999a) provide comprehensive reviews of univariate and multivariate techniques used to identify anomalous geochemical populations.

Many univariate methods exist for determining thresholds for identifying anomalous geochemical populations. A threshold, in this paper, follows the reasoning of Garrett (1991) and is defined as the outer (upper and/or lower) limit of local background variation. Methods used for determining anomalous populations include: exploration knowledge applied as *a rule of thumb* (e.g. it is known that within a given area rocks with > 1000 ppm Cu are anomalous); arbitrary methods using percentile or standard deviation values; visual inspection of probability plots (Stanley 1987; Grunsky 1997; Harris et al. 1997); visual inspection of area v. concentration plots (Cheng et al. 1994), and weights of evidence (WofE) (Bonham-Carter 1994). The WofE method requires knowledge of existing mineral prospects, whereas the other methods do not. The method presented by Cheng et al. (1994) requires geochemical data to be interpolated to a continuous surface map so that areas enclosed by different contour levels can be calculated.

Normal probability plots are arguably the most practical univariate method for determining thresholds, as they are easy to calculate and simple to interpret. Although percentile ranges

Cu anomalies in soil, till and humus based on the total population

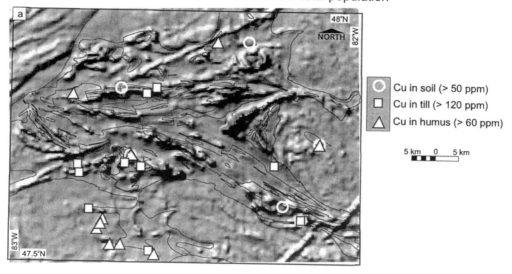

Cu anomalies in till, soil and humus - selective query by rock type

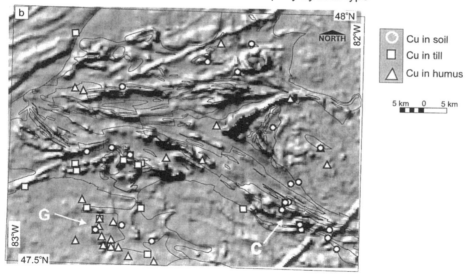

Fig. 6. Cu anomalies in surficial media, (**a**) based on total population irrespective of lithology, (**b**) based on selective query.

are often used to divide and display geochemical data distributions, percentile divisions (e.g. 50–75, 75–80, 80–90, 90–95) do not always correspond to breakpoints present on probability plots. Ideally, natural breakpoints divide geochemical data into separate populations that may reflect different geological or surficial processes (Grunsky 1997). Furthermore, upper or lower breakpoints on a probability plot may separate anomalous samples that may reflect mineralization.

Figure 5a shows normal probability plots for Cu and Zn concentrations in rock, for all samples within the SGB. Figure 5b shows a map (generated using the GIS) of anomalous samples (Cu, Zn) derived from selecting samples above the upper thresholds interpreted on the probability plots, overlaid on a shaded relief magnetic image. Elevated Cu and Zn concentrations characterize the Shunsby area (A), and an area in the south along the contact between mafic volcanic and granitoid rocks (B). Anomalous Zn and Cu concentrations characterize the southeastern and western portions of the SGB (C and D). Most of these anomalies fall on linear magnetic highs many of which are Fe formations, either contiguous (Old Woman Fe Formation (E)), or fragmented and broken up by later faulting (Rideout high strain zone – see Fig. 1) in the Shunsby area (A), and in the SE portion of the study area (C).

One of the advantages of GIS is the ability to divide geochemical data into separate populations on the basis of map-based spatial divisions. A 'point-in-polygon' operation a common GIS function, can be invoked that will intersect each geochemical sample point, over a polygon map. Thus, the geochemical samples can be grouped into different populations and analysed separately. This is particularly useful for: (1) evaluating geochemical data by lithological or surficial units; (2) normalizing lithogeochemical data to take into account lithological variations; and (3) identifying geochemical anomalies that are proximal to mineralization. Therefore, probability plots were calculated for Zn and Cu for each mapped lithological unit (see Fig. 1) and thresholds were selectively identified ('selective query' in GIS terminology) dividing the geochemical samples into background and anomalous. This has the effect of normalizing concentrations to lithology, thus accounting for variations in concentration as a function of rock type. Harris *et al.* (1999a, 2000) provide more details on normalization applied to rock samples in the SGB. The GIS database was exported to a statistical analysis package to perform statistical analysis and generate probability plots. Figures 5c and 5d show Cu and Zn anomalies in rock based on this selective query (normalizing) process by rock type. Comparison between Zn and Cu anomalies calculated from the entire study area (Fig. 5b) and anomalies based on selective querying (Fig. 5c, d), indicates that many more Cu and Zn anomalies are identified using selective querying. In particular, many more Cu anomalies in rock are found along the felsic volcanics that form the footwall to the overlying Old Woman Fe formation (E),

as well as in the central portion of the SGB along the length of the Brett River Synform (F) which comprises a package of ultramafic, mafic and intermediate volcanic rocks. Many more anomalous Zn samples occur in the southeastern portion of the belt within the Rideout high strain zone (Fig. 5c - C) than on Fig. 5a.

Figure 6a shows Cu anomalies in surficial media (humus, soil, till), calculated from thresholds on probability plots based on all the data irrespective of lithological unit whereas Fig. 6b show Cu anomalies based on selective querying by rock unit. Comparison of Figs. 6a and b reveals the presence of many more Cu anomalies on the image showing selective Cu anomalies, especially in the south over granitoid rocks (G) and in the SE over volcanic and sedimentary rocks (C).

Summary. Box and whisker plots are useful for characterizing and comparing geochemical distributions either between different elements or the same element in different media.

The major strength of the GIS is that the results of statistical analysis can be displayed spatially assisting in revealing patterns that may reflect lithological, surficial and mineralization processes. Furthermore, geochemical data can be broken into separate populations based on user-defined groupings (e.g. lithology, surficial unit, topography etc.) for detailed statistical and geostatistical analysis. Simple to complex queries can be undertaken by spatial unit, thus assisting to reveal geochemical anomalies that may be suppressed or entirely hidden when considering the entire geochemical population.

Visualization

Once geochemical distributions have been characterized, allowing appropriate methods to be selected with which to divide the data into meaningful intervals (e.g. percentiles or breakpoints on probability plots) and separating anomalous concentrations from background, the data may be displayed in a variety of ways. Most GIS provide the geologist with a wide range of software tools with which to visualize geochemical data. The process of visualization (see Harris *et al.* 1999b, for a detailed study) involves displaying and combining geochemical with other geoscience data for the purpose of exploratory data analysis. This is important in

Fig. 7. Proportional symbol (dot) maps, (**a**) Cu concentration in humus, (**b**) Cu concentration in soil, (c) Cu concentration in till.

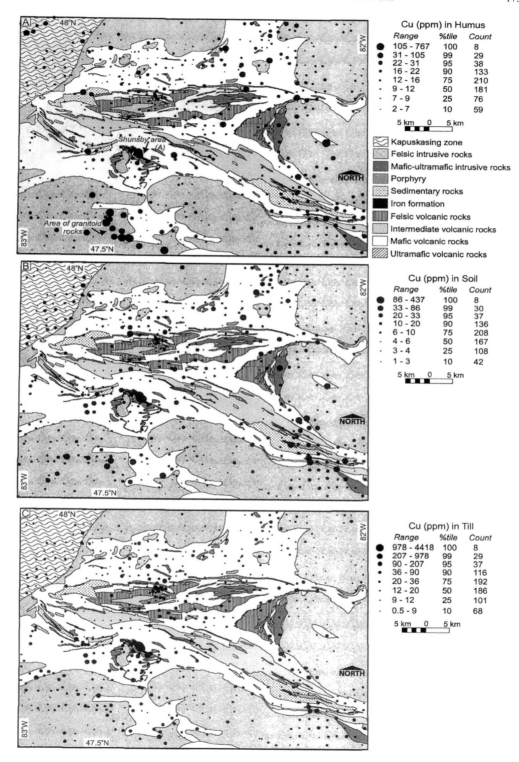

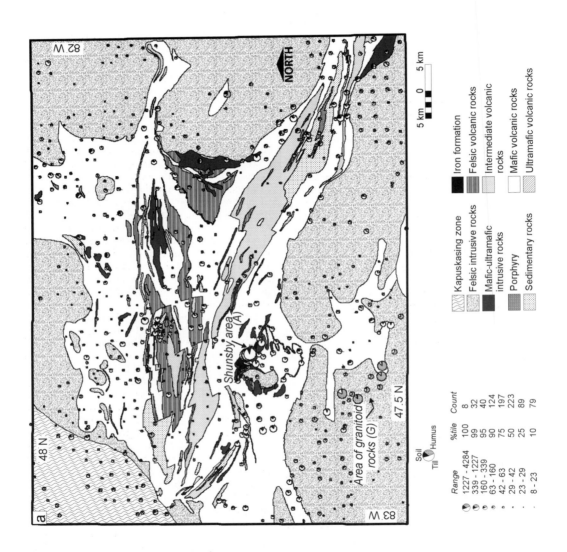

that non-traditional geochemical maps and images can assist the geologist in identifying spatial patterns that may be important for regional mineral exploration.

Proportional dot and pie plots. Traditional dot plots, in which the size of the symbol is proportional to element concentration, are straightforward to interpret and are effective for showing the spatial variation in concentration of elements. Figure 7 shows proportional dot plots of Cu concentration in humus, soil and till. Figure 8 is a variation of Figure 7, in that the size of the pie reflects the total Cu content in all three media (humus, soil and till), while the shading of the wedges comprising the pie reflects the proportion of the total Cu content by medium.

Areas of elevated concentration in all media, can be seen in the vicinity of the Shunsby base metal prospect (Fig. 7, A). Another area of elevated Cu concentration in especially humus and soil (but not till) is found within the area of granitoid rocks in the SE portion of the study area (G). Figure 8 shows that the area of anomalous Cu concentration in the granitoids (G) occurs dominantly in humus, whereas in the Shunsby area (A) till values are the highest, perhaps indicating different processes taking place. For example, scavenging of Cu by organics (humus) over the granitoids may have led to Cu enrichment in this area whereas mechanical breakdown of Cu-bearing minerals and dispersion in the clay + silt portion of glacial till predominates in the Shunsby area.

Interpolation. Establishing a zone of influence around a geochemical sample point, whether by artificial constructs or by interpolation, is often advantageous for visualization and modelling purposes within the GIS. A zone of influence around a geochemical sample can be as simple as setting a buffer zone of a given distance around each point. This distance can be determined based on exploration knowledge (i.e. alteration around a rock sample can be seen to extend for a radial distance of 50 m), or on geostatistical reasoning using variograms. Conversely, a natural zone of influence, such as a drainage basin, can be established around a lake or stream sediment sample point (see Bonham-Carter *et al.* 1987).

Geochemical data can be interpolated producing continuous surface maps if certain criteria are met:

(a) does the point data behave as a continuous random variable where the mean value is stationary over the entire area of interest;
(b) does the data behave isotropically or anistropically;
(c) is the point sampling dense enough to warrant meaningful interpolation?

The particular medium sampled for geochemistry often controls the first two factors. Spatially, lithogeochemical data differ from other geochemical media, such as till and lake sediments, in that rocks are a site-specific medium with a zone of spatial influence very close to the sample site. Geochemical data sampled from till and lake sediments reflect a much broader area (a drainage basin in the case of lake sediment data, and often a broad dispersal train in the case of till data), and therefore the concentration at a specific sample point represents an average of a much broader area than does a lithogeochemical sample. This has important implications for sampling strategies and for methods of data analysis and visualization.

Spatial behaviour of the data can be determined through analysis of a variogram. Experimental variograms can be calculated for all directions (omni-directional) or for any given direction (directional). Thus, directional effects, such as down-ice dispersion of metals in till, may be identified and accounted for in the interpolation process. Variograms are extremely useful, providing information such as maximum radius of influence around each point (range), which can be used to set the search radius (zone of influence) for various interpolation algorithms, as well as directional biases which can be used to set the shape and orientation of the search. If the variogram displays no spatial structure (i.e. pure nugget effect), then one should not interpolate the data or at best use an exact and simple interpolation algorithm (e.g. Delauney triangulation) that makes fewer spatial assumptions. Alternatively, one could choose a different method to present and visualize the data (proportional dots), or choose a natural zone of influence around each sample point (e.g. drainage basin for a lake sediment sample).

Fig. 8. Proportional symbol (pie) plot showing Cu concentration in soil, till and humus.

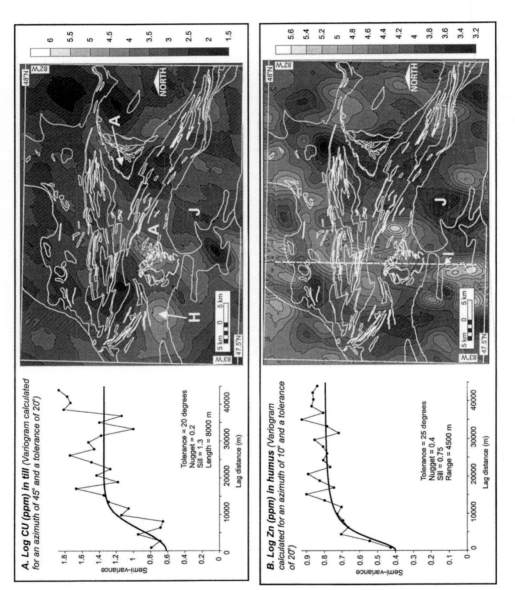

Fig. 9. Variograms and associated interpolated maps (kriged) of Cu and Zn data, (**a**) log Cu in till, (**b**) log Zn in humus.

Many different interpolation algorithms exist, such as kriging, minimum distance curvature and inverse-distance-weighted (IDW) interpolators. In this paper, kriging, using parameters derived from variograms, was used to interpolate the geochemical data. Further details on interpolation methods can be found in Davis (1986), and Issacks & Srivastava (1989). The advantages of interpolating geochemical data are that it: (1) facilitates overlay analysis within the GIS; (2) highlights regional geochemical patterns that may not be obvious when the data are displayed as points; and (3) facilitates the comparison of geochemical data from different media that are not collected at the same geographic location.

Figure 9 shows variograms calculated for log Cu in till and log Zn in humus. These orientations represent the best variograms (i.e. display the most coherent spatial structure between points), for each element in each medium. Better variograms were produced using log-transformed data, rather than raw data, as the transformed data were closer to a normal distribution, and the effects of outliers were thereby reduced. The associated kriged (interpolated) data are also shown on Figure 9. The variograms of Zn and Cu in rock have the smallest ranges, whereas the lake sediment, soil and humus data show larger ranges. This is reasonable given the effects of glacial transport and averaging over broad drainage basins. The best variogram for Cu was calculated from till (Fig. 9a) and consisted of a NE–SW anisotropy, reflecting the dominant, although not the youngest, direction of glacial transport (see Fig. 2). Range (length) of $c.$ 8 km potentially indicates the maximum detectable down-ice dispersion of Cu-bearing minerals above background. However, other factors such as variation in surficial and bedrock geology are important when interpreting information from variograms. The variogram for Zn in humus is spatially the most coherent, with a range of $c.$ 4.5 km (Fig. 9b). Generally, the poorest variograms were produced from Cu, in all media. The variograms for Cu were characterized by large nugget effects and large semivariance in the first lag (first distance interval used for calculation of semivariance) suggesting extreme variability between sample points at short distances.

Figure 9a shows high concentrations of Cu in till in an elongated zone centered over the Shunsby base-metal prospect (A) and a zone to the west at the intersection of the Wakami shear and Ridout high strain zones (H). Figure 9b shows a N–S trending zone (I) that is elevated in Zn in humus extending from the north of the SGB and terminating over the granitoid rocks in the south. The southern portion of this anomalous zone is centered on an area of organic and glaciolacustrine surficial deposits and the northern and central portions are roughly centered on N–S trending esker and glaciofluvial ice-contact deposits (see Fig. 2). This suggests that the anomaly may be surficial in origin, as opposed to reflecting bedrock. The area of apparent low concentration in both Cu in till and Zn (J on Figs. 9a & 9b) reflects an area of low sample density over a wide area of thick glaciolacustrine deposits. This area was not extensively sampled as the samples would not provide a representative picture of anomalies in the underlying bedrock (glaciolacustrine deposits would mask the geochemistry of the underlying bedrock).

Ternary images. Red–Green–Blue (RGB) ternary images (Broome *et al.* 1987; Harris *et al.* 1990), often used in the display of remotely sensed data, are also effective for visualizing interpolated geochemical data. These maps or images are constructed by taking three continuous surface (interpolated) geochemical maps, and displaying one in a blue, one in a green and the other in a red colour. The additive mixes of the primary colours, which results in a wide range of hues, can be interpreted with respect to the varying concentrations of each geochemical element, allowing the spatial associations between geochemical elements to be visually assessed.

Figures 10a and 10b show RGB ternary maps for total Zn and Cu concentration in till, soil and humus, respectively. Proportional circles (dots) in which the size of the circle reflects total base metal content and the wedges, the contribution of each medium to the total content (see Fig. 8), have been overlaid to facilitate comparison between the geochemical data visualized as points and the underlying interpolations. With respect to the Zn ternary image (Fig. 10a), it can be seen that humus is the dominant medium, as reflected by wide spatial variations in red hues. This is also verified by the box and whisker plot of Zn concentrations by medium (Fig. 4b). The Shunsby area (A) is elevated in Zn, in all media, as reflected by the white hue. Copper (Fig. 10b) shows much more variability in the different media, as shown by more variability in hue. The eastern portion of the study area is dominated by a NNW–SSE trending zone of yellow and green hues reflecting a predominance of Cu in soil and humus, whereas the western portion of the study area has a higher proportion of green hues, reflecting higher concentrations in soil, and blue hues

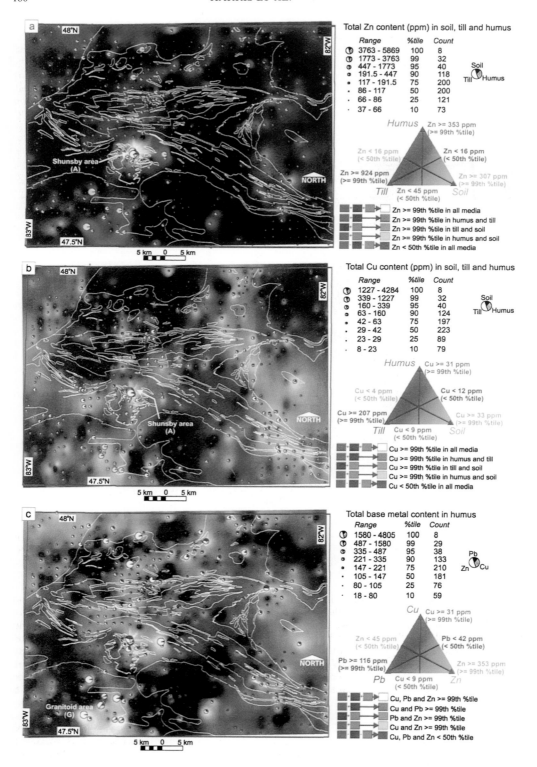

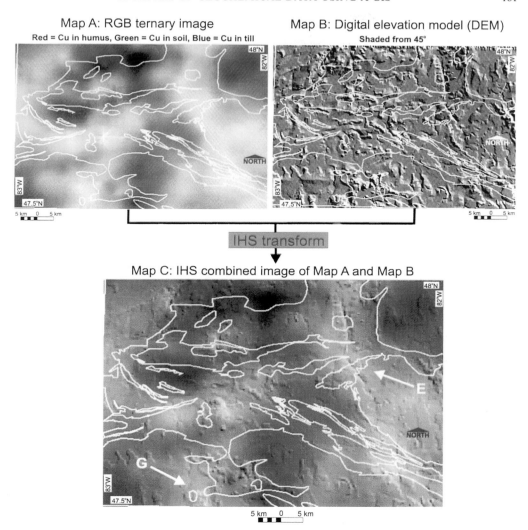

Fig. 11. Integrated geochemical image-maps. (**a**) RGB ternary map where Red, Cu in humus; Green, Cu in soil; Blue, Cu in till. (**b**) Digital elevation data (DEM) – artificial shading from 45°. (**c**) Ternary map shown in (a) and DEM shown in (b) combined using an IHS (intensity–hue–saturation) transform.

reflecting higher Cu concentrations in till. The central portion of the SGB over the Shunsby area (A) is dominated by elevated concentrations of Cu in till, as reflected by the bluish hue. Copper concentration is also relatively high in humus over granitoid rocks in the SW portion of the study area (G) as reflected by red hues.

Figure 10c is a ternary image that shows total base metal content (Cu + Zn + Pb) in humus. Total base metal content in humus over the study area is variable, reflected by a wide variation in hue. The northwestern portion of the study area is dominated by higher concentrations in Zn (green areas), whereas the eastern portion of the study area is dominated by higher concentrations in Pb and Cu, as indicated by a

Fig. 10. RGB (red–green–blue) ternary maps, (**a**) Zn concentration in till, soil and humus, (**b**) Cu concentration in till, soil and humus, (**c**) total base metal content in humus.

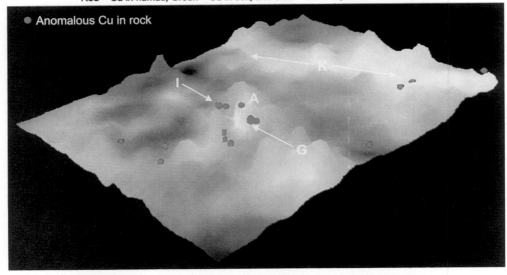

Fig. 12. 2.5 dimensional image-maps, **(a)** RGB ternary image showing Cu concentration in humus, soil and till (same as Fig. 11a) draped over a relief image representing Cu in lake sediments – Cu anomalies in rock have been overlayed as red dots, **(b)** RGB ternary image (as above) draped on topographical relief derived from a DEM.

magenta hue (red + blue). Local areas of elevated Cu concentration throughout the belt are reflected by a red hue. The black areas in Figure 10 reflect low element concentration and/or low sample density. Areas of low sample density can be masked out as demonstrated by Harris et al. (2000), resulting in an RGB ternary image in which areas of low sample density are not confused with areas of low element concentration.

Integrated images. The dispersion of elements in the surficial environment is often affected by topography (e.g. elevation, slope and aspect). Figure 11 shows an RGB ternary image in which log-transformed Cu concentrations in humus, soil and till are displayed in red, green and blue hues, respectively. This RGB ternary image uses the same data used in Figure 10b; however, the hues appear different due to the effect of a different contrast stretch algorithm applied to the data. This results in a different distribution of values, which in turn affects the overall distribution of hues. The RGB ternary image was then combined with a digital elevation model (DEM) to reveal possible associations between topographical and geochemical patterns. An intensity-hue-saturation (IHS) transform, available on most GIS, was used to combine the data (see Harris *et al.* 1990 and Harris *et al.* 1999*b* for a detailed description of the IHS transform).

The combined image (Fig. 11) shows that a N–S trending linear zone of Cu anomalies in humus and soil (greenish–yellow) is coincident with a N–S trending drainage channel in the eastern portion of the study area (see E) and that an area with high concentrations of Cu in humus is coincident with a zone of high relief over an area of granitoid rocks in the south (G).

2.5 dimensional images. The GIS can also make use of relief to portray geochemical data to advantage (Grunsky & Smee 1999). Figure 12a shows the RGB ternary map shown in Figure 11, painted over a 2.5 dimensional surface (artificial relief) representing Cu concentration in lake sediments. Anomalous Cu concentrations in rock are overlaid as red dots. The relationship between Cu in five media can therefore be seen on this one image. The areas of high relief representing anomalous Cu concentration in lake sediments are coincident with high concentrations of Cu in till, soil and humus (white hue) over the Shunsby area (A) and with high Cu concentrations in humus over granitoid rocks in the south (G). Elevated Cu concentration in lake sediments is focused in two N–S trending linear zones. The eastern zone (K) corresponds with anomalous concentrations in humus and soil (yellow hues) whereas the central linear zone (I) corresponds with high Cu in till in the north, all media in the central (white) and humus in the south (red) hues. These linear zones may reflect N–S glacial transport of Cu-rich debris as these zones crosscut regional stratigraphy. The Shunsby area (A) is also characterized by high Cu concentrations in rock.

Figure 12b is a similar image except that relief is represented by true topographic relief. The areas of elevated Cu in humus can clearly be seen to be coincident with an area of high topographic relief (G). The two distinct large N–S drainage channels are easily recognizable, the western-most (K) coincident with Cu anomalies in humus and soil, as previously mentioned.

Summary. Proportional dot and pie maps are simple to construct in a GIS and, since the points are not interpolated, no potential artifacts or generalizations due to the interpolation process are produced. However, only one element can be shown at a time (unless pie plots are used), and often percentile divisions, which is the most popular method of ranking and displaying the data, may not be sensitive to natural population breaks (thresholds) in the data. Furthermore, it is difficult to compare different elements both statistically and spatially.

Variograms provide information on spatial structure enabling the modelling of anisotropies such a glacial dispersion. Kriging generates an interpolated geochemical map that emphasizes regional geochemical patterns and also provides a measure of uncertainty in the interpolation process. This allows areas of high uncertainty in the kriging process to be excluded from the final interpolated map. Furthermore, producing a sample density map using the GIS can also identify areas of low sample density and these areas can also be excluded in the final map. Interpolated geochemical data often reveal both regional and subtle geochemical patterns to a greater extent than dot maps.

Using variations in the additive mixture of colours, RGB ternary images are useful for the multivariate comparison of geochemical patterns, either between different elements or between the same element sampled from different media. This not only highlights areas of anomalous concentrations but also the patterns identified, and the resulting colours may shed light on surficial processes at work over a given geographical area (e.g. processes of glacial dispersion, scavenging). Thus, RGB ternary images are not only useful for identifying anomalous concentrations of elements, but also for qualitatively identifying spatial correlations between three (or more) elements.

Combinations of geochemical data with other geoscience data such as topographical information using GIS technology (IHS transform) can reveal associations between geochemical data and other surficial factors that may be important for mineral exploration. Adding the third

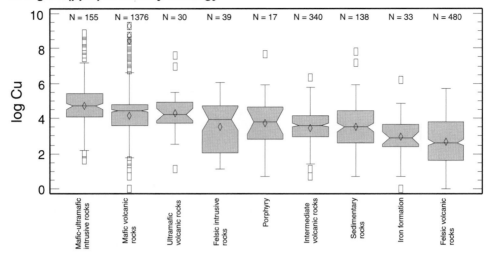

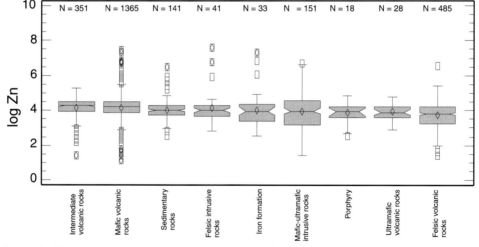

Fig. 13. Box and whisker plots showing log Cu and log Zn concentration by lithological unit, (**a**) log Cu from highest to lowest concentration, (**b**) log Zn from highest to lowest concentration. N, Number of samples.

dimension allows for the comparison of at least four variables in one image, facilitating the comparison between elements or media. These images can assist in revealing spatial associations between different elements and/or between elements and topography.

Lithological signatures in geochemical data sampled from surficial media

Relevant questions for mineral exploration when analysing geochemical data sampled from surficial media include:

(a) do element concentrations in surficial media vary over mapped surficial units;
(b) is there a bedrock signature in the geochemical data sampled from surficial media;
(c) do variations in element concentration in surficial media (till, humus, soil) reflect the concentration in the bedrock spatially and statisically;

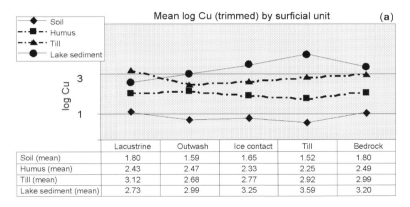

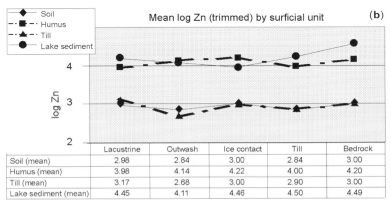

Fig. 14. Plots showing log Zn and log Cu concentration (mean values) in till, humus, soil and lake sediments by surficial unit (see Fig. 3), (**a**) log Cu by surficial unit, (**b**) log Zn by surficial unit.

(d) is there a spatial relationship between geochemical anomalies in bedrock and surficial media?

The GIS can be effectively utilized to help answer these questions by employing the 'point-in-polygon' procedure, to partition the geochemical data into separate groups based on mapped lithological and surficial units. A number of techniques can also be employed to assess the spatial relationships between geochemical signatures in different media. Examples of GIS analysis techniques used in concert with statistical analysis methods which address the questions listed above are illustrated in the following sections.

Statistical analysis of Zn and Cu by lithological and surficial units. Figures 13a and 13b are box and whisker plots showing log-transformed Cu and Zn concentrations in rock by mapped lithology (units shown on Fig. 1) ranked by concentration from highest on the left to lowest on the right. Outliers from each geochemical dataset have been removed (trimmed at approximately the 96th percentile for each data distribution) since, in this analysis, a mean signature for each lithological unit is required. Outliers may represent mineralization signatures as opposed to lithological signatures. Zinc concentration in rock is on average 50% higher than Cu, irrespective of lithological unit. Copper concentration is generally more variable between lithological units than Zn concentrations. This observation is also supported by generally more erratic variograms for Cu than Zn in the various media. Differences between Cu populations are also more significant than Zn populations between lithological units, as indicated by more variation in median values with less overlap between 95% confidence notches in the box and whisker plots. Copper is highest in ultramafic and mafic volcanic rocks and lowest in felsic volcanic rocks whereas Zn is highest in intermediate and mafic volcanic rocks and also lowest in felsic volcanic rocks.

Figures 14a and 14b are plots of mean values for log Cu and log Zn, respectively, for each surficial unit shown on the regional surficial geology map (Fig. 2). Once again the data have been trimmed. Error bars are not shown on these plots for the sake of clarity. However, both median and mean values where tested for significance using a Mann-Whitney test and t-test, respectively. Significant differences between median values were also assessed by using box and whisker plots (not shown). Both differences between mean values between log Cu and log Zn by medium (vertical direction on mean plots) and between values for each media with respect to different surficial units (horizontal direction on mean plots) were assessed for significance.

Many of the same trends seen on Figure 13 are also evident on Figure 14. Zinc concentration is appreciably higher than Cu in soil, by a factor of two and is c. 50% higher in humus. The concentrations of Cu and Zn in till are approximately the same. Zinc in lake sediments is greater than Cu in lake sediments by a factor of one. The higher levels of Zn is not surprising as Zn in rock is roughly 50% higher than Cu (see above). Zinc is highest in lake sediments and rock whereas Cu is highest in lake sediments over all surficial units except for lacustrine deposits.

The majority of differences between mean (also median) concentrations in each medium between surficial units are significantly different for Cu (Fig. 14a – compare means between media in a vertical direction on plot). With respect to Zn concentrations, the differences are significant between only the two obvious groupings seen on Figure 14b, indicating that between surficial units, lake sediments and humus have similar Zn concentrations, whereas till and soil have similar concentrations but are appreciably lower than the other media. Although there are differences between mean (and median) values in each medium between surficial units, many are not significant. Copper concentration is more variable than Zn concentration between surficial units as was the case for lithological units (Fig. 13). In general, Cu and Zn concentrations in lake sediments are highest in areas of till, which is generally thin in most parts of the SGB, and over areas of bedrock where weathering products of Cu- and Zn-bearing minerals would be deposited and dispersed locally within drainage basins. Most of the significant differences between populations for the surficial media across surficial units are between till, bedrock and glaciolacustrine deposits. Zinc and Cu concentrations are relatively high in glaciolacustrine deposits which may be acting as a *sink* for Cu and Zn through effects such as scavenging by Mn and Fe oxides.

The next problem is to determine whether the variability in Cu (and to a lesser extent Zn) concentrations between lithological and surficial units can be detected in the geochemical data obtained from surficial media. Figures 15a and 15b show log Zn and log Cu concentrations in rock and surficial media over each lithological unit. Ultramafic units are not shown due to a lack of geochemical samples. Again, error bars are not shown, but as before, differences between median and mean values were tested for significance using tests mentioned previously. Two groups are once again evident for Zn concentration (Fig. 15a): Zn is at higher concentrations in rock, lake sediments and humus and lower concentrations in till and soil. The differences in these two groups are significant statistically. A number of differences between media in the first group (rock, lake sediments, humus) are significant; however, there are no significant differences between lithology except between felsic volcanic rocks and other lithological units. The variation in mean (and median) Zn concentrations between lithology are similar for rock, lake sediment and humus except over felsic rocks where concentration is low in rock but relatively high in humus and lake sediments. This indicates that lake sediments and humus are varying sympathetically with bedrock and thus reflect bedrock signatures. Zinc in till and soil also vary sympathetically but at much lower concentrations. This divergence over felsic rocks may result from the geometry of the felsic units with respect to the dominant glacial dispersion direction. The felsic units trend generally in an E–W direction (see Fig. 1), perpendicular to the N–S glacial flow and are generally thin. Thus the relatively high Zn concentrations in surficial sediments may reflect the presence of mafic volcanic rocks north of each narrow felsic unit.

Copper concentrations (Fig 15b) between media for each lithology are significantly different except for till and lake sediments. Copper

Fig. 15. Plots of mean values for log Zn, Cu and Mg by lithological unit, (**a**) Zn in soil, humus, till, lake sediments and rock, (**b**) Cu in soil, humus, till, lake sediments and rock, (**c**) Mg in soil, humus, till, lake sediments and rock.

(a)
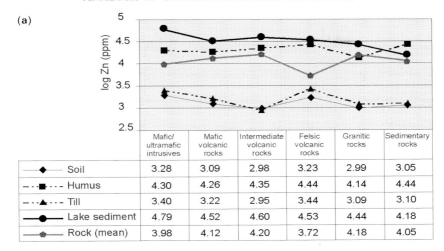

	Mafic/ultramafic intrusives	Mafic volcanic rocks	Intermediate volcanic rocks	Felsic volcanic rocks	Granitic rocks	Sedimentary rocks
◆ Soil	3.28	3.09	2.98	3.23	2.99	3.05
■ Humus	4.30	4.26	4.35	4.44	4.14	4.44
▲ Till	3.40	3.22	2.95	3.44	3.09	3.10
● Lake sediment	4.79	4.52	4.60	4.53	4.44	4.18
Rock (mean)	3.98	4.12	4.20	3.72	4.18	4.05

(b)
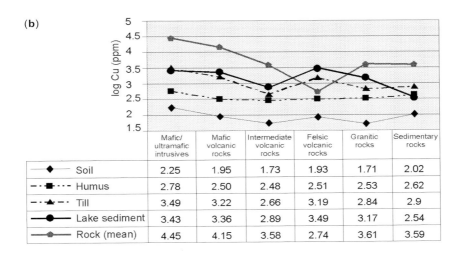

	Mafic/ultramafic intrusives	Mafic volcanic rocks	Intermediate volcanic rocks	Felsic volcanic rocks	Granitic rocks	Sedimentary rocks
◆ Soil	2.25	1.95	1.73	1.93	1.71	2.02
■ Humus	2.78	2.50	2.48	2.51	2.53	2.62
▲ Till	3.49	3.22	2.66	3.19	2.84	2.9
● Lake sediment	3.43	3.36	2.89	3.49	3.17	2.54
Rock (mean)	4.45	4.15	3.58	2.74	3.61	3.59

(c)
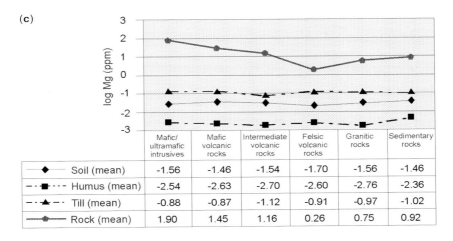

	Mafic/ultramafic intrusives	Mafic volcanic rocks	Intermediate volcanic rocks	Felsic volcanic rocks	Granitic rocks	Sedimentary rocks
◆ Soil (mean)	-1.56	-1.46	-1.54	-1.70	-1.56	-1.46
■ Humus (mean)	-2.54	-2.63	-2.70	-2.60	-2.76	-2.36
▲ Till (mean)	-0.88	-0.87	-1.12	-0.91	-0.97	-1.02
Rock (mean)	1.90	1.45	1.16	0.26	0.75	0.92

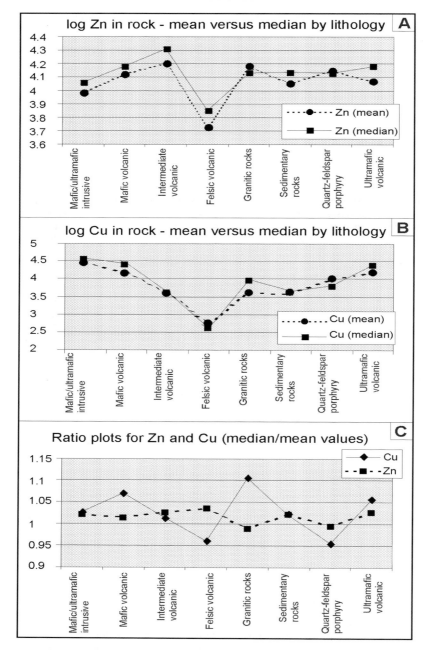

Fig. 16. Plots of the mean and median values and ratio between mean and median values for lithological units, (**a**) log Zn by lithology, (**b**) log Cu by lithology, (**c**) ratio of mean and median values for log Zn and log Cu by lithology.

concentrations between lithological units are more variable and the differences more significant than Zn concentrations, except for humus. Copper in till varies more sympathetically with rock than does Cu in soil and humus, except for felsic volcanic rocks for reasons discussed previously. A strong trend in decreasing Cu concentrations in rock can be seen across mafic

Table 5. *Correlation of Zn, in various media, and Cu in various media, based on interpolated geochemical maps*

	Humus	Soil	Till	Rock	Lake sediment
Humus		0.15 (**0.52**)*	0.04 (0.22)	-0.04 (0.01)	-0.01 (0.33)
Soil			**0.68** (**0.52**)	-0.02 (0.03)	0.18 (**0.41**)
Till				-0.01 (0.06)	**0.41** (0.30)
Rock					0.07 (0.01)
Lake sediment					

* Number in brackets is correlation for Cu, non-bracketed correlation is for Zn

rocks starting from ultramafic rocks, through mafic, intermediate and felsic rocks. This trend parallels a trend in Mg concentration in rock (Fig. 15c) and may reflect a fractionation trend as shown by a moderately strong Spearman rank correlation coefficient (0.5) between MgO and Cu in rock.

Figure 16 shows plots of mean and median values for log Zn (Fig. 16a) and Cu (Fig. 16b) concentrations in rock by lithology and a plot of the ratio between the median and mean values for Zn and Cu in rock by lithology (Fig. 16c). The data were not trimmed, as these plots can be used not only to assess the normality of each population but also as a mineralization indicator for each lithology. Larger ratio values (Fig. 16c) indicate median values greater than mean values reflecting a positively skewed distribution with the presence of many outliers (e.g. anomalously high values) possibly reflecting mineralization. Thus, with respect to Zn, although concentrations are not particularly high over felsic volcanic rocks (see Fig. 16a), the large ratio values for felsic volcanic rocks (Fig. 16c) indicate a higher percentage of outliers and therefore a higher prospectivity for Zn. Ultramafic and sedimentary rocks are also characterized by larger ratios. With respect to Cu concentrations, granitoid, mafic and ultramafic rocks have the largest ratios but not necessarily highest Cu concentrations (Fig. 16b). Opposite concentration trends for Cu and Zn are evident on Figures 16a and 16b for ultramafic, mafic and intermediate volcanic rocks: Zn concentration increases whereas Cu concentration decreases.

Summary. The 'point-in-polygon' technique automates the process of determining the map unit (e.g. lithology) in which a point (e.g. geochemical sample) occurs making it possible to analyse and compare geochemical populations by surficial or lithological units. This can result in more meaningful interpretation than when comparing total populations.

Copper and to a lesser extent Zn, show larger variations in concentration between media for each lithological unit than within each medium over the various lithologies. Copper concentration is generally more variable between lithological units than is Zn. Distinctive Zn and Cu signatures are evident in surficial media, but are suppressed with respect to the signatures in rock due to the homogenizing effect of weathering and erosion processes. The similarity in Zn and Cu concentrations in rock and lake sediments indicates that geochemical signatures in lake sediments (at least for Zn and Cu) reflect local bedrock concentrations. Till, soil and humus, with the exception of Zn concentrations in humus, do not reflect bedrock signatures as closely.

The effect of glacial dispersion on Zn and Cu concentrations in till and soil

One concern for exploration is whether elevated levels of Cu and Zn in rock are related, statistically and spatially, to elevated levels in surficial media. To address this issue, a nearest-point search algorithm (available on many GIS) was employed. The advantage of the nearest-point method is that it takes into account spatial variations in Cu and Zn concentration over distance and direction within the study area. Thus, assessment of varying patterns by direction and distance is possible, as opposed to a strictly global statistical measure which correlation coefficients provide.

Table 5 presents global correlation coefficients (i.e. coefficients between the total data population over the entire study area) between Zn and Cu in rock, till, humus, soil, and lake sediments. The correlations are based on spatially interpolated data as the surficial media were sampled at different geographical locations than rock and lake sediment samples, making point-to-point comparisons impossible. The highest correla-

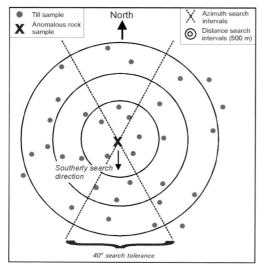

Fig. 17. Cartoon diagram of nearest-point search strategy showing distance radii and directional search wedges.

tions are shown by: (1) Zn in soil and till; (2) Cu in soil and till; (3) Zn in till and lake sediments; (4) Cu in soil and humus; and (5) Cu in soil and lake sediment. The results for both Zn and Cu suggest that soil is directly derived from the underlying thin till over much of the belt. The correlation between Zn in till and lake sediments is reasonable, as both lake sediments and till would present an average signature of Zn over broad zones (i.e. drainage basin and glacial dispersion area). In addition, given the greater mobility of Zn in the surficial environment and the higher background Zn levels (in a range of rock-forming minerals), this is not an unexpected result. The moderately strong correlation between Cu in soil and humus is also not unexpected given the well-known ability of organic matter to scavenge Cu.

There is little correlation between Zn and Cu concentrations in rock with other media. This is not entirely unexpected, as correlation coefficients are global in nature, and do not account for spatial factors such as glacial dispersion and variations due to lithology. Furthermore, the zone of influence for a rock sample is small, whereas the zone of influence for till, lake sediments, and to a lesser extent soil and humus, are broader.

With respect to the nearest-point search process, the data distributions for Cu and Zn concentrations in rock were divided into separate percentile rankings (>95th and >75th percentiles), and the samples above these percentiles were used as a basis for comparison with Zn and Cu in surficial media. The identified Zn and Cu samples in rock can be regarded as anomalous (> 95%) and possibly anomalous (>75%). The nearest-point method involves searching from each rock sample with anomalous Zn and Cu concentration to the closest Zn and Cu anomaly in the another media (till, soil, humus) out to a maximum distance of 3 km in all directions, or optionally in specific directions chosen by the geologist. The mean and median concentration of Zn and Cu are then plotted against distance. Figure 17 is a cartoon example illustrating how the nearest-point algorithm searches by distance and direction, or by a combination of the two (i.e. search by distance intervals in a specific direction).

Figures 18 and 19 shows cumulative plots of median and mean concentration of log Zn and log Cu values in till and soil plotted by distance (using an omni-directional search) to the nearest Zn and Cu anomaly in rock, for rock samples above the 95th percentile. The same plots were produced for rocks samples above the 75th percentile (graphs not shown). Generally, Cu and Zn concentrations, for the 95th percentile ranking, in till and soil decrease with distance from an anomaly in rock suggesting that Zn and Cu anomalies in till reflect anomalies in bedrock within a distance of 1 km. The plots for humus and soil (not shown) were more variable, and did not show a clear trend of decreasing concentrations of Zn and Cu concentration with distance from an anomalous rock sample. Box and whisker plots are also included in Figures 18 and 19 to determine whether concentrations are statistically different with distance. Differences in Zn and Cu median values in till and soil, between the closest distance to a rock anomaly (0–0.5 km) and other distances (e.g >0.5 km), are significant (median notches do not overlap). The exception is for Cu concentration in soil (see Fig. 19). A significant difference was also noted between Zn concentrations in both till and soil with distance from a Zn sample in rock (>75%) (graph not shown). However, no difference in

Fig. 18. (**a**) Cumulative plot of log-Zn concentration (mean and median values) in till and soil by distance to nearest anomalous rock sample (values for rock are above the 95th percentile), (**b**) Box and whisker plot (not cumulative by distance) of log Zn in till by distance to Zn anomaly in rock, (**c**) Box and whisker plot (not cumulative by distance) of log Zn in soil by distance to Zn anomaly in rock.

(a)

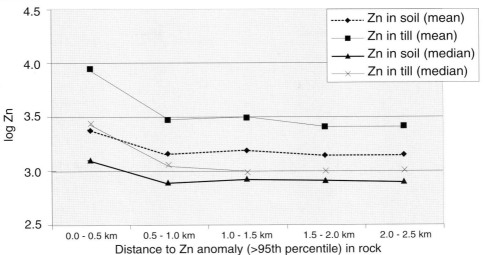

A. Distance/concentration plot: log Zn in till and soil versus Zn anomaly (>95th percentile) in rock

(b)

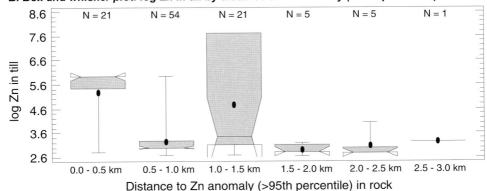

B. Box and whisker plot: log Zn in till by distance to Zn anomaly (>95th percentile) in rock

(c)

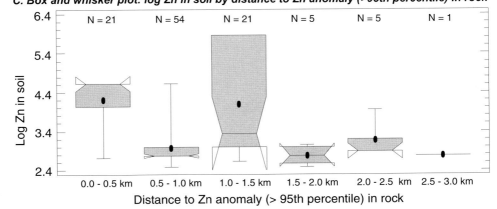

C. Box and whisker plot: log Zn in soil by distance to Zn anomaly (>95th percentile) in rock

(a)

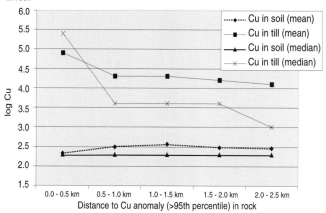

A. Distance/concentration plot: log Cu in till and soil versus Cu anomaly (>95th percentile) in rock

(b)

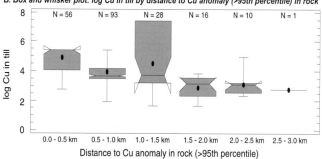

B. Box and whisker plot: log Cu in till by distance to Cu anomaly (>95th percentile) in rock

(c)

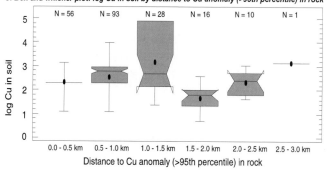

C. Box and whisker plot: log Cu in soil by distance to Cu anomaly (>95th percentile) in rock

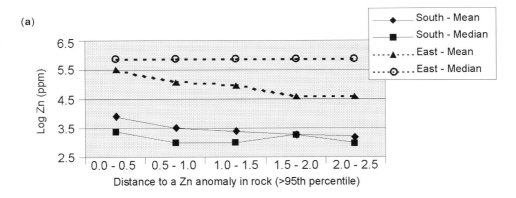

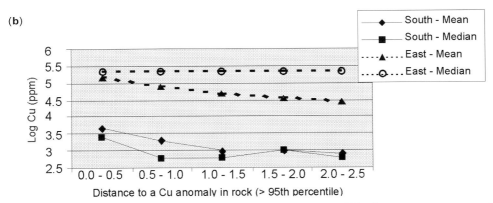

Fig. 20. (a) Cumulative plot of log Zn concentration (mean and median) in till by distance to nearest anomalous rock sample (values for rock are above the 95th percentile)- southerly and easterly search directions, (b) Cumulative plot of log Cu concentration (mean and median) in till by distance to nearest rock sample (values for rock are above the 95th percentile)- southerly and easterly search directions.

Cu concentration at the 75th percentile was evident.

The search was then divided by direction (four cardinal directions) with a 40° tolerance (see Fig. 17). Only the south and east searches resulted in the trends seen in Figure 20. The decrease (albeit weak) in Cu and Zn concentration in till in a southerly direction suggests that this may be a down-ice dispersion effect, whereas the easterly trend may reflect the E–W strike of bedrock lithological units. It is interesting to note that concentration levels (as indicated by mean values) are higher in an easterly search than a southerly search direction, again reflecting the general E–W strike of lithological units.

Summary. The GIS can be used to compare the spatial relationship between geochemical anomalies in rock and anomalies in surficial media by using a nearest-point search algorithm which can be programmed to search by distance and in any given direction. This can direct an explorationist's search from a given element anomaly in rock in a direction in which the concentration of the same element in surficial media is the highest. From a process point of

Fig. 19. (a) Cumulative plot of log Cu concentration in till and soil (mean and median) by distance to nearest anomalous rock sample (values for rock are above the 95th percentile), (b) Box and whisker plot (not cumulative by distance) of log Cu in till by distance to Cu anomaly in rock, (c) Box and whisker plot (not cumulative by distance) of log-Cu in soil by distance to Cu anomaly in rock.

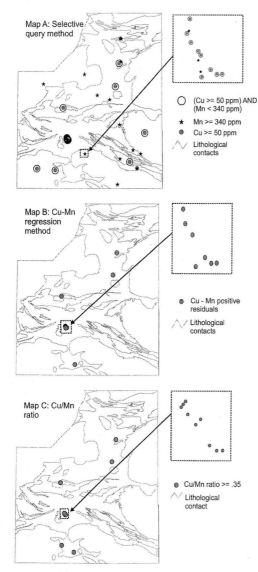

Fig. 21. Maps of Cu samples from soil screened for Mn scavenging, (**a**) selective query, (**b**) regression – Cu v. Mn, (**c**) Cu : Mn ratio.

data can be detected using the nearest-point technique. The E–W bias can be attributed to the general E–W strike of stratigraphy whereas the N–S bias is a reflection of down-ice glacial dispersion of Zn- and Cu- bearing minerals.

Identifying geochemical anomalies due to mineralization

GIS/statistical methods can be used to screen or filter out anomalies in soil, humus or lake sediment data that may be due to factors not related to mineralization such as lithological variations and/or metal scavenging due to organic matter and Fe and Mn oxides. Many methods for screening geochemical data for scavenging effects exist. Four related methods that are easily implemented in a GIS are reviewed in this section: (a) selective query; (b) linear regression; (c) ratios; and (d) RGB ternary maps. These methods can be applied to multi-media data from the same or different geographical locations. The former involves processing the data as discrete points whereas the latter necessitates interpolating the data to produce continuous surface maps. Copper concentration in soil is used as an example to demonstrate the advantages and disadvantages of each method.

The selective query method involves inspecting normal probability plots of Cu, Mn, Fe and LOI (as a measure of organic matter) and selecting upper breakpoints that divide the geochemical population into background and anomalous groups. The GIS is then used to select anomalous Cu samples that are not coincident with anomalous Fe, Mn or LOI concentrations. This can be done using a SQL statement as follows:

Select (Cu > = 80 ppm) AND (Mn < 40 ppm) AND (Fe < 30 ppm) AND (LOI < 5%)

Figure 21a shows the results of the selective query process in which anomalous Cu concentrations has been screened for Mn scavenging using the following query:

Select (Cu > 50 ppm) AND (Mn < 340 ppm)

Conversely, Cu may be regressed against Fe, Mn or LOI as explanatory variables and only the residual values (i.e. values that do not fit the linear model within a certain prediction interval) selected. Residuals represent samples where there is little correlation between the elements (Cu v. Fe, Mn and LOI) under study. The regression may be univariate, taking one ex-

view this type of analysis can detect down-ice dispersal patterns in till in the absence of an orientaion survey or detailed mapping, and also identify the effects of lithology on geochemical trends and patterns that may be present in surficial media. Zinc and, to a lesser extent, Cu concentration in till is related to Zn and Cu concentration in bedrock within distances of c. 1 km. An E–W and N–S directional bias in the

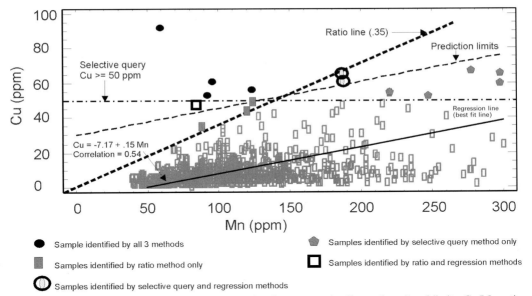

Fig. 22. Scatterplot between Cu and Mn in soil showing linear regression line and predicted limits, Cu/Mn ratio (threshold line –0.35) and selective query threshold.

planatory variable at a time, or multivariate where all three explanatory variables are used simultaneously.

Figure 22 is a scatterplot of Mn and Cu concentrations in soil; a moderate correlation of 0.54 exists. The best fit line, as determined by linear regression, and lines showing the predicted limits are included on this plot. The lines showing the prediction limits represent the limits for the forecast of the fitted value (Cu) associated with a given value of the independent variable (Mn). The positive residuals that are elevated in Cu (above predicted line) and low in Mn (<340 ppm), but greater than the predicted line, are highlighted as square and circle symbols on the plot. These are the samples shown in Figure 21b.

Ratios, where Cu is divided by Fe, Mn or LOI can be used as a method to correct for scavenging effects. Ratios can be calculated within the GIS (or statistical package) and since only high ratio values (e.g. high Cu and low Mn for example) are important, a probability plot of the ratio values can be constructed and anomalous upper thresholds can be selected which presumably reflect samples free from scavenging effects. Figure 21c shows samples with high ratios (>0.35) between Cu and Mn. High ratio values were determined by identifying an upper break (threshold) on a normal probability plot of the ratio values.

Figure 22 indicates that all three methods are similar, differing only slightly on how the samples with elevated Cu and low Mn are identified. Over a 75% overlap in samples identified by all three methods exists, although there are some unique combinations of samples identified by only one or two methods. The selective query method is simple and straightforward as samples which fall above 50 ppm Cu and below 340 ppm Mn (threshold not shown on Fig. 22) are selected. With respect to the regression method, samples above the predicted line are selected (Fig. 22) and for the ratio method, samples above a ratio threshold (>=0.35) are highlighted. The selective query method resulted in 16 samples high in Cu and low to moderately low in Mn. The regression method identified 12 samples whereas the ratio method identified 14 samples.

Figure 23 presents a RGB ternary map, which is another method of visually screening anomalies due to scavenging. This map shows concentrations of Cu, Mn and Fe in lake sediments as red, blue and green colours, respectively. Thus, intense red anomalies represent elevated Cu concentrations whereas intense yellow hues represent coincident Cu and Mn anomalies, magenta hues represent coincident Cu and Fe anomalies and white represent coincident Cu, Mn and Fe anomalies. Thus, anomalies displayed in yellow, magenta and white might be considered as lower priority, as they may represent Cu anomalies due to scavenging. One must be cautious when visually interpreting the colours in Figure 23 as the white area, which reflects coincident Cu, Mn and Fe anomalies is also coincident with mineralized Fe formations and therefore may be more a reflection of the underlying geology as opposed to scavenging processes.

Summary. All methods screened for the effects of scavenging yield similar results. The selective query technique is perhaps the simplest and most effective method, as it can be applied directly in the GIS using simple query statements. Caution must be used when using the ratio method as high ratio values can result from Cu concentrations that are not necessarily anomalous and are below upper thresholds present on probability plots.

Discussion and conclusions

Zinc and Cu concentrations in the various media are summarized from Figure 4 as follows:

(a) Zn – lake sediments > (humus > rock) > (till > soil)
(b) Cu – rock > lake sediments > till > humus > soil

The media enclosed in brackets above do not have significantly different concentrations of Zn. Zn concentrations in all media are higher than Cu. This may be due to generally higher background concentrations, as a function of the predominance of mafic and intermediate volcanic rocks, and greater mobility. Concentrations of Cu and Zn in till are similar. The greatest amount of mixing would occur in till thus averaging out total concentration (similar to the effect of lake sediments) resulting in similar Zn and Cu concentrations at least in the clay + silt fraction of glacial till. Concentration of Zn and Cu in lake sediments and rock are the highest and typified by the highest contrast anomalies. Soil and, to a lesser extent, till (especially for Zn) have the lowest concentrations of Zn and Cu as well as the lowest contrast between anomalies and background.

The highest concentration for Zn is found in lake sediments. This is not unexpected given the relative mobility of Zn and the large catchment areas involved in lake sediment sampling. In addition, mafic volcanic rocks dominate the study area and in these rocks Zn can readily substitute in mafic-forming minerals resulting in higher background levels. The relatively high Zn concentration in humus is somewhat anomalous. The following factors may be important in accounting for this observation: (1) scavenging of Zn by organic matter and colloidal Fe and Mn oxides still present in the humus; (2) vegetation tapping mineralized bedrock directly (in thin soil/till conditions) resulting in higher concentrations in humus than in soil or till; and (3) higher background levels of Zn (see above) combined with preferential uptake of Zn by vegetation and subsequent recycling in humus. It is also noteworthy that Zn concentration in humus resulted in the best-structured variogram. This is also somewhat surprising, given that humus is generally considered to reflect element concentrations local to the sample point (i.e small zone of influence). This result may suggest that humus has a homogenizing effect (e.g. scavenging) on more mobile elements such as Zn, thus averaging out concentrations over space, or that trees (from which the humus is derived) are efficient in gathering Zn over broad areas.

The concentration of Zn in rocks is generally less variable than Cu. Zinc exhibits both chalcophile and lithophile characteristics and can readily substitute in major-rock forming silicate minerals such as biotite, pyroxene and amphiboles; oxide minerals such as magnetite; and in sulphides (sphalerite). Copper displays chalcophile characteristics and prefers sulphides to silicates or oxide minerals. Thus, Cu will be found predominately in sulphides (e.g. chalcopyrite, covellite) whereas Zn can be found both in sulphides, and in common rock-forming silicate minerals. Therefore, high Cu concentrations may reflect primarily sulphide mineralization whereas high Zn concentrations may represent both sulphide mineralization and lithological variations. The more homogeneous distribution of Zn between lithological units suggests that Zn concentration is more a reflection of lithological variations (e.g. more Zn occurring ubiquitously as a substitute mineral in mafic-forming minerals). This may reflect a general trend (or difference) between lithophile and chalcophile elements.

Copper concentration is highest in rock and given that most Cu will be found in sulphides, anomalies in Cu are more likely to reflect mineralization or extensive weathering of sulphides. High concentrations in lake sediment again may represent a mineralization signature rather than primarily a lithological signature, as is the case for Zn. This illustrates the importance of accounting for lithological variations using normalization techniques. Copper concentration in humus is lower than Zn in humus.

Global correlation of Cu and Zn in the various media is highest between soil and till, suggesting that in the generally thin till conditions across the belt, soil is derived directly from the till, and that the till is for the most part residual (travelled short distances generally $<=3$ km). This suggests that either till or soil can be used as an effective medium for exploration purposes in Archean greenstone belts covered by a thin veneer of till.

With respect to Zn concentration between lithological units, rock, lake sediments and humus have higher Zn concentrations than soil and till. Zinc concentration in humus varies sympathetically with lake sediments across lithology as does soil and till. Humus is similar to rock except over felsic volcanic rocks. With respect to Cu concentrations between lithological units, all media vary sympathetically with lithology across ultramafic, mafic and intermediate rocks and to a lesser extent over felsic, granitoid and sedimentary rocks. Copper concentrations in lake sediments and till are similar and vary sympathetically over all rock units. These results suggest that lake sediments and humus (at least in thin till conditions) are good media for sampling for base metal exploration as they reflect a broad geochemical signature that is reflective of bedrock variations.

Copper and, even more so, Zn concentrations in till (and to a lesser extent soil) are related to anomalous concentrations in rock within a distance of 1 km. This, again, may reflect a stronger bedrock signature for Zn than Cu concentration in till. The GIS, using a nearest-point algorithm, can be used to identify directional anisotropies in geochemical data sampled from till, thus assisting in geographically narrowing down prospective targets. This is especially important for exploration in areas that are not well mapped (e.g. in areas where information on surficial units and glacial movement direction is lacking).

With regard to visualizing geochemical data, RGB ternary images are effective in presenting interpolated geochemical data. However, continuous surface (interpolated) geochemical maps can only be generated if certain criteria, discussed in this paper, are met (e.g. sufficient sample density, well-behaved variograms). Otherwise, the data should be visualized and analysed as discrete points or with a natural zone of influence (e.g. drainage basin for lake sediment data) or using small zones of influence (e.g. buffers). Variogram analysis in conjunction with kriging is the most powerful method of interpolating geochemical data. Variograms provide critical information on spatial associations between points, directional biases and spatial structure. The degree of smoothing desired in the interpolation process obviously depends on what the data will be used for. If the goal is to preserve all high frequency variability, which may be important for regional mineral exploration, then less smoothing, as obtained from the IDW interpolations, is desirable. However, if the goal is to produce a geochemical map useful for characterizing surficial or lithological patterns then a certain degree of smoothing is warranted.

The hue variations in RGB ternary images can not only indicate anomalous areas (both in concentration and spatial pattern) worthy of exploration follow-up, but also suggest differences in surficial/glacial processes operating between different elements in the different media. For example, the dominance of Zn in the humus (Fig. 10a) may reflect a direct bedrock signature, greater mobility of Zn, greater scavenging of Zn by organic matter, preferential uptake of Zn by vegetation and subsequent recycling in humus and the relative homogenizing influence of humus with respect to Zn especially over a N–S linear trending zone in the central portion of the belt. Copper concentration is much more variable in the different media (Fig. 10b) reflecting more heterogeneous concentrations in bedrock and variability in surficial processes ranging from scavenging by humus in the southern granitoid rocks, mechanical distribution of Cu-bearing minerals in the central portions of the study area and a combination of scavenging and chemical breakdown of Cu-bearing minerals in soil and humus over primarily granitoid rocks in the eastern portion of the study area.

In processing geochemical data, normalization (e.g. selective query) to lithology is important for especially for lithogeochemical data (also see Harris et al. 1999a, 2000). The selective query process can identify anomalous elemental concentrations by lithology. Thus, a sample that may fall between the 70th and 80th percentile for the total population of samples (and not considered anomalous), may fall above the 95th percentile range for a specific lithology and thus would be selected as anomalous. The success of the normalization process in identifying anomalies depends on the element, and in this case, because Zn concentration is both reflective of lithological variations as well as mineralization, selective querying by rock unit may be too liberal in that anomalies due to lithological variations are also being identified. The majority of Cu anomalies reflect mineralization.

Correcting (screening) geochemical data (soil, humus, lake sediments) for extraneous effects

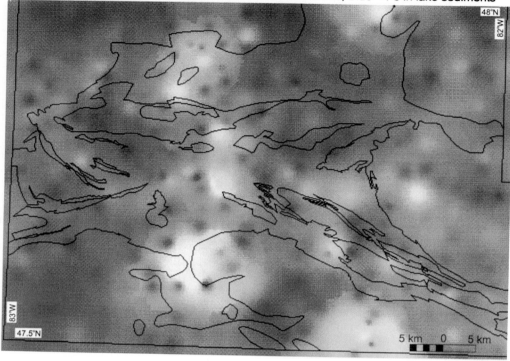

Fig. 23. RGB (red–green–blue) ternary map where R, Cu concentration in soil, G, Mn concentration in soil, B, Fe concentration in soil.

such as scavenging is important to help identify anomalies more likely due to mineralization. Providing the geochemical data pertain to the same geographical location, the selective query method presented in this paper is the simplest and most straightforward method for screening scavenging effects. An RGB ternary map showing the concentration of a given element (Cu, Zn) and elements that may be responsible for scavenging (Fe, Mn) assists in identifying areas where anomalies may be due to strong scavenging processes rather than mineralization. However, consideration of the underlying geology is crucial in the interpretation process.

In conclusion, the GIS offers geologists a powerful tool with which to manage, evaluate, analyse and visualize geochemical data sampled from a variety of media. GIS, in concert with statistical analysis and geostatistical analysis packages, can be used to fully assess the statistical and spatial characteristics of geochemical data using techniques ranging from simple SQL queries to more complex multivariate methods. Furthermore, analysis of geochemical data using a GIS can be conducted in a timely fashion which is an important factor when working under exploration deadlines. The GIS, in conjunction with other software tools, assists the geologist in extracting the most information from their data. More time can be spent on the analysis and visualization of the data rather than the laborious task of overlaying maps manually. Analysis of geochemical data using a GIS, in addition to providing information on geochemical anomalies, can often reveal subtle associations between different geochemical elements and/or different media through the integration and overlay of multiple layers of geochemical data. Lastly, but not insignificantly, the GIS provides an organized and geo-referenced archive for large volumes of geochemical data that can be accessed and updated quickly and efficiently.

The authors would like to thank Graeme Bonham-Carter, Gwendy Hall and Harvey Thorleifson, all from the Geological Survey of Canada, for thorough and thoughtful reviews of this paper. Additionally, Eric Grunsky (Government of Alberta) and one anonymous reviewer are also thanked for providing helpful reviews. All reviewers contributed to making this a much better (and succinct) paper. Falconbridge are thanked for contribution of large volumes of lithogeochemical data. Geological Survey of Canada No. 1999211.

References

AYER, J. 1995a. *Precambrian geology, northern Swayze greenstone belt, District of Sudbury*. Ontario Geological Survey Report **297**.

AYER, J. 1995b. *Precambrian geology, northern Swayze greenstone belt, District of Sudbury*. Ontario Geological Survey Map **2627**.

AYER, J. A. & THERIAULT, R. 1993. *Geology of Keith and Muskego Townships, Northern Swayze Greenstone Belt*. 1992-1993 Northern Ontario Development Agreement (NODA) Summary Report **26-33**.

BERNIER, M. A. 1994. Quaternary Geology and Surface Till Sampling Program, Rush Lake and Opeepeesway Lake Map Areas, Eastern Swayze Greenstone Belt. *In: Summary of Field Work and Other Activities 1994*. Ontario Geological Survey, Misc. Paper **163**, 241-243.

BERNIER, M. A. & GOFF, J. R. 1993. Quaternary Mapping and Surface Drift Sampling Program, Western Swayze Greenstone Belt. *In: Summary of Field Work and Other Activities 1993*. Ontario Geological Survey, Misc. Paper **162**, 250-255

BERNIER, M. A. & KASZYCKI, C. A. 1995. *Till, humus and B-horizon soil geochemical database, Surficial Sediment Sampling Program, Swayze greenstone belt, Northern Ontario, 1992-94*. Ontario Geological Survey, Miscellaneous Release – Data 15

BERNIER, M. A., HARRIS, J. R. & JOHNSTON, T. D. 1996. *Recent Developments and Digital Integration of the Surficial Sediment Compositional Database, Swayze Greenstone Belt, Northern Ontario*. NODA Summary Report, 97-101.

BONHAM-CARTER, G. F. 1994. *Geographic Information Systems for Geoscientists: Modeling with GIS*. Pergamon, Elsevier Science Inc., New York.

BONHAM-CARTER, G. F., ROGERS, P. J. & ELLWOOD, D. J. 1987. Catchment Basin Analysis Applied to Surficial Geochemical Data, Cobequid Highlands, Nova Scotia. *Journal of Geochemical Exploration*, **29**, 259-278

BROOME, J. H., CARSON, J. M., GRANT, J. A. & FORD, K. L. 1987. *A modified ternary radioelement mapping technique and its application to the South Coast of Newfoundland*. Geological Survey of Canada Paper **87-14**.

CHENG, Q., AGTERBERG, F. P. & BONHAM-CARTER, G. F. 1994. A Spatial Analysis Method for Geochemical Anomaly Separation. *Journal of Geochemical Exploration*, **56**, 183-195.

DAVIS, J. C. 1986. *Statistics and Data Analysis in Geology (2nd Edition)*. John Wiley and Sons, New York.

FRISKE, P. W. B., HORNBROOKE, E. H. W. & MCCURDY, M., 1988. *Regional lake sediment and water geochemical data, Ontario*. Geological Survey of Canada Open File **1640**, NGR 105

FUMERTON, S. & HOULE, K. 1993. *Mineral showings, occurrences, deposits and mines of the Swayze greenstone belt, interim report. Volumes 1 and 2.* Ontario Geological Survey Open File Report **5871**.

FUMERTON, S., HOULE, K. & ARCHIBALD, G., 1993. *Digital data on the mineral showings, occurrences, deposits and mines of the Swayze greenstone belt, plus a computer application to update and edit the data using Foxpro, interim report*. Ontario Geological Survey Open File Report **5872**.

GARRETT, R. G. 1991. *The Management, Analysis and Display of Exploration Geochemical Data, Exploration Geochemistry Workshop*. Geological Survey of Canada Open File **2390**, Paper **9**, 9–1 to 9–4.

GEOLOGICAL SURVEY OF CANADA (GSC), 1987. *Regional lake sediment and water geochemical data, Ontario*. GSC Open File **1357**, NTS 41J and 41O.

GEOLOGICAL SURVEY OF CANADA (GSC), 1988. *Regional lake sediment and water geochemical data, Ontario*. GSC Open File **1640**, NGR 105-1988, NTS 41P, part of NTS 31M.

GRUNSKY, E. C. 1986. Recognition of alteration in volcanic rocks using statistical analysis of lithogeochemical data. *Journal of Geochemical Exploration*, **25**, 157–183.

GRUNSKY, E. C. 1997. *Numerical Techniques and Strategies for the Interpretation of Geochemical Data*. Current Topics in GIS and Integration of Exploration Datasets Short Course., Exploration 97 Workshops, Ottawa.

GRUNSKY, E. C. & SMEE, B. W. 1999. The differentiation of soil types and mineralization from multi-element geochemistry using multivariate methods and digital topography. *Journal of Geochemical Exploration*, **67** (1–3), 287–299.

GRUNSKY, E. C., EASTON, R. M., THURSTON, P. C. & JENSEN, L. S. 1992. Characterization and Statistical Classification of Archean Volcanic Rocks of the Superior Province using Major Element Geochemistry in Geology of Ontario. Ontario Geological Survey, Special Volume **4**, Part 2, 1347–1438.

HARRIS, J. R., MURRAY, R. & HIROSE, T. 1990. IHS Transform for the integration of radar imagery and other remotely sensed data. *Photogrammetric Engineering and Remote Sensing*, **56**(12), 1631–1641.

HARRIS, J. R., GRUNSKY, E. C. & WILKINSON, L. 1997. Developments in the effective use and interpretation of lithogeochemistry in regional exploration programs: application of GIS technology. *In: Proceedings of Exploration 97: Fourth Decennial International Conference on Mineral Exploration*. Toronto, Ontario, Canada, 285–292.

HARRIS, J. R., GRUNSKY, E. C. & WILKINSON, L. 2000. Effective use and interpretation of lithogeochemical data in regional exploration programs. *Ore Geology Reviews*, **16**, 107–143.

HARRIS, J. R., WILKINSON, L. & VANDERKAM, R. 1999. *A compilation of OGS, GSC and industry geoscience data over the Swayze greenstone belt, Ontario.*

Geological Survey Open File **D3770** (CD-ROM), Ontario Geological Survey Miscellaneous release data (MRD) **47**.

HARRIS, J. R., WILKINSON, L., GRUNSKY, G., HEATHER, K. & AYER, J. 1999a. Techniques for analysis and visualization of lithogeochemical data with applications to the Swayze greenstone belt, Ontario. *Journal of Geochemical Exploration*, **67**(1–3), 301–334.

HARRIS, J. R., VILJOEN, D. & RENCZ, A. 1999b. Integration and Visualization of Geoscience Data. *In*: RENCZ, A. N. (ed.) *Remote Sensing for the Earth Sciences*. John Wiley and Sons Inc., New York, 707.

HEATHER, K. B. 1993. *Regional geology, structure, and mineral deposits of the Archean Swayze greenstone belt, southern Superior Province, Ontario*. Current Research Geological Survey of Canada Paper **93-1C**, 295–305.

HEATHER, K. B. & SHORE, G. T. 1999. *Geology of the Swayze Greenstone Belt, Ontario; eight (8) 1.50,000 scale maps and accompanying legend/marginal notes*. Geological Survey of Canada Open File Maps **3384a-i**.

HEATHER, K. B. & VAN BREEMEN, O. 1994. *An interim report on geological, structural and geochronological investigations of granitoid rocks in the vicinity of the Swayze Greenstone Belt, southern Superior Province, Ontario*. Current Research, Geological Survey of Canada Paper **1994-C**, 259–268.

HEATHER, K. B, SHORE, T. G. & VAN BREEMAN, O. 1995. *The convoluted 'layer-cake': An old recipe with new ingredients for the Swayze Greenstone Belt, Southern Superior Province, Ontario*. 1994-1995 NODA Summary Report, 94–103.

HEATHER, K. B., SHORE, G. T. & VAN BREEMEN, O. 1996. Geological investigations in the Swayze greenstone belt, southern Superior Province, Ontario. Current Research Geological Survey of Canada Paper **1996-C**, 125–136.

HOWARTH, R. J. 1983. Mapping. *In*: HOWARTH, R. J. (ed.) *Statistics and data analysis in geochemical prospecting. Handbook of exploration geochemistry*. Elsevier, Amsterdam, 111–205.

ISAAKS, E. H. & SRIVASTAVA, R. M. 1989. *An Introduction to Applied Geostatistics*. Oxford University Press, New York.

ONTARIO GEOLOGICAL SURVEY (OGS). 1997. *Ontario airborne magnetic and electromagnetic surveys: Archean and Proterozoic greenstone belts, Swayze area*. ERLIS dataset **1015** (CD-ROM). Ontario Geological Survey, Sudbury, Ontario

STANLEY, C. F. 1987. Instruction Manual for PROB-PLOT: An interactive computer program to fit mixtures of normal (or log normal) distributions with maximum likelihood optimization procedures. *Association of Exploration Geochemists*, Special Volume **14**, 40.

THOMPSON, M. & HOWARTH, R. J. 1976. Duplicate Analysis in Geochemical Practice: Part 1. Theoretical Approach and Estimation of Analytical Reproducibility. *Analyst*, **101**, 690–698.

THOMPSON, M. & HOWARTH, R. J. 1978. A New Approach to the Estimation of Analytical Precision. *Journal of Geochemical Exploration*, **9**, 23–30

TUKEY, J. W. 1977. *Exploratory Data Analysis*. Addison-Wesley, Reading, Massachusetts.

WILKINSON, L., HARRIS, J. R. & GRUNSKY, E. C. 1999. *Building a lithogeochemical database for GIS analysis of the Swayze Greenstone Belt; Methodology, problems and solutions*. Geological Survey of Canada Open File Report **3788**

Regional and local-scale gold grain and till geochemical signatures of lode Au deposits in the western Abitibi Greenstone Belt, central Canada

M. BETH McCLENAGHAN

Geological Survey of Canada, 601 Booth Street, Ottawa, Ontario K1A 0E8, Canada
(e-mail: bmcclena@nrcan.gc.ca)

Abstract: This paper is an overview of drift exploration methods for lode Au deposits in areas of thin and thick cover of glacial sediments within the Abitibi Greenstone Belt of central Canada. It summarizes a large volume of data produced by government regional surveys and case studies as well as that from industry-led gold exploration programs. Regional till surveys can be used as targeting mechanisms for further Au exploration. Anomalies are defined by a series of samples with elevated Au concentrations that lie along significant bedrock structures, occurring in clusters or as isolated samples in areas of low sample density. Thresholds between background and anomalous Au grain abundances or Au concentrations are variable and depend on location within the Abitibi Greenstone Belt. Case studies around known deposits provide examples of geochemical and mineralogical signatures of Au deposits that can be expected in till down-ice. These serve as sources of information on appropriate sampling methods and size fractions to analyse, and on ice flow patterns, local glacial stratigraphy and suitable till units for sampling. Two methods for measuring the Au content of till are commonly used: (1) a count of visible Au grains and (2) geochemical elemental analysis. Close to source, till contains thousands to hundreds of thousands ppb Au and several hundred Au grains. The Au grains vary from coarse sand to silt sizes and have pristine shapes. The presence of high Au concentrations in till indicates that the ore zones subcrop and that glacial processes have produced Au dispersal trains down-ice.

The Abitibi Greenstone Belt (AGB) in central Canada is one of the world's largest Archean greenstone belts, covering an area of c. 100 000 km² (Fig. 1). The belt is dominated by metavolcanic rocks, but also includes metasedimentary and intrusive rocks. Many of the Au deposits are spatially related, in a broad sense, to major east–west trending deformation zones such as the Destor–Porcupine and Larder Lake–Cadillac fault zones (Robert 1996). The main ore mineral in most deposits is native Au, which typically contains some Ag. Gold occurs as isolated grains and fracture fillings in quartz as well as coatings on, or as inclusions and fracture fillings within, sulphide grains, most commonly pyrite (Robert 1996). World class Au mining camps within the AGB include Timmins, Kirkland Lake, Larder Lake, Val d'Or and Cadillac. The first Au mines opened in 1910, and mining continues today at numerous deposits. During the past 90 years, c. 130 million ounces of Au worth, at $300 US per ounce, $39 billion have been mined from the AGB (M. Doggett, pers. comm. 1999), making it one of the most important Au mining regions in the world.

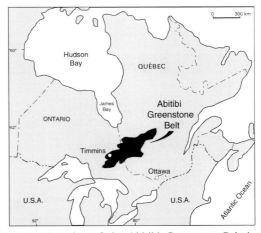

Fig. 1. Location of the Abitibi Greenstone Belt in central Canada.

The AGB is one of the most challenging regions in which to use drift exploration methods because: (1) bedrock outcrop is rare; (2) till units were deposited by a complex

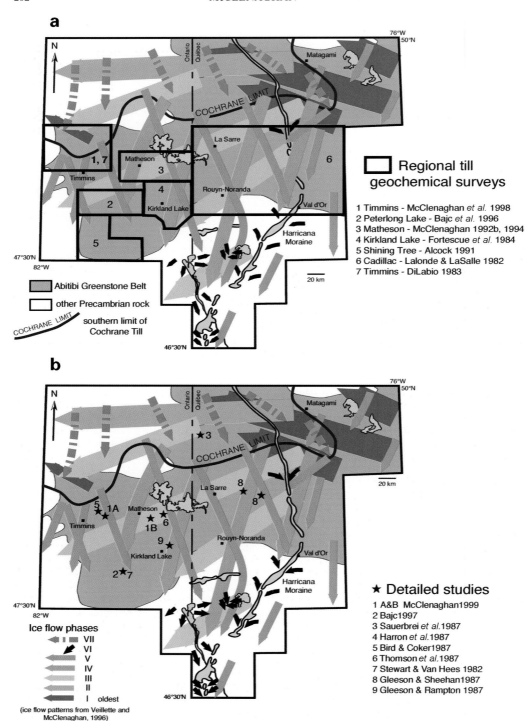

Fig. 2. Regional ice flow patterns across the Abitibi Greenstone Belt, southern limit of clay-rich Cochrane Till, and location of: (**a**) regional geochemical surveys summarized in Table 2; and (**b**) detailed Au deposits studies summarized in Table 3 (ice flow patterns from Veillette & McClenaghan 1996).

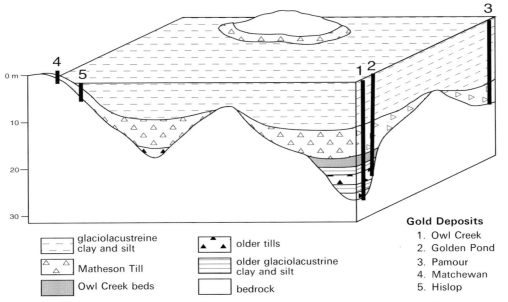

Fig. 3. Stylized block diagram of the Timmins region showing till distribution and thickness and the stratigraphic setting of glacial sediments overlying Au deposits (1 to 5) described in this paper.

sequence of ice flow; (3) tills often are capped by thick (5 to 30 m) glaciolacustrine clay and silt deposits and thus there are few natural exposures of the till for sampling and determining till stratigraphy; and (4) the flat lying and poorly drained glaciolacustrine sediments are often covered by extensive peatlands which limit access for sampling. It was in response to these challenges that techniques such as reverse circulation drilling (McMartin & McClenaghan 2001) were developed to collect till samples in the thick drift-covered, inaccessible terrain of the AGB.

The earliest Au deposits discovered in the Abitibi region were in areas of abundant bedrock outcrop or thinly covered glacial sediments. As production from these initial deposits declined, exploration expanded to areas with similar bedrock geology covered by thick glacial sediments. In these areas, geophysical methods are used in combination with overburden drilling and till sampling to explore for Au. The greatest challenge in Au exploration continues to be in the thick drift areas, which contain multiple till units related to several ice flow phases. Mineral exploration in these areas is assisted by conducting three dimensional till geochemical surveys in conjunction with studies of regional till stratigraphy and ice flow patterns.

In the past 20 years, till geochemistry and Au grain methods have evolved into sophisticated exploration techniques in the AGB. This paper reviews the glacial history of the Abitibi region, the use of these methods, including general descriptions of field and laboratory methods, and provides examples of regional geochemical surveys and deposit-scale studies.

Quaternary geology

The Abitibi region was glaciated during the Wisconsinan and was affected by a complex sequence of ice flows (Fig. 2) (Veillette & McClenaghan 1996). Initially, ice flowed northwestward across the area (Phase I, Fig. 2) during an early growth phase of the Laurentide Ice Sheet (Veillette 1995). Ice flow subsequently shifted counterclockwise, to the W–SW (220° to 240°, Phase II and III) during the main phase of the Laurentide Ice Sheet, and then south (170° to 200°, Phase IV), and finally SE (160° to 180°, Phase V) and SW (180 to 200°, Phase V) during the final phases of the ice sheet. Surface till in areas of thin glacial sediment cover was deposited by later phases of ice flow to the SW, south and SE. Thicker till units underlying glaciolacustrine sediments and in buried bedrock valleys

(Fig. 3) were deposited during older phases of ice flow.

Across the AGB, a silty sand till, referred to as Matheson Till (Hughes 1965; McClenaghan 1992a), occurs at the surface in thin wedges around bedrock outcrops (Fig. 3) and as thick (>3 m) deposits in isolated patches. More commonly, Matheson Till occurs as till sheets beneath thick (3 to 30 m) glaciolacustrine silt and clay (Fig. 3) deposited in glacial Lake Ojibway near the end of glaciation c. 8000 years ago (Veillette 1995). Matheson Till is a carbonate-rich (10 to 30% matrix carbonate) till containing 10 to 25% clasts, of which c. 40 to 80% are locally derived bedrock (Hughes 1965; McClenaghan 1992a; McClenaghan et al. 1998). Because of its high sand content and local bedrock component, Matheson Till is the primary sampling medium for Au exploration. Older tills and associated sediments occur beneath Matheson Till (Fig. 3) in buried bedrock valleys or low areas on the bedrock surface (Bird & Coker 1987; DiLabio et al. 1988; Smith 1992; McClenaghan & DiLabio 1995). The older silty sand tills are also useful sample media for Au exploration (e.g. Bird & Coker 1987; Sauerbrei et al. 1987). North of Timmins (Fig. 2), the surface is covered by Cochrane Till (Hughes 1965; Paulen & McClenaghan 1998a, b) which is a clast-poor, silty clay till produced by glacial erosion of underlying fine-grained glaciolacustrine sediments during a late glacial surge into Lake Ojibway. The paucity of sand-sized material in the matrix, and its derivation from the underlying fine-grained glaciolacustrine sediments makes this till unsuitable for drift prospecting.

Till sampling methods

General sampling procedures used to collect till for Au exploration have been summarized by Hirvas & Nenonen (1990), Kauranne et al. (1992), Plouffe (1995) and McMartin & McClenaghan (2001). In the AGB, till is collected in areas of thin glacial sediments from hand-dug holes, backhoe excavator trenches, natural sections along river or lake shorelines, road cuts, and open pits. Where glacial sediment thickness exceeds 5 m and extensive peatlands limit access, reverse circulation or rotasonic drilling is used to collect till below the surface cover of other sediments, to characterize the till stratigraphy and to determine lateral and vertical variations in till geochemistry. Availability and cost are major factors in the choice of drilling method (Coker & DiLabio 1989; Coker 1991). Reverse circulation drills (McMartin & McClenaghan 2001) recover a slurry of cuttings up to 1 cm in size, down to and into bedrock. The more expensive rotasonic drilling method (McMartin & McClenaghan 2001) produces 9 cm diameter continuous core of glacial sediments and bedrock. It has been used in the initial stages of exploration programs and in government regional geochemical surveys to determine the types and distribution of glacial sediments and to document the characteristics of different till units, enabling much more detailed geological observation, stratigraphical interpretation and till sampling (Averill et al. 1986; Averill 1990; McClenaghan 1994).

Laboratory methods

This section describes procedures for Au grain recovery and geochemical analysis most commonly used in regional till geochemical surveys and in local, property-scale Au exploration programs in the AGB. The presence of significant numbers of visible Au grains in the heavy mineral fraction of till is of interest to Au exploration because it can be a direct indication of bedrock Au mineralization (Averill 1988; Bajc 1996). Visible Au grains in till typically vary in size from 10 μm to 2000 μm; 10 μm is the lower limit of Au visibility under a binocular microscope (Averill 1988). Visible Au grains are routinely recovered from till samples and examined to determine their abundance, size and shape, to detect dispersal trains from Au deposits and to predict the size, grade and character of the bedrock mineralization (Averill 1990, 2001).

Geochemical analysis of till also provides important information about the presence of bedrock mineralization. Two different size fractions of till are most commonly analysed for Au and pathfinder elements: the non-magnetic heavy mineral fraction, and the <0.063 mm (silt + clay) fraction (Shelp & Nichol 1987; McClenaghan 1992b, 1994; Bajc 1997). Both size fractions should be used in Au exploration programs because Au in bedrock deposits in the AGB varies from coarse-grained free Au in quartz-carbonate veins to fine-grained Au associated with sulphides (Robert 1996). The additional information gained by analysing both size fractions will help to identify and confirm Au anomalies and provide insights into the nature (grain size) of Au mineralization. Analysis of the finer fraction is especially important where till is weathered (Nichol et al. 1992; Bajc 1996). However, geochemical analysis of the fine fraction of till samples collected by reverse circulation drilling has limited value because of

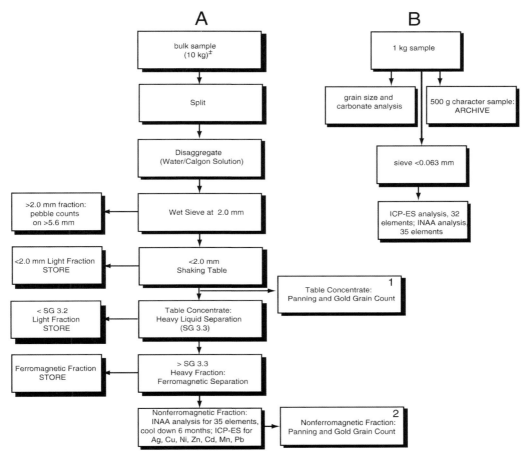

Fig. 4. Sample processing flow diagram for till sample preparation: (**a**) preparation of heavy mineral concentrates for Au grain counting and geochemical analysis; and, (**b**) preparation of the < 0.063 mm fraction for geochemical analysis. Au grains may be examined and counted at two different stages of the processing: (1) during sample processing; or (2) after nondestructive geochemical analysis.

significant losses of this fraction during drilling (Shelp & Nichol 1987) and because of cross-contamination of samples by recirculated water (Coker & Shilts 1991).

Sample preparation

Samples collected for Au grain recovery are usually 10 to 12 kg (5 to 6 l), of which small 1 kg subsamples are set aside for fine fraction geochemical analysis and for archiving (Fig. 4). The remaining material is disaggregated, typically by agitation in a dispersant, and the gravel fraction (> 2 mm) is screened off for lithological analysis of the pebble-sized (> 5.6 mm) fraction. The < 2 mm fraction is then pre-concentrated by density separation, using a shaking table, to recover Au grains between 10 and 2000 μm and to obtain a heavy mineral fraction for geochemical analysis. Final density concentration is completed using heavy liquids such as methylene iodide (specific gravity 3.3 g/cm^3) to recover heavy minerals. The ferromagnetic minerals (and drill steel) are then removed from the heavy mineral fraction using a hand magnet or roll separator, weighed and archived, leaving on average 15 to 30 g of non-ferromagnetic heavy mineral concentrate for geochemical analysis (Averill 1988). This non-ferromagnetic heavy mineral fraction is commonly referred to as the 'heavy mineral fraction' or HMC. Au grains are panned, counted, and classified with the aid of optical or scanning electron microscopy at one

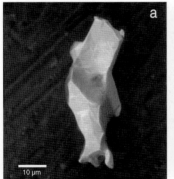 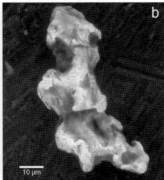 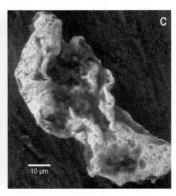

Fig. 5. Secondary electron images of Au grains from till in the Timmins region showing the three 'conditions' routinely documented for Au grains: (**a**) *pristine* Au grain with equant molds suggestive of former quartz–feldspar–carbonate–sulphide gangue; (**b**) *modified* Au grain with vestiges of equant gangue molds and edges that are slightly curled; and (**c**) *reshaped* Au grain showing pitted surfaces and well curled edges.

of two stages during the processing: after density pre-concentration, or after non-destructive geochemical analysis, using the Au concentrations as a guide (Fig. 4).

Till samples which are to be used for geochemical analysis of the fine fraction are air dried below 40°C, to prevent the loss of volatile elements such as Hg, and sieved using stainless steel or plastic screens to recover specific size fractions, most commonly the <0.063 mm fraction, for geochemical analysis (Fig. 4). Averill (1990) noted the tendency for Au grains to pass through sieves preferentially, and therefore complete screening of a sample rather than the rapid recovery of a small portion is advisable.

Gold grains

Condition

The degree of rounding, polishing and bending of the Au grains in till may provide information about glacial transport distance or insight into the style of Au mineralization (Averill 2001). DiLabio (1990, 1991) proposed a graphically descriptive classification scheme (pristine–modified–reshaped) for describing conditions and surface textures of Au grains that builds on Averill's (1988) descriptions of Au grain shape related to glacial transport distance. The progression from pristine to reshaped grains can be interpreted to represent increasing distance of glacial transport. However, caution should be used when utilizing Au grain condition as an indication of transport distance because Au grain morphology can be variable in the bedrock source, and Au grains can be released from mineralized bedrock fragments at any distance down-ice during glacial transport or during subsequent post-glacial weathering (Coker & Shilts 1991; Henderson & Roy 1995).

Pristine gold grains

Pristine Au grains (Fig. 5a) retain primary shapes and surface textures and appear not to have been damaged in glacial transport. They occur as angular wires, rods and delicate leaves that formed as fracture fillings, as crystal faces and grain molds, and as inclusions in sulphides. The transport history of pristine grains may be interpreted in two ways. Au grains were eroded from a bedrock source nearby and glacially transported to the site with little or no surface modification, and transport distance is generally short. Alternatively, Au grains were liberated from rock fragments during *in situ* weathering of glacially transported sulphide grains containing Au, wherein the pristine shape and surface texture provide little information on transport distance, but do provide important information on the style of Au mineralization, i.e. sulphide-hosted Au mineralization (Henderson & Roy 1995).

Modified gold grains

Modified Au grains (Fig. 5b) retain some primary surface textures but all edges and protrusions have been damaged during transport and they are commonly striated. Irregular edges and protrusions are crumpled, folded and curled. Grain molds and primary surface textures are preserved only on protected faces of grains. Till samples that contain elevated con-

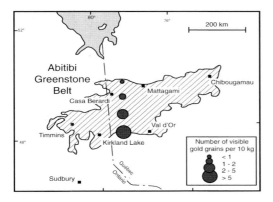

Fig. 6. Southward increasing background concentrations of Au grains in the heavy mineral fraction of till across the Abitibi Greenstone Belt (modified from Averill & Huneault 1991).

centrations of modified grains are generally proximal to the bedrock source.

Reshaped gold grains

Reshaped Au grains (Fig. 5c) have undergone sufficient glacial transport that all primary surface textures have been destroyed and the original grain shape is no longer discernible. Reshaped grains are flattened to rounded resulting from repeated folding of leaves, wires and rods. Grain surfaces may be pitted from impact marks from other grains, but surfaces are not leached of Ag. Although these grains can have a complex transport history, the presence of large numbers of reshaped grains in discrete areas should be considered significant (Bajc 1996). Most background Au grains have a reshaped morphology (Averill 1988).

Abundance

Exploration programs and government regional surveys conducted in the AGB routinely report Au grain abundance in the <2.0 mm HMC of 10 kg till samples. In general, background concentrations of Au grains increase from north to south across the AGB, along the general direction of ice flow (Fig. 6) and are related to the amount of mineralized metavolcanic bedrock present up-ice (Averill 1988; Averill & Huneault 1991). For specific areas in the AGB, government regional till geochemical surveys (Fig. 2a) have defined more precise thresholds between background and anomalous concentrations of Au grains using the 95th percentile. For

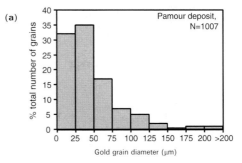

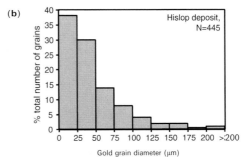

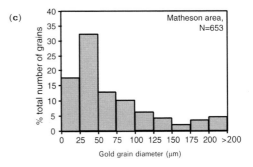

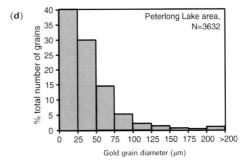

Fig. 7. Distribution of Au grain size (largest dimension) in till samples from: (**a**) the Pamour gold mine No. 5 open pit in Timmins; (**b**) the Hislop Au deposit near Matheson; (**c**) from surface till in the Matheson region; and, (**d**) from surface till in the Peterlong Lake-Radisson Lake region.

example, the threshold between background and anomalous concentrations of Au grains in the west part of the AGB (Timmins) is five grains (McClenaghan et al. 1998). South of Timmins, in the Peterlong Lake area, the threshold between background and anomalous Au grain abundances is ten grains (Bajc 1996). The thresholds identified by these surveys serve as guidelines for exploration companies.

Size

Most Au grains recovered from regional studies in the western Abitibi are silt-sized (McClenaghan 1992b, 1994; Bajc 1996; McClenaghan et al. 1998). For example, Au grains recovered from regional till samples collected around Timmins vary between 10 µm and 250 µm, and most are < 50 µm. This small size may reflect glacial transport history (i.e. folding and crumpling of grains to small sizes) or may reflect the nature of bedrock Au mineralization (i.e. fine- v. coarse-grained Au) (Bajc 1997; Averill 2001). In contrast, visible Au grains recovered from till immediately down-ice of Au deposits (Fig. 7) have greater ranges in size, typically between 10 µm to 800 µm (e.g. Bajc 1997; McClenaghan 1999).

Gold grain size, condition and abundance are all important factors to consider when interpreting Au grains counts and Au analyses for heavy mineral concentrates. A till sample containing numerous small, pristine Au grains may yield only low to moderate Au concentrations (500 to 1000 ppb) although the sample is considered to be anomalous. In contrast, a till sample containing a couple of large (> 100 µm) reshaped Au

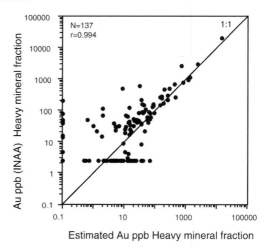

Fig. 8. Calculated Au concentration based on visible Au grain counts v. actual Au concentration determined by INAA for heavy mineral fraction of regional till samples from the Timmins area (from McClenaghan et al. 1998).

grains can yield what appears to be an anomalous Au concentration (Table 1) when in fact the sample is not.

Geochemical analysis

Heavy mineral concentrates are the most commonly analysed till fraction for Au exploration. They are analysed for Au and other elements by non-destructive instrumental neutron activation analysis (INAA) (Hoffman et al. 1999). For unoxidized till, a small representative split of the

Table 1. *Comparisons of visible gold grain abundance, size, and shape to gold concentration in the heavy mineral fraction of anomalous and background till samples*

Location and sample number	Number of visible gold grains	Gold grain shape and size (µm)	Gold (ppb) in heavy mineral concentrate	Interpretation
Timmins 96MPB6076 (McClenaghan et al. 1998)	50	all < 150 µm, 70% pristine shape	487 ppb	anomalous: large number of small gold grains, low gold concentration
Matheson 843707 (OGS 1986)	2	100 × 100, modified shape 550 × 250, reshaped	4100 ppb	not anomalous: two very large gold grains, high concentration of gold
Peterlong Lake 96AFB4032 (Bajc 1996)	1024	all < 200 µm 99% pristine shape	1250 ppb (calculated)	anomalous: large number of small gold grains, high concentration of gold

Table 2. *Regional till geochemical surveys conducted in the Abitibi Greenstone Belt by provincial and federal geological surveys*

	Location	Reference	Till sampling methods	Sample size	Gold Grains counted	Till size fractions analysed	Pebble count data	Other media analysed
1	Timmins	McClenaghan et al. 1998	surface sampling	10 kg and 2 kg	yes	< 2.0 mm HMC and < 0.063 mm	yes	humus
2	Peterlong Lake/ Matachewan	Bajc 1996, 1997; Bajc et al. 1996	surface sampling	10 kg and 2 kg	yes	< 0.063 mm	no	lake sediment, lake water
3	Matheson	McClenaghan 1990, 1991, 1992b, 1994; Nichol et al., 1992	surface sampling and overburden drilling	10 kg and 2 kg	yes	< 2.0 mm HMC, < 2.0 mm and < 0.063 mm	yes	none
4	Kirkland Lake	Fortescue et al. 1984; McClenaghan & Wyatt 1990	surface sampling and overburden drilling	5 to 8 kg	no	> 0.125 mm HMC, < 0.125 HMC, < 0.063 mm	no	none
5	Shining Tree	Alcock 1991	surface sampling	10 kg and 2 kg	yes	< 0.063 mm	yes	none
6	Noranda-Val d'Or	Lalonde & LaSalle 1982	percussion drilling	not reported	no	< 0.177 mm, > 0.177 mm HMC, > 0.177 mm light minerals	no	none
7	Timmins	DiLabio 1983	overburden drilling	5 to 8 kg	no	< 2.0 mm HMC	no	none

HMC, heavy mineral concentrate.

HMC may also be analysed for base metals and other elements by aqua regia/inductively coupled plasma-emission spectrometry (ICP-ES). Destructive analytical methods for the HMC, such as fire assay, are not commonly used so that concentrates may be examined later for Au grains, kimberlite indicator minerals or other indicator minerals.

The concentration of visible Au in the HMC can be estimated using the number and dimensions of Au grains recovered. Averill & Zimmerman (1987) describe the method that has been used for most government surveys and case studies conducted in the AGB. Estimated concentrations are generally a good proxy (Fig. 8) for actual values determined by geochemical analysis of the HMC (Averill 1988; McClenaghan et al. 1998). However, predicted Au values will be lower than actual geochemical results when not all Au grains have been identified and counted. Au grains may be missed where they are very small (< 10 μm), present in sulphide minerals, or covered with secondary minerals such as limonite or goethite. The latter case is an important consideration when using surface till samples (< 2 m from surface) for Au exploration, as they have been affected by oxidation. Therefore, geochemical analysis of the HMC is strongly recommended.

For the < 0.063 mm fraction, Au is determined by INAA or by fire assay followed by aqua regia/graphite furnace-atomic absorption spectrometry (GF-AAS) or /ICP-ES on a minimum of 30 to 50 g (Hoffman et al. 1999). Au pathfinder elements, such as As, Ag, Sb, Cu, Bi, Mo, W, Te, B, and Co, are commonly determined by aqua regia/ICP-ES. For both the HMC and < 0.063 mm fractions, inexpensive multielement analytical packages, based on ICP-ES and ICP-mass spectrometry, now offer determination of large suites of elements that allow for exploration of other commodity types simultaneously.

Thresholds (95th percentile) between background and anomalous Au concentrations in till have been determined for large parts of the western AGB by government regional geochemical surveys (Fig. 2; McClenaghan 1991, 1994;

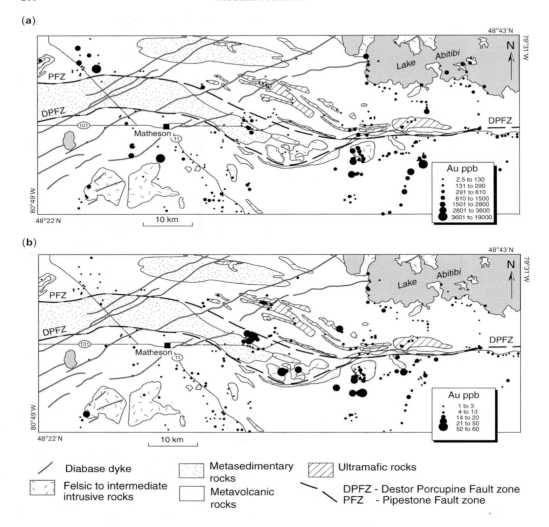

Fig. 9. Au concentration in surface till samples in the Matheson area: (**a**) the heavy mineral fraction; and (**b**) <0.063 mm fraction (modified from McClenaghan 1992b).

Bajc 1996; McClenaghan et al. 1998). Thresholds vary from 500 to 1500 ppb Au in the HMC and 5 to 25 ppb Au in the <0.063 mm fraction depending on location within the AGB and the amount of mineralized volcanic bedrock up-ice. At the property scale, threshold values may be higher. For example, Harron et al. (1987) used a threshold of 2000 ppb Au in the HMC of till to identify anomalous values on a property NE of Timmins. Bird & Coker (1987) used a cut-off of 5300 ppb Au to identify highly anomalous Au concentrations in the HMC around the Owl Creek Au deposit.

Regional geochemical surveys

Regional till geochemical surveys dramatically increase our knowledge of the distribution, composition and origin of till. They establish regional thresholds between background and anomalous Au concentrations in till, and, using these thresholds, identify geochemical anomalies that warrant further investigation. Regional surveys also provide the framework in which other regional geochemical surveys or smaller, property-scale geochemical studies can be interpreted. Several surveys (Table 2) have been

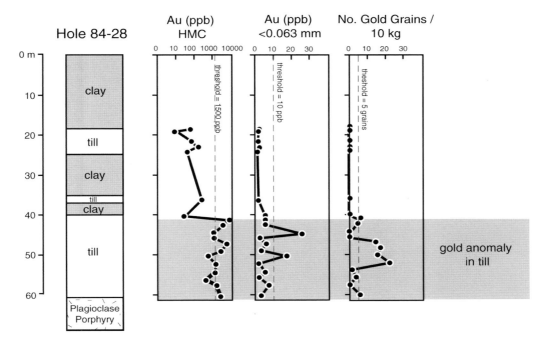

Fig. 10. Comparison of Au grain abundance and Au concentration in the heavy mineral and <0.063 mm fractions of thick till in Ontario Geological Survey hole 84–28 (modified from McClenaghan 1994).

carried out in the past 20 years (Fig. 2) and results from two of them are highlighted below.

Matheson area

Regional scale till sampling was carried out across a 4000 km^2 area centered on the town of Matheson (survey 3, Fig. 2) by the Ontario Geological Survey between 1984 and 1988 (McClenaghan 1990, 1991, 1992b, 1994). Glacial sediments in this area are generally thick (5 to 40 m) and include a cover of 5 to 30 m of glaciolacustrine sediments (Fig. 3). Bedrock outcrop in the area is rare. Till was deposited by ice flowing west, then southwest and finally south or southeast. In areas of thin (<5 m) drift, till was deposited by the most recent ice flow to the south or southeast. In areas underlain by thicker glacial sediments, till deposited by older ice flows as well as younger, south or southeast ice flows may be present. Using surface till sampling in areas of thin drift and sonic overburden drilling (McMartin & McClenaghan 2001) in areas of thick drift (Steele et al. 1989), 1100 till samples were collected from 282 surface till sites (site spacing of 0.5 to 2 km) and 225 drill holes (drill hole spacing of 2 to 5 km). In thick till sequences, vertically adjacent samples were collected from drill holes, typically over 1 to 2 m intervals. From each sample site or interval, a large (10 kg) sample was collected for Au grain counting, pebble counting and multielement geochemical analysis of the HMC, and a 2 kg sample was collected for geochemical analysis of the <2.0 mm and <0.063 mm fractions.

Gold grain counts combined with multielement geochemical analysis of the HMC and <0.063 mm fraction of till were the most cost-effective and valuable for identifying Au anomalies. The threshold between background and anomalous concentrations of Au grains in till is five grains per 10 kg of till. Most Au grains in the till samples are silt-sized (Fig. 7c), which may reflect either the glacial transport history (reducing grain size through folding, crumpling and flattening) or the fine-grained nature of Au in many of the deposits in the area. Numerous Au geochemical anomalies were identified using thresholds of 1000 to 1500 ppb Au in the HMC and 10 ppb in the <0.063 mm fraction. Varying combinations of elevated As, Sb, and, to a lesser extent, Cu, Co, Ni and Zn concentrations accompany the anomalous Au values.

On a regional scale, the distribution of anomalous Au concentrations is very different for the HMC (Fig. 9a) and <0.063 mm fraction

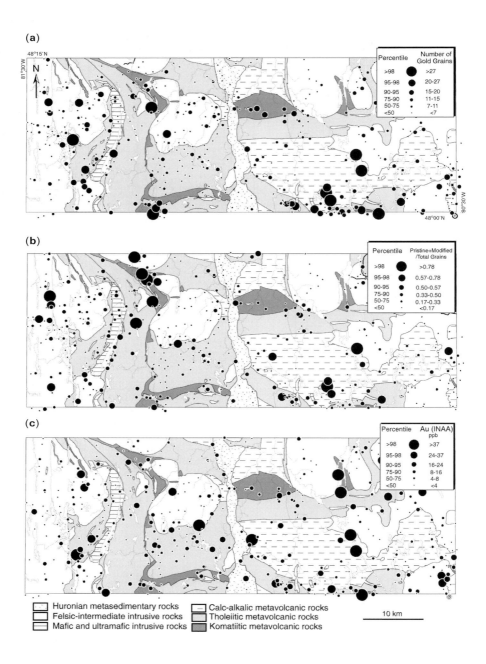

Fig. 11. Regional till geochemistry of the Peterlong Lake area: (**a**) Au grains in heavy mineral fraction; (**b**) ratio of pristine + modified Au grains to total number of grains; and (**c**) Au in < 0.063 mm fraction (from Bajc 1996).

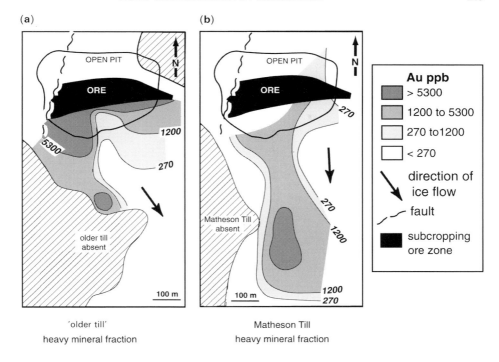

Fig. 12. Au abundance in the heavy mineral fraction of: (**a**) older till; and (**b**) younger till (Matheson Till) dispersal trains down-ice from the Owl Creek Au deposit (modified from Bird & Coker 1987).

(Fig. 9b) of till for both surface and subsurface till samples. These differences are likely to be related to the nature of Au in the bedrock and the distance down-ice from the bedrock source; they show the importance of analysing both size fractions for Au. In contrast, comparison of multiple till samples from a specific anomalous site reveals similar distribution patterns for the HMC and < 0.063 mm fraction (Fig. 10). Where these similarities occur, the bedrock source is likely to contain both coarse- and fine-grained Au.

Peterlong Lake–Radisson Lake area

The Ontario Geological Survey also conducted a regional till sampling survey in the Peterlong Lake–Radisson Lake area (survey 2, Fig. 2), 25 km south of Timmins (Bajc 1996, 1997). Lake sediment and lake water samples were also collected as part of the survey (Bajc et al. 1996). The area is characterized by upland regions of bedrock outcrop and thin discontinuous till cover with isolated pockets of thicker till. Most till was deposited by the last ice flow event to the south-southeast. Till samples were collected from shallow hand-dug pits at 414 sites at a sample spacing of about 3 km. A 10 kg sample was collected for visible Au grain counting only and a 2 kg sample was collected for geochemical analysis of the < 0.063 mm fraction.

Most till samples contain visible Au, with the threshold between background and anomalous concentrations at ten grains per 10 kg sample (Fig. 11a). Some Au grains in till are likely from the Timmins Au camp 25 km to the north. Most Au grains are in pristine condition and silt-sized (Fig. 7d), which may reflect glacial transport history and/or the nature of Au mineralization in the area (predominantly fine-grained Au). The ratio of pristine plus modified grains to the total number of Au grains (Fig. 11b) provides a means of identifying anomalies for further investigation. Using a threshold of 24 ppb Au (95th percentile) for the < 0.063 mm fraction, Bajc (1996) identified numerous Au anomalies in till (Fig. 11c). Varying combinations of elevated Ag, Mo, As, and Sb concentrations accompany the anomalous Au values (Bajc 1996). Similar to

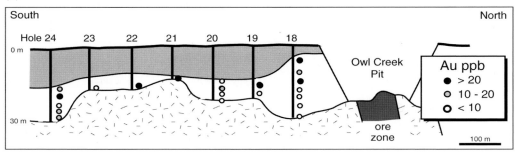

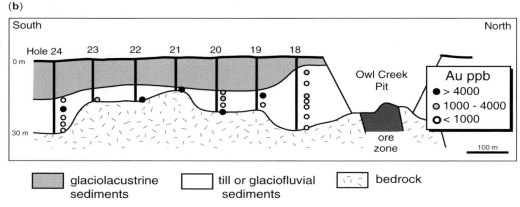

Fig. 13. Comparison of Au abundance in the (**a**) <0.063 mm and (**b**) heavy mineral fractions of till at the Owl Creek Au deposit (modified from Shelp & Nichol 1987).

the Matheson regional survey described above, geochemical anomalies in the <0.063 mm fraction do not always coincide with the anomalous areas identified using Au grain counts in the HMC. Au grain and geochemical anomalies are associated with significant bedrock structures or particular bedrock lithologies and many are coincident with known Au mineralization (Bajc 1996).

Property scale studies

Several property scale geochemical studies have been conducted in the AGB to evaluate properties for potential Au mineralization (Gleeson & Rampton 1987; Harron *et al.* 1987; Sauerbrei *et al.* 1987) or to document the nature of Au dispersal in till from known Au deposits (Stewart & Van Hees 1982; Bird & Coker 1987; Gleeson & Sheehan 1987; Bajc 1997; McClenaghan 1999). These studies are listed in Table 3 and their locations are shown in Figure 2. Highlights from five detailed studies in both thin- and thick-drift covered terrain (Fig. 3) within the AGB are described below.

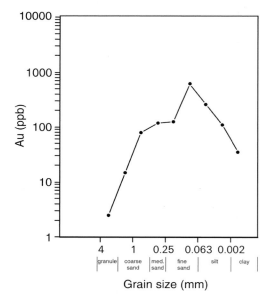

Fig. 14. Abundance of Au v. grain size of analysed fractions of unoxidized till from the Owl Creek Au deposit (from DiLabio 1995).

Table 3. Detailed, deposit-scale till geochemical studies conducted in the Abitibi Greenstone Belt

	Location	Reference	Till sampling methods	glacial sediment thickness	Dispersal train (length × width)	Sample size	Till size fractions analyzed	Highest gold grain count	Highest Au value (size fraction)	Pathfinder elements	Pebble counts
1	Timmins (Pamour Mine, Hislop Deposit)	McClenaghan et al. 1998	open pit section; backhoe	thick thin moderate	not determined; >10 m length	10 kg and 2 kg	<2.0 mm HMC and <0.063 mm	880 grains; 371 grains	20,200 ppb (HMC), 250 ppb (<0.063 mm; 23,700 ppb (HMC)	As, Ag, Cd Ag	yes yes
2	Matachewan (Matachewan Con. Mine)	Bajc 1997	open pit and surface sampling	thin to moderate	200 to 400 m length	10 kg and 2 kg	<0.063 mm	1024 grains	510 ppb (<0.063 mm)	As, Sb	no
3	Casa Berardi (Golden Pond)	Sauerbrei et al. 1987	RC overburden drilling	thick	400×200 m, but truncated by bedrock ridge	10 kg	<2.0 mm HMC	700 grains	106,000 ppb (HMC)	As	yes
4	Timmins	Harron et al. 1987	RC overburden drilling	thick	ribbon, 2 km × 500 m	10 kg	<2.0 mm HMC	not reported	19,000 ppb (HMC)	none	no
5	Timmins (Owl Creek Mine)	Bird & Coker 1987; Shelp & Nichol 1987	RC and sonic overburden drilling	thick	two trains at different stratigraphic levels, ribbon, 500×200 m	10 kg	<2.0 mm HMC; <0.063 mm	not reported	250,000 ppb (HMC)	none	yes
6	McCool Township (Belore deposit)	Thomson et al. 1987	RC overburden drilling	thick	300 m length	10 kg	<2.0 mm	not reported	>3000 ppb (<2.0 mm)	As	yes
7	Matachewan	Stewart & Van Hees 1982	backhoe and auger	thin	ribbon 200 to 800 m ×60 to 150 m	2 to 5 kg	<2.0 mm and <0.063 mm	not counted	2760 ppb (<2.0 mm)	none	no
8	Malarctic (Doyon Mine)	Gleeson & Sheehan 1987	RC overburden drilling	thin to moderate	200 m length	not known	<2.0 mm HMC and <2.0 mm	not counted	>15,000 ppb (HMC); 3250 ppb (<2.0 mm)	none	no
	(Bosquet Mine)		percussion drill	thin to moderate	15 to 30 m length	not known	<2.0 mm HMC and <0.15 mm	not counted	9600 ppb (HMC); 420 ppb (<0.15 mm)	none	no
9	Kirkland Lake (Morrissette Creek)	Gleeson & Rampton 1987	surface sampling	thin to thick	none	not known	<0.063 mm	not counted	61 ppb (<0.063 mm)	none	no

HMC = heavy mineral fraciton

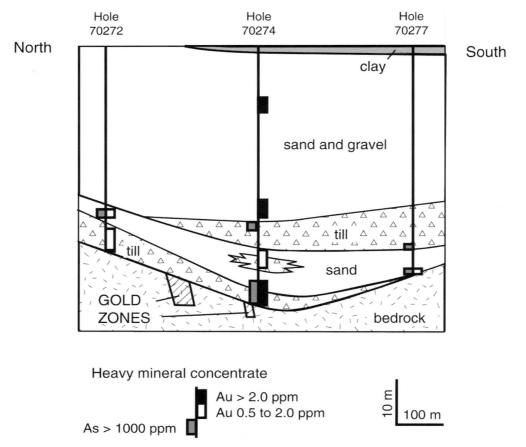

Fig. 15. North–south section at the Golden Pond Au deposit showing stratigraphy and heavy mineral concentrate geochemical anomalies c. 100 m down-ice of Au mineralization (modified from Sauerbrei et al. 1987).

Owl Creek Au deposit

The Owl Creek Au deposit is located 18 km E of Timmins (deposit 5, Fig. 2) in an area covered by thick (10 to 20 m) glacial sediments. Mineralization is hosted in mafic tholeiitic rocks bounded to the north and south by greywacke and argillite and occurs as free Au in quartz veins and as fine-grained Au associated with pyrite (Jonasson et al. 1999; Pressacco 1999). Bird & Coker (1987) reported on a detailed till sampling program carried out around the deposit using sonic and reverse circulation overburden drilling. Multiple till units capped by 5 m of glaciolacustrine clay and silt were intersected in the drill holes (Fig. 3) and all till units were sampled for Au grain counting, heavy mineral geochemistry and pebble counting.

Anomalous Au concentrations in the HMC of till define two dispersal trains down-ice of the Owl Creek Gold Mine (Bird & Coker 1987). The lower train in an older till is 400 to 500 m long and oriented towards the southeast (Fig. 12a). Au concentrations range up to 250 000 ppb, with the highest values in till occurring from 0 to 200 m down-ice. The upper train in younger till is 700 m long and is oriented to the south (Fig. 12b). Au concentrations in this train are highest 400 to 500 m down-ice from subcropping mineralization. The displacement of the anomaly in upper till down-ice is likely to be due to recycling of the Au-rich lodgment till of the lower dispersal train into the upper till and not to direct glacial erosion of the ore zone (Bird & Coker 1987; Coker & Shilts 1991). Au values are much lower in the upper dispersal train, which is likely due to the recycling and to dilution by more distally derived material that dominates the upper till. Pebble counts confirm that the lower dispersal train contains more locally

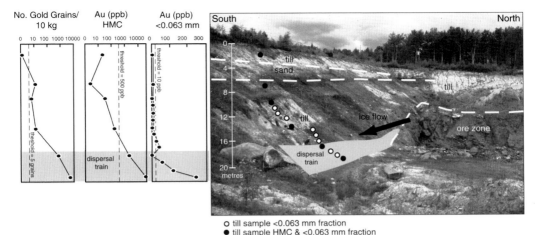

Fig. 16. Au grain abundance and till geochemistry for the heavy mineral and <0.063 mm fraction of till from the No. 5 open pit at the Pamour Au deposit (modified from McClenaghan 1999). Threshold between anomalous and background concentrations indicated by dashed lines.

derived bedrock fragments than the upper train. Shelp & Nichol (1987) showed that the Au content in the HMC and <0.063 mm fraction of till (Fig. 13) clearly depicts glacial dispersal from the deposit, with similar dispersal distances but lower contrasts and absolute concentrations of Au in the <0.063 mm fraction. The highest Au concentrations in till overlying the Owl Creek deposit are in the fine sand to silt-sized ranges (Fig. 14). This is not unexpected as much of the Au in the Owl Creek deposit is hosted in pyrite, and most pyrite in the till is sand-sized (DiLabio 1995).

Golden Pond Au deposit

The Golden Pond Au deposit is 150 km north of Rouyn-Noranda (deposit 3, Fig. 2), in an area of extremely thick glacial sediments. Mineralization consists of pyrite–arsenopyrite and native Au-bearing quartz-carbonate veins in metasedimentary and volcanic rocks. Sauerbrei et al. (1987) conducted a till sampling program using reverse circulation drilling to test for glacial dispersal from the deposit and to evaluate till sampling as an exploration method in this area. Following the success of their orientation survey, till sampling was used to explore for additional bedrock mineralization along strike and to identify diamond drill targets, and ultimately the Golden Pond East deposit.

Overburden drilling revealed the presence of two till units (lower and upper) overlain by thick (20 to 30 m) glaciofluvial sand and gravel and glaciolacustrine clay (Fig. 3). Till samples were collected from both till units for Au grain counting, heavy mineral fraction geochemistry and pebble counting. Au anomalies associated with the Golden Pond deposit predominate in the lower till, up to 400 m down-ice (Fig. 15). The highest concentrations in till occur down-ice from the Golden Pond East deposit with the most anomalous till sample containing 106 ppm Au and c. 700 Au grains. Most of the Au grains in till are 50 to 150 μm in size. Unusually high As concentrations (up to 140 000 ppm) in till accompany the high Au values, indicating that As is a strong pathfinder element for Au mineralization here.

Pamour Au deposit

The Pamour deposit, 20 km E of Timmins (deposit 1A, Fig. 2), has been actively mined since 1936. Au mineralization occurs north of Destor–Porcupine fault zone at the unconformity between Timiskaming metasedimentary rocks to the south and mafic and ultramafic volcanic rocks to the north. Au mineralization occurs as free Au in quartz-carbonate veins along with traces of sphalerite, galena and arsenopyrite and as fine-grained Au associated with pyrite (Tyler et al. 1999; Poulsen et al. 2000). Mineralization in the No. 5 pit is covered by 20 m of Matheson Till.

McClenaghan (1999) collected till samples from a section of till exposed in the west wall of the No. 5 open pit, a few metres down-ice

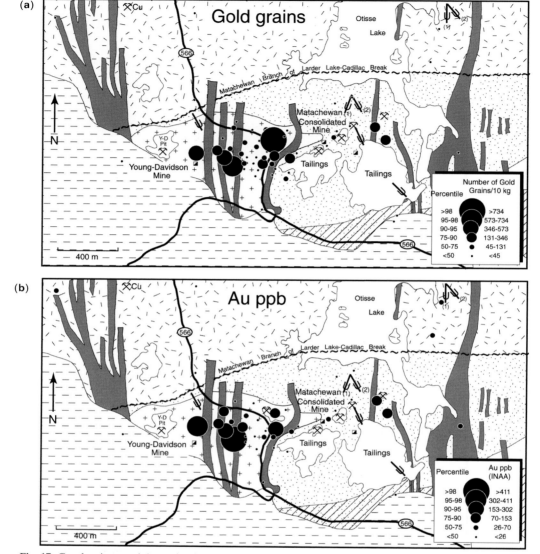

Fig. 17. Geochemistry and Au grain abundance in till around the Matachewan Consolidated gold mine: (**a**) Au grains, (**b**) Au ppb (from Bajc 1997).

from mineralized bedrock (Fig. 16). This thick vertical section was sampled to simulate an overburden drill hole in thicker drift covered terrain. Till samples were collected for Au grain counting, geochemical analysis of the HMC and <0.063 mm fraction and pebble counts. The lowermost till sample above bedrock contains 880 Au grains in 10 kg of till (Fig. 16). Au content decreases exponentially up the section to a background concentration of two grains at the top of the section. Approximately 95% of the Au grains in the anomalous till are in pristine condition with reshaped grains becoming more abundant higher in the section. Au grains vary from 10 μm to 300 μm; however c. 70% of the grains are silt-sized (<50 μm) (Fig. 7a). The HMC of till contains elevated concentrations of Au (Fig. 16), As, Cd, Ag, Sb and Zn, indicating the presence of sulphide minerals in the till. The <0.063 mm fraction contains anomalous concentrations of Au and As (Fig. 16).

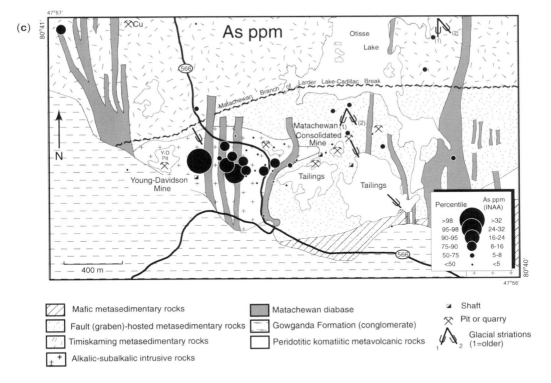

Fig. 17. Geochemistry and Au grain abundance in till around the Matachewan Consolidated gold mine: (c) As ppm in <0.063 mm fraction (from Bajc 1997).

Matachewan Consolidated Mines Au deposits

Property scale till geochemical studies were conducted around the Matachewan Consolidated and Young-Davidson Au deposits (deposit 2, Fig. 2), 70 km SE of Timmins, by Stewart & Van Hees (1982) and more recently by Bajc (1997). Au mineralization in the two deposits is hosted by sheared alkalic to subalkalic intrusive rocks as well as sheared, fault- or graben-hosted metasedimentary rocks (Fig. 17). Au is present as both coarse visible Au and fine-grained Au within sulphides. Glacial sediments in the area are generally thin (<3 m), and consist mainly of till from surface to bedrock which was deposited by ice flowing southwest and south–southeast (Fig. 17).

Bajc (1997) collected 40 till samples from hand-dug pits in a 6 km² area around the deposits for Au grain counting in the HMC and geochemical analysis of <0.063 mm fraction. Regional background concentrations of Au grains in till are between ten to 15 Au grains per 10 kg sample (Bajc 1996). However, around the deposits most till samples contain more than 20 grains, up to a maximum of 1024 grains at one site (Fig. 17a). Au grains are in pristine condition and vary in size from 10 μm to 800 μm; however, 95% of the grains are silt-sized (<50 μm). As a result of the predominantly small Au grains, the estimated Au concentrations for the till samples are low. For example, a till sample containing 1024 grains is predicted to yield only 1250 ppb Au (Bajc 1997). The highest Au (Fig. 17b), As (Fig. 17c) and Sb concentrations in the <0.063 mm fraction define an east-trending anomalous zone over the mineralized bedrock, with the highest values over the subalkalic intrusive rocks in the west. Arsenic and Sb are elevated compared to regional background values (Bajc 1997) in samples that contain anomalous Au concentrations. Glacial dispersal appears to be short, in the range of 200 to 400 m (Bajc 1997).

Hislop Au deposit

A second example from an area of thin to

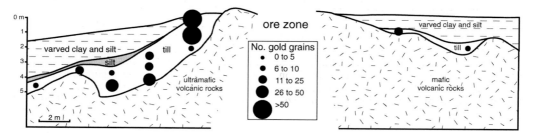

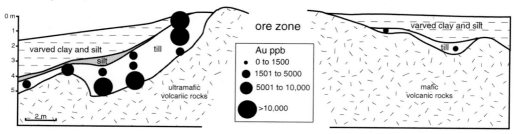

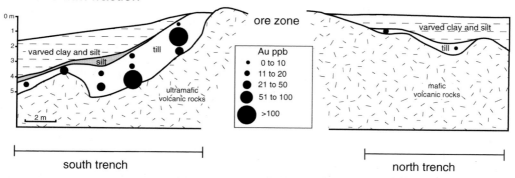

Fig. 18. Comparison of Au abundance in the heavy mineral and <0.063 mm fractions of till north and south of the Hislop Au deposit (from McClenaghan 1999).

moderate glacial sediment cover is the Hislop Au deposit, 70 km E of Timmins (deposit 1B, Fig. 2). Mineralization occurs at the contact between mafic metavolcanic rocks to the north and ultramafic metavolcanic rocks to the south within the Destor–Porcupine fault zone. Visible Au occurs in quartz veins and as inclusions in pyrite. Pyrite is the main sulphide mineral present, with chalcopyrite and galena present in minor quantities. The Hislop deposit outcrops at the surface and its flanks are overlain by highly oxidized silty-sand till which is in turn overlain by glaciolacustrine clay and silt. Bedrock striations on the subcropping ore zone indicate ice flowed to the southwest and then to the southeast.

McClenaghan (1999) collected till samples from two 3 m deep backhoe trenches excavated up-ice (north) and down-ice (south) of the mineralized outcrop (Fig. 18) for Au grain counting, geochemical analysis of the HMC and <0.063 mm fractions, and pebble counting. The threshold between background and anomalous concentrations of Au grains in 10 kg of till is five grains in this area. Till down-ice (south) of the deposit contains between 20 and 371 Au grains. Most Au grains are in pristine condition. They vary in size from 15 µm to 625 µm; however 70% of the grains are silt-sized (<50 µm) (Fig. 7b).

The HMC contains between 1000 and 23 700 ppb Au (0.06 g per tonne in a 10 kg till sample) and anomalous Au values are accompanied by varying combinations of elevated Sb, Ag, Cd, Zn, and Cu, indicating the presence of sulphide minerals in the local bedrock. However, no one element is consistently anomalous in till samples with elevated Au. For most till samples, the estimated Au concentration in the HMC based on the visible Au grain count is less than the actual Au concentration determined by INAA, some up to 15 times lower, due to masking of Au by Fe-oxide or gangue minerals. Iron-oxide coatings on Au grains formed during weathering of pyrite in the till. Au concentrations in the <0.063 mm fraction of till down-ice vary between 9 and 1324 ppb. No other elements in the <0.063 mm fraction display anomalous concentrations.

Conclusions

This paper has presented an overview of drift exploration for Au in areas of thin and thick glacial sediment cover within the AGB, for both regional geochemical surveys and deposit-scale studies. The Abitibi Greenstone Belt is a unique glaciated terrain in which to explore for lode Au deposits because it is one of the largest greenstone belts in the world and it hosts several world class Au camps. It has been continuously explored for Au over the past 90 years, the last 15 of which have included the use of drift prospecting methods. A significant volume of published literature is now available on drift exploration in the region. Most significantly, direct comparisons of data from government regional geochemical surveys and exploration programs over the past 15 years are possible because consistent heavy mineral concentrating and Au grain counting procedures have been used.

Regional geochemical surveys completed in the AGB in the past 12 years provide baseline datasets and establish thresholds for identifying anomalous Au concentrations in till. These thresholds can be used to identify anomalies in government datasets and in new geochemical data collected in the region by exploration companies. Thresholds between background and anomalous Au grain abundance or Au concentration are variable and depend on location north to south within the AGB with the lowest thresholds at the northern edge of the AGB. Au deposits in the region may contain both coarse and fine-grained Au. This fact, coupled with the observation that the HMC and <0.063 mm fractions of till display different Au anomaly patterns, highlights the need to analyse both size fractions for Au. The regional till geochemical survey results may be used to focus further Au exploration. Targets for further study are defined by anomalous samples that lie along significant bedrock structures, occur in clusters or are isolated in areas of low sample density. Surface and subsurface mapping of glacial deposits and studies of ice flow history carried out in conjunction with regional geochemical surveys provide the glacial geological framework in which to interpret the geochemical results.

Orientation or case studies around known Au deposits are important sources of information on: the geochemical and mineralogical signatures of Au deposits; appropriate sampling methods and size fractions to analyse; and local glacial stratigraphy and ice flow patterns. Case studies around Au deposits covered by thick glacial sediments (Owl Creek, Golden Pond, Pamour) demonstrate the need for overburden drilling to collect till samples from thick till units in order to detect glacial dispersal down-ice. Case studies in areas of thin glacial sediments (Matachewan, Hislop) show the value of shallow till sampling in detecting Au mineralization. In the case studies described here, till down-ice from Au deposits contains ppm levels of Au and several hundred Au grains. These high Au concentrations indicate that the ore zones subcrop and glacial processes have produced Au dispersal trains down-ice. Lower, yet still anomalous, Au concentrations in till will be found farther down-ice (more distal parts of dispersal trains), where till has been reworked and diluted by younger ice flows (e.g. Owl Creek deposit) or where subcropping Au mineralization is of much lower grade. Dispersal trains vary in length from a few tens of metres to >1 km. Close to source, visible Au grains vary in size from silt- to coarse sand-size, are in pristine condition and vary in concentration from tens of grains to hundreds of grains. In contrast to regional geochemical surveys, Au concentrations in both the HMC and <0.063 mm fraction of property-scale studies display similar anomaly patterns. These similar patterns indicate that geochemical analysis of the <0.063 mm fraction may be sufficient when exploring for fine-grained Au deposits. Geochemical analysis for pathfinder elements such as those described here (As, Sb, Ag and Cd) is also useful. Where till is strongly affected by surface weathering (e.g. Hislop deposit and surface till samples from regional surveys), the Au content of till should be determined using both Au grain counting and geochemical methods. Estimating Au content of weathered till using only visible Au grain counts may result in under-estimation because a significant proportion of the Au grains may be masked by coatings of goethite formed during weathering of pyrite.

The gold grain and till geochemical methods described here also have been used successfully to explore for lode Au deposits in other parts of Canada (e.g. Pronk & Burton 1988; Bernier & Webber 1989; Chapman et al. 1990). The major limitation of these methods, however, is their ability to detect only Au mineralization that subcrops at the bedrock/glacial sediment interface. Recently, selective leach methods applied to B-horizon soil and organic matter have been tested in the AGB (Jackson 1994; Bajc 1998; Hamilton & McClenaghan 1998) as a potential exploration tool for Au mineralization covered by thick glacial sediments. These selective leach methods may be helpful in detecting Au deposits that do not subcrop.

A. F. Bajc and S. A. Averill were extremely helpful to the author by providing geological information and figures. Secondary electron (SEM) images of Au grains were produced by Overburden Drilling Management Ltd. M. Fedikow, W. B. Coker, R. N. W. DiLabio and D. Kerr are thanked for their comments and suggestions which improved the manuscript, and G. E. M. Hall for her editorial comments. Geological Survey of Canada Contribution No. 199185.

References

ALCOCK, P. W. 1991. *Quaternary geology of the Shining Tree area*. Ontario Geological Survey, Open File Report **5810**.

AVERILL, S. A. 1988. Regional variations in the gold content of till in Canada. *In:* MACDONALD, D. R. & MILLS, K. A (eds) *Prospecting in Areas of Glaciated Terrain – 1988*. Canadian Institute of Mining and Metallurgy, 271–284.

AVERILL, S. A. 1990. Drilling and sample processing methods for deep till geochemistry surveys: making the right choices. *In:* AVERILL, S. A., BOLDUC, A., COKER, W. B., DILABIO, R. N. W., PARENT, M. & VEILLETTE, J. *Application de la géologie du Quaternaire à l'exploration minérale*. Association professionelle des géologues et des géophysiciens du Québec, 139–173.

AVERILL, S. A. 2001. The application of heavy indicator minerals in mineral exploration. *In:* MCCLENAGHAN, M. B., BOBROWSKY, P. T., HALL, G. E. M. & COOK, S. J. (eds) *Drift Exploration in Glaciated Terrain*. Geological Society, London, Special Publications, **185**, 201–224.

AVERILL, S. A. & HUNEAULT, R. 1991. Using silt-sized visible gold grains to explore for Au deposits concealed by Quaternary overburden: Nevada vs. Canada. *Explore* (Association of Exploration Geochemists Newsletter), **72**(1), 10–12.

AVERILL, S. A. & ZIMMERMAN, J. R. 1986. The Riddle resolved: the discovery of the Partridge gold zone wing sonic drilling in glacial overburden at Waddy Lake, Saskatchewan. *Canadian Geology*, **1**, 14–20.

AVERILL, S. A., MACNEIL, K. A., HUNEAULT, R. G. & BAKER, C. L. 1986. *Rotasonic drilling operations (1984) and overburden heavy mineral studies, Matheson Area, District of Cochrane*. Ontario Geological Survey, Open File Report **5569**.

BAJC, A. F. 1996. *Regional distribution of gold in till in the Peterlong Lake-Radisson Lake area, southern Abitibi subprovince; potential exploration targets*. Ontario Geological Survey, Open File **5941**.

BAJC, A. F. 1997. *A regional evaluation of gold potential along the western extension of the Larder Lake-Cadillac Break, Matachewan area: results of regional till sampling*. Ontario Geological Survey, Open File Report **5957**.

BAJC, A. F. 1998. A comparative analysis of enzyme leach and Mobile Metal Ion selective extractions: case studies from glaciated terrain, northern Ontario. *Journal Geochemical Exploration*, **61**, 113–148.

BAJC, A. F., HAMILTON, S. M., AYER, J. & JENSEN, L. S. 1996. *New exploration targets in the Peterlong Lake-Radisson Lake area, Southern Abitibi Subprovince; till, lake sediment and lake water sampling program*. Ontario Geological Survey Open File Report **5942**.

BIRD, D. J. & COKER, W. B. 1987. Quaternary stratigraphy and geochemistry at the Owl Creek gold mine, Timmins, Ontario, Canada. *Journal of Geochemical Exploration*, **28**, 267–284.

BERNIER, M. A. & WEBBER, G. R. 1989. Mineralogical and geochemical analysis of shallow overburden as an aid to gold exploration in southwestern Gaspésie, Quebec, Canada, *Journal of Geochemical Exploration*, **34**, 115–145.

CHAPMAN, R., CURRY, G. & SOPUCK, V. 1990. The Bakos deposit discovery – a case history. *In:* BECK, L. S., HARPER, C. T. (eds) *Modern Exploration Techniques*. Saskatchewan Geological Society, Special Publications, **10**, 195–212.

COKER, W. B. 1991. Overburden geochemistry in mineral exploration. *In: Exploration Geochemistry Workshop*. Geological Survey of Canada, Open

File **2390**, 3-1-3-60.

COKER, W. B. & DILABIO, R. N. W. 1989. Geochemical exploration in glaciated terrain: geochemical responses. *In:* GARLAND, G. D. (ed.) *Proceedings of Exploration '87.* Ontario Geological Survey, Special Volume, **3**, 336–383.

COKER, W. B. & SHILTS, W. W. 1991. Geochemical exploration for gold in glaciated terrain. *In:* FOSTER, R. P. (ed.) *Gold Metallogeny and Exploration.* Blackie, London, 336–359.

DILABIO, R. N. W. 1983. *Gold content of overburden samples in the Abitibi Clay Belt.* Geological Survey of Canada, Open File **945**.

DILABIO, R. N. W. 1990. Classification and interpretation of the shapes and surface textures of gold grains from till on the Canadian Shield. *In: Current Research, Part C.* Geological Survey of Canada, Paper **90-1C**, 323–329.

DiLabio, R. N. W. 1991. Classification and interpretation of the shapes and surface textures of gold grains from till. *In:* HÉRAIL, G. & FORNARI, M. (eds) *Alluvial Gold Placers.* Proceedings of the International Symposium on Alluvial Gold Placers, Institut Francais de Recherche Scientifique pour le Développment en Coopération, 297–313.

DILABIO, R. N. W. 1995. Residence sites of trace elements in oxidized tills. *In:* BOBROWSKY, P. T., SIBBICK, S. J., NEWELL, J. M. & MATYSEK, P. F. (eds) *Drift Exploration in the Canadian Cordillera.* British Columbia Ministry of Energy, Mines and Petroleum Resources, Paper **1995-2**, 139–148.

DILABIO, R. N. W., MILLER, R. F., MOTT, R. J. & COKER, W. B. 1988. The Quaternary stratigraphy of the Timmins area, Ontario, as an aid to mineral exploration by drift prospecting. *In: Current Research, Part C.* Geological Survey of Canada, Paper **88-1C**, 61–65.

FORTESCUE, J. A. C., LOURIM, J., GLEESON, C. F., JENSEN, L. & BAKER, C. 1984. *A synthesis and interpretation of basal till geochemical and mineralogical data obtained from the Kirkland Lake (KLIP) area (1979-1982).* Ontario Geological Survey, Open File Report **5506**.

GLEESON, C. F. & RAMPTON, V. N. 1987. The use of near surface materials in gold exploration-Kirkland Lake, Ontario. *Canadian Institute of Mining and Metallurgy Bulletin*, **80**(904), 62–68.

GLEESON, C. F. & SHEEHAN, D. G. 1987. Humus and till geochemistry over the Doyon, Bosquet and Williams Au deposits. *Canadian Institute of Mining and Metallurgy Bulletin*, **80**(898), 58–66.

HAMILTON, S. M. & MCCLENAGHAN, M. B. 1998. Field data in support of electrochemical transport of elements through thick glacial overburden. *In:* AYER, J. A., BAKER, C. L., IRELAND, J. C., KELLY, R. I. & THURSTON, P. C. (eds) *Summary of Field Work and Other Activities-1998.* Ontario Geological Survey, Miscellaneous Paper **169**, 267–274.

HARRON, G. A., MIDDLETON, R. S., DURHAM, R. B. & PHILIPP, A. 1987. Geochemical and geophysical gold exploration in the Timmins area, Ontario; a case history. *Canadian Institute of Mining and Metallurgy Bulletin*, **80**, 52–57.

HENDERSON, P. J. & Roy, M. 1995. Distribution and character of gold in surface till in the Flin Flon greenstone belt, Saskatchewan. *In: Current Research 1995-E.* Geological Survey of Canada, 175–186.

HIRVAS, H. & NENONEN, K. 1990. Field methods for glacial indicator tracing *In:* KUJANSUU, R. & SAARNISTO, M. (eds) *Glacial Indicator Tracing.* A. A. Balkema, Rotterdam, 217–248.

HOFFMAN, E. L., CLARK, J. R. & YEAGER, J. R. 1999. Gold analysis-fire assaying and alternative methods. *Exploration and Mining Geology*, **7**, 155–160.

HUGHES, O. L. 1965. Surficial geology of part of the Cochrane District, Ontario, Canada. *In:* WRIGHT Jr., H. E. & FREY, D. G. (eds) *International Studies on the Quaternary.* Geological Society of America, Special Paper **84**, 535–565.

JACKSON, R. G. 1994. *The application of water and soil geochemistry to detect blind mineralization in areas of thick overburden.* Ontario Geological Survey, Open File Report **5927**.

JONASSON, I. R., KINGSTON, D. M., WATKINSON, D. H., ELLIOTT, S. R. 1999. Role of pyrite in the formation and localization of gold mineralization at the Owl Creek Mine, Timmins, Ontario. *Economic Geology*, Monograph **10**, 627–660.

KAURANNE, K., SALMINEN, R. & ERIKSSON, K. (eds) 1992. *Handbook of Exploration Geochemistry, Volume 5: Regolith Exploration Geochemistry in Arctic and Temperate Terrains.* Elsevier, Amsterdam.

LALONDE, J-P. & LASALLE, P. 1982. *Atlas géochimique de l'argile et du till de base de l'Abitibi.* Ministere de l'Energie et des Ressources, **DVP-830**.

MCCLENAGHAN, M. B. 1990. *Summary of results from the Black River-Matheson (BRiM) reconnaissance surface till sampling program.* Ontario Geological Survey, Open File Report **5749**.

MCCLENAGHAN, M. B. 1991. *Geochemistry of tills from the Black River-Matheson (BRiM) sonic overburden drilling program and implications for exploration.* Ontario Geological Survey, Open File Report **5800**.

MCCLENAGHAN, M. B. 1992a. *Quaternary geology of the Matheson-Lake Abitibi area.* Ontario Geological Survey, Open File Report **5836**.

MCCLENAGHAN, M. B. 1992b. Surface till geochemistry and implications for exploration, Black River-Matheson area, northeastern Ontario. *Exploration and Mining Geology*, **1**, 327–337.

MCCLENAGHAN, M. B. 1994. Till geochemistry in areas of thick drift and its application to gold exploration, Matheson area, northeastern Ontario. *Exploration and Mining Geology*, **3**, 17–30.

MCCLENAGHAN, M. B. 1999. *Till geochemical signatures associated with Au deposits in the Timmins-Matheson area, Western Abitibi Greenstone Belt, northeastern Ontario.* Geological Survey of Canada, Open File **3707**.

MCCLENAGHAN, M. B. & DILABIO, R. N. W. 1995. *Overburden drill hole compilation, Timmins, Ontario (NTS 42A/11, 12, 13, 14).* Geological Survey of Canada, Open File **3086**.

MCCLENAGHAN, M. B. & WYATT, P. H. 1990. *Summary of overburden and bedrock trace element data from* the Kirkland Lake Initiatives program (KLIP) reconnaissance till sampling program. Ontario Geological Survey, Open File Report **5737**.

McCLENAGHAN, M. B., PAULEN, R. C., AYER, J. A., TROWELL, N. F. & BAUKE, S. 1998. *Regional till and humus geochemistry of the Timmins-Kamiskotia (NTS 42A/11, 12, 13, 14) area, northeastern Ontario*. Geological Survey of Canada, Open File **3675**.

McMARTIN, I. & McCLENAGHAN, M. B. 2001. Till geochemistry and sampling techniques in shield terrain. *In*: McCLENAGHAN, M. B., BOBROWSKY, P. T., HALL, G. E. M. & COOK, S. J. (eds) *Drift Exploration in Glaciated Terrain*. Geological Society, London, Special Publications, **185**, 19–43.

NICHOL, I., LAVIN, O. P., McCLENAGHAN, M. B. & STANLEY, C. R. 1992. The optimization of geochemical exploration for gold using glacial till. *Exploration and Mining Geology*, **4**, 305–326.

ONTARIO GEOLOGICAL SURVEY 1986. *Sonic drillhole 84-37, Stock Township, District of Cochrane*. Ontario Geological Survey Map **80790**.

PAULEN, R. C. & McCLENAGHAN, M. B. 1998a. *Surficial geology of the Buskegau River (NTS 42A/14) area, northeastern Ontario*. Geological Survey of Canada, Open File Map **3619**.

PAULEN, R. C. & McCLENAGHAN, M. B. 1998b. *Surficial geology of the Manning Lake (NTS 42A/13) area, northeastern Ontario*. Geological Survey of Canada, Open File Map **3618**.

PLOUFFE, A. 1995. Drift prospecting sampling methods. *In*: BOBROWSKY, P. T., SIBBICK, S. J., NEWELL, J. M. & MATYSEK, P. F. (eds) *Drift Exploration in the Canadian Cordillera*. British Columbia Ministry of Energy, Mines and Petroleum Resources, Paper **1995-2**, 43–52.

POULSEN, K. H., ROBERT, F. & DUBÉ, B. 2000. *Geological classification of Canadian Au deposits*. Geological Survey of Canada, Bulletin **540**.

PRESSACCO, R. 1999. Economic geology and mineralization of the Owl Creek Mine. *In*: PRESSACCO, R. (ed.) *Special Project: Timmins Ore Deposits Descriptions*. Ontario Geological Survey, Open File Report **5985**, 169–200.

PRONK, A. G. & BURTON, D. M. 1988. *Till geochemistry as a technique for gold exploration in northern New Brunswick*. Canadian Institute of Mining and Metallurgy Bulletin, **81**, 90–98.

ROBERT, F. 1996. Quartz-carbonate vein gold. *In*: ECKSTRAND, O. F., SINCLAIR, W. D. & THORPE, R. I. (eds) *Geology of Canadian Mineral Deposit Types*. Geological Survey of Canada, Geology of Canada, **8**, 350–366.

SAUERBREI, J. A., PATTISON, E. F. & AVERILL, S. A. 1987. Till sampling in the Casa Berardi gold area, Québec: a case history in orientation and discovery. *Journal of Geochemical Exploration*, **28**, 297–314.

SHELP, G. S. & NICHOL, I. 1987. Distribution and dispersion of gold in glacial till associated with gold mineralization in the Canadian Shield. *Journal of Geochemical Exploration*, **27**, 315–336.

SMITH, S. L. 1992. *Quaternary stratigraphic drilling transect, Timmins to the Moose River Basin, Ontario*. Geological Survey of Canada, Bulletin **415**.

STEELE, K. G., BAKER, C. L. & McCLENAGHAN, M. B. 1989. Models of glacial stratigraphy determined from drill core, Matheson area, northeastern Ontario. *In*: DILABIO, R. N. W. & COKER, W. B. (eds) *Drift Prospecting*. Geological Survey of Canada, Paper **89-20**, 127–138.

STEWART, R. A. & VAN HEES, E. H. 1982. Evaluation of past-producing gold mine properties by drift prospecting: an example from Matachewan, Ontario, Canada. *In*: Evenson, E.B., Schluter, Ch. & Rabassa, J. (eds) *Tills and Related Deposits*. A. A. Balkema, Rotterdam, 179–193.

THOMSON, I., BURNS, J. G. & FAULKNER, F. H. 1987. Gold exploration in deep glaciated overburden: experiences from the Belore case history. *Journal of Geochemical Exploration*, **29**, 435–436.

TYLER, R. K., SIMUNOVIC, M., WILSON, S., PENNA, D., GERTH, D., KILBRIDE, B., COAD, P. & HARVEY, P. 1999. Economic geology and mineralization of the Pamour Mine. *In*: Pressacco, R. (ed.) *Special Project: Timmins Ore Deposits Descriptions*. Ontario Geological Survey, Open File Report **5985**, 201–221.

VEILLETTE, J. J. 1995. New Evidence for northwestward glacial ice flow, James Bay region, Quebec. *In: Current Research 1995-C*. Geological Survey of Canada, 249–258.

VEILLETTE, J. J. & McCLENAGHAN, M. B. 1996. *The sequence of ice flow in Abitibi-Timiskaming; implications for mineral exploration and dispersal carbonates from the Hudson Bay Basin, Québec and Ontario*. Geological Survey of Canada, Open File **3033**.

Application of composite glacial boulder geochemistry to exploration for unconformity-type uranium deposits in the Athabasca Basin, Saskatchewan, Canada

STEVEN EARLE

Grasswood Geoscience Ltd., 696 Western Acres Rd., Nanaimo, British Columbia V9R 5W9, Canada (e-mail: geosci@home.com)

Abstract: Sampling glacially transported boulders is effective for mapping subcrop clay alteration patterns associated with deeply-buried unconformity-type uranium deposits in the Athabasca Basin of northern Saskatchewan, Canada. The technique works well because the subcrop alteration haloes show significant geochemical contrast with background, because the glacial deposits are consistently distributed and rich in boulders, and the altered boulders are typically well-represented in the boulder population. Boulder sampling is preferred over some other techniques because it is rapid and inexpensive, and the use of a lithological sampling medium provides a direct measure of subcrop clay-mineralogy, in an area where outcrop exposure is very restricted.

A large composite boulder sample data set, comprising c. 20 000 samples from within the eastern part of the Athabasca Basin, has been compiled. Sampling and analytical techniques are described, and results are presented at both regional and semi-regional scales. The distributions show clear correlations with both basin-scale features, such as stratigraphical variation and regional alteration patterns, and with deposit-related features, such as hydrothermal illite, dravite and chlorite alteration. Composite boulder lithogeochemistry may be equally effective for mineral exploration in other glaciated regions with restricted bedrock exposure.

The Athabasca Basin of northern Saskatchewan (Fig. 1) is the world's most important uranium province. Large high-grade uranium deposits such as Key Lake, Cigar Lake, Cluff Lake, Rabbit Lake and McArthur River, plus numerous smaller but similarly rich deposits, are situated at the unconformity between the Proterozoic Manitou Falls Formation (Athabasca Group), and the underlying Aphebian and Archean gneisses and schists (Hoeve *et al.* 1980; Kirchner *et al.* 1980; Fouques *et al.* 1986; McGill *et al.* 1993). The Manitou Falls Formation, which is the basal unit of the Athabasca Group in the eastern part of the basin, is a medium- to coarse-grained fluviatile quartz arenite with a total thickness of c. 1000 m (Ramaekers 1990). Quartz-pebble conglomerate units are common near to the base of the formation. The Manitou Falls Formation is overlain in this area by the Wolverine Point Formation (interbedded sandstone and mudstone) and the Locker Lake Formation (pebbly quartz arenite).

Most of the known uranium deposits are

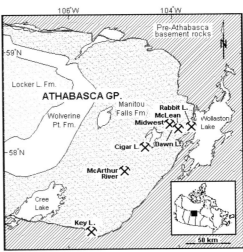

Fig. 1. Location of the Athabasca Basin in northern Saskatchewan, and locations of some important unconformity-type uranium deposits. Geology after Ramaekers (1990).

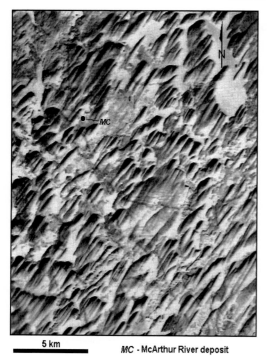

Fig. 2. Winter-time Landsat-TM image of part of the Athabasca Basin around the McArthur River uranium deposit. The direction of glacial transport is from NE to SW. Drumlins and smaller flutings are clearly evident as elongate and elliptical ridges. An esker complex is visible as a sinuous line crossing the southeastern part of the image area. Dark lines crossing the glacial direction are related to forest-fires. Lakes are snow-covered and appear white.

located in a part of the basin where the sandstone ranges in thickness from a few metres to just over 500 m. Basement-rock structural and lithological features typically associated with the mineralization, such as reverse faulting and graphite-bearing pelitic rocks, have been traced successfully using geophysical techniques (e.g. Matthews *et al.* 1997). It has also been shown that the clay-mineral component of the sandstone has been pervasively and extensively altered by the mineralizing fluids and that the alteration haloes, which extend upward from the deposits to the present bedrock surface, can be clearly delineated by mineralogical and lithogeochemical methods (Golightly *et al.* 1983; Sopuck *et al.* 1983; Hoeve 1984; Earle & Sopuck 1989; Earle *et al.* 1999). The challenge for geochemical exploration for unconformity-type uranium deposits in the Athabasca Basin is that there is very little exposure of bedrock because of a consistent cover of thick (>5 m) glacial sediments.

Northern Saskatchewan has been completely and repeatedly affected by Quaternary glaciation (Schreiner 1979). While there is considerable bedrock exposure in areas underlain by granitic and gneissic basement rocks, the relatively soft sandstone of the Athabasca Basin is characterized by a nearly continuous cover of morainal and glaciofluvial deposits. The predominant drift type is morainal, and as shown in Figure 2, drumlinoid features display a well developed NE–SW trend. This southwesterly ice-flow direction is assumed to represent the most recent ice advance event in northern Saskatchewan. Multiple till sheets have been observed within the study area (Schreiner 1979). Analysis of satellite imagery by the author has shown evidence of previous ice-advance events with ice-flow directions, which differ from the most recent one by no more than a few tens of degrees.

Since the average drift thickness within the Athabasca Basin is >5 m, and bedrock exposure is typically <5%, systematic sampling of outcrop to detect alteration haloes is not practical. Some attempts have been made to delineate the subcrop expression of alteration haloes by drilling shallow holes through the drift (cf. Clark 1987), but because of the high cost of drilling through thick bouldery till, this has not proven to be a practical alternative. Sampling the till itself is relatively easy, however the strongest and most clearly defined anomaly haloes are those of the hydrothermal alteration products, such as illite, chlorite and dravite, and these are not generally detected by geochemical analysis of the finer fractions (<0.063 mm) of the till. On the other hand, clay mineral signatures are clearly defined by analysis of the coarse (rock-fragment) fraction of the till, and it has been shown that there is a close correlation between the geochemistry of composite samples from Manitou Falls Formation boulders and that of the underlying bedrock (Earle *et al.* 1990).

Since 1986, over 30 000 composite boulder samples have been collected within the Athabasca Basin. In fact the technique has been used by virtually every exploration company working within the basin during this period. The geochemical data for most of the samples collected are available from government assessment files. The sampling and analytical techniques have been sufficiently consistent to allow compilation of the results into a single regional data set. Clay mineral distributions based on composite-sample boulder lithogeochemical data for 20 000 samples from the eastern part of the Athabasca

Basin are presented here, along with an interpretation of the results, and suggestions for application of this technique in other glaciated areas.

Composite boulder sampling

Glacially transported material is comprised of a mixture of particles which have been moved varying distances. The silt- or clay-sized material within a till includes grains which have been transported from a few centimetres to tens of kilometres down-ice. Similarly, a till includes cobbles and boulders which have been transported only a short distance, and others which have moved greater distances. It would be impossible to separate the grains of proximal and distal origin from a typical fine-fraction sub-sample, and in most cases it is reasonable to accept that while some of the material is far-travelled, the sample as a whole is generally representative of an area some distance up-ice. In this respect the fine-fraction of a till sample is a composite sample. A composite rock-fragment sample can also be collected from till, either by sieving out some of the coarse-grained material, or by selecting larger fragments such as cobbles or boulders.

The till overlying the eastern part of the Athabasca Basin is rich in cobble- and boulder-sized clasts. Based on lithological counts made at thousands of locations, it has been found that, on average, c. 85% of the larger clasts in the till are Manitou Falls Formation sandstone and most of the remaining 15% are basement-derived granite and gneiss which have been transported for tens of kilometres, from outside (north and east) of the Athabasca Basin. The Manitou Falls Formation contains c. 3% aluminosilicate minerals (illite, kaolin etc.). The surrounding basement rocks (typically gneisses, schists and granitic rocks) consistently contain several tens of percent aluminosilicate minerals (mica, feldspar, clays etc.). Basement rock-fragments must be excluded from composite samples because geochemical techniques are routinely used for determination of clay mineral types in the Manitou Falls Formation, and thus a single basement rock fragment can have a significant effect on the estimate of the amount and proportions of clay minerals. A composite sample comprised of sub-samples from small rock fragments (pebble-sized or smaller) would not be useful because it would be very difficult to identify and remove all basement material. Instead, composite samples are collected from boulders exposed at the surface of the till.

In order to determine reasonable sample sizes

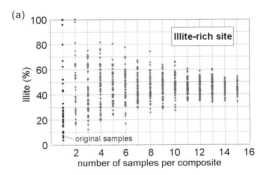

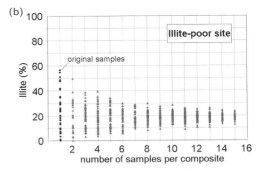

Fig. 3. Simulations of differing sizes of composite boulder samples based on actual data ('original samples') from two different sites (A & B). Fifty synthetic composite samples were randomly selected for each sample size from 2 to 15, and each point represents the mean illite content in one of those 'composite' samples. Composites comprised of fewer than five boulders tend to have highly variable illite levels, while those comprised of more than ten boulders are much more consistent.

(number of boulders) for composite boulder sampling, a study of individual boulder geochemistry was carried out adjacent to a known uranium deposit in the central part of the study area. At each of two sites 40 individual sandstone boulders were collected and analysed. Site A is situated 100 m down-ice (SW) from a uranium deposit covered with 200 m of sandstone and 10 to 20 m of till. The 40 samples collected from this location have between 4 and 100% illite. Site B is situated 1500 m to the SE of the same deposit, and therefore should not contain any debris eroded from the deposit, i.e. background conditions. The 40 samples collected at this location contain between 0 and 56% illite. The actual data for these samples have been used to generate a series of synthetic composite boulders sample values, where the synthetic composites are comprised of between 2

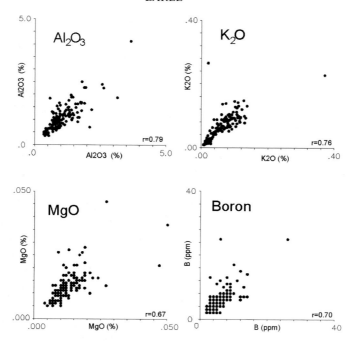

Fig. 4. Comparative data for 275 pairs of composite boulder sample field duplicates.

and 15 sub-samples selected randomly from the group of 40 original data values. Fifty such synthetic composite samples were derived for each number of sub-samples, from 2 to 15.

As shown on Figure 3, composites consisting of 4 sub-samples contain 20 to 80% illite at site A, and from 3 to 40% illite at site B. Composites of 8 sub-samples have illite content ranging from 25 to 75% at site A, and from 8 to 30% at site B. Composites of 12 sub-samples contain 35 to 55% illite at site A, and 10 to 25% illite at site B. Results such as these have been used to show that while it might be desirable to collect composite samples from dozens of boulders, it is both practical and adequate to collect a composite sample from between 10 and 15 boulders per site.

Composite boulder samples comprise chips (c. 50 g) taken from between 10 and 15 of the population of relatively large and angular Manitou Falls Formation boulders observed at surface within a radius of a few tens of metres of each sample site. The chips are taken from the surfaces of the boulders, as experience has shown that the elements of interest have not been significantly affected by postglacial surface weathering. Chips collected from 10 to 15 boulders at each site are combined and analysed as one sample. Sampling spacing between sites is typically 100 to 200 m, on lines spaced at between 500 and 2000 m apart. In most cases the sampling lines are orientated perpendicular to the ice-advance direction.

Analysis

The c. 20 000 composite boulder samples included in the area covered by this compilation have been analysed for both major and trace elements. This paper is focused on the clay mineral alteration patterns because the trace element analytical techniques have not been sufficiently consistent to allow comparison between data sets. For this reason only the major element analytical techniques are described here.

Major-element analytical techniques have varied slightly over the period of sampling (1986 to 1999) but the resulting values are closely comparable. A 'total' acid digestion (HF, $HClO_4$, HNO_3 and HCl) has been used in all cases, and for most of the samples the analysis of Al_2O_3, MgO and K_2O has been by inductively-coupled plasma atomic emission spectrometry (ICP-AES). Boron has also been analysed by ICP-AES, following fusion with NaOH and dissolution in HCl.

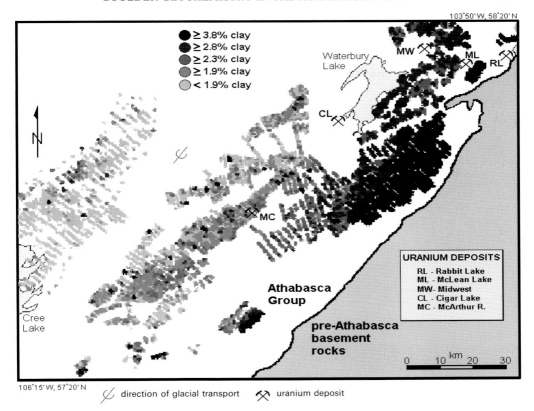

Fig. 5. Total clay contents of composite boulder samples from till in the southeastern part of the Athabasca Basin. The distribution has been re-sampled to a 500 × 500 m grid. Shading intervals are based on the 90, 70, 50 and 30th percentiles.

Some examples of the repeatability for 275 pairs of field duplicate boulder samples are shown in Figure 4. The duplicates were collected from separate sets of boulders at each site, and were submitted blind to the laboratory. This field duplicate sampling was part of a separate project carried out by the author at several different locations within the eastern Athabasca Basin. The variability shown here represents combined sampling, sample preparation and analytical variability. For all of the elements except B, the total variability is < 10%, while for B, it is close to 12%. (Variabilities are determined as the difference between a duplicate pair expressed as a percentage of the average of the duplicate pair. The values given above are the averages of these between-pair differences for the 275 pairs.)

The term 'clay' in this paper refers to all aluminosilicate minerals in the sandstone matrix, including the clay minerals kaolin and illite, as well as chlorite and dravite. The Al_2O_3, MgO and K_2O and B data have been used to estimate clay mineral proportions in these samples using the simplified normative technique first described by Earle & Sopuck (1989). The technique is based on the well-established observation that the Manitou Falls Formation is devoid of significant amounts of feldspar. The only important aluminosilicate minerals observed within the Manitou Falls Formation by the author using short-wave infrared reflectance spectra are illite, chlorite (sudoite), kaolin (both dickite and kaolinite) and dravite (Earle et al. 1999). These observations are supported by x-ray diffraction data (Hoeve 1984) and by petrographic data (Ramaekers 1990).

In the normative technique used here, the proportion of dravite is estimated from the B content; the proportion of chlorite is estimated from the remaining amount of MgO; the proportion of illite is estimated from the remaining amount of K_2O and the proportion of kaolin is estimated from the remaining

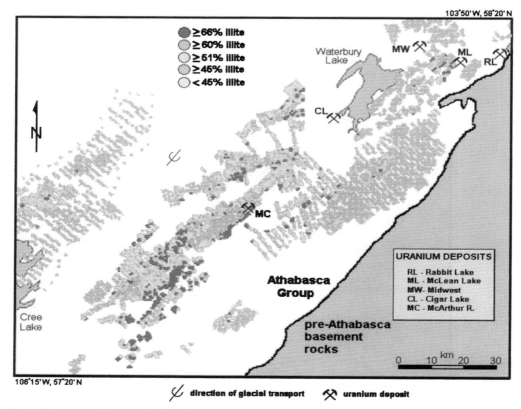

Fig. 6. Illite contents of composite boulder samples from till in the southeastern part of the Athabasca Basin. The distribution has been re-sampled to a 500 × 500 m grid. Shading intervals are based on the 97, 95, 90 and 80th percentiles. Illite is expressed as a proportion of the total clay fraction.

amount of Al_2O_3. The total clay content is determined as the sum of the proportions of these minerals, and is reported as a weight percentage of the whole rock. The contents of the individual clay minerals are reported as a weight percentage of the total clay content.

Regional lithogeochemical trends

Lithogeochemical data for composite boulder samples from c. 20 000 sites from the southeastern part of the basin are shown on Figures 5 to 7. In order to facilitate presentation of the results at a relatively small scale, the original distributions have been digitally resampled to a 500 m square grid. For each grid node a value has been calculated as the arithmetic average of all points which fall within a 375 m radius circular window. No distance-weighting has been applied.

There is a general decrease in the total clay content from the bottom to the top of the Manitou Falls Formation. In the lower part of the formation clay is typically > 5%, while in the upper part the content is typically < 2%. This trend is clearly reflected in the regional distribution of clay contents in the eastern part of the basin (Fig. 5). The highest clay levels (> 4%) are concentrated along the eastern margin of the basin in areas underlain by the lower part of the Manitou Falls Formation. Clay levels decrease towards the centre of the basin, and are lowest in areas underlain by the upper part of the formation.

The Manitou Falls Formation is characterized by illite levels which are generally < 25% of the total clay fraction. Illite contents > 50% are most commonly a product of hydrothermal alteration related to unconformity-type uranium mineralization (cf. Hoeve 1984), although there is evidence of regional illitization in the eastern part of the basin (Earle & Sopuck 1989). The boulder data show an extensive trend of

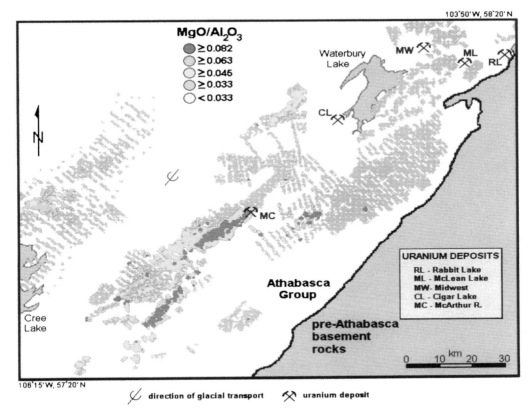

Fig. 7. MgO/Al$_2$O$_3$ contents of composite boulder samples from the southeastern part of the Athabasca Basin. The distribution has been re-sampled to a 500 × 500 m grid. Shading intervals are based on the 97, 95, 90 and 80th percentiles.

significant illitization in the central part of the study area (Fig. 6). This trend coincides closely with the trend of regional illite enrichment in bedrock which extends from the Key Lake deposit area (see Fig. 1) in a northeasterly direction towards the McArthur River deposit area, as was first described by Earle & Sopuck (1989) on the basis of sandstone drill-core. Smaller zones of boulder-sample illite enrichment are evident to the SW of the area of the McLean deposit, and to the N of Cree Lake (at the western edge of the study area).

The Manitou Falls Formation contains minor amounts of detrital tourmaline (as dravite) (Ramaekers 1990). Elevated levels of dravite (of apparent hydrothermal origin) are spatially associated with uranium deposits such as Key Lake (Earle & Sopuck 1989) and McArthur River (McGill et al. 1993), however dravite is also enriched in some areas without known uranium deposits (Earle & Sopuck 1989). In most of the eastern Athabasca Basin, the Manitou Falls Formation contains essentially no chlorite (not measurable geochemically), however there is significant enrichment of magnesium chlorite (sudoite) in the sandstone around deposits such as McArthur River (McGill et al. 1993). There is also some enrichment of chlorite in a few areas without known uranium deposits. The MgO/Al$_2$O$_3$ ratio of the sandstone provides a general indication of combined chlorite and dravite enrichment (since dravite is a B- & Mg-bearing tourmaline), and is a useful indicator of magnesium alteration. The background level for the ratio is around 0.025, and there is an extensive area to the southwest of the McArthur River deposit for which the MgO:Al$_2$O$_3$ ratios are consistently >0.082 (Fig. 7). This trend of Mg enrichment is c. 20 km in extent. Another separate strong Mg anomaly trend occurs to the S–SW of the McArthur River deposit, and there are others to the east, to the north (W of the Cigar Lake deposit) and also farther to the west (N of Cree Lake).

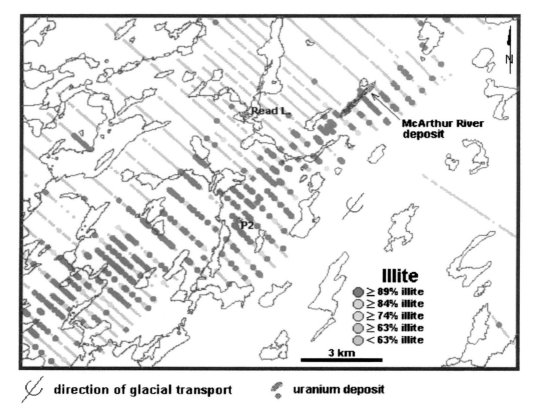

Fig. 8. Illite contents of composite boulder samples from till in the McArthur River deposit area. The distribution has been re-sampled to a 250 × 250 m grid. Illite is expressed as a proportion of the total clay fraction.

Deposit-related lithogeochemical trends

McArthur River is the only major Athabasca Basin uranium deposit for which relatively detailed boulder geochemical data are publicly available. Data for a 20 by 15 km area around McArthur River are shown in Figures 8 to 10. In this case, the original distributions have been resampled to a 250 m square grid, with 190 m radius circular smoothing windows. The location of the large McArthur River deposit is shown, as are locations of two smaller mineralized zones, Read Lake and P2. The lithological and structural fabric of the basement rocks of this area (as in much of the rest of the eastern Athabasca Basin) follows a consistent northeasterly trend, and the P2 mineralization is situated on the same litho-structural trend as the McArthur River deposit (McGill et al. 1993). This trend includes a clearly-defined zone of basement rock conductivity (related to graphitic pelitic rocks), and a reverse fault (the P2 Fault) which is characterized by an offset in the order of tens of metres (McGill et al. 1993).

A clearly-defined zone of boulder illite enrichment exists in the McArthur River area (Fig. 8). This zone is part of the regional illite anomaly shown above in Figure 6. In this area, many of the boulder samples contain >80% illite. The McArthur River deposit is situated at the northwestern edge of the illite zone, while the P2 deposit is entirely within the illitic area. Evidence of strong illitization in the sandstone above the McArthur River Deposit has been found in drill core (McGill et al. 1993) and, in particular, in the sandstone from holes drilled immediately to the east of the deposit (Matthews et al. 1997), directly beneath the area of the boulder illite anomaly. The Read Lake area is not characterized by illite-rich boulders, although there is a zone of illitic boulders along strike to the southwest.

There is a trend of very strong dravitization in boulders which has its up-ice limit in the

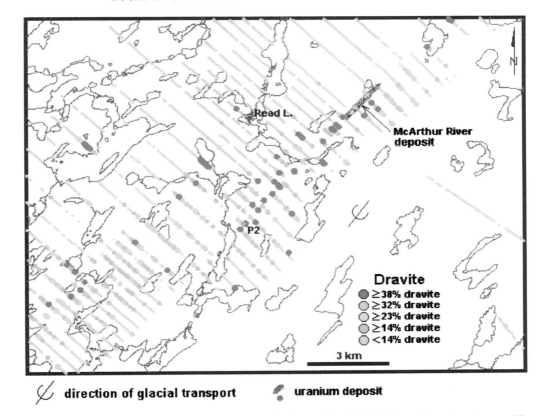

Fig. 9. Dravite contents of composite boulder samples from till in the McArthur River deposit area. The distribution has been re-sampled to a 250 × 250 m grid. Dravite is expressed as a proportion of the total clay fraction.

immediate area of the McArthur River deposit (Fig. 9). Many of these samples have > 30% dravite in the clay fraction, as compared with a background level in the 1 to 3% range for the eastern part of the basin as a whole. These boulder results are consistent with those from drill core, in which sandstone contains B concentrations of several hundred ppm (McGill et al. 1993), values which are equivalent to tens of % dravite in the clay fraction. The McArthur River boulder dravite zone extends along strike to the P2 area. There is no evidence of strong dravitization at Read Lake from the boulder data, but there is a persistent boulder-sample dravitic zone c. 3 km to the southwest. Apart from these two main sub-parallel trends of strong dravitization, there is an area of moderate dravite enrichment (with levels consistently > 14%) which coincides generally with the area of illite-rich boulders shown in Figure 8.

The McArthur River area is characterized by a generally high Mg background (Fig. 10). MgO/Al$_2$O$_3$ ratios are consistently > 0.10, as compared with the background level of < 0.05 (see Fig. 7). There is moderate to strong enrichment of Mg in boulders from the area of the McArthur River deposit, and also the P2 area. These results are consistent with chlorite data for drill core presented by McGill et al. (1993), which show > 10% chlorite in the sandstone above the McArthur River deposit. The Mg enrichment in boulders also extends along the structural trend to the northeast of the deposit. The Read Lake area is also characterized by strong and persistent Mg enrichment in boulders, especially to the southwest of the mineralized zone.

Conclusions

The examples described here have demonstrated that the geochemistry of composite boulder samples from till is an effective technique for mapping both regional and semi-regional scale

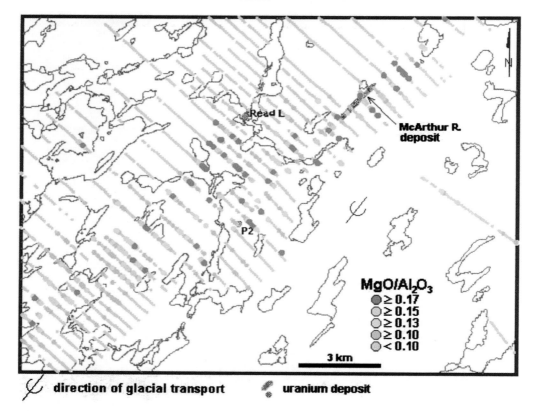

Fig. 10. MgO/Al$_2$O$_3$ contents of composite boulder samples from till in the McArthur River deposit area. (The distribution has been re-sampled to a 250 × 250 m grid.)

variations in bedrock geochemistry in the eastern part of the Athabasca Basin. Sampling is rapid and relatively inexpensive, and the results are consistent with observed patterns from bedrock lithogeochemical studies.

On a regional scale, the boulder data reveal broad patterns, such as variations in the total clay content, which are consistent with the observed stratigraphical features of the Manitou Falls Formation. The data also show major regional alteration features, such as a very extensive zone of illitization, which is consistent with observations based on bedrock data.

In the McArthur River area, the boulder results are consistent with observed patterns based on both detailed deposit drilling, and more regional exploration drilling. The most prominent boulder lithogeochemical feature at McArthur River is a very strong dravite anomaly, which is a reflection of the strong dravitization of the sandstone directly above the deposit, and which extends along the structural trend towards the P2 mineralized zone.

The effectiveness of boulder geochemistry in this environment is dependant on a number of factors: (1) the extent and contrast of the primary lithogeochemical (or mineralogical) haloes; (2) the relatively even distribution of boulder-bearing till deposits; (3) the consistent pattern of glacial erosion and dispersion; and (4) the premise that the hydrothermally altered rock is not under-represented in the boulder population.

The first three of these criteria apply consistently in the eastern part of the Athabasca Basin. All of the known unconformity-type deposits have extensive and strong lithogeochemical haloes which subcrop. The distribution of morainal material is consistent, although there are also areas covered by glaciofluvial deposits, and the most recent glacial transport direction is remarkably consistent throughout the region.

The last criterion is not as consistently met. In several cases, including McArthur River, the altered sandstone is at least as competent as the surrounding unaltered sandstone, and altered sandstone appears to be well represented in the population of boulders within the area of alteration. At some other deposits, at least part of the altered sandstone is less competent than the unaltered sandstone, and there is evidence that the softer altered boulders are under-represented in the boulder population in till. With only one exception, the unconformity-type deposits studied by the author show some lithogeochemical anomalies in boulder samples, although not all of the anomaly patterns are as well-defined as for the McArthur River Deposit.

It is reasonable to expect that boulder lithogeochemistry should be applicable in other glaciated regions, and in exploration of other types of deposits, providing that the criteria listed above are met. The technique has been applied with success in orientation studies and exploration programs for porphyry Cu–Au deposits in the glaciated terrain of central British Columbia, Canada (Earle et al. 1998).

The composite boulder sampling technique was developed by the author and Jim Murphy (formerly of Uranerz Exploration and Mining Ltd.) with the logistical and financial assistance of Uranerz Exploration and Mining Ltd. and Cameco Corporation. The paper benefited from reviews by L. Clark, R. E. Lett and M. B. McClenaghan

References

CLARK, L. 1987. Near-surface lithogeochemical halo as an aid to discovery of deeply buried unconformity-type uranium deposits, Athabasca Basin, Saskatchewan. In: GARRETT, R. G. (ed.) Geochemical Exploration 1985. Journal of Geochemical Exploration, **28**, 71–84.

EARLE, S. & SOPUCK, V. 1989. Regional lithogeochemistry of the eastern part of the Athabasca Basin uranium province, Saskatchewan, Canada. In: MULLER-KAHLE, E. (ed.) Uranium Resources and Geology of North America, Proceedings of I.A.E.A. meeting held in Saskatoon, 1987. International Atomic Energy Agency, **TECDOC-500**, 263–298.

EARLE, S., MCGILL, B. & MURPHY, J. 1990. Glacial boulder lithogeochemistry: an effective new uranium exploration technique in the Athabasca Basin, Saskatchewan. In: BECK, L. & HARPER, C. (eds) Modern Exploration Techniques. Saskatchewan Geological Society, Special Volume, **10**, 94–114.

EARLE, S., SKETCHLEY, D. & LEVSON, V. 1998. Application of composite boulder lithogeochemistry to exploration for porphyry deposits in British Columbia. Proceedings, Pathways '98 Symposium, B.C. & Yukon Chamber of Mines and The Society of Economic Geologists.

EARLE, S., WHEATLEY, K. & WASYLIUK, K. 1999. Application of reflectance spectrometry to assessment of alteration mineralogy at the Key Lake uranium deposit, Saskatchewan. In: ASHTON, K. & HARPER, C. (eds) MinExpo'96 Symposium - Advances in Saskatchewan Geology and Mineral Exploration. Saskatchewan Geological Society, Special Publication, **14**, 109–123.

FOUQUES, J., FOWLER, M., KNIPPING, H. & SCHIMANN, K. 1986. The Cigar Lake uranium deposit: discovery and general characteristics. In: EVANS, L. (ed.) Uranium Deposits of Canada. Canadian Institute Mining and Metallurgy, Special Volume, **33**, 218–229.

GOLIGHTLY, J., GLEESON, C., BRUMMER, J. & SARACOGLU, N. 1983. Geochemical zoning around the McLean Uranium deposits, Saskatchewan, Canada. In: PARSLOW, G. (ed.) Geochemical Exploration 1982. Journal of Geochemical Exploration, **19**, 101–102.

HOEVE, J. 1984. Host rock alteration and its application as an ore guide at the Midwest Lake uranium deposit, northern Saskatchewan. Bulletin of the Canadian Institute of Mining and Metallurgy, **77**, 63–72.

HOEVE, J., SIBBALD, T., RAMAEKERS, P. & LEWRY, J. 1980. Athabasca Basin unconformity-type uranium deposits: a special class of sandstone-type deposits?. In: FERGUSON, J. & GOLEBY, A. (eds) Uranium in the Pine Creek Geosyncline. International Atomic Energy Agency, Publication **555**, 575–594.

KIRCHNER, G., LEHNERT-THIEL, K., RICH, J. & STRNAD, J. 1980. The Key Lake U-Ni Deposits: A model for Lower Proterozoic uranium deposition. In: FERGUSON, J. & GOLEBY, A. (eds) Uranium in the Pine Creek Geosyncline. International Atomic Energy Agency, **555**, 617–630.

MATTHEWS, R., KOCH, R. & LEPPIN, M. 1997. Advances in integrated exploration for unconformity-type deposits in western Canada. In: GUBINS, A. (ed.) Proceedings of Exploration 97. Fourth Decennial International Conference on Mineral Exploration, 993–1024.

MCGILL, B., MARLATT, J., MATTHEWS, R., SOPUCK, V., HOMENIUK, L. & HUBREGTSE, J. 1993. The P2 North uranium deposit, Saskatchewan, Canada. Exploration and Mining Geology, **2**, 321–331.

RAMAEKERS, P. 1990. Geology of the Athabasca Group (Helikian) in northern Saskatchewan. Saskatchewan Energy and Mines Report **195**.

SCHREINER, B. 1979. Reconnaissance Quaternary geology mapping, northeastern shield., In: CHRISTOPHER, J. & MACDONALD, R. (eds) Summary of Investigations 1979. Saskatchewan Geological Survey, Miscellaneous Report **79-10**, 68–72

SOPUCK, V., DE CARLE, A., WRAY, E. & COOPER, B. 1983. The application of lithogeochemistry in the search for unconformity-type uranium deposits, northern Saskatchewan, Canada. In: PARSLOW, G. (ed.) Geochemical Exploration 1982. Journal of Geochemical Exploration, **19**, 77–99.

An evolutionary model of glacial dispersal and till genesis in Maritime Canada

RUDOLPH R. STEA & PHILLIP W. FINCK

Nova Scotia Department of Natural Resources, PO Box 698, Halifax, Nova Scotia B3J 2T9, Canada (e-mail: rrstea@gov.ns.ca)

Abstract: Glacier process models of till genesis cannot fully explain the properties of tills in Maritime Canada. A succession of local ice caps, called the Appalachian Ice Complex, developed during the last glaciation and were drained by ice streams into the submarine channels bordering the region. The migration of these centres produced areas with widely differing flow patterns, landform assemblages and deposits. Early regional phases of ice flow were characterized by wide, rapidly-flowing ice-streams that formed thick exotic, silty tills. In later phases, ice divides developed over highland areas underlain by metamorphic and igneous rocks, forming stony local tills. Terrain zones characterized by distinct transport histories and depositional sequences were produced by the interplay and migration of regional ice sheets and local ice caps. The interaction of local glaciers and previously deposited tills formed hybrid tills through two reworking processes: inheritance and overprinting. Inheritance is incorporation of till components and/or fabric into a younger till by erosion and entrainment of material from an older till. Overprinting is the injection or imprint of matrix, clasts or fabric on older tills by overriding ice. Glacial dispersal of rocks, minerals and trace elements in this complex glaciated terrain is controlled by the location of former ice divides. Simple unidirectional trains are preserved in relict terrains under divides. In ice marginal areas, reworking processes result in complex dispersal fans produced by smearing and widening of previously formed trains and fans. These dispersal fans can be modelled by vector addition of discrete flow events within each dispersal zone. A simple empirical model of glacier dispersal is presented using exponential decay and uptake algorithms, and incorporating the reworking of older till material.

Maritime Canada has been prospected for several hundred years, and most outcrops have been visited at least once by someone looking for mineral wealth. In Nova Scotia, prospecting was given a major boost by the gold rush of the 1800s and during a subsequent mining resurgence in the twentieth century both in Au and base metals. The complex terrains of the Appalachian Orogen (Fig. 1) host a wide variety of mineral deposits. Prospectors and geologists are now faced with the daunting task of finding ore bodies in the new frontier, the vast area of rock hidden underneath glacial drift. Till and other glacial deposits form the surface cover over 90% of Nova Scotia and can exceed 70 m in thickness. In Nova Scotia, mineral exploration using glacial drift has a long history. In 1896, the naturalist W. H. Prest was given the task of finding the source of auriferous boulders in a till-covered area that had eluded prospectors for decades. He traced a Au dispersal train using modern techniques of drift prospecting to the lode source (Fig. 2). More recently, the discovery of North America's first primary Sn mine at East Kemptville, Nova Scotia can be attributed to a combination of basic prospecting and advanced drift exploration using till as a sampling medium (McAuslan *et al.* 1980; O'Reilly *et al.* 1992). The new frontier continues to give up its hidden secrets to those willing to explore in it.

This paper summarizes the paradigms developed during surficial mapping and till geochemistry programs of the 1970s and 1980s that were initiated to provide basic data about glacial deposits, and to aid mineral exploration in Nova Scotia (e.g. Stea & Fowler 1979; Graves & Finck 1988; Finck & Stea 1995). The Meguma Zone, particularly the South Mountain Batholith, was targeted for reconnaissance till sampling because

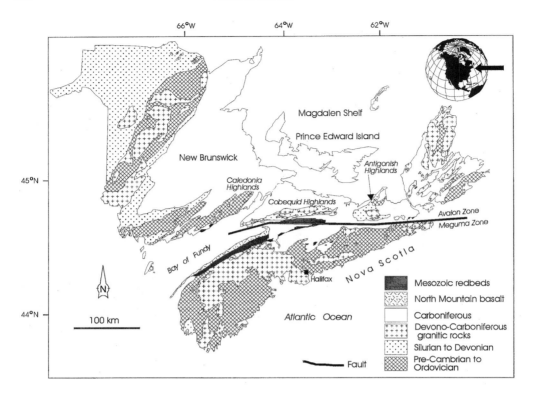

Fig. 1. Generalized bedrock geology of Maritime Canada, showing the main geological terrain boundaries and location of highland regions.

of the newly discovered tin deposits and the potential for base metals and Au. Establishing a framework of glacial history will enable the prospector to link glacial deposits with ice flow events, enhancing the interpretation of surface geochemistry and giving clues to the sources of dispersed ore boulders. Unfortunately, due to the downturn of the mineral industry, much of these data are lying dormant, with till, stream sediment and lake sediment metal anomalies waiting to be followed up, using the techniques that will be described.

Over the past half-century a comprehensive and very useful genetic till classification scheme has been developed whereby till properties are explained by the complex modes of erosion, entrainment and transport (Dreimanis 1989). Explorationists applying glacial geology to the search for mineral deposits are primarily concerned with transport directions and distances and the dynamics controlling till composition and provenance (i.e. where is the source of mineralized boulders and till geochemical anomalies?). The fashionable lodgement v. flow till debates in drift exploration workshops of the 1970s and 1980s seemed strangely superfluous to most concerned with mineral exploration. The present process-oriented classifications and models are not sufficient to fully explain the compositional variations of till. The trend of time-independent process-model thinking has recently been supplanted by evolutionary concepts on the significance of glaciation histories and interactions between glaciers and previously formed tills and landforms (Stea & Brown 1989; Clark 1993; Kleman 1994; Stea 1994). This is fertile ground for explorationists as glacial dispersal and the composition of the resulting till deposits are controlled as much by ice flow history as ice dynamics (Klassen 1997, 1999). In this paper a few more necessary terms to the till lexicon have been added to explain how till composition and dispersal can vary as a result of the shifting ice divides, ice stream activity and reworking of old tills. An attempt is made to model erosion and transport in section and in plan view, working towards a comprehensive model of glacial erosion, transport and deposi-

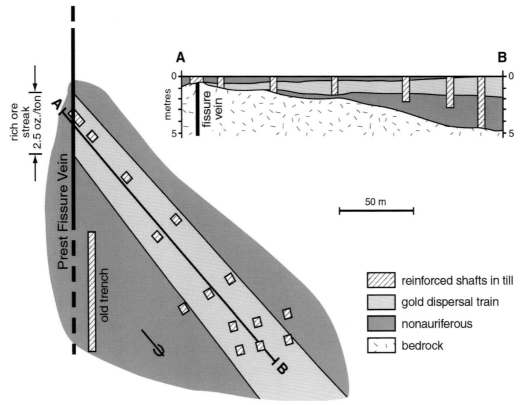

Fig. 2. Example of early drift prospecting by Prest (1911). The map is reproduced from the report, the cross section was inferred from his account. This may be the first account of a dispersal 'plume' in the literature.

tion in the multi-glaciated terrains of Appalachian Canada. Regional till pebble lithology data sets and detailed stratigraphical sections from Finck & Stea (1995) and Stea (1995) will be utilized in the examples described here. Methods for sample collection and preparation are described in those publications. The examples will focus on pebble lithology as a proxy for bulk till composition. It is easier to demonstrate till compositional variation from pebble lithology rather than geochemistry because of the added problems of sampling and analytical variability.

Glaciation of Maritime Canada

It is no longer considered self-evident that landforms deposited by previous glaciations have been erased by the latest glacial advance (Clark 1993; Kleman 1994). In Maritime Canada, the history of glaciations has been unravelled from palimpsest erosional and depositional glacial landforms (Prest *et al.* 1972; Grant 1977, 1994; Rampton *et al.* 1984; Stea 1984; Pronk *et al.* 1989; Stea *et al.* 1998). Palimpsest glacial landforms are defined as 'landforms that have been overprinted and reoriented by later flows and bear evidence of the trends of the older and younger ice flows' (Stea 1994, p. 141). The 'palimpsest' concept has been applied to dispersal trains in areas of complex glacial geology (e.g. Parent *et al.* 1996). Drumlins in Nova Scotia are largely relict and palimpsest landforms, formed during Early and Late Wisconsinan ice flows from both external and local ice centres (Stea & Brown 1989).

Bedrock surfaces reveal many palimpsest erosional forms including, cross-striated outcrops (cf. Veillette & Roy 1995), 'beveled' facets (Prest 1983; Grant 1989) and multi-stossed outcrops (Stea & Finck 1984). The relationships between older, weathered striations on lee-side surfaces, and freshly striated surfaces has been termed 'erosional stratigraphy' (term suggested by W. W. Shilts, pers. comm., 1986; Stea *et al.* 1992*b*, 1998). Flow events defined by discrete,

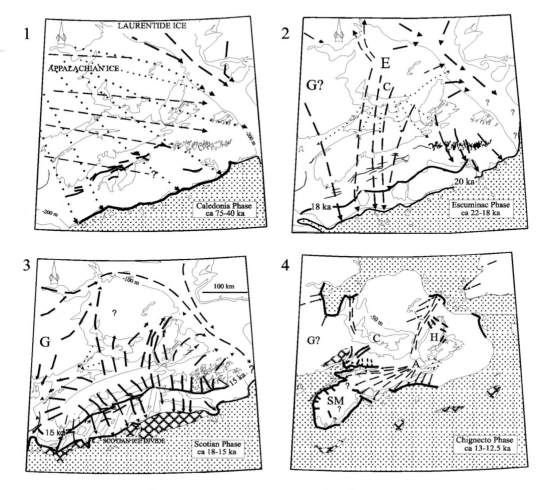

Fig. 3. Evolution (advance and retreat) of ice divides and domes over Maritime Canada during the Wisconsinan (75–10 ka). Note: white, ice; stipple, ocean; x-hatch, emergent land areas; diagonal lines, possible ice shelf (Scotian Phase); dashed arrows, flowlines; dotted arrows, later flowline during same phase of flow: (**1**) Caledonia Phase (Phase 1) in Atlantic Canada and margins (early eastward flow designated phase 1a); (**2**) Escuminac Phase (Phase 2) from the Escuminac Ice Centre and Divide (E) on the Magdalen Shelf and the Gaspereau ice centre in New Brunswick (G). Chignecto glacier (C) active for a short time prior to Phase 3; (**3**) Scotian Phase (Phase 3) (advance and retreats) (Scotian Ice Divide); cross-hatched area, emergent marine landscapes; (**4**) Chignecto Phase (Phase 4) from local centres over the Antigonish Highlands and Chedabucto Bay (A); the South Mountain (SM); Cape Breton Highlands (H) and Prince Edward Island (C). A minor phase of ice flow (Collins Pond Phase) during the Younger Dryas Chronozone not represented.

regionally mappable trends of striations of the same relative age are termed ice flow phases (cf. Rampton *et al.* 1984). The patterns of ice flow are determined by linking flowlines through striae and the orientation of streamlined glacial landforms (cf. Lowell *et al.* 1990). These flowlines are later utilized in the reconstruction of former ice sheets.

In order to develop the evolutionary till model, it is necessary to summarize the glacial history of Maritime Canada that was developed largely through glacial landform and erratic dispersal mapping. Maritime Canada is a collage of tectonic terranes (Fig. 1) that is relatively well-known, allowing the bedrock source of glacial erratics to be accurately located (e.g. Stea & Pe-Piper 1999). Ice centres collectively termed the Appalachian Ice Complex (Prest & Grant 1969), shifted in a clockwise manner across the complex bedrock terrains of the region, producing cross-striated bedrock outcrops and compositionally-distinct till sheets. Linking these superimposed till sheets with regional erosional flow patterns helps to verify the established erosional stratigraphy and allows for a more accurate reconstruction of glacial flow patterns (Stea 1984; Graves & Finck 1988; Stea et al. 1989; McClenaghan & DiLabio 1995) and correlation with offshore end moraines (Stea et al. 1998).

Caledonia Phase (Phase 1)

The oldest observed ice flow patterns on land in Maritime Canada are eastward and southeastward (Caledonia Phase, Fig. 3). Eastward flow patterns (Phase 1a), prevalent along the Northumberland Strait region (Fig. 3), were originally assigned by Chalmers (1895) to the Appalachian-based Northumberland Glacier. The ice configuration later changed to a long-lived ice centre in New Brunswick (Gaspereau Ice Centre; Rampton et al. 1984) which produced southeastward striation patterns on the Caledonia highlands in southern New Brunswick. These southeast flow patterns relate to an extensive and thick ice sheet that eroded the tops of all highland areas in Maritime Canada (Grant 1977, 1989). Overconsolidated, matrix-rich tills (McCarron Brook Till, Hartlen Till, East Milford Till; cf. Williams et al. 1985) found throughout Nova Scotia have been linked with this southeastward flow pattern through till fabric and provenance studies, (Grant 1963; Prest et al. 1972; Nielsen 1976; Alcock 1984; Stea 1984; Stea et al. 1992a, b).

The source of the Caledonia Phase glacier is not known. Little striation or erratic evidence exists for the passage of the southeastward-flowing Laurentide ice sheet sourced in the Canadian Shield across central and northeastern New Brunswick (Rampton et al. 1984; Pronk et al. 1989; Lamothe 1992), although Lamothe (1992, p. 29) indicates that shield erratics may be indistinguishable from local high-grade metamorphic rocks. Striae orientations in northeastern New Brunswick indicate local glaciers were funnelled eastward into Baie des Chaleurs from Appalachian sources to the west and southwest (Rampton et al. 1984; Pronk et al. 1989). Anorthosite boulders derived from the Canadian Shield and found in western Prince Edward Island (Prest & Nielsen 1987) and in the Gulf of St. Lawrence (Loring & Nota 1973; Stea 1991) indicate that the Laurentide Ice Sheet was at one time confluent with, or superseded, ice from the Appalachian Ice Complex in the Gulf of St. Lawrence (Fig. 3).

Escuminac Phase (Phase 2)

Southward and southwestward ice flow paths that post-date the Caledonia Phase are recorded by striated outcrops in Nova Scotia and Prince Edward Island (Prest 1973; Fig. 3). These flow trajectories were a result of an ice centre on the Magdalen Shelf that funnelled ice southwestward into Chignecto Bay and the Bay of Fundy (Chalmers 1895; Goldthwait 1924; Prest 1973; Stea 1983, Rampton et al. 1984; Foisy & Prichonnet 1991) and radiated north into the Chaleur trough and possibly over the Magdalen Islands (Pronk et al. 1989; Parent & Dubois 1990). Josenhans & Lehman (1999) describe a till formed by ice stemming from the Magdalen Shelf, overlying a Laurentide derived till in the Laurentian Channel. The ice dome or divide, roughly situated north of Prince Edward Island, has been termed the Escuminac Ice Centre (Rampton & Paradis 1981; Rampton et al. 1984; Fig. 3). During the Escuminac Phase, the red drumlin Lawrencetown Till (Fig. 1) was formed by southward transport of red, hematitic mud from the Magdalen Shelf onto the Meguma Zone, largely underlain by grey metamorphic and igneous rocks (Stea 1994; Figs 4 and 5). This till is also characterized by consistently high percentages of allochthonous pebbles derived mainly from the Cobequid Highlands of the Avalon Zone (Finck & Stea, 1995; Fig. 4). The pebble source areas of the Lawrencetown Till defined by Stea (1994) using visual identification have been confirmed and narrowed with the application of petrological and whole-rock geochemical techniques (Stea & Murphy 1994; Stea & Pe-Piper 1999).

Ice outflow from the Escuminac Ice Centre during the Escuminac Phase was in the form of rapidly-flowing ice streams or ice 'currents', first envisioned by Grant (1963). These streams converged into inter-bank channels on the outer shelf, drawn by calving of the ice margin in deeper water (Fig. 3). Rapid ice stream flow can also be inferred from the properties and distribution of the Lawrencetown Till. The Lawrencetown Till is found only in relatively narrow drumlin fields in low-lying areas of Nova Scotia and has a consistently high erratic clast content, without dilution with local debris along its flow path (Grant 1963; Stea et al. 1992a; Finck & Stea 1995; Fig. 4).

Escuminac Phase ice streams crossed the inner Scotian Shelf to the outer banks and slope, culminating between 18 and 21 ka as determined by dating of marginal glaciomarine deposits (Mosher et al. 1989). Throughout most of the Escuminac Phase and later flow phases the

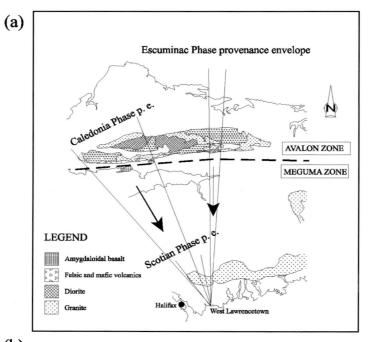
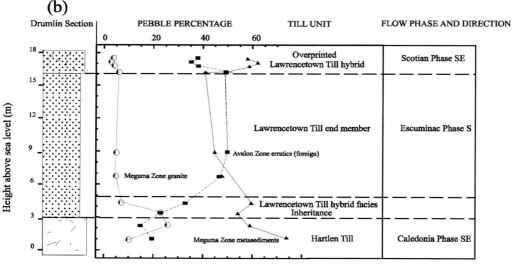

Fig. 4. (a) Bedrock source areas or 'provenance envelopes' (cf. Stea & Pe-Piper 1999) for the till units in the drumlin section at West Lawrencetown established using the overlapping ranges of source areas for erratics identified using petrology and whole-rock geochemistry; (b) three till units exposed at the West Lawrencetown section, showing the vertical variations in abundance of three pebble lithologies from the source areas shown in (a). The uppermost Lawrencetown 'hybrid' Till formed by overprinting of the Lawrencetown Till during the Scotian Phase. (Location 6 in Fig. 6).

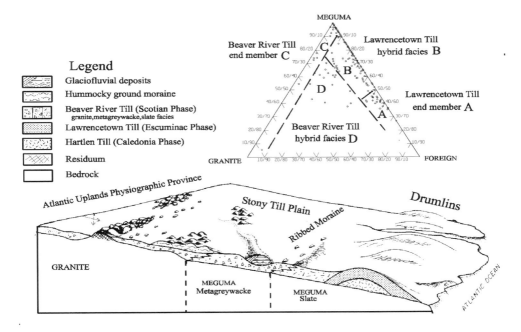

Fig. 5. Generalized cross-section through the 'stony' till plain and drumlins along the Eastern Shore of Nova Scotia (east of Halifax, Fig. 1) and pebble composition of terrestrial diamictons from the eastern shore (data from Stea 1995). The autochthonous, clast-dominated Beaver River Till (Scotian and Chignecto Phases 3 and 4), which makes up most of the stony till plain, slightly overlaps or completely overlies drumlins. Drumlins are mantled with an allochthonous reddish, silty till (Lawrencetown Till-Escuminac Phase 2) and cored by a grey to greyish red silty till (Hartlen Till-Caledonia Phase 1). A triangular plot shows the ratio of local Meguma metasedimentary pebbles (MEGUMA), Meguma Zone granitic pebbles (GRANITE) and 'foreign' or erratic pebbles (FOREIGN) from the Avalon Zone, (Fig. 1). The end member of the Lawrencetown Till (A) contains > 50% allochthonous components, but hybrid facies (B) shows a wide range of compositions due largely to overprinting by later ice movements. The Beaver River Till produced during the Scotian Phase is almost entirely composed of local Meguma Zone (C) metamorphic and granitic rocks.

Magdalen Shelf was an ice rise (cf. Hughes 1995), drawn into deep channels west of Cape Breton and Chaleur troughs (Fig. 3). Ice divides and cold-based ice located over the Magdalen Shelf may explain thin till cover and lack of glacier erosion over Prince Edward Island (Prest 1973) and the Magdalen Islands (Dredge et al. 1992; Fig. 1).

Scotian Phase (Phase 3)

An ice divide (Scotian Ice Divide) formed over Nova Scotia (Fig. 3) as ice thinned rapidly, due to calving into the marine channels bordering Nova Scotia including the Gulf of Maine-Bay of Fundy and Cape Breton Channels (Prest & Grant 1969). In this manner, Nova Scotia ice was cut off from the Escuminac and Gaspereau ice sources (cf. Denton & Hughes 1981). Flow from the Scotian Ice Divide was northwestward into the Bay of Fundy and southeastward over the Eastern Shore, veering southward east of Country Harbour. It was also funnelled northward into Georges Bay and the Cape Breton Channel from Cape Breton and the mainland. At an early stage of development, the ice flow from the Scotian Ice Divide merged with glaciers in New Brunswick to flow northward through the Shediac Channel into the Laurentian Channel. The ice dome off Cape Breton Island proposed by Grant (1977) was probably part of this divide which can be traced from mainland Nova Scotia eastward through Chedabucto Bay onto the continental shelf. Under this divide a stony, local till was produced (Fig. 5).

Chignecto Phase (Phase 4)

Cross-striated bedrock exposures along the coast of the Bay of Fundy record an intermittent

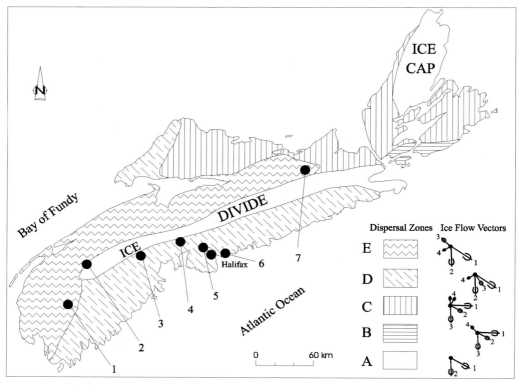

Fig. 6. Zonal concept of glacial dispersal for Nova Scotia, showing the ice flow vectors associated with each zone. Locations of dispersal studies indicated by black dots: (1) East Kemptville-Davis Lake (Zone E-Fig. 8); (2) Kejimikujik Lake (Zone E-Figs. 13, 17); (3) New Germany (Zone D-Fig. 12); (4) Head of St. Margarets Bay (Zone D-Fig. 15); (5) Halifax Peninsula (Fig. 7b); (6) West Lawrencetown section (Fig. 4); and, (7) Garden of Eden dispersal train (Fig. 10).

shifting of ice flow from northeastward during the Scotian Phase, to northwestward and finally westward during the Chignecto Phase (Prest *et al.* 1972; Stea & Finck 1984; B. McClenaghan, pers. comm., 1994). Small ice caps were generated from the melting Scotian Phase glacier during drawdown into the Bay of Fundy or a brief late-glacial climatic reversal (Fig. 3). These small ice caps or glaciers formed over southern Nova Scotia, (South Mountain Ice Cap; MacNeill & Purdy 1951) the Northumberland Strait area (Chignecto Glacier; Chalmers 1895; Northumberland Strait glacier; Rampton & Paradis 1981) and the Antigonish Highlands in northern Nova Scotia (Antigonish-Chedabucto Bay Glacier Complex; Stea *et al.* 1989; Fig. 3).

Zonal Concept of Glacial Dispersal

The interplay of the regional ice sheets and local ice caps coupled with the physiographical variability of Nova Scotia produced distinct zones of glacial erosion and deposition characterized by discrete transport histories (Fig. 6). The main controlling factor of dispersal is the Late Wisconsinan Scotian Ice Divide which straddled the province (Fig. 3). Mapped ice flow directions shift abruptly across the divide from north to south. Zone A is located underneath the Late Wisconsinan Scotian Phase ice divide across Nova Scotia, and underneath the Cape Breton Highlands, long-lived ice cap (Grant 1994). The main flow trends or 'vectors' are southeastward and southward, relict from earlier regional Wisconsinan ice flows (cf. Kleman *et al.* 1999). The presence of bedrock residuum and saprolite in Zone A attests to limited Late Wisconsinan erosion and deposition (McKeague *et al.* 1983; Grant 1994). Zone D on the Atlantic coast side of the divide also features south and southeastward ice flow trends, where flow from the Scotian Ice Divide is parallel to earlier flow events with only a minor overprint of southwest flow from the last Chignecto Phase glaciers

Lake Pluton southeast for at least 8 km (Fig. 8a). There is also evidence of south–southwest and northwest dispersal of Sn in the sand-sized heavy mineral fraction of till. This pattern mimics the dominant southeastward granite clast dispersal direction (Fig. 8b).

The zones shown in Figure 6 are a result of position and migration of ice centres and divides over time. The number, magnitude and direction of the ice flow vectors and associated dispersal patterns are controlled by: (1) position relative to ice divides; and (2) non-erosive, cold-based ice in highland regions and adjacent to ice streams (ice rises) (Denton & Hughes 1981; Hughes 1995). Ice movement is minimal in these cold-based zones and little water is produced at the ice-bed interface, suppressing the processes of erosion and deposition. The opposite occurs in the ice stream areas which are zones of enhanced erosion and deposition, as evidenced by drumlin fields and heavily striated bedrock surfaces. As the centres of flow of the Appalachian Ice Complex shifted, so did zones of erosion and deposition. In areas under new ice divides, deposits from earlier glaciations were preserved. In other areas of warm-based ice, the older glacial deposits were eroded and transported. Mineral exploration companies using drift prospecting methods within these various zones have had to contend with widely differing patterns and magnitudes of dispersal. Some areas show relatively simple, and local (Fig. 2) patterns of southeastward dispersal and one till sheet, while others are complicated by additional ice flows, and multiple tills. The zonal concept is a method to integrate the net effects of the flow phases, allowing the explorationist to instantly assess the complications he/she will face in the regions under assessment.

The concept of vector addition (Fig. 9) can be used to model palimpsest glacial dispersal in these zones. Each dispersal event has a definite magnitude and direction, in essence a vector quality. For example, hypothetical mineralized boulders are dispersed along flow lines during the Caledonia Phase with vector magnitude and direction (Vector 0-1). These boulders are re-entrained during the Escuminac Phase and moved along Vector 1-2. The resultant Vector (0-2) is the net effect of glacial dispersal during two phases of ice flow. Finally during the Scotian Phase, the boulders are moved along Vector 2-3, and the resultant dispersal Vector 0-3, represents the vector sum of dispersal during three glaciations. The shaded areas in Figure 9 represent the hypothetical dispersal fans resulting from glacial smearing as flow directions shift. Re-entrainment, or reworking occurs from mineralized debris along the entire original Caledonia Phase dispersal vector. In the model total reworking of the previous 'fan' is assumed, but in reality, the dispersal fan would probably encompass the entire area bounded by the dispersal vectors. Reworking of a dispersal train into a fan would not produce as well defined a train as that produced directly from glacial erosion of the source along one ice flow only and the amount of train reworking can be quite variable. Parent et al. (1996) describe similar palimpsest dispersal fans in till defined by trace element geochemistry where a significant amount of the fan area is due to reworking. Pebble dispersal fans in Zone E (Fig. 6) can be deciphered using the vector addition concept. An example in northern Nova Scotia is the diffuse 'amoeboid-shaped' (cf. Shilts 1995) fan developed during four phases of ice flow around a distinctive mafic intrusive body in northern Nova Scotia (Turner & Stea 1987; Fig. 10). Dispersal 'fans' are likely to be the result of ice flow shifts and smearing of previously deposited dispersal trains rather than ice dynamics during a single ice flow event (Flint 1971; Batterson 1989; Parent et al. 1996; Klassen 1997). The narrow 'ribbon-shaped' dispersal train may be the result of laminar flow along a single flow line (DiLabio 1990).

'End Member' Tills

The third dimension of glacial dispersal comprises the constituent till sheets that make up the dispersal fans. They are discussed in this section. In the simplest case the texture, provenance and dispersal characteristics of a till are controlled by bedrock source areas and ice sheet dynamics. This is true where there is no interaction between temporally distinct ice flow events and their related deposits. Such 'simple' processes produce what the authors refer to as 'end member' till and dispersal within this till unit can be relatively straightforward (Fig. 2). In thick multiple till sections, end member tills are found above and below till contacts which are the zones of interaction. Where end member tills are found at the surface: (1) they were the last or only till sheet deposited; (2) there are no underlying tills available for reworking; and (3) ice sheet dynamics (cold-based ice, ice divides) have mitigated interactions between glaciers and previously formed till.

In this paper, focus is placed on tills on the Atlantic coast of Nova Scotia underlain by Ordovician to Carboniferous igneous and metamorphic rocks (Meguma Zone), where most of

Fig. 11. (a) Section through a drumlin at Smiths Cove east of Halifax showing three tills (1-Hartlen, 2-Lawrencetown, and 3-Beaver River Till); (b) A dispersal plume of auriferous quartz (boulder and cobbles) from a vein 50 m to the north (to the right) of the photograph at the Beaver Dam Gold deposit. Note the fracture and shattering of the large quartz boulder in the Beaver River Till. Photograph courtesy of M. Parkhill, New Brunswick Dept. Natural Resources.

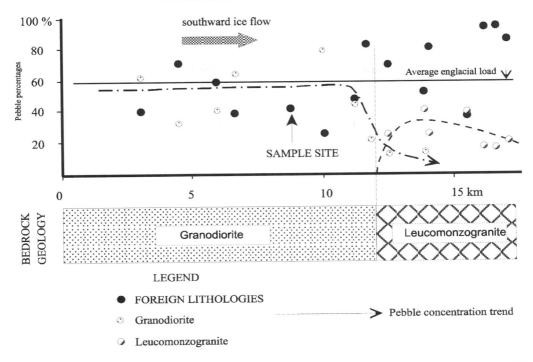

Fig. 12. Dispersal (decay and uptake) of local granite in the pebble fraction (>2.5 cm) of the Lawrencetown Till of a transect over the South Mountain Batholith and the persistence of the englacial load (Location 3 in Fig. 6).

the detailed till sampling work has been done (Fig. 1).

Silty 'foreign' drumlin tills

Drumlins are the dominant glacial landform of the Atlantic coast of Nova Scotia comprising four major drumlin fields and thousands of individual drumlins. The long axis orientations of these drumlins are parallel to local striations, with occasional lobate, 'palimpsest' drumlins parallel to both earlier and later ice flow phases (Stea & Brown 1989; Stea 1994). Drumlins are generally drift-cored and can exceed 30 m in total thickness at the stoss summit.

Sections through drumlins along the Atlantic coast reveal thick, predictable drift sequences which can be correlated regionally (Stea & Fowler 1979). At the core of many drumlins is a grey and grey-red overconsolidated, matrix-supported, silty diamicton (Hartlen Till) formed during the Caledonia Phase and overlain by a red-brown, matrix-supported, silty diamicton (Lawrencetown Till) formed during the Escuminac Phase (Figs 4 and 5). The contact between tills is sharp, and is often accompanied by striated, bullet-shaped boulder horizons or pavements (Grant 1963; Nielsen 1976). The red matrix of the Hartlen and Lawrencetown Tills is distinctive, and entirely foreign to the Meguma Zone, which is comprised largely of grey and black slates, metawackes and white granites. In some drumlin sections a third till, the Beaver River Till, overlies the Lawrencetown Till along an erosive contact (Fig. 11a). Grant (1963) originally interpreted these diamicton units as tills formed by separate glaciations, with contacts (including boulder pavements and thin waterlain units) representing interstadial intervals. Nielsen (1976) interpreted this widespread three-till sequence as a single depositional package from one ice flow encompassing lodgement, and englacial and supraglacial melt-out tills. The authors' present view is a compromise of the two theories with each till unit deposited by separate phases of ice flow from evolving ice centres and shifting divides during one major glaciation (Wisconsinan) (Stea et al. 1992a, b, 1998).

Tills forming drumlins, especially the Lawrencetown Till, have a high content of far-travelled or exotic bedrock components (Figs 4 and 5). As the Escuminac Phase glacier flowed southward over the Cobequid Highlands, Car-

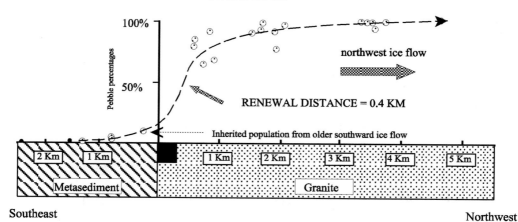

Fig. 13. Exponential uptake of granitic clasts in till in the Beaver River Till near Kejimikujik National Park. In this area (Zone E – Location 2, Fig. 6, Fig. 7), the Scotian Phase flow which produced the Beaver River Till was towards the northwest. Renewal distances (distance for 50% uptake of local rocks) are 0.4 km. The till contains an inherited (reworked) population of granitic clasts (c. 10%) from an earlier southeastward ice flow.

boniferous red-beds from the Magdalen Shelf (Fig. 1) were eroded and entrained in an englacial position. In a similar manner, many igneous bedrock erratics derived from the Cobequid Highlands were transported southward, *en masse*, and deposited in drumlins along the Atlantic coast (Stea & Pe-Piper 1999). The Escuminac Phase glacier overrode stagnant basal ice on the lee of the Cobequid Highland plateau shifting its debris load higher in the ice profile, where it was transported by rapid ice stream flow (e.g. Clark 1987) into marine channels at the continental shelf edge (Stea *et al.* 1998). An analogous process is upshearing at the margin of polythermal glaciers where basal debris in warm-based sliding ice is elevated over the cold-based ice at the edge (Kirkbride 1995). In areas down-ice of the Cobequid Highlands, there was little interaction between basal and englacial ice. 'Skip Zones' as much as 20 km wide were formed where local bedrock debris was only sparingly incorporated into the Lawrencetown Till. Figure 12 is a transect of the Lawrencetown Till across granite facies of the South Mountain Batholith showing the changing pebble composition. A high percentage of foreign englacial components is maintained across the 'skip zone' masking local bedrock variations.

Stony, local tills (ice divide tills)

The regions away from drumlin fields are characterized by irregular, hummocky topography or ribbed moraine with large quartzite or granite boulders (from <1 m up to 20 m in diameter) strewn on the surface, and are referred to as the Stony Till Plain (Fig. 5; Stea *et al.* 1992*a*). The till that makes up the Stony Till Plain is called the Beaver River Till (Grant 1980; Finck & Stea 1995). Beaver River Till thicknesses vary widely, from 1 to 5 m in upland areas (>200 m elevation) to >7 m in lower elevations. It consists of a stony, sandy, olive-grey diamicton with a high percentage of angular-subangular cobbles and boulders (Fig. 11b). Recent excavations of a natural gas pipeline corridor across Nova Scotia in the Beaver River Till revealed increasing clast abrasion (striae and rounding) in a down-ice direction towards the coast. The pebble lithology of an 'end member' Beaver River Till closely reflects underlying bedrock geology (Figs 7 and 13). End member Beaver River Tills are dominated by a single local bedrock-derived, clast lithology, generally >90%. Deviations from this end member clast lithology content reflect reworking and incorporation of underlying till units. Because of its strong local component, the clast lithology of the Beaver River Till can be used to map bedrock geology in areas of little bedrock outcrop (MacDonald & Horne 1987). Although originally interpreted as an ablation till because of its looseness and coarse texture (e.g. Nielsen 1976) an englacial or melt-out origin for the Beaver River Till is unlikely because of local derivation and rapid lithological transitions across bedrock contact zones (e.g. Fig. 11b).

The renewal distance, as defined by Peltoniemi (1985), is the distance down-ice over which the proportion of a new rock type increases in the till from 0% to 50%. Renewal distances

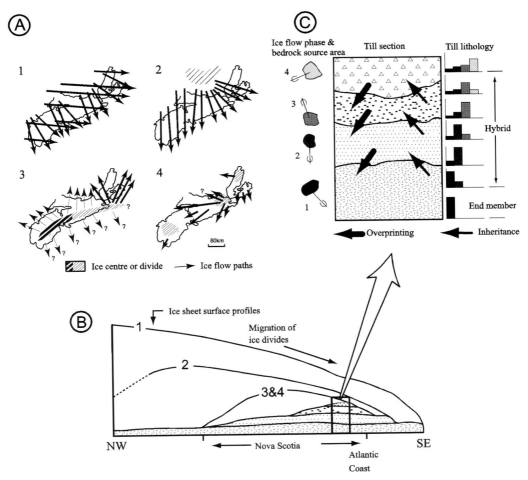

Fig. 14. Spatial and temporal variations in ice dynamics affecting the distribution, properties and provenance of tills. As ice divides migrated towards local areas in Maritime Canada (A) during four phases of Wisconsinan ice flow, till sheets became texturally 'immature' with an increasing clast/matrix ratio and autochthonous clast components (B). Till matrix and clast composition is controlled by the basal thermal regimes of the evolving glaciers and reworking processes. Local glaciers reworked earlier tills and produced hybrid facies through inheritance and overprinting. The hypothetical evolution of till lithological composition through four ice flow phases is shown (C).

calculated for the Beaver River Till average about 0.4 km (Fig. 13). In some areas an accurate assessment of renewal distances is possible because the granite/metasediment contact is nearly perpendicular to ice-flow (Fig. 7). These extremely low renewal distances (e.g. Scandinavian Tills average 5 to 20 km; Peltoniemi 1985) are indicative of low-velocity basal ice conditions in proximity to an ice divide (Stea et al. 1989). The angular nature and lack of striae on the clasts as well as the predominance of clast over matrix modes, suggests that glacial quarrying and fracturing are dominant mechanisms in the formation of the Beaver River Till rather than abrasion (e.g. Iverson 1995). Low sliding velocities are implied by the negligible dispersal, reinforcing the idea that these tills are deposited near or under the Scotian Ice Divide. The Beaver River Till can be considered an 'immature' till (cf. Dreimanis & Vagners 1971) implying that it was formed in a relatively short period of time, and not given the chance to develop a matrix component through comminution and abrasion. Stony 'local' tills are common near former ice centres in central New Brunswick (Broster et al. 1997) and throughout the shield areas of Canada

(Dredge 1983). Stea et al. (1989) used the term 'ice divide till' for these till facies.

'Hybrid' tills

'Hybrid' tills (cf. Grant 1963) are created in the areas where there are interactions between glaciers and previously deposited tills through reworking. Reworking can modify the basic characteristics of the 'end member' tills to the extent that the resulting hybrid tills no longer reflect (to greater or lesser degrees) the properties of the original tills (Fig. 14). Two distinct reworking processes can be defined as: inheritance and overprinting (Finck & Stea 1990).

Inheritance is the incorporation of till components and/or fabric into a younger till by erosion and entrainment of material from an older till. If assimilation of the older till is incomplete the stratigraphical relationship between the materials will be preserved in section. However, where complete reworking has taken place the older material will be assimilated and will not exist as a regionally mappable till unit. In this case, erosional evidence such as striations, may be all that is left of an earlier ice flow.

Overprinting is the injection or mixing of matrix and clast components and imprint of fabric on older tills by overriding ice. Where a till unit is associated with the later ice flow, the stratigraphical relationships and overprinting of the older till will be observed in section. If the younger ice flow does not form a recognizable deposit, the only evidence for its existence will be its effect on the older till(s) and/or the preservation of erosional evidence such as cross cutting striations.

Each till unit starts with an 'end member' clast composition which is altered by processes of either inheritance, overprinting, or both. The original composition reflects the dynamics of the depositing glacier and bedrock sources. The resultant clast composition of a particular sample will be the sum of end members and reworking processes, in effect the compositional evolution of the tills through several glaciations (Fig. 14). The reworked components will form a background population of clast types, generally subordinate to the influx of fresh bedrock lithologies, provided there is access to outcrop by ensuing glaciers.

Hybrid tills by inheritance

Many of the hybrid tills in Nova Scotia were produced from the interactions of the Scotian and Chignecto Phase glaciers and previously deposited Lawrencetown and Hartlen Tills. A hybrid Beaver River Till surrounds and occasionally covers Lawrencetown Till drumlins. The older Lawrencetown Till (or any other older till) has acted as a source for material reworked into the hybrid facies of the Beaver River Till, which inherited some red silt and clay (as inclusions or spread through the matrix) and foreign lithologies from the Lawrencetown Till. Grant (1963) was the first to recognize the significance of hybrid clast lithology of the Beaver River Till and its spatial association with Lawrencetown Till.

Inheritance is obvious in the regional till data sets compiled for Nova Scotia. For example, till overlying the South Mountain Batholith (Figs 7 and 13) has inherited a population of distally derived granite clasts from the earlier southeastward Caledonia Phase ice flow, when flow reversed towards the northwest during the Scotian Phase. Another example of inheritance can be found in two till sections on the South Mountain Batholith where Beaver River Till overlies Hartlen Till (Fig. 15). The correlative till facies of the Hartlen Till over granitic rocks in this region is an indurated sandy facies observed under Escuminac Phase Lawrencetown Till in coastal sections (Graves & Finck 1988). The inferred flow directions for the Hartlen and Beaver River Tills are southeastward (Fig. 15). In Section A, the lower till (Hartlen Till) contains elevated concentrations of local bedrock biotite monzogranite (lithology 2). Near surface samples of the overlying younger till (Beaver River Till) are dominated by medium-grained leucomonzogranite (lithology 4) which is proximal but up-ice (Fig. 15). The uptake of foreign lithologies and the relative increase in lithology 2 in the lower part of the Beaver River Till represents inheritance, because the material was eroded or reworked from the older till. The inheritance relationship is typical of superimposed till units in stratigraphical section. The surface till is largely a hybrid facies of the Beaver River Till.

Hybrid tills by overprinting

Overprinting of the Beaver River and older drumlin-forming Lawrencetown Tills is widespread across the Atlantic Uplands. Areas covered by Beaver River Till west of Halifax on the South Mountain Batholith (Fig. 7b) contain elevated concentrations of Meguma Group lithologies with very few erratic (foreign) clasts commonly associated with the process of inheritance. Either the till was formed during a separate southwest flow phase (Chignecto Phase) or the Meguma Group clasts were

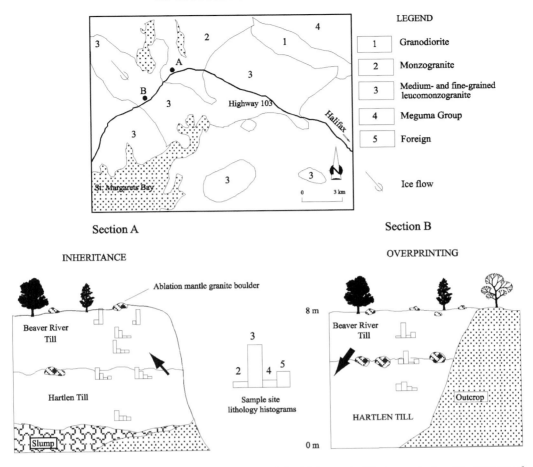

Fig. 15. Bedrock geology and location of sections A and B which contain tills demonstrating the concepts of inheritance and overprinting. (Location 4, Fig. 6).

overprinted on the end member Beaver River Till from source areas to the northeast by southwest flowing ice. The homogeneity of till sections suggests the latter hypothesis is correct.

Overprinting, like inheritance is best observed in multi-till drumlin sections such as that depicted in Section B (Fig. 15), which shows Beaver River Till overlying Hartlen Till without an intervening Lawrencetown Till. In this example, the till sample collected below the contact between the two units shows the effect of overprinting. The overriding Scotian Phase glacier has added leucomonzogranite clasts to the Hartlen Till, thereby increasing the background concentration of local leucomonzogranite pebbles in the underlying Hartlen Till. If the upper part of the Hartlen Till is a meltout facies, then the englacial/supraglacial debris would be expected to contain a higher percentage of monzogranite clasts from sources further up-ice (e.g. Shilts 1976; Dreimanis 1989).

A drumlin section east of Halifax reveals three till units: the silty Hartlen Till; red clay-rich Lawrencetown Till; and, an uppermost hybrid, brown sandy facies of the Lawrencetown Till (Fig. 4b). Local Meguma Zone metasedimentary clasts abruptly increase in abundance in the hybrid facies of the Lawrencetown Till at the top of the section (Fig. 4b). A process-oriented model of till genesis would predict an increase in foreign Avalon Zone lithologies towards the top of the drumlin section, as they are carried higher and further in the englacial zone (e.g. Shilts 1976). The paradoxical increase in local metasedimentary clasts can best be explained by compositional modification by an overriding glacier with a local source area. A southeastward-trending fabric of the uppermost hybrid

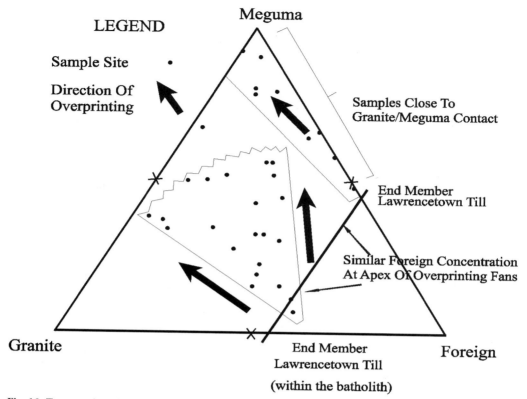

Fig. 16. Ternary plot of % granite, Meguma Zone metasedimentary and foreign (largely Avalon Zone) clast content in Lawrencetown Tills in the peninsula west of Halifax (Location shown in Fig. 7b). Overprinting of Lawrencetown Till by Scotian and Chignecto Phase glaciers results in the increasing percentages of local granites or Meguma metasediments. The apex of these fans represents the 'end member' composition.

till suggests an overprinting by Scotian Phase ice as the end member Lawrencetown Till exhibits a southward-trending fabric (Stea 1994). The southward transport direction that emplaced the Lawrencetown Till was confirmed using whole-rock geochemistry of granitic clasts that are from known point-source plutons in the Cobequid Highlands (Stea & Pe-Piper 1999). This overprinting effect can also be inferred from results for regional till samples collected over the Eastern Shore and the South Mountain Batholith (Figs 10 and 16). The mean foreign clast composition of end member Lawrencetown Till in drumlin sections in these two areas is about 50%. Most surface samples of Lawrencetown Till are hybrid tills reflecting a compositional bias towards local bedrock, the effect of overprinting of locally-eroded clasts by Scotian Phase ice. The effects of the overprinting process may be significantly underestimated. For example, in Nova Scotia older erratic-rich tills serve as sources for material inherited by younger, local tills. The inherited material (erratics) are lithologically distinct from locally-derived rocks in the younger tills. The younger, local glaciers which overprinted erratic-rich tills were eroding and entraining local Meguma Zone lithologies from adjacent bedrock source areas. Therefore overprinting of locally-derived granite clasts during the Chignecto Phase on a pre-existing locally-derived granitic till from the Scotian Phase is difficult or impossible to recognize. If the overprinted granitic material is mineralized, then the significance of overprinting to exploration becomes apparent.

Mechanisms of reworking

The mechanisms of reworking are unclear, but a general hypothesis can be formed from empirical data. Previous work on erosion and entrainment has focussed on a bedrock base (e.g. Iverson 1995). Hybrid tills described in this paper are presumed to have formed by glaciers flowing over till substrates. Till below a glacier can be a deforming layer (e.g. Hart & Boulton 1991), which for fast-moving ice streams, is likely to be the case (Alley 1991). The Appalachian ice complex however, consisted largely of local ice caps and divides developed on metamorphic and igneous bedrock substrates with variable thermal regimes. Much of the erosion and reworking of till substrates involved freeze–thaw mechanisms. Some of the older till components were entrained (inherited) through freeze-on and later sheared and homogenized by changing compressional and tensional flow regimes down-ice. Till wedges of the Beaver River Till injected into the Lawrencetown Till have been described by Nielsen (1976, p. 47) suggesting a frozen till substrate. In the case of overprinting, the till substrate, consisting of finer-grained older tills, is partially remobilized and fresh bedrock components carried in the ice above are added through basal melting. Subglacial deformation along the flow line then mixes the freshly eroded bedrock material and the older tills. The thickness of the lithologically mixed or overprinted layer in a deforming till in Nova Scotia is in the order to 2 to 3 m (e.g. Fig. 4b). Evidence for overprinting of till fabric, to depths of 6 to 7 m, has been documented at the Joggins section in northern Nova Scotia (Stea et al. 1986).

Empirical modelling of glacial dispersal in complex flow terrains

Individual clast components in a till sample can reliably be traced to source areas by understanding their genesis and the glacial history of the region. Dispersal parameters such as renewal distance can be estimated for a given till, and predictions of the distance to up-ice bedrock or mineralized zones quantitatively made based on the changes in clast percentages of till samples. Renewal distance and its corollary, half-distance, are dependant on the following factors (e.g. Salonen 1992): (1) grain size; (2) lithology; (3) width of outcrop; (4) glacial dynamics; and (5) glacial history.

Some studies have shown that tills close to a source area are dominated by coarser-sized clasts of that component and further away become enriched in the finer modes of that component (Drake 1972; Szabo et al. 1975). In other words, half-distance values increase with decreasing grain size of the indicator commodity. Others have demonstrated the opposite effect (Salminen & Hartikainen 1985). The glacier base, called the traction zone or 'glacial mill', produces fine-grained material by abrasion and crushing with 'terminal grades' in the silt sizes (Dreimanis & Vagners 1971). Some of the finer material can also be inherited from older tills or regolith. The 'grain-size effect' may be significant only in the traction zone as material in englacial transport is suspended above the base and undergoes little comminution (Boulton 1978). In the case of melt-out tills formed of englacial debris, the down-ice decrease in source-rock percentages is largely due to dilution rather than comminution.

Lithology affects the renewal or half-distance length (Gillberg 1967; Bouchard et al. 1984; Salonen 1986). Salonen (1986) suggests that the resistance of a rock type to impact shattering is a major determining factor in transportability. Impact-susceptible rocks, such as limestone, are transported furthest because they are readily crushed, enhancing entrainment whereas phyllites, resistant to impact, are not as readily crushed/shattered by the glacier. Rock hardness may also be a factor if the transport is limited to the basal zone of the glacier.

Till stratification and genesis are of primary importance in clast dispersal. Tills derived from englacial or supraglacial zones, where clast contact and comminution are minimized, contain higher percentages of far travelled rock types than basal tills (Harrison 1960; Shilts 1973; Dreimanis 1989). In the basal zone the effect of local variations in bedrock lithology are enhanced, whereas in englacially or supraglacially transported tills these variations are often overshadowed by allochthonous clast components (Fig. 12). In a region of complex ice flow history, reworking (inheritance and overprinting) is the other major factor in controlling the percentages of clasts in a given sample.

The abundance of clasts of a particular rock type in any till sample is the result of uptake (transported from underlying bedrock sources) or comminution (decay) down-ice from distant source. The decrease or decay of rock types with distance is a result of dilution with newly eroded rock types (Boulton 1984) and comminution of fragments into finer grain sizes (Dreimanis & Vagners 1971). An empirical exponential decay relationship has been well defined in the literature (Krumbein 1937; Shilts 1976; Salonen 1986; Bouchard & Salonen 1989). Decay is given as:

$$Y_0 = Y_i (e^{-ax}) \qquad (1)$$

where, Y_0 is the percentage of the up-ice bedrock lithology in the clast fraction; Y_i is the initial concentration of lithology Y_0 when x equals 0, i.e. the y–intercept; x is the distance down-ice from the source area; $x_{1/2}$ is the half distance when half of the original lithologies are gone; and a is the coefficient of particle distribution $(0.693)/x_{1/2}$. Stea (1991), Finck & Stea (1995), Parent et al. (1996) and Klassen (1999) modified the decay equation in order to approximate uptake. This yielded the equation:

$$Y_0 = Y_a (1-e^{-ax}) \qquad (2)$$

In this case, the constant 'a' is the renewal distance; Y_0 is the percentage of a particular clast type being incorporated from underlying bedrock; and Y_a is the asymptotic value or saturation point of the sample for the particular clast lithology being incorporated into the till (Fig. 17).

Both the englacial (allochthonous) load (e.g. Lawrencetown Till, Fig. 12) and reworked components of a till sample can be accounted for in equations (1) and (2) through initial (Y_i) and final saturation percentages (Y_a). 'Final' or saturation percentages of a particular rock type will not always be 100%. This value depends on the width of the source rock formation and amount of allochthonous components. It is obvious that a narrow source rock area will provide less material for entrainment than a wide one, and when the glacier flows across it, uptake does not achieve a maximum value (100%). In addition, mixing with englacial layers and/or inheritance from stratigraphically older tills serving as sources of material will dilute the till and prevent Y_0 from ever reaching 100% (e.g. Fig. 12). The asymptotic value is then determined as:

$$Y_a = 100\% - E$$

where, E is the Sum $(E_1 + E_2 + \ldots)$ and E_1, E_2, etc. are the percentages of the various englacially transported (*en masse*) and inherited lithologies in each sample. Since uptake and decay work in unison, the percentage of a particular clast type that represents underlying bedrock geology in a till sample down-ice of a bedrock contact is represented as:

$$Y_0 = Y_i (e^{-a1x}) + Y_a (1 - e^{-a2x}) \qquad (3)$$

here, a_1 is 0.693/half-distance and a_2 is 0.693/renewal-distance.

The renewal-distance as defined by Peltoniemi (1985) assumes that the percentage of a particular rock type being incorporated in a till starts at 0%. If there is a pre-existing concentration level of the particular clast (rock) type in the till from another source then Equation (3) will give the half-distance or renewal-distance at which the clast percentage increases by 50% minus the percentage that its original concentration decays or decreases. It is important to remember that Y_0 and Y_i are the same rock type. In other words clasts of the particular type are moved and then reworked by ice movements from different directions and hence give 'false' initial percentages which are corrected by assuming a decay. Klassen (1999) used a similar approach to model a two component clast system without reworking.

In stony, locally-derived Beaver River Tills without a substantial foreign englacial component, or Lawrencetown Tills where locally eroded material entrained in the traction zone dominates over englacial debris and transport, renewal distance can be assumed to equal half-distance, i.e. if $a_1 = a_2$ thus solving Equation (3) for renewal distance or half-distance gives:

$$Y_0 = Y_i (e^{-ax}) + Y_a - Y_a e^{-ax}$$
$$Y_0 - Y_a = (Y_i - Y_a)(e^{-ax})$$
$$\ln[(Y_0 - Y_a)/(Y_i - Y_a)] = -ax = -0.693(x)/x_{1/2}$$
$$x_{1/2} = -0.693(x)/\ln[(Y_0 - Y_a)/(Y_i - Y_a)]$$

or solving for transport distance down-ice 'x' gives:

$$x = (x_{1/2}/-0.693)(\ln[(Y_0 - Y_a)/(Y_i - Y_a)])$$

Equations (1), (2) and (3), and the resulting derivations allow estimates of renewal-distances, half-distances and transport distance 'x' to be calculated for a particular pebble count from a till sample. An exponential function between transport distance 'x', the observed clast percentage(s) is assumed. Transport in a known direction is also assumed (degrees of inheritance or overprinting can be accounted for in the Y_i and Y_a factors). For additional up-ice lithologies, Equations (1) and (3) can be summed (Stea 1991; Parent et al. 1996). The Y_i parameter in Equation (3) is an approximation of inheritance which can be quantified using vector addition. Including vector quantities (Fig. 9) in two and three dimensions, would enable complete modelling of ore dispersal plumes through several ice movements.

Testing the model

The dispersal model was tested in an area where: (1) the till type was suitable and consistent; (2) the location of bedrock contacts was rigorously constrained; (3) ice-flow directions was known or could be approximated; and (4) till samples were appropriately located with respect to the bedrock contacts (i.e. not to close and not to far; if a sample is located on the flat part of the dispersal curve experimental error in determining the clast percentage will yield

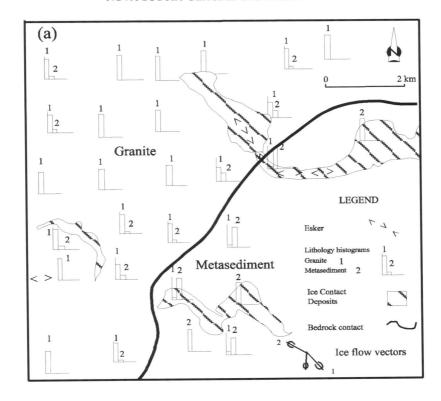

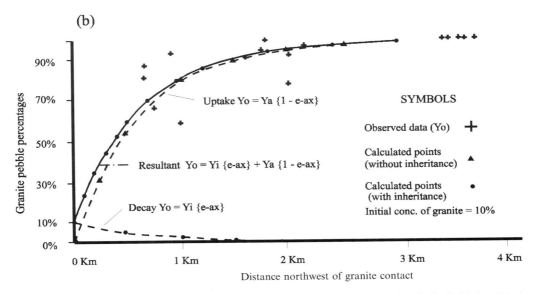

Fig. 17. (a) Regional clast geology of the Beaver River Till in the area northwest of Kejimikujik National Park (Location 7, Fig. 6). Lithologies 1-granite, 2-metasedimentary rock; (b) dispersal data compared with model curves, for the uptake without inheritance, starting at 0% and the resultant model taking into account inheritance of a pre-existing population of granite with initial percentages Y_i of 10%. Solid line represents dispersal curve from a lower Y_i value.

disproportionate changes in the value of 'x' or '$x_{1/2}$').

Table 1. *Dispersal parameters for Beaver River Till, northwest corner of Kejimikujik Park*

Sample #	Distance 'x' (km)	Initial % (Y_i)	Observed % (Y_o)	Renewal dist. '$x_{1/2}$'
3	0.45	10	100	0.35
8	3.75	10	100	0.38
10	1.8	10	100	0.18
11	0.9	10	93	0.24
12	0.65	10	81	0.65
19	0.65	10	87	0.23
21	1	10	69	0.65
22	3.5	10	100	0.36
27	3.6	10	100	0.37
28	2.15	10	98	0.39
29	0.75	10	66	0.53
33	1.6	10	91	0.48
34	2	10	93	0.54
35	3.65	10	94	0.87
39	4.75	10	100	0.48
40	2	10	79	0.95
49	1.75	10	95	0.42

The average $x_{1/2}$ value equals 0.44 km, calculated by ignoring the largest and smallest $x_{1/2}$ values.

The test area (Figs 7 and 17) is the NW corner of Kejimikujik National Park, at the contact between Meguma Group metasedimentary rocks to the east and granitic rocks to the west (Horne & Corey 1994). The bedrock contact is oriented approximately southwest–northeast. This contact is perpendicular to early Caledonia Phase ice flow that transported granitic material southeastward over metasedimentary bedrock. The contact is also approximately perpendicular to later Scotian Phase northeastward ice-flow. Till containing Meguma Group metasedimentary clasts and granitic clasts deposited during the Caledonia Phase was subsequently transported northwest over the underlying granite bedrock (Fig. 17) during the Scotian Phase. Thus the younger Beaver River Till contains a pre-existing concentration of granite pebbles (Y_i) equal to c. 10% (Fig. 17). Table 1 gives the transport distances (x), dispersal function parameters (Y_0, Y_i, Y_a) and the calculated $x_{1/2}$ value for each sample. The decay function Y_i (e^{-ax}) accounts for the pre-existing concentration of granitic pebbles. Y_a (the asymptote) was set to 100%, based on observations of pebble lithology. The actual values of Y_0 are plotted in Figure 17b along with the model uptake and decay curves based on an average calculated renewal distance of 0.44 km. Observed values of Y_0 and the calculated curve display a strong positive correlation.

Prediction of the location of a bedrock contact

Figure 18 shows simplified bedrock geology of the western portion of the New Germany map sheet (Horne 1993), the distribution of bedrock outcrop, the distribution of till samples and their corresponding clast lithologies. Using Equation (3) an average renewal distance of 0.33 km was calculated for Beaver River Till in this locality (Table 2). Using the average $x_{1/2}$ value of 0.33 km, the transport distances to up-ice bedrock contacts 'x' were calculated for till samples where bedrock outcrops were sparse or non-existent (Fig. 18). These calculated contacts are more undulatory than the contacts of Horne (1993), but are consistent with other areas on the South Mountain Batholith where this method was applied and confirmed by information from continuous outcrop or diamond drilling. Of particular significance is that the predicted contact based on the clast lithology agrees with the known bedrock outcrop distribution and acts as a refinement rather than as a contradiction. Where bedrock outcrop is scarce and till is used to predict bedrock contacts, this mathematical approach using till clast geology eliminates bias and allows for a more consistent and rational interpretation.

Table 2. *Dispersal parameters for Beaver River Till on the western part of the New Germany map sheet*

Sample #	x (km)	Y_i(%)	Y_o(%)	Y_a(%)	$x_{1/2}$(km)
35	0.25	0	45	100	0.290
146	1.1	0	93	100	0.287
153	1.25	0	93	100	0.326
154	1.43	0	90	100	0.429

The average $x_{1/2}$ value equals 0.33 km.

Conclusions

Maritime Canada was under the influence of the Appalachian Ice Complex, a series of autonomous local ice caps, for most of the last glaciation. Ice divides shifted across the region producing varied zones of erosion, deposition and non-deposition. Zones characterized by discrete ice flow histories can be delineated. Tills were largely deposited at the ice margins, and in ice stream areas of these local ice caps. Fine-grained 'foreign' tills were produced during early

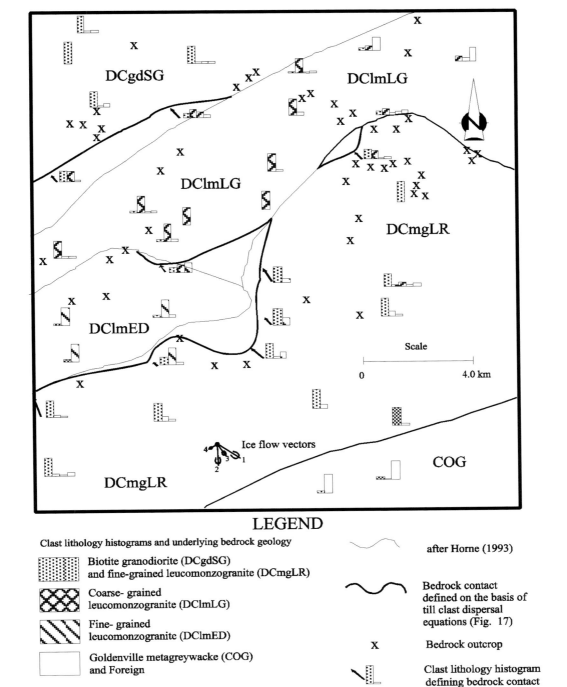

Fig. 18. Bedrock contacts estimated using dispersal Equation (3) compared with contacts mapped by Horne (1993) in the New Germany area (Location 3, Fig. 6; Fig. 7).

regional ice flows across Carboniferous and Triassic sedimentary basins. Stony, local tills were produced in proximity to ice divides over metasedimentary and igneous rocks during later, local ice flow events. Hybrid tills were formed by the interaction of successive glacier flow phases with previously deposited tills. Reworking processes include inheritance and overprinting (defined below), and serve to significantly alter the physical and chemical properties of resultant tills. Erosion, entrainment and deposition of mineralized bedrock during multiple ice flows produces, in plan view, a complex 'palimpsest' dispersal fan (e.g. Turner & Stea 1987; Parent et al. 1996), which can be simply modelled by vector addition of the ice flows. Modification of drumlins and moraines also produced 'palimpsest' glacial landforms (Stea 1994).

A simple exponential model of the uptake and decay of till pebbles down-ice from a bedrock contact was presented with emphasis on the end member and hybrid stony, ice divide tills. This model can be used to map bedrock in till-covered terrains as well as locate buried ore deposits, and it takes into account the reworking of older tills. The model can be modified for auriferous quartz veins for example, that produce linear trains of pebbles and associated gold grains (e.g. Fig. 2). Dispersal trains defined by trace element geochemistry can also be modelled in a similar fashion but analytical and sample variability have to be taken into account (Klassen 1999). Mineral exploration in the more complex, multi-till lowland areas of Maritime Canada may be simplified by using pebble lithologies. Turner & Stea (1990) demonstrated that trace element levels can be related to lithological variations in Nova Scotia tills. For the englacial components of end member tills (such as the Lawrencetown Till) Stea & Pe-Piper (1999) developed whole-rock geochemical techniques to analyse the source rock distributions and refine the 'provenance envelope' (e.g. Fig. 4a) to map up-ice bedrock and find ore bodies. For hybrid tills, a concomitant glacial mapping program is always necessary to establish the ice flow parameters of the vector dispersal model. The empirical dispersal model presented assumes a predominant ice flow direction. Glacier dispersal fan production during multiple glaciations would incorporate vector quantities to model changing ice flow directions and in the third dimension, uptake through the till sheets to the source. The 'provenance envelope' (e.g. Fig. 4a) techniques can be utilized to map up-ice bedrock and find ore bodies.

The model assumes a predominant ice flow direction and is the first step towards a unified empirical model of glacier dispersal fan production during multiple ice flows which would incorporate vector quantities to model changing ice flow directions and in the third dimension.

Glossary of terms

A few terms introduced and recycled in this paper are summarized as follows:

Zonal concept of glacial dispersal:
Zones characterized by distinct transport histories produced by the interplay and migration of regional ice sheets and local ice caps.

Ice divide till:
Local stony tills with renewal distances < 1 km showing little evidence of clast abrasion and produced near ice divides.

Ice stream till:
Exotic, silty tills with renewal distances > 10 km, and associated with drumlin fields along the Atlantic coast.

Skip zones:
Zones where local bedrock is slightly represented in the clast fractions of overlying melt-out tills due to rapid ice stream flow and englacial transport, where little or no mixing in the traction zone has occurred.

End member till:
A till whose properties are controlled by bedrock source areas and the dynamics of the depositing ice flow.

Hybrid till:
A till whose properties are controlled by reworking of older tills by younger ice sheets.

Inheritance:
The incorporation of till components and/or fabric into a younger till by erosion and entrainment of material from an older till.

Overprinting:
The injection or imprint of matrix, clasts or fabric on older tills by overriding ice.

Vector addition model:
Modelling glacial dispersal in terms of the direction and magnitude of transport, with the resultant dispersal taken to be the vector sum of constituent dispersal events.

References

ALCOCK, P. W. 1984. *A sedimentologic and stratigraphic study of Wisconsin tills in an area of the Cobequid Highlands, northwestern Nova Scotia.* MSc Thesis, University of Toronto.

ALLEY, R. B. 1991. Deforming bed origin for southern Laurentide ice sheets? *Journal of Glaciology*, **37**, 41–50.

BATTERSON, M. 1989. *Quaternary geology and glacial dispersal in the Strange lake area, Labrador.* Geological Survey Branch, Department of Mines and Energy Report **89-3**.

BOUCHARD, M. A. & SALONEN, V.-P. 1989. Glacial dispersal of boulders in the James Bay lowlands of Quebec, Canada. *Boreas*, **18**, 189–199.

BOUCHARD, M. A., CADIEUX, B. & GOUTIER, F. 1984. L'origine et les characteristiques des lithofacies du till dans le secteur nord du Lac Albanel, Québec: une etude de la dispersion glaciare clastique. *In:* GUHA, J. & CHOWN, E. H. (eds) *Chibougamou-Stratigraphy and Mineralization*. CIM Special Volume, **34**, 240–260.

BOULTON, G. S. 1978. Boulder shapes and grain-size distributions as indicators of transport paths through a glacier and till genesis. *Sedimentology*, **25**, 773–799.

BOULTON, G. S. 1984. Development of a theoretical model of sediment dispersal by ice sheets. *In: Prospecting in Areas of Glaciated Terrain*. Institute of Mining and Metallurgy, 213–224.

BROSTER, B. E., MUNN, M. D. & PRONK, A. G. 1997. Inferences on glacial flow from till clast dispersal: Waterford area, New Brunswick. *Géographie physique et Quaternaire*, **51**, 29–39.

CHALMERS, R. 1895. *Report on the surface geology of eastern New Brunswick, northwestern Nova Scotia and a portion of Prince Edward Island.* Geological Survey of Canada Annual Report **1894, 1:7: pt. M**.

CLARK, P. U. 1987. Subglacial sediment dispersal and till composition. *Journal of Geology*, **95**, 527–541.

CLARK, C. D. 1993. Megascale glacial lineations and cross-cutting ice flow landforms. *Earth Science Processes and Landforms*, **18**, 1–29.

DENTON, G. H. & HUGHES, T. J. 1981. *The Last Great Ice Sheets*. John Wiley and Sons, Inc., Toronto.

DILABIO, R. N. W. 1990. Glacial dispersal trains. Chapter 7 *In:* KUJANSUU, R. & SAARNISTO, M. (eds) *Glacial Indicator Tracing*. Balkema, Rotterdam, 109–122.

DRAKE, L. D. 1972. Mechanisms of clast attrition in basal till. *Geological Society of America Bulletin*, **83**, 2159–2165.

DREDGE, L. A. 1983. *Surficial geology of the Sept Iles area, Quebec north shore*. Geological Survey of Canada Memoir **408**.

DREDGE, L., MOTT, R. J. & GRANT, D. R. 1992. Quaternary stratigraphy, paleoecology, and glacial geology, Iles de la Madelaine, Quebec. *Canadian Journal of Earth Sciences*, **29**, 1981–1996.

DREIMANIS, A. 1989. Genetic classification of tills. *In:* GOLDTHWAIT, R. P. & MATSCH, C. I. (eds) *Genetic Classification of Glacigenic Deposits*. Balkema, Rotterdam, 17–84.

DREIMANIS, A. & VAGNERS, U. 1971. Bimodal distribution of rock and mineral fragments in basal tills. *In:* GOLDTHWAIT, R. P. (ed.) *Till: A Symposium*. Ohio State University Press, 237–250.

FINCK, P. W. & STEA, R. R. 1990. Inheritance and overprinting: controlling factors of till genesis in Nova Scotia. *In: Programme and Abstracts*, Canadian Quaternary Association-American Quaternary Association, First Joint Meeting, 18.

FINCK, P. W. & STEA, R. R. 1995. *The compositional development of tills overlying the South Mountain Batholith*. Nova Scotia Department of Natural Resources, Mines and Minerals Branch Paper **95-1**.

FLINT, R. F. 1971. *Glacial and Quaternary Geology*. John Wiley and Sons, Toronto, 892

FOISY, M. & PRICHONNET, G. 1991. Reconstruction of glacial events in southeastern New Brunswick. *Canadian Journal of Earth Sciences*, **28**, 1594–1612.

GILLBERG, G. 1967. Further discussion of the lithological homogeneity of till. *Geologiska Foreningens i Stockholm Forhandlingar*, **89**, 29–49.

GOLDTHWAIT, J. W. 1924. *Physiography of Nova Scotia*. Geological Survey of Canada Memoir **140**.

GRANT, D. R. 1963. *Pebble lithology of the tills of southeast Nova Scotia*. MSc Thesis, Dalhousie University, Halifax, Nova Scotia.

GRANT, D. R. 1977. Glacial style and ice limits, the Quaternary stratigraphic record, and changes of land and ocean level in the Atlantic Provinces, Canada. *Géographie physique et Quaternaire*, **31**, no. 3–4, 247–260.

GRANT, D. R. 1980. *Quaternary stratigraphy of southwestern Nova Scotia: glacial events and sea level changes*. Geological Association of Canada and Mineralogical Association of Canada Guidebook.

GRANT, D. R. 1989. Quaternary Geology of the Atlantic Appalachian region of Canada. *In:* FULTON, R. J. (ed.) *Quaternary Geology of Canada and Adjacent Greenland*. Geological Survey of Canada, **1**, 393–440.

GRANT, D. R. 1994. *Quaternary Geology, Cape Breton Island*. Geological Survey of Canada Bulletin **482**.

GRAVES, R. M. & FINCK, P. W. 1988. The provenance of tills overlying the eastern part of the South Mountain Batholith, Nova Scotia. *Maritime Sediments and Atlantic Geology*, **24**, 61–70.

HARRISON, W. 1960. Original bedrock composition of Wisconsin till in central Indiana. *Journal of Sedimentary Petrology*, **30:3**, 432–446.

HART, J. K. & BOULTON, G. S. 1991. The interrelation of glaciotectonic and glaciodepositional processes within the glacial environment. *Quaternary Science Reviews*, **10**, 335–350.

HORNE, R. J. 1993. *Geological map of New Germany (NTS 21A/10), Nova Scotia*. Nova Scotia Department of Natural Resources, Mines and Minerals Branch Map **93-01**.

HORNE, R. J. & COREY, M. C. 1994. *Geological map of Kejimikujik (NTS 21A/02 and 21A/07), Nova Scotia*. Nova Scotia Department of Natural Resources, Mines and Minerals Branch Map **94-05**.

HUGHES, T. J. 1995. Ice sheet modelling and the reconstruction of former ice sheets from geo(morpho)logical field data. *In:* MENZIES, J. (ed.) *Modern Glacial Environments: Processes, Dynamics and Sediments*. Butterworth-Heinemann, 77–99.

IVERSON, N. 1995. Processes of erosion. *In:* MENZIES, J. (ed.) *Modern Glacial Environments: Processes, Dynamics and Sediments*. Butterworth-Heineman, 241–247.

JOSENHANS, H. & LEHMAN, S. 1999. Late glacial stratigraphy and history of the Gulf of St. Lawrence Canada. *Canadian Journal of Earth*

Sciences, **36**, 1327–1345.

KIRKBRIDE, M. P. 1995. Processes of transportation. In: MENZIES, J. (ed.) *Modern Glacial Environments: Processes, Dynamics and Sediments*. Butterworth-Heineman, 261–293.

KLASSEN, R. A. 1997. Glacial history and ice flow dynamics applied to drift prospecting and geochemical exploration. In: GUBINS, A. G. (ed.) *Proceedings of Exploration 97: Fourth Decennial International Conference on Mineral Exploration*. 221–232.

KLASSEN, R. A. 1999. The application of glacial dispersal models to the interpretation of till geochemistry in Labrador, Canada. *Journal of Geochemical Exploration*, **67**, 245–269.

KLEMAN, J. 1994. Preservation of landforms under ice sheets and ice caps. *Geomorphology*, **9**, 19–32.

KLEMAN, J., HÄTTESTRAND, C. & CLARHäLL, A. 1999. Zooming in on frozen patches: scale dependant controls on Fennoscandian ice sheet basal thermal zonation. *Annals of Glaciology*, **28**, 189–194.

KRUMBEIN, W. C. 1937. Sediments and exponential curves. *Journal of Geology*, **64**, 577–601.

LAMOTHE, M. 1992. *Pleistocene stratigraphy and till geochemistry of the Miramichi Zone, New Brunswick*. Geological Survey of Canada Bulletin **433**.

LORING, D. H. & NOTA, D. J. G. 1973. *Morphology and sediments of the Gulf of St. Lawrence*. Fisheries Research Board of Canada Bulletin **182**.

LOWELL, T. V., KITE, J. S., CALKIN, P. E. & HALTER, E. F. 1990. Analysis of small scale erosional data and a sequence of late Pleistocene flow reversal, northern New England. *Geological Society of America Bulletin*, **102**, 74–85.

MACDONALD, M. A. 1994. *Geological map of the South Mountain Batholith, Western Nova Scotia*. Nova Scotia Department of Mines and Energy Map **1994-1**.

MACDONALD, M. A. & HORNE, R. J. 1987. *Geological map of Halifax and Sambro (NTS sheets 11D/05 and 1D/12), Nova Scotia*. Nova Scotia Department of Mines and Energy Map **87-06**.

MACNEILL, R. H. & PURDY, C. A. 1951. A local glacier in the Annapolis-Cornwallis Valley (abstract). *Proceedings of the Nova Scotia Institute of Science*, **23**, 111.

MCAUSLAN, D. A., DICKIE, G. B., SARKAR, P. K., SINCLAIR, P. E. & WILSON, B. H. 1980. The history of discovery and the description of the East Kemptville tin deposit southwest Nova Scotia. In: *82nd Annual General Meeting of the Canadian Institute of Mining and Metallurgy*. Shell Canada Resources Limited. 1–11.

MCCLENAGHAN, M. B. & DILABIO, R. N. W. 1995. Till geochemistry and its implications for mineral exploration: southeastern Cape Breton Island, Nova Scotia, Canada. *Quaternary International*, **20**, 107–122.

MCKEAGUE, J. A., GRANT, D. R., KODAMA, H., BEKE, G. J. & WANG, C. 1983. Properties and genesis of soil and the underlying gibbsite-bearing saprolite, Cape Breton Island, Canada. *Canadian Journal of Earth Sciences*, **20**, 37–38.

MOSHER, D. C., PIPER, D. J. W., VILKS, G. V., AKSU, A. E. & FADER, G. B. 1989. Evidence for Wisconsinan glaciations in the Verrill Canyon area, Scotian Slope. *Quaternary Research*, **31**, 27–40.

NIELSEN, E. 1976. *The composition and origin of Wisconsinan tills in mainland Nova Scotia*. PhD Thesis, Dalhousie University, Halifax, Nova Scotia.

O'REILLY, G. A., MACDONALD, M. A., KONTAK, D. J. & COREY, M. C. 1992. Granite and metasedimentary-hosted mineral deposits of southwest Nova Scotia. In: *Wolfville '92, Geological Association of Canada-Mineralogical Association of Canada*, Geological Association of Canada-Mineralogical Association of Canada Field Excursion Guidebook C-3.

PARENT, M. & DUBOIS, J. M. M. 1990. Les dépots glaciares et marins des ils de la madelaine (Québec), temoins d'une englaciation rapide dans les appalaches septentrionales au Pleistocene Superieur. In: *Programme and Abstracts, Canadian Quaternary Association – American Quaternary Association, First Joint Meeting*, 28.

PARENT, M., PARADIS, S. & DOIRON, A. 1996. Palimpsest glacial dispersal trains and their significance to drift prospecting. *Journal of Geochemical Exploration*, **56**, 123–140.

PELTONIEMI, H. 1985. Till lithology and glacial transport in Kuhmo, eastern Finland. *Boreas*, **14**, 67–74.

PREST, V. K. 1973. *Surficial deposits of Prince Edward Island*. Geological Survey of Canada Map **1366A**.

PREST, V. K. 1983. *Canada's heritage of glacial features*. Geological Survey of Canada Miscellaneous Report **28**.

PREST, V. K. & GRANT, D. R. 1969. *Retreat of the last ice sheet from the Maritime Provinces-Gulf of St. Lawrence region*. Geological Survey of Canada Paper **69-33**.

PREST, V. K., GRANT, D. R., MACNEILL, R. H., BROOKS, I. A., BORNS, H. W., OGDEN, J. G., III, JONES, J. F., LIN, C. L., HENNIGAR, T. W. & PARSONS, M. L. 1972: Quaternary geology, geomorphology and hydrogeology of the Atlantic Provinces. In: *24th International Geological Congress*, Excursion Guidebook A61-C61.

PREST, V. K. & NIELSEN, E. 1987. The Laurentide ice sheet and long distance transport. In: KUJANSUU, R. & SAARNISTO, M. (eds) *INQUA Till Symposium, Finland 1985*. Geological Survey of Finland, Special Paper, **3**, 91–102.

PREST, W. H. 1911. Prospecting in Nova Scotia. *Nova Scotia Mining Society Journal*, **16**, 73–91.

PRONK, A. G., BOBROWSKY, P. T. & PARKHILL, M. A. 1989. An interpretation of the late Quaternary glacial flow indicators in the Baie des Chaleurs region, northern New Brunswick. *Géographie physique et Quaternaire*, **43:2**, 179–190.

RAMPTON, V. N., GAUTHIER, R. C., THIBAULT, J. & SEAMAN, A. A. 1984. *Quaternary geology of New Brunswick*. Geological Survey of Canada Memoir **416**.

RAMPTON, V. N. & PARADIS, S. 1981. *Quaternary geology of Amherst map area (21H)*. New Brunswick Department of Natural Resources Map Report **81-3**.

SALMINEN, R. & HARTIKAINEN, A. 1985. *Glacial trans-*

port of till and its influence on interpretation of results in North Karelia, Finland. Geological Survey of Finland Bulletin **335**.

SALONEN, V.-P. 1986. *Glacial transport distance distributions of surface boulders in Finland.* Geological Survey of Finland Bulletin **338**.

SALONEN, V.-P. 1992. Glacigenic dispersion of coarse till fragments. *In:* KAURANNE, L. K., SALMINEN, R. & ERIKSSON, K. (eds) *Handbook of Exploration Geochemistry, Volume 5, Regolith Exploration in Arctic and Temperate Terrains.* Elsevier, Rotterdam, 127–142.

SHILTS, W. W. 1973. *Glacial dispersal of rocks, minerals and trace elements in Wisconsinan till, southeastern Quebec, Canada.* Geological Society of America, Memoir **136**, 189–219.

SHILTS, W. W. 1976. Glacial till and mineral exploration. *In:* LEGGETT, R. F. (ed.) *Glacial Till: An Interdisciplinary Study.* Royal Society of Canada, Special Publication, **12**, 205–224.

SHILTS, W. W. 1995. Glacial drift exploration. *In:* MENZIES, J. (ed.) *Modern Glacial Environments: Processes, Dynamics and Sediments.* Butterworth-Heineman, 411–438.

STEA, R. R. 1983. Surficial geology of the western part of Cumberland County, Nova Scotia. *In: Current Research, Part A.* Geological Survey of Canada, Paper **83-1A**, 197–202.

STEA, R. R. 1984. The sequence of glacier movements in northern mainland Nova Scotia determined through mapping and till provenance studies. *In:* MAHANEY, W. C. (ed.) *Correlation of Quaternary Chronologies.* Geo Books, Norwich England, 279–297.

STEA, R. R. 1991. *Clast dispersal in the Laurentian Channel off Cape Breton Island and mainland Nova Scotia: test of a glacial dispersal model.* Dalhousie University, Centre of Marine Geology Technical Report **13**.

STEA, R. R. 1994. Relict and palimpsest glacial landforms in Nova Scotia, Canada. *In:* WARREN, W. P. & GROOT, D. G. (eds) *Formation and Deformation of Glacial Deposits.* Balkema, Rotterdam, 141–158.

STEA, R. R. 1995. *Late Quaternary glaciations and sea-level change along the Atlantic Coast of Nova Scotia.* PhD Thesis, Dalhousie University, Halifax, Nova Scotia.

STEA, R. R. & BROWN, Y. 1989. Variation in drumlin orientation, form and stratigraphy relating to successive ice flows in southern and central Nova Scotia. *Sedimentary Geology,* **62**, 223–240.

STEA, R. R. & FINCK, P. W. 1984. Patterns of glacier movement in Cumberland, Colchester, Hants, and Pictou Counties, northern Nova Scotia. *In: Current Research, Part A.* Geological Survey of Canada, Paper **84-1A**, 477–484.

STEA, R. R. & FOWLER, J. H. 1979. *Minor and trace element variations in Wisconsinan tills, Eastern Shore region, Nova Scotia.* Nova Scotia Department of Mines and Energy Paper **79-4**.

STEA, R. R. & MURPHY, B. 1994. The source of erratics in Lawrencetown Till along the Eastern Shore of Nova Scotia: petrology and whole-rock geochemistry. *In: Mines and Minerals Branch, Report of Activities 1993.* Nova Scotia Department of Natural Resources, Mines and Energy Branches, Report **94-1**, 93–103.

STEA, R. R. & PE-PIPER, G. 1999. Using whole rock geochemistry to locate the source of igneous erratics from drumlins on the Atlantic coast of Nova Scotia. *Boreas,* **28**, 308–325.

STEA, R. R., CONLEY, H. & BROWN, Y. (compilers) 1992a. *Surficial geology of the Province of Nova Scotia.* Nova Scotia Department of Natural Resources Map **92-1**.

STEA, R. R., FINCK. P. W. & WIGHTMAN, D. M. 1986. *Quaternary geology and till geochemistry of the western part of Cumberland County, Nova Scotia (sheet 9).* Geological Survey of Canada Paper **85-17**.

STEA, R. R. MOTT, R. J., BELKNAP, D. F. & RADTKE, U. 1992b. The Pre Late Wisconsinan chronology of Nova Scotia, Canada. *In:* CLARK, P. U. & LEA, P. D. (eds) *The Last Interglaciation/Glaciation Transition in North America.* Geological Society of America, Special Paper **270**, 185–206.

STEA, R. R., PIPER, D. J. W., FADER, G. B. J. & BOYD, R. 1998. Wisconsinan glacial and sea-level history of Maritime Canada, a correlation of land and sea events. *Geological Society of America Bulletin,* **110**, 821–845.

STEA, R. R., TURNER, R. G., FINCK, P. W. & GRAVES, R. M. 1989. Glacial dispersal in Nova Scotia: a zonal concept. *In:* DILABIO, R. N. W. & COKER, W. B. (eds) *Drift Prospecting.* Geological Survey of Canada, Paper **89-20**, 155–169.

SZABO, N. L., GOVETT, G. J. S. & LAYTAI, E. Z. 1975. Dispersion trends of elements and indicator pebbles in glacial till around Mt. Pleasant, New Brunswick, Canada. *Canadian Journal of Earth Sciences,* **12**, 1534–1556.

TURNER, R. G. & STEA, R. R. 1987. The Garden of Eden dispersal train. *In: Mines and Mineral Branch Report of Activities 1986.* Nova Scotia Department of Mines and Energy, Report **97-1**, 165–169.

TURNER, R. G. & STEA, R. R. 1990. Interpretation of till geochemical data in Nova Scotia, Canada using mapped till units, multi-element anomaly patterns, and the relationship of till clast geology to matrix geochemistry. *Journal of Geochemical Exploration,* **37**, 225–254.

VEILLETTE, J. J. & ROY, M. 1995. The spectacular cross-striated outcrops of James Bay, Quebec *In: Current Research 1995, Part C.* Geological Survey of Canada, 243–248.

WILLIAMS, G. L., FYFFE, L. R., WARDLE, R. J., COLMAN-SADD, S. P., BOEHNER, R. C. & WATT, J. A. 1985. *Lexicon of Canadian Stratigraphy, Volume VI, Atlantic Region.* Canadian Society of Petroleum Geologists.

Contrasting styles of glacial dispersal in Newfoundland and Labrador: methods and case studies

MARTIN J. BATTERSON & DAVID G. E. LIVERMAN

Geological Survey of Newfoundland and Labrador, Department of Mines and Energy, P.O. Box 8700, St. John's, Newfoundland A1B 4J6, Canada, (e-mail: mjb@zeppo.geosurv.gov.nf.ca)

Abstract: A review of practical approaches to drift exploration intended for use by exploration geologists working in drift covered areas is presented. The contrasting styles of glacial dispersal between Labrador, dominated by the effects of the Laurentide ice sheet, and the Island of Newfoundland, affected by small, coalescing ice caps at the glacial maximum and smaller topographically-controlled ice centres during deglaciation, are described. The effect has been to produce longer, ribbon-shaped dispersal trains in Labrador, except in the Labrador Trough near the centre of the Labrador sector of the Laurentide Ice Sheet, and shorter more diffuse dispersal patterns in Newfoundland.

This paper is intended for use by exploration geologists working in glacial drift covered areas, and focuses on practical approaches to drift exploration through the illustration of methods used, and dispersal patterns observed, in Newfoundland and Labrador by the Geological Survey of Newfoundland and Labrador (GSNL). Labrador, which is part of the mainland of Canada, has a long history of mineral exploration, beginning in the 1800s. The advent of air travel in the 1920s opened up Labrador for exploration, and the iron ore deposits of western Labrador were discovered in 1929. Exploration in the supracrustal rocks of the Central Mineral Belt of Labrador (Fig. 1) in the 1950s and 1960s resulted in the discovery of base metal (mostly Cu), and rare earth element (Be and Nb) deposits. Reserves generally were insufficient for mining. Deposits of U however, were considered to be of minable quantities. Environmental concerns and plummeting prices stalled uranium production indefinitely. A surge of exploration activity occurred in the late 1970s to early 1980s following discovery, through drift prospecting, of a Y–Zr–REE deposit at Strange Lake on the Labrador–Québec border. Recently, intense exploration activity followed the discovery of the Voiseys Bay nickel deposit near Nain (Fig. 1).

Mineral development on the Island of Newfoundland has a long history, initiated by Viking attempts to smelt bog iron about AD 1000 (Wallace 1990). The earliest mines were started in the 1500s (Martin 1983), representing some of the earliest mining initiatives in North America. Most exploration has occurred in the central part of the Island, resulting in the discovery of a major Cu–Pb–Zn deposit at Buchans that was mined between 1906 and 1984, and the copper deposits around Notre Dame Bay resulting in the mines at Tilt Cove (1864–1967), Betts Cove (1874-1886) and Little Bay (1878–1902) (Fig. 1). Western Newfoundland contains past-producing base metal (e.g., Zn at Daniels Harbour), chromite (e.g., Lewis Hills) and precious metal (e.g., Hope Brook gold) mines. The largest mine in eastern Newfoundland was the Bell Island iron ore mine that operated between 1894 and 1966, although several smaller mines, e.g., near St. John's (Cu), La Manche (Pb) and Brigus (Au), existed for shorter periods.

A major contribution to mineral exploration activity has been the province-wide multi-element lake sediment geochemical surveys, and follow-up till geochemical programs supported by detailed ice flow indicator mapping. Several mineral occurrences have been discovered through the use of drift prospecting techniques,

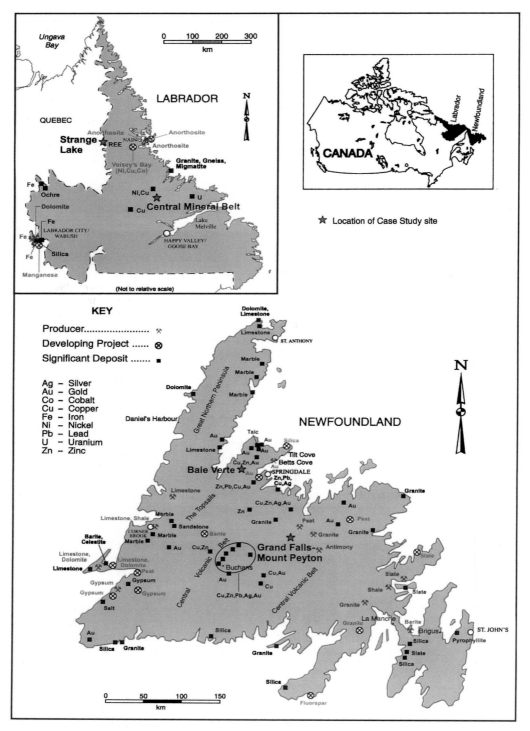

Fig. 1. Location of Newfoundland and Labrador on the east coast of Canada, and location of significant mineral prospects, place names mentioned in the text and case study sites.

including the Strange Lake Y–Zr–REE deposit. Recent staking activities are commonly the result of the release of till geochemical data by the government (e.g. Batterson *et al.* 1998), and mineral explorationists in Newfoundland and Labrador are increasingly venturing into drift-covered areas where surface bedrock exposures are scarce. Sampling of surficial materials, interpretation of ice flow directional indicators, and definition of dispersal trains thus are becoming necessary skills for the successful prospector operating in Newfoundland and Labrador.

Glacial history of Newfoundland and Labrador

The two parts of the Province of Newfoundland and Labrador have very different glacial histories, and styles of glaciation. Labrador was glaciated by the continental-scale Laurentide Ice Sheet, whereas the island of Newfoundland supported independent ice caps, on a much smaller scale. The tip of the Great Northern Peninsula was the only area covered by the Laurentide Ice Sheet during the last glaciation (Grant 1987). Evidence of pre-Late Wisconsinan glaciation is meagre (e.g., Brookes *et al.* 1982; Klassen & Thompson 1993), and in both parts of the Province multiple till sheets are rare or absent.

Labrador

The Laurentide Ice Sheet covered most of Canada during the last glaciation. All of Labrador was covered by the eastern sector of the Laurentide Ice sheet, except for the highest peaks of the northern Torngat Mountains and the Mealy Mountains south of Lake Melville. Ice flowed out from the centre of the Labrador Peninsula in all directions. In northern Labrador, the Torngat Mountains presented a barrier to eastward ice movement. Glaciers reached the coast via the major valleys through the mountain range. End moraines and kames mark the farthest extent of the ice sheet, although in places the ice sheet may have terminated in the sea.

The major valleys leading to the coast contain thick sequences of glaciofluvial outwash, deposited as the ice sheet retreated. These are commonly overlain by marine sediments, which were deposited during periods of higher sea level following deglaciation. On the uplands, recessional moraines and eskers mark the pattern of retreat. As the ice melted, lakes formed where the ice sheet blocked river drainage. The largest of these (glacial lakes Naskaupi and McLean) were trapped between the westward retreating ice margin and the drainage divide between the Atlantic Ocean and Ungava Bay. The main Laurentide Ice Sheet finally melted in the Schefferville area of western Labrador about 6500 years ago. Small cirque glaciers are still found in the Torngat Mountains today. In the central part of the Labrador Peninsula ice flow is complex and difficult to interpret, because records of several past glaciations may be preserved. Klassen & Thompson (1993) outline the ice flow history, which is particularly complicated in the Labrador Trough. In coastal areas, ice flow history is generally simpler.

Newfoundland

The tip of the Great Northern Peninsula was the only area covered by the Laurentide Ice Sheet during the last glaciation (Grant 1987). The Appalachian ice complex, characterized by smaller ice caps independent of the Laurentide Ice Sheet, covered the rest of the island. Separate centres of glacier accumulation existed on the Avalon Peninsula, in central Newfoundland, and on the Long Range Mountains. Complicated ice flow patterns resulted where an area was, at different times, covered by ice from more than one centre (cf. Catto 1998). As the ice melted, these accumulation areas became isolated from each other, and as many as 15 smaller ice caps probably existed for a short time (Grant 1974). The complex interaction of numerous small ice caps resulted in a complicated ice flow history. The farthest limit of glacial advance in many areas was onto the continental shelf. As the glaciers melted, coastal areas became ice free between 11 000 and 14 000 years ago. The climate cooled again about 11 000 to 10 000 years ago, and some glaciers re-advanced for a short time (Grant 1989).

Sea-level history

Understanding the distribution of marine sediments is important in planning geochemical sampling because, unlike primary glacigenic sediments marine deposits are more difficult to link to bedrock source. The Newfoundland and Labrador coastline contains numerous raised beaches and deltas that mark the position of the coast as sea level fell during deglaciation. Today, much of the Newfoundland coast is sinking as a result of continued settling of the crust, although the coast of Labrador continues to rise as a result of isostatic adjustment (Liverman 1994).

In Newfoundland, landforms that mark the highest level of the sea (marine limit) are found at higher elevations toward the northwest, with the highest marine limits found on the tip of the Great Northern Peninsula. Most of the island shows raised beaches apart from small areas on the Avalon Peninsula. In Labrador, the highest raised beaches are in the southeast where beaches up to about 150 m above sea level occur. The marine limit decreases northward, to about 55 m in the Torngat Mountains, and 17 m at Cape Chidley, the northernmost part of Labrador.

Methodology

The definition of glacial dispersal patterns may be accomplished through a combination of field and laboratory techniques, many of which are amply discussed in other papers in this volume. Surficial geology mapping completed as part of till geochemical programs provide an essential means by which to evaluate geochemical anomalies or trends. The reader is referred to other papers in this volume that deal explicitly with methodology.

Striation mapping

Bedrock outcrop is common in most of the province, and the favoured method of delineating ice-flow is by mapping striations. Striations are excellent indicators of ice flow as they are formed by the direct action of moving ice. The information from striations should be treated with caution, as regional ice flow patterns can show considerable local variation where ice flow was deflected by local topography. Regional flow patterns can only be deduced after examining numerous striation sites. The orientation of ice flow can easily be discerned from a striation by measuring its azimuth. Determination of the direction of flow can be made by observation of the striation pattern over the outcrop; where areas in the lee of ice flow may not be striated; by the presence of such features as 'nail-head' striations, and miniature crag and tails (rat-tails); and by the morphology of the bedrock surface, which may show the effects of sculpturing by ice (Iverson 1991). At many sites, the direction of ice flow is unclear, and only the orientation of ice flow can be deduced. Where striations representing separate flow events are found, the age relationships are based on cross cutting of striation sets, and preservation of older striations in the lee of younger striations.

Klassen & Thompson (1993) initiated systematic striation mapping in Labrador. Techniques developed in their project were refined in subsequent widespread systematic striation mapping of the Island of Newfoundland (e.g. Liverman & St. Croix 1989; St. Croix & Taylor 1991). To standardize striation observations, a basic form is used by government geologists listing location, elevation, dimensions and surface orientation of outcrop; striation direction and range of measurements; confidence in the identification of the striations; confidence in assigning a flow direction or orientation; and age relationships and basis. These are entered into a database that currently lists over 10 000 observation sites for the Island of Newfoundland (Taylor et al. 1994). Striation data can be readily plotted at a variety of scales, or imported into GIS applications or viewers.

Clast fabric analysis

The trend of clast fabrics may be used to indicate the direction of ice flow at the time the sediment was deposited, and is thus a useful method for explorationists in heavily drift-covered areas. Clast fabric analysis is also critical in determining diamicton genesis, although such studies must be supplemented by examination of the sedimentary structures of the diamicton. (Diamicton is a descriptive term referring to any sediment with a wide range of grain sizes.) Detailed examination of a sediment is required to determine if it is a till (a sediment deposited directly by the action of glaciers), or secondary products such as sediment gravity flow deposits. Strong unimodal fabrics are typical of basal melt-out and lodgement tills, and less oriented fabrics are typical of supraglacial deposits and diamictons produced by ice-rafting of clasts in lakes or oceans.

The orientation of 25 pebble- to cobble-sized clasts with an a-axis to b-axis ratio of 3:2 or greater is measured at each selected site, with clasts taken from a 1m^2 area on a vertical outcrop face. The results are plotted on a stereogram, analysed statistically using the Stereo™ package for the Apple Macintosh® computer (MacEachran 1990), and the principal eigenvectors and eigenvalues calculated. The principal eigenvalue divided by the sample size is known as the normalized eigenvalue (S_1; Mark 1973, 1974; Woodcock 1977). This is a measure of the strength of orientation of the clasts, and can range between 0.33 and 1.00. A sample with most clasts aligned in similar orientations will have a value close to 1.00, whereas a random sample will have a value close to 0.33. Only fabrics with $S_1 > 0.6$ are considered to be possibly indicative of basal glacial transport and deposition. A second statistical parameter

Table 1. *Elements analysed by the GSNL and commercial laboratories.*

Method	Elements analysed
AAS	Ag
Gravimetric analysis	LOI
ICP-ES	Al, Ba, Be, Ca, Ce, Co, Cr, Cu, Dy, Fe, Ga, K, La, Li, Mg, Mn, Mo, Na, Nb, Ni, P, Pb, Sc, Sr, Ti, V, Y, Zn, Zr
INAA	As, Au, Ba, Br, Ca, Ce, Co, Cr, Cs, Eu, Fe, Hf, La, Lu, Na, Nd, Rb, Sb, Sc, Sm, Tb, Th, U, Yb,

(K) indicates whether the distribution is unimodal, or girdled. Low values of K (< 1.0) suggest a girdle distribution, atypical of basally deposited tills, and therefore only those fabrics with a K value > 1.0 are interpreted here as being due to glacial transport and deposition. Appropriate statistical analysis allows strong, unimodal fabrics to be identified, but the relation of these fabrics to ice flow is dependent on examination and interpretation of diamicton sedimentary structures.

Clast rock types

The pattern of glacial dispersal may be determined from plotting the distribution of mineralized or indicator clasts on a map. In this manner, the gross shape of glacial dispersal trains may be shown. Numerous examples of this approach have been used in Newfoundland and Labrador (e.g. Liverman 1992; Batterson 1994, 1998). Average glacial transport distances can be estimated from plots of clast lithology against distance from source (Salonen 1986; Bouchard & Salonen 1990). The half distance method is also used to estimate transport distance from clast lithology. The half distance is the distance at which the frequency of a particular rock type in till decreases to half its original abundance (Krumbein 1937; Gillberg 1965; Bouchard & Salonen 1990).

Till geochemistry

Analytical work on the silt-clay fraction of diamictons is carried out at the GSNL Geochemical Laboratory, with additional analyses from commercial laboratories. Analytical methods in the GSNL laboratory include atomic absorption spectrometry (AAS), gravimetric analysis, and inductively coupled plasma emission spectrometry (ICP-ES) using an aqua regia digestion. External analyses by commercial laboratories are by instrumental neutron activation analysis (INAA). These methods and the elements analysed are summarized in Table 1. Data quality is monitored using laboratory duplicates (analytical precision only), and standard reference materials.

Till samples collected for geochemical analysis ideally are sampled on an irregular grid. Regional surveys in Newfoundland and Labrador sample at about 2 km intervals in areas of poor access, where helicopter support is required. Areas with an adequate road network allow 1.5 km sample spacing. Sampling at these densities provides a gross indication of glacial dispersal patterns, and allows the GSNL to sample 1 to 2 1:50 000 NTS map sheets per year. Detailed sampling at the property-scale is largely the responsibility of individual exploration companies or prospectors.

Surficial geology mapping

Mapping of Quaternary sediments in association with sampling for geochemistry is critical in the interpretation of geochemical data. Adequate mapping prior to sampling allows identification of till, glaciofluvial, glaciomarine, and modern fluvial or modern marine deposits, all of which should be treated separately in the interpretation of the data. Mapping the surficial geology commonly will guide the sampling. Comparison of geochemical sampling locations with surficial geology maps during the data compilation stage allows modification of the data-set to reflect similar sediment types, commonly tills. The recognition of sediment types is particularly important in coastal areas, such as Newfoundland and Labrador, because many parts of these areas were submerged by higher seas following deglaciation and till is commonly covered by marine sediments. In Newfoundland, calving glaciers produced glaciomarine sediments, exposures of which are common below marine limit. Simple visual examination of these exposures produced in this environment is insufficient to differentiate between tills deposited directly by glaciers, and those secondary deposits produced by, for example, sediment gravity flow.

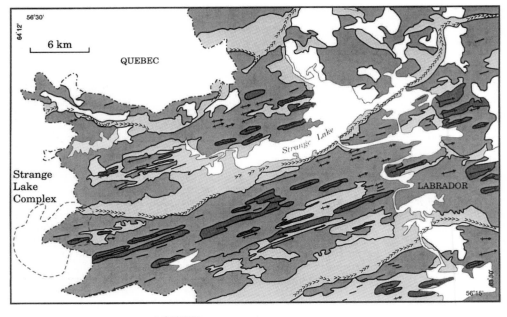

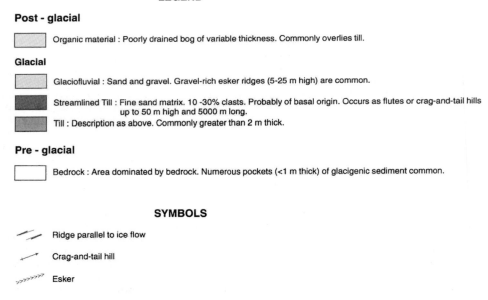

Fig. 2. Surficial geology of the area down-ice of the Strange Lake Complex.

Case studies

Numerous glacial dispersal trains and patterns in Newfoundland and Labrador have been documented. In Labrador, Klassen & Thompson (1993) described amoeboid-shaped dispersal of clasts and till geochemistry in the Labrador Trough, produced by multiple ice flows in an area close to a Laurentide Ice Sheet dispersal centre. In the Central Volcanic Belt of Newfoundland (Fig. 1), ribbon-shaped dispersal trains from the Buchans Cu–Pb–Zn ore-body

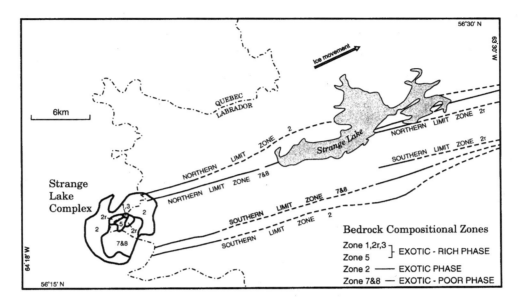

Fig. 3. Distribution of mineralized clasts from sub-units of the Strange Lake Complex (from Batterson 1989b).

have been described by James & Perkins (1981), and from the Gullbridge Cu deposit by O'Donnell (1973). Klassen & Murton (1996) provide a more complete description of glacial dispersal patterns for the Central Volcanic Belt. The case studies presented below are from work conducted by the GSNL.

Strange Lake, northern Labrador

The Strange Lake study in northern Labrador defined geochemical and clast dispersal trains from a highly mineralized source over plateau terrain formed by a single ice flow event. The resulting pattern is atypical of dispersal in most of Newfoundland and Labrador. The Strange Lake complex was discovered in 1979 through boulder tracing during follow-up of unrelated lake sediment anomalies. A peralkaline granite intruded into gneiss and rapakivi granite hosts Zr–Y–Nb–Be–REE mineralization. The deposit is situated on the Québec–Labrador border on the low relief Nain plateau. The area was glaciated during the Late Wisconsinan by the Labrador sector of the Laurentide Ice Sheet, flowing coastward (070°). Most of the area is drift covered, commonly masking the underlying bedrock geology. Several large crag-and-tail hills, in excess of 5 km long, are found on the plateau, and several shallow valleys host glaciofluvial sediments and well-defined esker ridges (Fig. 2).

Till samples were collected by GSNL to characterize glacial dispersal from the ore-body. Over 500 sites were sampled from an average depth of 40 to 60 cm (i.e., mostly BC- to C-horizon soil samples), which provided a nominal sample density of one site per 2 km^2, although sample density was increased to one site per 0.5 km^2 in the vicinity of the mineralization. The lithology of at least 50 clasts were identified at each sample site, and the silt-clay (<0.063 mm) fraction of the till analysed for a wide-range of elements using atomic absorption spectrophotometry and x-ray fluorescence.

Clast (2–6 cm diameter) distribution shows a long, ribbon-shaped dispersal train. Adjacent to the mineralization, high concentrations (maximum 52%) of Strange Lake complex clasts occur, with this clast content generally decreasing to <10% within 1 km down-ice of the complex. At 40 km down-ice from the complex, 1–4% of surface clasts are from the Strange Lake complex. Within the dispersal train are a series of inter-related and over-lapping clast dispersal trains that originate from the sub-units within the Strange Lake complex (cf. Miller 1986). A transect anywhere across the dispersal train perpendicular to ice flow reveals a sequence that mimics the outcrop distribution of the sub-units in the bedrock source (Fig. 3). The concentration of clasts within the dispersal

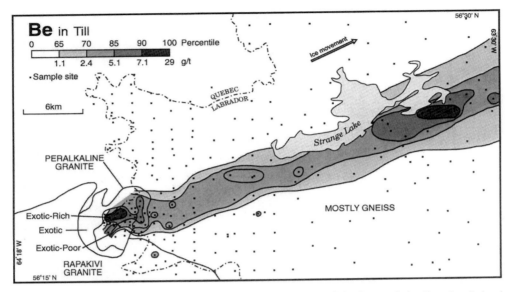

Fig. 4. Distribution of Be in the <0.063 mm fraction of till down-ice of the Strange Lake Complex, Labrador (from Batterson 1989b).

trains is highest on the crag-and-tail hills, representing up to 10% clasts at distances of 25 km down-ice of the complex.

The long, ribbon-shaped dispersal train of mineralized clasts is also defined by the till geochemistry (McConnell & Batterson 1987; Batterson 1989a,b). The geochemical patterns for a group of elements including Be, Pb, Th, U, Y and Zr match the dispersal train delineated by the mineralized clasts, whereas the patterns for other elements (e.g., Co, Cu and Ni) reflect dispersal from the gneissic terrain to the north of the complex or the rapakivi granites to the south (McConnell & Batterson 1987).

Several element associations are shown by the data. Co–Ni–Cu associations reflect a mafic component in the gneiss complex. Be–La–Nb–Pb–Th–U–Y–Zr associations represent lithophile elements associated with the ore minerals of the Strange Lake complex. Geochemical patterns of these lithophile elements are broadly similar. The highest values occur over the peralkaline complex, particularly over the mineralized zone (e.g. Be, Fig. 4). The dispersal train extending down-ice from the source is ribbon-shaped and is clearly delineated for at least 40 km. The geochemical contrast between the peralkaline-related dispersal train and local background over the surrounding gneissic and rapakivi granite terrain to the north is sharp. Crag-and-tail hills within the Strange Lake dispersal train have anomalously high geochemical and complex-related clast concentrations compared to the surrounding lowlands, somewhat independent of distance from source. At 40 km down-ice of the complex, Be and Y concentrations are anomalous (95th percentile of values) (Fig. 4), indicating that these topographical highs acted as 'interceptors' of complex-derived sediment transported either englacially or high in the basal debris layer (Batterson 1989a, b).

Central Mineral Belt, Labrador

This case study illustrates the effects on dispersal where a regional ice flow event was followed by a topographically-controlled event. Clast dispersal trains show a distinct 'dog-leg' pattern. The Central Mineral Belt of Labrador is an east-trending belt of Proterozoic supracrustal sedimentary and volcanic rocks with associated granites (Ryan 1985). Aphebian and Palaeohelikian rocks contain significant mineralization. Rocks of the Aphebian Moran Lake Group contain base metal, precious metal and U showings, many of which were found during exploration in the 1950s, particularly by AMCO (American Metals Company) and BRINEX (British Newfoundland Exploration). The Palaeohelikian rocks, particularly the Bruce River Group, host both base metals and U. Base metal exploration produced disappointing results,

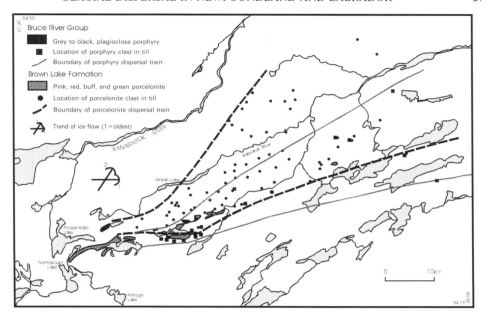

Fig. 5. Dispersal of indicator erratics from the Bruce River Group and Brown Lake Formation.

although significant U showings were discovered near Moran Lake. Further east, U deposits at Kitts and Michelin were of minable grades and volumes, although development was stalled due to unfavourable market and environmental conditions.

The region has a diverse physiography. Valleys adjacent to the modern coast are commonly filled with fine-grained marine sediments, in some places capped by post-glacial fluvial sediments deposited as relative sea level fell. Valleys above the marine limit, about 130 m above modern sea level (Batterson et al. 1988), are commonly filled with sand and gravel deposited either by glaciofluvial meltwater from waning glaciers inland or post-glacial fluvial sediments.

Two separate ice flows from the Laurentide Ice Sheet were identified. The earliest flow was a regional event towards about 020°, crossing major valleys such as that of the Kaipokok River. The most recent ice flow event was eastward. In the area west of Moran Lake, stossed bedrock forms show evidence of this flow with evidence of earlier flows commonly preserved on the lee-side of outcrops. East of Moran Lake the youngest flow is not found on hilltops, being restricted to valleys. Striations from the youngest flow are commonly found cross-cutting striations produced by the earlier ice flow. This topographically-controlled event affected glacial dispersal, likely cutting dispersal trains derived from the earlier ice flow event, e.g., at Melody Lake (Batterson et al. 1987). The influence of the two ice flows on dispersal patterns is shown by the distribution of indicator erratics in the Central Mineral Belt. These clasts are visually distinctive and have a unique source area. The distribution of a grey to black plagioclase porphyry clasts from the Bruce River Group, and pink, red, green to buff porcelonite and pink, red, green to buff volcaniclastic sandstone clasts, both from the Brown Lake Formation, was used by Batterson et al. (1988) to illustrate dispersal patterns (Fig. 5). Dispersal was mapped on the presence or absence of the indicator clasts from mudboil-dominated uplands or test pits in till. Porcelonite clasts form a fan-shaped dispersal train within the study area, which is 7 km wide at its source and 20 km wide down-ice (east). A similar pattern of dispersal is displayed by plagioclase porphyry clasts. The patterns are consistent with eastward and northeastward ice flow. Near and west of Moran Lake, glacial dispersal is eastward, (i.e. parallel to the most recent flow). East of Moran Lake (Fig. 5), the orientation of the dispersal train is northeastward. This 'dog-leg' is likely the result of increasing influence of topography on ice flow and dispersal patterns, and is an example of a

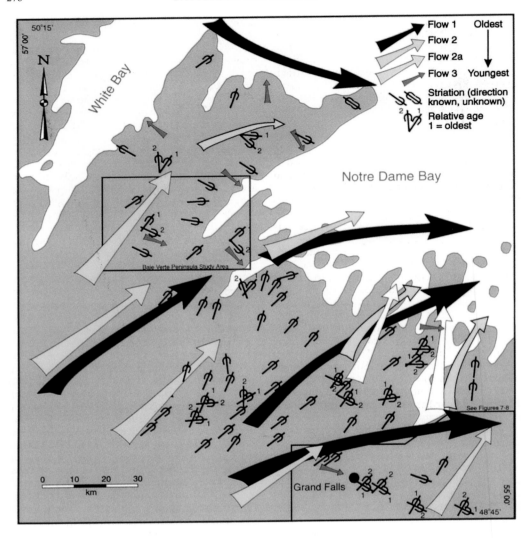

Fig. 6. Ice flow directions in northeast Newfoundland. See Figure 1 for place name location.

palimpsest glacial dispersal train (Parent et al. 1996).

The dispersal train was originally produced by the regional northeast ice flow. As ice began to thin it became topographically-controlled through the east-trending valleys, producing the eastward trending part of the train. The northeast part of the dispersal train is largely confined to highland areas and were unaffected by the eastward ice flow. This changing pattern of dispersal during deglaciation, as ice thins and is increasingly topographically influenced, is an important consideration in areas with considerable topographical relief close to ice sheet margins, such as coastal Labrador.

Central Newfoundland

This case study illustrates the short, diffuse geochemical dispersal patterns associated with small island-centred ice caps characteristic of the Appalachian ice complex, which are in contrast to the generally longer, well-defined dispersal trains produced by, and previously illustrated for, the Laurentide Ice Sheet in Labrador. The effect of till geochemical surveys on staking activity in central Newfoundland is also discussed.

Mineral exploration activity in the 1980s in central Newfoundland has shifted focus from the traditional base metal exploration programmes that followed discovery of the Buchans Cu–Pb–Zn deposit, to precious metals, particularly Au. While the Au-bearing base metal volcanogenic massive sulphide (VMS) deposits are well known (e.g., at Buchans and Rambler), exploration over the last two decades has resulted in the recognition of numerous epigenetic Au showings in the province. Important recent discoveries include the Hope Brook deposit, near Burgeo, and the Pine Cove and Nugget Pond deposits near Baie Verte (Fig. 1). Other areas of interest include SW Newfoundland, the Avalon Peninsula, Burin Peninsula, and the Grand Falls area. The latter area was the focus of a GSNL till geochemistry and surficial mapping project. The area extends from Grand Falls eastward to the southwest Gander River, and is underlain by Cambrian to Middle Ordovician mafic to felsic volcanic rocks of the Victoria Lake Group, south of Diversion Lake and north of West Lake (Kean & Mercer 1981). To the east are late Ordovician and early Silurian sedimentary rocks and associated submarine mafic volcanic rocks. The youngest rocks in the area are those of the Silurian–Devonian Mount Peyton intrusive suite. These rocks constitute the Mount Peyton batholith. The western and northern parts of the batholith are composed of grey, fine- to medium-grained gabbro, and the eastern and southern parts are a pink biotite granite (Dickson 1992). Mineral exploration activity has been centred on areas underlain by volcanic and sedimentary rocks, and along the western margin of the Mount Peyton batholith. Much of the area between Lemotte's Lake and Diversion Lake was staked in 1997, with Au as the main target.

The area has a generally rolling topography dissected by northeast–southwest oriented river valleys, which lead into the eastward-flowing Exploits River valley. The physiography is largely controlled by bedrock structure, and hills commonly are oriented northeastward. Mount Peyton (482 m asl) is the highest point in the study area. The ice flow history was defined by striation mapping. Three separate phases of ice flow were identified (Fig. 6), similar to those first described by St. Croix & Taylor (1991): (1) An early eastward ice flow is recorded by striations over the northern part of the area, although bedrock outcrops were rarely stossed by this flow. Evidence of the early eastward flow is found across much of northeastern Newfoundland (Batterson & Vatcher 1991; Scott 1994); (2) a regional north to northeast ice flow is shown by striations and bedrock stossing on outcrops. Flow directions are generally more northward towards the coast, produced by draw-down into the Bay of Exploits (Batterson et al. 1998); and, (3) the most recent ice flow event was eastward, and is recorded by fine striations overprinting those striations produced by the regional north to northeastward flow event. This late eastward flow did not mould bedrock outcrops.

A detailed till sampling program was conducted across the area. Sample spacing was roughly one sample per $2\,km^2$ in areas of road access. In areas where helicopter support was required, sample spacing was roughly one sample per $4\,km^2$. Approximately 850 samples were collected, mostly from the BC- or C-horizon. Samples were sieved in the laboratory and the silt-clay ($<0.063\,mm$) fraction was retained for geochemical analysis.

The patterns of dispersal between Grand Falls and Glenwood suggest the regional north to northeastward ice flow event controlled dispersal. Till clasts from the Botwood Group were found over the northwest margins of the Mount Peyton Batholith up to 5 km from source, and red conglomerate clasts from the Rogerson Lake conglomerate found just south of Diversion Lake were found up to 10 km to the northeast. Glacial dispersal patterns are also apparent from the trace element concentrations. Dispersal trains are commonly short (generally $<5\,km$). The proximity to ice divides is considered the main explanation for the short dispersal, although the broadly-spaced sample spacing (one per $4\,km^2$) may be a contributing factor in the poor definition of dispersal. Principal component analysis of the geochemical data identified several groupings of elements (Batterson et al. 1998). A mafic component contained elements that clearly define the gneiss component of the Mount Peyton Batholith, e.g., Cr (Fig. 7). The low Cr concentration in till over the southern part of the batholith, and high concentrations over the northern edge indicate that glacial dispersal was towards the northeast ($020°$). Low Cr contents in the south reflect the dispersal of Cr-poor till derived from the adjacent Botwood Group sediments southwest of the batholith, and similarly the dispersal of Cr-rich till from the Mount Peyton batholith over Botwood Group sedimentary rocks in the north. In both cases, dispersal directions are parallel to the northward ice flow (ice flow 2, Fig. 6), and dispersal distances are less than 5 km. Similar trends are found in the distribution of the other elements defined as the mafic component (Fe, Cr, Cu, Mg, Ti etc). Incompatible elements suggest an association with specialized or peralkaline granites (Be, K, Nb, Y, Zr etc), and are highest in till over the granite component of the Mount Peyton complex. Glacial dispersal is illustrated by northward transport from the northern edge of the granite, and from low potassium tills derived from the gneiss component of the Mount Peyton batholith that are dispersed over the southwestern edge (cf. Batterson et al. 1998).

Au in till (Fig. 8) does not display discernable dispersal patterns. Instead, Au anomalies occur as single anomalous values or clusters overlying sedimentary rocks in the west part of the area, or over the Mount Peyton Batholith. Nevertheless, the release of the Au data in late 1998 resulted in the staking of 377 km^2 of new claims, several over geological terrains previously little explored, including the Mount Peyton batholith, where Au concentrations in till are up to 120 ppb.

Baie Verte

The Baie Verte Peninsula is an area of active Au exploration (Fig. 1). It is also an area affected by multiple ice flow events. This case study illustrates the use of clast fabric data to assist exploration activities by defining those ice flows with the greatest influence on sediment dispersal. The relationship between ice flow indicators, particularly striations, and sediment dispersal can be unclear, as striations are produced by glacial erosion, and till sampled in drift prospecting studies is produced by glacial erosion and deposition. Clast fabric analysis uses evidence from the same medium being sampled; reflects the ice flow conditions at the time the sediment was deposited, and allows the relationship between erosional evidence and dispersed sediment to be evaluated.

The use of this method in exploration is illustrated with two examples from the southern Baie Verte Peninsula (Fig. 6), an area where two ice flow directions separated by c. 90° are possible agents of dispersal and must be considered in interpretation of geochemical anomalies (Fig. 6). In the first example, clast fabric analysis from 44 exposures allows the dominant sediment transport event to be identified. Exposures in the area show a sandy diamicton, containing dominantly local clasts, some of which are striated. The diamicton generally is structureless, but contains lenses and laminae of sand and gravel that are internally stratified, and lie horizontally, with no evidence of deformation. On the basis of sedimentary structures present, the genesis of most of the diamicton in the area is basal melt-out till (Halderson & Shaw 1982; Shaw 1982). Using the clast fabric criteria of strong fabrics having an $S_1 > 0.6$, only 12 out of 44 clast fabrics completed are considered to indicate ice flow direction. Considerable variation is found at a single site, with analyses at different depths showing different orientation patterns, or strengths of clast alignment, possibly due to post-depositional re-sedimentation of the diamicton. Of the twelve sites, only one does not show a mean orientation within 30° of at least one striation found within 3 km of the sample site. At several sites, local ice flow directions, as indicated by striations, diverge by over 90°. In contrast, clast fabric analysis is successful in identifying the ice flow direction to the northeast and it is this orientation that should be used in tracing geochemical anomalies, or mineralized clasts up-ice to their source.

Table 2. *Data from thirteen clast fabrics measured from diamicton exposed in exploration trenches NW of Flatwater Pond, Baie Verte Peninsula*

Site	S_1	Mean orientation (direction°/dip°)
88-26	0.80	151/07
88-27	0.63	240/14
88-28a	0.60	322/00
88-28b	0.84	150/18
88-28c	0.92	140/07
88-28d	0.82	340/09
88-60	0.52	305/07
88-61	0.89	333/16
88-62	0.54	055/13
88-63a	0.48	261/25
88-63b	0.72	356/22
88-64	0.67	002/02
88-65	0.85	096/09

In the second example, 13 fabrics were measured from diamicton exposed in exploration trenches northwest of Flatwater Pond (Fig. 1, Table 2). Two striation sites in the Flatwater Pond vicinity show evidence of: (1) an older ice flow between 325° and 350°; (2) subsequent ice flow at 310° to 320°; and (3) a younger ice flow at 290°. Four other sites in the area show single sets of striations oriented between 340° and 350° produced by the earliest ice flow (Liverman & St. Croix 1989). Diamicton exposed in the trenches has a sub-horizontal fissility but is otherwise structureless. It contains clasts that are mostly angular, and of local provenance. Pebbles to boulders constitute 60 to 80% of the diamicton,

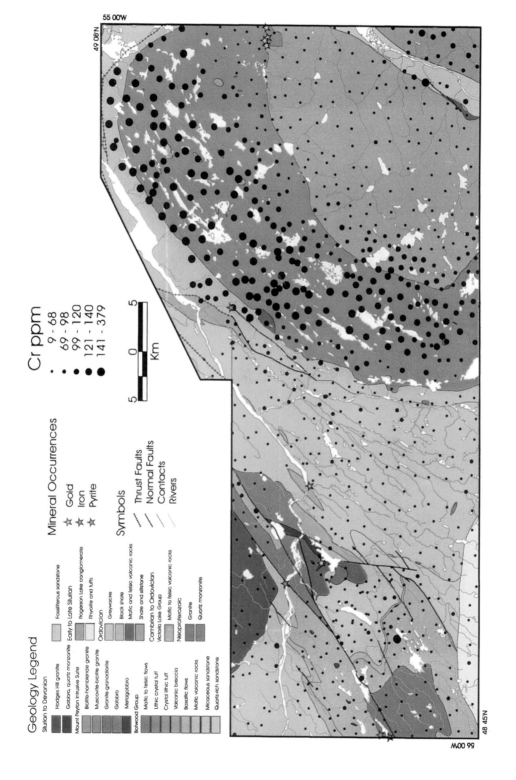

Fig. 7. Distribution of Cr in the <0.063 mm fraction of till, Grand Falls – Mount Peyton area (from Batterson et al. 1998). See Figure 6 for location.

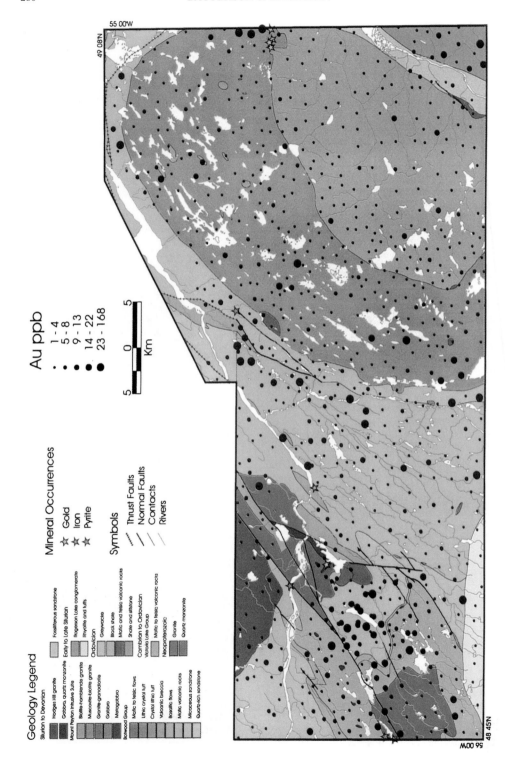

Fig. 8. Distribution of Au in the <0.063 mm fraction of till, Grand Falls – Mount Peyton area (from Batterson *et al.* 1998). See Figure 6 for location.

with sandy matrix forming the remaining 20 to 40%. The diamicton contains 'fragile' clasts of highly weathered serpentinite that have been 'smeared' along horizontal planes. Rarely, clasts are aligned along distinct sub-horizontal lines. Resistant clasts of volcanic rocks are commonly striated. Diamicton thickness is variable and controlled by bedrock topography. In topographical lows, it is up to 2.5 m thick, but is < 1 m over bedrock highs.

The results of fabric analysis of the diamicton at 13 sites are variable (Table 1). Seven sites show strong to very strong unimodal fabrics (S_1 values between 0.72 and 0.92). Such fabrics are considered typical of basal tills (Shaw 1982, 1987; Dowdeswell & Sharp 1986). The absence of sorted strata or lenses, the dominance of clasts with a local provenance material, and the presence of fissility and 'smeared' clasts suggests that the diamicton is a lodgement till (Kruger 1979; Muller 1983). Six of these seven sites show a mean orientation suggesting ice flow to be between 320° and 356°. Thus it is likely that the dominant sediment-moving ice flow was towards approximately 330°. This direction is similar to the dominant striation direction observed in the surrounding area, and the earliest flow recorded in multiple striation sites.

Since the flow direction suggested by the fabric analysis is similar to the earliest recorded in the area (as deduced from striations), it is possible that the variability in fabric orientations is due to the re-orientation of clasts by subsequent ice flows. This would produce moderately-oriented fabrics, with a secondary mode to the west of the primary mode. This is not the case here. Alternatively, Dowdeswell & Sharp (1986) suggested that typical lodgement tills are composed of two layers, with the upper layer displaying moderately- to poorly-oriented fabrics. In the Flatwater Pond example, where two fabrics were taken at different vertical positions at the same site (88-28 and 88-63), the strongest clast orientations were found at the higher site. The diversity found is more likely to be due to mass movement rather than either primary depositional processes or re-orientation by later ice flow. Relief on the bedrock surface is considerable (30-40 m), and slopes are steep (up to 20°, generally dipping to the east). Mass movement of saturated sediment down the bedrock surface could take place either at the base of the glacier, or post-glacially. The vertical variation in fabric strength and orientation is therefore due to a series of minor mass movements taking place as cohesive debris flows, with most movement taking place along a zone of dislocation at the base of the sediment. The upper parts of the diamicton would be transported in-tact relatively undisturbed, but clasts at the base would be re-oriented. Sediment at sites 88-27, 88-28A, 88-60, 88-62, and 88-63A have relatively weak girdle fabrics, typical of debris flow diamictons (Nardin *et al.* 1979; Lawson 1982), suggesting that at these sites, the diamicton is re-sedimented secondary tills.

These examples demonstrate that clast fabric analysis is a useful tool for drift prospecting areas of complex ice flow. Clearly, clast orientation measurements from one site are not sufficient to define ice flow in an area. Instead, multiple site measurements are useful in determining the directions of glacial transport. It is also important to use appropriate statistical criteria in evaluating the results of fabric analysis. Clast fabric analysis is also useful in determining the genesis of glacigenic diamictons, and thus identifying appropriate sampling media.

The use of clast lithology has proven to be an important tool in defining glacial dispersal on the Baie Verte Peninsula. As outlined above, ice flow history in the Baie Verte area is complex, with widely divergent flows identified. In attempting to apply this method, only sites with strong unimodal clast fabrics paralleling local striation measurements were used. The complex bedrock geology of the area makes precise determination of origin of a given clast difficult. Many of the units contain similar rock types and precise lithological identification is difficult from a pebble-sized clast. When using this method, a very simplified interpretation of the bedrock geology, based on the dominant rock type within a group or formation is used. In some cases there is more than one potential source area for a given rock type, so the nearest source up-ice along the projected ice flow vector was used.

Results for one site on the southern Baie Verte Peninsula are plotted in Figure 9. Similar plots for other sites in the area provide half-distance transport distances for pebbles derived by this method in the range of 0.7 to 2.9 km. The effects of reworking are ignored, but may well be significant. Calculation of half-distance assumes that all granitic clasts are derived from bedrock at least 8 km west of the site. However, earlier northward ice flow also would have dispersed large volumes of granitic rocks from extensive bedrock exposures 1 to 3 km south to southeast of the site, and these granitic rocks to the south and southeast may well be the source of granitic clasts in diamicton. If reworking is considered, the effect is generally to decrease the estimates of half-distances.

Estimates of half-distance values were made

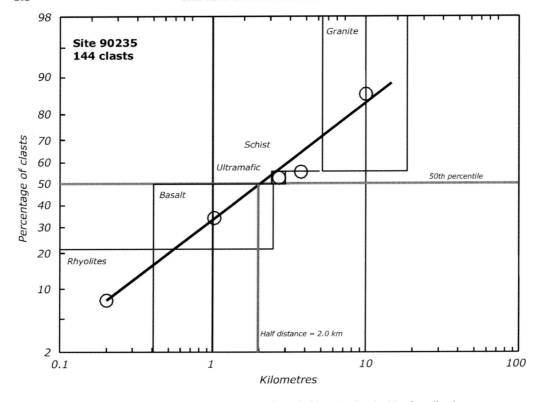

Fig. 9. Mean transport distances for clasts at a site on the Baie Verte Peninsula, Newfoundland.

on sites in the Springdale area, southeast of the Baie Verte Peninsula (Fig. 1), where in most cases the distribution of only two rock types could be traced. The fitting of a straight line to two points makes considerable assumptions concerning the distribution of the data, but results for six sites range between 5.8 and 8.8 km, apart from a single site at 2.8 km. Despite the limitations of the data for this area, these results suggest that transport distances for pebbles are generally less for the southern Baie Verte Peninsula as compared to the Springdale area. Thus, geochemical anomalies in glacial diamictons are likely to be closer to a bedrock source, as are mineralized boulders.

Estimates of half-distance value for the two areas discussed above were also made by choosing distinctive rock types that could be related to an up-ice source area. Ultramafic rocks were chosen for the southern Baie Verte Peninsula, and Topsails granite for the Springdale area. For individual sample sites within each area, the distance up-ice to the proximal contact of the chosen rock type was estimated, and plots made of percentage of that rock type v. distance from bedrock source (Fig. 10). Simple linear regression was used to fit straight lines to the data. For both areas investigated, the data are 'noisy', but F-tests suggest that the regression analyses are significant. Once the relationship between distance from source and concentration is established, then the half distance can be calculated as follows. The intercept on the Y-axis is the concentration when distance is zero (or logarithm of the concentration in log-linear plots). Half this value can thus be substituted into the equation of the line to estimate the half-distance. The mean half-distance estimates for the southern part of the Baie Verte Peninsula are 6.0 km, and 17 km for the Springdale area. Bouchard & Salonen (1990) suggest that the areal size of the bedrock unit is a major influence on half-distance, and in this case the source area underlain by granite covers a much greater area than that ultramafic rock. However, results generated by this method suggest that there is a significant difference in glacial transport distance between the Baie Verte Peninsula and Springdale.

The differences in transport distance estimates

(a)

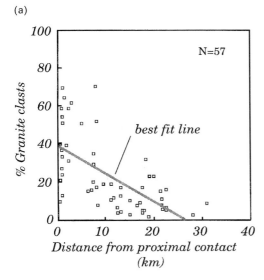

(b)

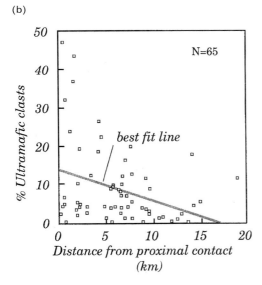

Fig. 10. Graphs showing distance of selected rock types in till to the proximal bedrock source for sites on the (a) Springdale area and (b) Baie Verte Peninsula.

can be explained in terms of proximity to late-glacial ice divides. The ice flow history for most of the Springdale area is comparatively simple, and consistent throughout the last glaciation (Fig. 6). Ice flowed from a major ice accumulation centre in The Topsails, 25 km to the southwest. The southern Baie Verte Peninsula has a complex ice flow history, and the transport directions identified for sites are related to late, deglacial flow events, caused by flow from a small, local ice cap (Fig. 6). The maximum transport distance can be no more than the distance from the site being investigated to the centre of the ice cap. In the case of the southern Baie Verte Peninsula, this is in the order of 15 to 25 km, whereas in the Springdale area, sites are located 20 to 50 km from the centre of the likely source of ice. Thus average glacial transport distances will be greater if sediment is transported by ice flow from a major ice centre, and the site is relatively distant from the accumulation centre.

These results imply that in areas dominated by local ice flow and short transport distances the till will contain a greater proportion of local material and dispersal trains from mineralized bedrock will be easier to identify and trace. Greater glacial transport distances will produce longer, but less well defined, dispersal trains.

Discussion

The adoption of a multi-faceted approach to the collection of geochemical data has proven useful in defining glacial dispersal in Newfoundland and Labrador. A combination of clast identification, striation mapping, surficial mapping and till geochemistry is employed by the GSNL. Clast lithology analysis rarely focuses on mineralized material, but instead uses visually distinctive rock types with a discrete source area to define general patterns of dispersal. Surficial geological mapping is essential in guiding sampling for geochemistry, and to the interpretation of geochemical data. Striation mapping also provides exploration companies with ice flow data that should be used when sampling till and interpreting results.

Newfoundland has a distinct glacial history from that of Labrador. The Island was covered by a series of coalescing ice caps during the Late Wisconsinan, with only limited coverage by Laurentide ice. Here, glacial dispersal is short, and dispersal trains are diffuse, such as those described from the Baie Verte Peninsula and from Grand Falls. Labrador was covered by the Labrador sector of Laurentide Ice Sheet, with a dispersal centre in western Labrador. Dispersal patterns have an amoeboid shape in the Labrador Trough area, shown by a series of dispersal trains trending in several directions, commonly of limited extent (Klassen & Thompson 1993). Away from the ice dispersal centres dispersal trains are longer, better developed and, ribbon-shaped (e.g. Strange Lake) or fan-shaped (e.g. Central Mineral Belt). Other factors contribute to the length and definition of dispersal trains. The Strange Lake dispersal train, for instance, is narrow, ribbon-shaped, and well-defined for more than 40 km. This results from a fortuitous combination of unique geology, topography, and an simple glacial transport history. Dispersal trains on the Island of Newfoundland are commonly less well-defined, geochemically, due to chemical similarities in adjacent rock units.

Recent releases of geochemical data in central Newfoundland have shown that geochemical sampling on a regional basis, taking into account the Quaternary history, provide vital base-line data for use by mineral exploration companies. Results from government survey activities also point to successful strategies for explorationists working at the property scale. In particular, the results emphasize the importance of understanding local surficial geology before planning a soil sampling programme, the usefulness of striation mapping in boulder tracing, and the importance of clast fabric studies in deciphering the source of anomalies in areas of complex ice-flow. The methods described require some expertise, but cost little and greatly increase the effectiveness of drift prospecting in areas of drift cover.

Numerous student assistants contributed to the case studies presented in this paper, and some of the ideas generated have come from discussions with many colleagues past and present. T. Sears is thanked for drafting most of the diagrams, and D. Taylor contributed diagrams generated using ArcView. An earlier version of this paper was prepared for the 1999 Association of Exploration Geochemists workshop on Drift Prospecting in Glaciated Terrain held in Vancouver, Canada. The manuscript was improved considerably through reviews by R. R. Stea and E. Nielsen. Thanks also to M. B. McClenaghan for her editorial skills.

References

BATTERSON, M. J. 1989a. *Quaternary geology and glacial dispersal in the Strange Lake area, Labrador.* Newfoundland Department of Mines and Energy, Geological Survey Branch Report **89-3**.

BATTERSON, M. J. 1989b. Glacial dispersal from the Strange Lake alkalic complex, northern Labrador. *In:* DILABIO, R. N. W. & COKER W. B. (eds) *Drift Prospecting.* Geological Survey of Canada, Paper **89-20**, 31–40.

BATTERSON, M. J. 1994. *Ice flow indicators, Pasadena map sheet, 12H/04.* Newfoundland Department of Mines and Energy, Geological Survey Branch Open file **12H/1284**.

BATTERSON, M. J. 1998. *Quaternary history, palaeogeography and sedimentology of the Humber River basin and adjacent areas.* PhD Thesis, Memorial University of Newfoundland.

BATTERSON, M. J. & VATCHER, S. V. 1991. Quaternary geology of the Gander (NTS 2D/15) map area. *In:* *Current Research.* Geological Survey, Department of Mines and Energy, Report **91-1**, 1–12.

BATTERSON, M. J., SCOTT, S. & SIMPSON, A. 1987. Quaternary mapping and drift exploration in the eastern part of the Central Mineral Belt, Labrador. *In:* HYDE, R. S., WALSH, D. G. & BLACKWOOD, R. F. (eds) *Current Research.* Mineral Development Division, Newfoundland Department of Mines and Energy, Report **87-1**, 1–9.

BATTERSON, M. J., SIMPSON, A. & SCOTT, S. 1988. Quaternary mapping and drift exploration in the Central Mineral Belt (13K/7 and 13K/10), Labrador. *In:* HYDE, R. S., WALSH, D. G. & BLACKWOOD, R. F. (eds) *Current Research.* Mineral Development Division, Newfoundland Department of Mines and Energy Report **88-1**, 331–341.

BATTERSON, M. J., TAYLOR, D. M. & DAVENPORT, P. H., 1998. *Till Geochemistry of the Grand Falls – Mount Peyton area (NTS 2D/13, 2D/14 and part of 2E/03).* Geological Survey of Newfoundland and Labrador, Department of Mines and Energy Open File NFLD **2664**.

BOUCHARD, M. A. & SALONEN, V.-P. 1990. Boulder transport in shield areas. *In:* KUJANSUU, R. & SAARNISTO, M. (eds) *Glacial Indicator Tracing.* A. A. Balkema, Rotterdam, 87–107.

BROOKES, I. A., MCANDREWS, J. H. & VON BITTER, P. H. 1982. Quaternary interglacial and associated deposits in southwest Newfoundland. *Canadian Journal of Earth Sciences*, **19**, 410–423.

CATTO, N. R. 1998. The pattern of glaciation on the Avalon Peninsula of Newfoundland. *Géographie physique et Quaternaire*, **52**, 23–45.

DICKSON, W. L. 1992. *Geology of the Mount Peyton map area, central Newfoundland (NTS 2D/14).* Department of Mines and Energy, Geological Survey Branch, Open File **002D/14/0273**. Map **92-022**.

DOWDESWELL, J. A. & SHARP, M. J. 1986. Characterization of pebble fabrics in modern terrestrial glacigenic sediments. *Sedimentology*, **33**, 699–710.

GILLBERG, G. 1965. Till distribution and ice movements on the northern slopes of the south Swedish highlands. *Geologiska Foreningens I Stockholm Förhandlingar*, **86**, 433–484.

GRANT, D. R., 1974. Prospecting in Newfoundland and the theory of multiple shrinking ice caps. *In:* *Report of Activities.* Geological Survey of Canada, Paper **74-1B**, 215–216.

GRANT, D. R. 1987. *Quaternary geology of Nova Scotia and Newfoundland (including Magdalen Islands).* International Union for Quaternary Research, XII INQUA Congress, Ottawa, Excursion Guidebook A-3/C-3, National Research Council of Canada, Publication **27525**.

GRANT, D. R. 1989. Quaternary geology of the Atlantic Appalachian region of Canada. *In:* FULTON, R. J. (ed.) *Quaternary Geology of Canada and Greenland.* Geological Survey of Canada, Geology of Canada, **1**, 391–440.

HALDORSEN, S. & SHAW, J. 1982. The problem of recognising melt-out till. *Boreas*, **11**, 261–277.

IVERSON, N. R. 1991. Morphology of glacial striae: Implications for abrasion of glacier beds and fault surfaces. *Geological Society of America Bulletin*, **103**, 1308–1316.

JAMES, L. D. & PERKINS, E. W. 1981. Glacial dispersion from sulphide mineralisation, Buchans area, Newfoundland. *In:* SWANSON, E. A., STRONG, D. F. & THURLOW, J. G. (eds) *The Buchans orebodies: Fifty years of geology and mining.* Geological Association of Canada, Special Paper **22**, 269–283.

KEAN, B. F. & MERCER, N. L. 1981. *Grand Falls, Newfoundland.* Department of Mines and Energy, Mineral Development Division Map **81-099**.

KLASSEN, R. A. & MURTON, J. B. 1996. Quaternary geology of the Buchans area, Newfoundland: Implications for mineral exploration. *Canadian Journal of Earth Sciences*, **33**, 363–377.

KLASSEN, R. A. & THOMPSON, F. J. 1993. *Glacial history, drift composition, and mineral exploration, central Labrador.* Geological Survey of Canada Bulletin **435**.

KRUGER, J. 1979. Structures and textures in till indicating subglacial deposition. *Boreas*, **8**, 323–340.

KRUMBEIN, W. C. 1937. Sediments and exponential curves. *Journal of Geology*, **45**, 577–601.

LAWSON, D. E. 1982. Mobilisation, movement, and deposition of active sub-aerial sediment flows, Mantanuska Glacier, Alaska. *Journal of Geology*, **90**, 279–300.

LIVERMAN, D. G. E. 1992. Application of Regional Quaternary Mapping to Mineral Exploration, Northeastern Newfoundland, Canada. *Transactions of the Institution of Mining and Metallurgy Section B – Applied Earth Science*, **101**, B89-B98.

LIVERMAN, D. G. E. 1994. Relative sea-level history and isostatic rebound in Newfoundland, Canada. *Boreas*, **23**, 217–230.

LIVERMAN, D. G. E. & ST. CROIX, L., 1989. *Ice flow indicators on the Baie Verte Peninsula.* Newfoundland Department of Mines and Energy Open File Map **89-36**.

MACEACHRAN, D. B. 1990. *Stereo,© the sterographic projection software program for the Apple Macintosh® computer.* Distributed by Rockware Inc, Wheat Ridge, Colorado, USA.

MARK, D. M. 1973. Analysis of axial orientation data, including till fabrics. *Bulletin of the Geological Society of America*, **84**, 1369–1374.

MARK, D. M. 1974. On the interpretation of till fabrics. *Geology*, **2**, 101–104.

MARTIN, W., 1983. *Once Upon a Mine: Story of Pre-Confederation Mines on the Island of Newfoundland.* Canadian Institute of Mining and Metallurgy, Special Volume **26**.

MCCONNELL, J. W. & BATTERSON, M. J. 1987. The Strange Lake Zr–Y–Nb–Be–REE deposit, Labrador: A geochemical profile in till, lake and stream sediment and water. *Journal of Geochemical Exploration*, **29**, 105–127.

MILLER, R. R. 1986. Geology of the Strange Lake Alkalic Complex and the associated Zr–Y–Nb–Be–REE mineralization. *In:* BLACKWOOD, R. F., WALSH, D. G. & GIBBONS, R. V. (eds) *Current Research.* Newfoundland Department of Mines and Energy, Mineral Development Division, Report **86-1**, 11–20.

MULLER, E. H. 1983. Dewatering during lodgement of till. *In:* EVENSON, E. B., SCHLUCHTER CH. & RABASSA, J. (eds) *Tills and Related Deposits.* A. A. Balkema, Rotterdam, 13–18.

NARDIN, T. R., HEIN, F. J., GORSLINE, D. S. & EDWARDS, B. D. 1979. A review of mass movement processes, sediment and acoustic characteristics, and contrasts in slope and base-of-slope systems versus canyon-fan-basin systems. *In:* DOYLE L. J. & PILKEY, JR., O. H. (eds). *Geology of Continental Slopes.* Society of Economic Paleontologists and Mineralogists, Special Publication, **27**, 2761–2773.

O'DONNELL, N. D. 1973. *Glacial indicator trains near Gullbridge, Newfoundland.* MSc Thesis, University of Western Ontario.

PARENT, M., PARADIS, S. J. & DOIRON, A. 1996. Palimpsest glacial dispersal trains and their significance for drift prospecting. *Journal of Exploration Geochemistry*, **56**, 123–140.

RYAN, A. G. 1985. *Regional geology of the central part of the Central Mineral Belt, Labrador.* Newfoundland Department of Mines and Energy, Mineral Development Division Memoir **3**.

ST. CROIX, L. & TAYLOR, D. M. 1991. Regional striation survey and deglacial history of the Notre Dame Bay area. *In:* PEREIRA, C. P. G., WALSH, D. G. & BLACKWOOD, R. F. (eds) *Current Research,* Newfoundland Department of Mines and Energy, Geological Survey of Newfoundland, Report **91-1**, 61–68.

SALONEN, V. P. 1986. *Glacial transport distances of surface boulders in Finland.* Geological Survey of Finland Bulletin **338**.

SCOTT, S. 1994. Surficial geology and drift exploration of Comfort Cove - Newstead and Gander River map areas (NTS 2E/7 and 2E/2). *In:* PEREIRA, C. P. G., WALSH, D. G. & BLACKWOOD, R. F. (eds) *Current Research.* Geological Survey, Department of Mines and Energy, Report **94-1**, 29–42.

SHAW, J. 1982. Meltout till in the Edmonton area, Alberta, Canada. *Canadian Journal of Earth Sciences*, **19**, 1548–1569.

SHAW, J. 1987. Glacial sedimentary processes and environmental reconstruction based on lithofacies. *Sedimentology*, **34**, 103–116.

TAYLOR, D. M., ST. CROIX, L. & VATCHER, S. V. 1994. *Newfoundland Striation Database.* Newfoundland Department of Mines and Energy, Geological Survey Branch Open File Nfld **2195** (version 3).

WALLACE, B. L. 1990. L'anse aux Meadows: Gateway to Vinland. *In: The Norse of the North Atlantic. Acta Archaeologica*, **61**, 166–197.

WOODCOCK, N. H. 1977. Specification of clast fabric shapes using an eigenvalue method. *Bulletin of the Geological Society of America*, **88**, 1231–1236.

The glacial transport and physical partitioning of mercury and gold in till: implications for mineral exploration with examples from central British Columbia, Canada

A. PLOUFFE

Geological Survey of Canada, 601 Booth Street, Ottawa, Ontario K1A 0E8, Canada
(e-mail: aplouffe@nrcan.gc.ca)

Abstract: Mercury glacial dispersal was measured in the clay-sized fraction (<0.002 mm) and heavy mineral concentrate (0.063–0.250 mm, specific gravity >3.3 g/cm^3) of till in a region of bedrock cinnabar occurrences, in central British Columbia, Canada. Most of the Hg in till occurs as sand-sized cinnabar (HgS) grains. A longer dispersal train was measured with the heavy mineral concentrates because Hg concentrations in heavy minerals yielded a higher ratio between anomalous and background concentrations when compared to the clay-sized material. It is proposed that geochemical or mineralogical analyses on a specific grain size fraction or density fraction of till, where the desired metal resides, result in a higher contrast between anomalous and background concentrations. Such a great contrast translates into a longer detectable dispersal train and hence, a larger target for mineral exploration. Therefore, in drift exploration programs, it is crucial to identify the mode of occurrence of a sought commodity in till; this can be achieved in part with a simple partitioning study whereby metal concentrations are measured in specific grain size fractions of till. Physical partitioning results for Au in the study area indicate that close to the bedrock source, large metal concentrations in some cases are present in the sand- (0.063–2 mm) and granule-sized (2–4 mm) fractions. Therefore, the significance of a regional Au anomaly, commonly defined in the silt plus clay-sized fraction of till could be evaluated by further determining the Au content of coarser size fractions (sand and granule).

As part of the Canada–British Columbia Agreement on Mineral Development (1991–1995), a regional till sampling program was completed in central British Columbia over two 1:250 000 scale NTS map sheets: 93 K and N (Fig. 1). The objectives of this survey were to: (1) provide data on till geochemistry in support of mineral exploration; (2) provide baseline information on surficial sediment geochemistry that could ultimately be used in environmental assessments; and, (3) develop drift prospecting methodologies for the Canadian Cordillera by studying and modelling glacial transport from known sources.

Preliminary results on regional till geochemistry obtained after the first field season (1991) indicated that the abundant known Hg occurrences in bedrock, located along one of the major faults of the region (Pinchi Fault), containing cinnabar (HgS) as the main mineral, were reflected by high levels of Hg in till. Although a wealth of information has been published on Hg in rocks and soils, little has been published on the glacial transport and

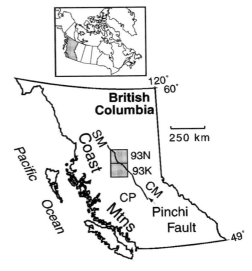

Fig. 1. Study area (shaded box) and Pinchi Fault locations. CM, Cariboo Mountains; CP, Chilcotin Plateau; SM, Skeena Mountains.

From: McClenaghan, M. B., Bobrowsky, P. T., Hall, G. E. M. & Cook, S. J. (eds) 2001. *Drift Exploration in Glaciated Terrain*. Geological Society, London, Special Publications, **185**, 287–299.
0305-8719/01/$15.00 © The Geological Society of London 2001.

mode of occurrence of Hg in till. To fill part of this gap, a detailed investigation of the glacial transport of Hg and its physical partitioning in till was undertaken. Results of this research deal dominantly with the environmental aspect of Hg distribution in the surficial environment (Plouffe 1995a, 1997a, 1998). This paper condenses and summarizes this work with respect to its implication for drift exploration. In addition, new data on the partitioning of Au in different grain size fractions of till are presented to emphasize the relationship between the significance of an isolated Au anomaly in a regional till geochemistry survey and the presence of Au in the sand- and granule-sized fractions.

Study area

The study area is located in central British Columbia, Canada and is defined by the extent of the 93 K and N map sheets (Fig. 1). It straddles the Omineca Mountains, the Nechako Plateau, and the Fraser Basin (Holland 1976) (Fig. 2). The Omineca Mountains consist of rugged mountainous terrain with cirques, peaks, and arêtes which contrast with the Nechako Plateau, characterized by rolling topography and rounded hills. The Fraser Basin is generally flatter and lower than the Nechako Plateau.

Bedrock geology

Central British Columbia is underlain by large crustal fragments (terranes) of island arc and oceanic crust affinities that were accreted to the North American Craton in the Mesozoic (Monger 1984; Gabrielse et al. 1991). Following accretion, dextral strike slip motion took place along major fault systems, such as the Pinchi Fault which extends over a distance of approximately 450 km in central British Columbia (Fig. 1). Within the study area, the Pinchi Fault separates dominantly oceanic sedimentary bedrock of the Cache Creek terrane to the west from dominantly volcanic rocks of Quesnellia to the east (Fig. 2). Plutonic rocks occur on both sides of the fault and porphyry mineralization has been found in association with some intrusions (e.g. Mt. Milligan, Delong et al. 1991).

As a result of deep circulation of meteoric water during the faulting event, low temperature Hg mineralization formed along Pinchi Fault (Nesbitt et al. 1987). During regional bedrock mapping of the area, several cinnabar (HgS) occurrences were detected in the bedrock (Armstrong 1949) which led to the discovery and exploitation of the Pinchi Lake and Bralorne Takla Hg deposits, the only two commercial Hg

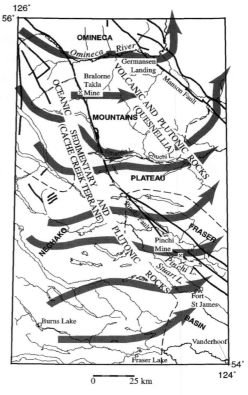

Fig. 2. Generalized ice-flow directions (grey arrows). Bold lines define the major faults of the region (Armstrong 1949; Bellefontaine et al. 1995), and the dashed lines mark the boundaries between physiographic divisions (modified from Plouffe 1995c).

mines which have been active in Canada (Fig. 2). Cinnabar mineralization at the Pinchi Lake mine occurs in brecciated fault zones in limestone in association with stibnite (Sb_2S_3), pyrite (FeS_2), marcasite (FeS_2) and chalcopyrite ($CuFeS_2$) (Freeze 1942; Armstrong 1949).

Glacial geology

During the Late Wisconsinan glaciation, ice lobes advanced over the plateau area of central British Columbia from accumulation zones located in the Coast, Skeena and Cariboo mountains (Tipper 1971; Plouffe 1991, 1995b). Because the Coast and Skeena mountains were the principal source of ice to the area, regional ice flow was predominantly east to southeast (Fig. 2). During ice build-up, minor local fluctuations in ice flow took place because of topographical influences. In the eastern sector of the study area, at the climax of glaciation, ice

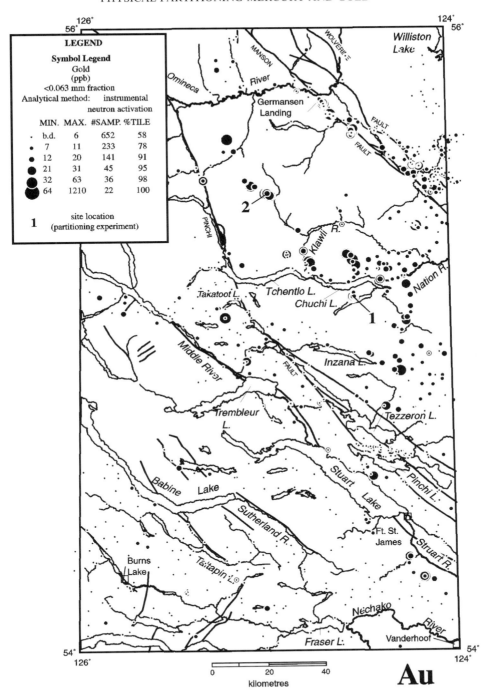

Fig. 3. Au content of the silt plus clay-sized fraction of till (modified from Plouffe 1995*b*). Thicker lines represent faults (b.d., below detection limit).

was deflected to the northeast due to pressure exerted by an ice lobe flowing north from the Cariboo Mountains (Fig. 2) (Tipper 1971; Plouffe 1991, 1995b). Till is the most abundant surficial sediment within the study area. In places, till is overlain by glaciofluvial sand and gravel, well-sorted fine sand, silt and clay of glaciolacustrine origin, and sporadic organic deposits. Till sampling was hindered in areas where the overlying sediments are thick.

Methods

Till samples were collected from road side sections and hand-dug pits at a minimum depth of 1 m, well below the zone of intense soil weathering. For the regional till sampling survey, the spacing between sample sites averaged 5 km along roads. The sample spacing was reduced to 1 km in the vicinity of Pinchi Lake mine to define the extent of the Hg dispersal train. In addition, property-scale sampling (five to ten samples per km^2) was completed near selected areas of bedrock Au mineralization: south of Chuchi Lake and north of Tchentlo Lake (sites 1 and 2, respectively; Fig. 3). For practical reasons, till sampling was dominantly limited to areas accessible by road.

For the regional till geochemistry study, Hg and Sb analyses were conducted on the clay-sized fraction (<0.002 mm) of till samples because Hg and other metals such as Sb have a tendency to be adsorbed onto clay particles and phyllosilicates (Jonasson 1970; Shilts 1975, 1984, 1995; Nikkarinen et al. 1984). In the case of Au, analyses were performed on the silt plus clay-sized fraction (<0.063 mm or −250 mesh) because in addition to its presence in the clay-sized fraction, Au can occur abundantly in the silt-sized material of till near Au bedrock mineralization (DiLabio 1985, 1988, 1995; Coker et al. 1988; Coker & Shilts 1991; Sibbick & Fletcher 1993; Delaney & Fletcher 1995; Sibbick & Kerr 1995). Restricting geochemical analyses to a particular size fraction, as opposed to analysing an entire bulk till sample, reduces the effect of textural variation on geochemistry and allows for better comparison between regions (Shilts 1975). Separations of the clay- and silt plus clay-sized fractions were done following procedures outlined in Lindsay & Shilts (1995). For the Pinchi Lake mine site only, geochemical analyses and visual cinnabar grain counts were conducted on non-magnetic heavy mineral concentrates (0.063–0.250 mm), which were prepared from a 5 kg sample using a shaker table and methylene iodine (specific gravity = 3.3 g/cm^3). Cinnabar grains were identified by their disctinct red colour and were counted using a binocular microscope.

For the partitioning study, till samples (<4 mm) were separated into six different grain size fractions as described in Plouffe (1997a). All size fractions coarser than 0.063 mm, including heavy mineral concentrates, were pulverized to <0.063 mm prior to their geochemical analyses.

Mercury determinations were done by cold-vapour atomic absorption spectrometry preceded by a hot aqua-regia digestion. This analytical method has a detection limit of 10 ppb. Sb content was determined by inductively coupled plasma-atomic emission spectrometry (detection limit of 2 ppm) following an aqua-regia digestion. Gold analyses were performed by instrumental neutron activation on 30 g samples, with a detection limit of 2 ppb. Analytical precision at the 95% confidence level computed following Garrett's (1969) method are: ±8% for Hg, ±35% for Sb and ±39% for Au. Each pair of duplicate samples was prepared from the same field sample. The lower precision for Au and Sb analyses is ascribed to metal concentrations near detection limit. The low precision of Au analyses is also attributable to the heterogeneous distribution of Au particles, i.e. the nugget effect. Even with the lower analytical precision of Sb and Au analyses, the analytical error is smaller than the overall data variability.

Regional till geochemistry

Mercury

The highest Hg concentrations in till are found over and down-ice (generally to the east) from cinnabar occurrences located along the Pinchi Fault (Plouffe 1997b, p.224, fig. 3). Note, down-ice refers to down the direction of paleo-glacier flow. Within the study area, Hg concentrations in the clay-sized fraction of till average 180 ppb, except within 2 km down-ice from Pinchi and Manson faults where Hg levels range from 600 to 21 500 ppb.

The average Hg concentration in till is slightly more elevated over the volcanic bedrock on the east side of the Pinchi Fault (220 ppb) as opposed to the west side (140 ppb) which is underlain by oceanic sedimentary rocks. This difference is attributed to a combination of eastward glacial transport of debris enriched in Hg derived from cinnabar mineralization in bedrock along Pinchi Fault and, the potentially higher Hg background concentrations in the volcanic rocks (Plouffe 1998).

Gold

Gold concentrations do not display as strong a relationship with a bedrock lithology or structure as Hg, but rather show a more sporadic distribution of high levels (Fig. 3). Most of the high concentrations can be related to known bedrock mineralization, notably in the areas north and south of Chuchi Lake, c. 25 km N of Tchentlo Lake, and E of Inzana Lake. The high Au content of till in the Germansen Landing area is attributed to reworking of Au from placer deposits that underlie till of the last glaciation (Plouffe 1995*b*). If an anomaly threshold is arbitrarily set at the 90th percentile (20 ppb), several Au anomalies, unrelated to known mineral occurrences, are evident: one anomaly N of Trembleur Lake (30 ppb); one S of Takatoot Lake (47 ppb); one NE of Stuart Lake (21 ppb); and, one 10 km N of Tezzeron Lake (21 ppb) (Fig. 3). These anomalies could be of two types: (1) true anomalies related to proximal concealed bedrock mineralization or buried placer deposits reworked during the last glaciation; or (2) false anomalies related to the heterogeneous distribution of Au, i.e. the nugget effect. Clearly, the first anomaly type is the most attractive and significant from a mineral exploration perspective. A potential methodology to differentiate between both types of anomalies for a follow-up survey is suggested below.

Mercury dispersal train

The detailed sampling conducted in the Pinchi Lake mine region was hampered by the presence of thick deposits of glaciolacustrine and glaciofluvial sediments, and organic deposits overlying till (Plouffe 1994). In the Pinchi Lake mine region, ice predominantly flowed to the east with some local fluctuations (Plouffe 1994, 1997*b*). Glacial transport of Hg was identified by geochemical analyses of the clay-sized fraction and the heavy mineral concentrates, and by the quantification of cinnabar grains in the heavy mineral concentrates.

Clay-sized fraction geochemistry

Mercury concentrations in the clay-sized fraction of till are highest (21 500 ppb) in two samples collected 100 m down-ice from the mineralization at Pinchi Lake mine (Fig. 4a). Further east of these two sites, concentrations generally decrease. Using a threshold of 600 ppb, Plouffe (1998) estimated that the Hg dispersal train in the clay-sized fraction of till extends a distance of 12 km. Fluctuations in Hg concentrations within the dispersal train could be related to the heterogeneous distribution of Hg in till or the presence of concealed cinnabar mineralization along secondary faults (Armstrong 1966) that contributed Hg enriched debris into till. Plouffe (1998) suggested that the bedrock mineralization at Pinchi Lake mine is not the sole source of Hg in till and that the Hg dispersal train is most likely a 'cumulative dispersal train', i.e. a series of dispersal trains derived from more than one source.

Heavy mineral concentrates geochemistry

Higher Hg levels were detected in pulverized heavy mineral concentrates than in the clay-sized fraction (note the change in concentration units in Fig. 4a, b) because of the large abundance of cinnabar in the heavy mineral concentrates (cinnabar specific gravity = 8.09g cm^{-3}; Berry *et al.* 1983). In other words, light minerals with low Hg content which act as diluents were removed from the samples prior to analysis. As indicated previously, geochemical analyses of heavy mineral concentrates and cinnabar grain counts were only performed on the samples of the Pinchi Lake mine region. Hence, fewer samples are reported on Figure 4b and c than on Figure 4a. Because of the smaller number of heavy mineral samples, the threshold between background and anomalous concentrations for the clay and the heavy mineral concentrates are not the same percentile values.

Using a threshold of 10 ppm, Plouffe (1998) estimated that the Hg dispersal train defined by the heavy mineral concentrates has a minimum length of 24 km. As was interpreted for the clay-sized fraction, the dispersal train defined by the heavy mineral concentrates might be a 'cumulative dispersal train', with more than one bedrock source.

Cinnabar grain count

Using cinnabar grain counts normalized to the weight of the heavy mineral concentrates (to account for slight variations in the weight of the heavy mineral concentrates), a Hg dispersal train with a minimum length of 24 km was defined (Fig. 4c). This train should also be described as a 'cumulative dispersal train' (Plouffe 1998).

Antimony dispersal train

Stibnite (Sb_2S_3) occurs with cinnabar in Pinchi Lake deposit and its presence is reflected by high Sb concentrations in the clay-sized fraction of

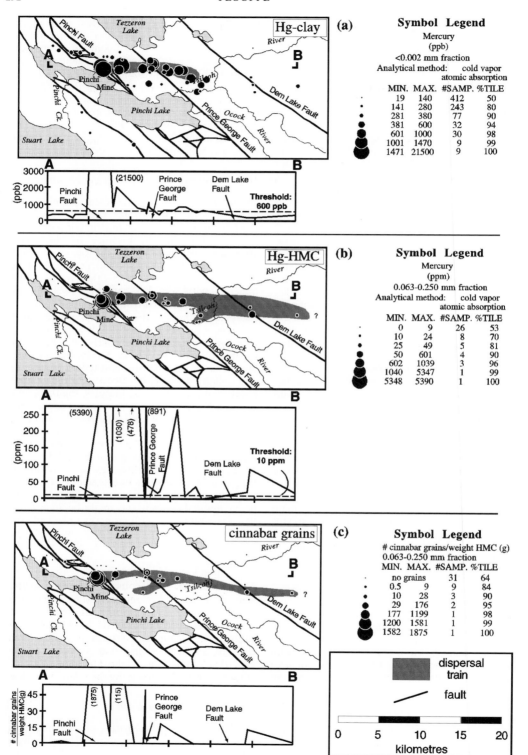

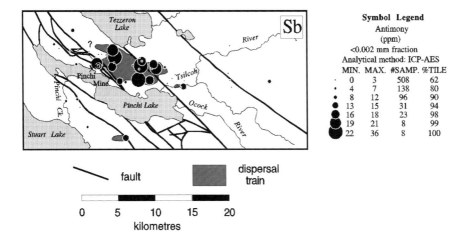

Fig. 5. Sb concentrations in the clay-sized fraction of till in the Pinchi Lake region. Percentiles in the legend were calculated using the regional data (modified from Plouffe 1998).

till down-ice from the mine (Fig. 5). However, the highest Sb concentrations occur to the east of the highest Hg concentrations. This distribution pattern for Sb could be related to: (1) the presence of concealed stibnite mineralization east of Pinchi Lake mine; or (2) the low analytical precision for Sb. Using a threshold of 14 ppm, the Sb dispersal train extends to a maximum length of 9 km E of Pinchi Lake mine (Plouffe 1998).

Factors controlling the length of Pinchi Lake mine dispersal trains

As illustrated here and other publications on the subject of glacial transport, the net effect of glaciation in the vicinity of bedrock mineralization generally results in dispersal trains which can have a much greater aerial expression than the bedrock source. Several factors control the length of dispersal trains including: (1) the selection of the threshold, (threshold being defined as the boundary between anomalous and background concentrations); (2) the position of the bedrock source with respect to the regional topography; (3) the hardness and structure (jointed v. massive) of the source material (i.e. the ease of glacial erosion); (4) the geochemical, lithological or mineralogical contrast between the bedrock source and the surrounding region; (5) the size and number of sources available for glacial erosion; (6) the size fraction and the mineralogical phase of the glacial sediment on which analyses are being conducted; and, (7) the physics of the glacier (e.g. flow rate, cold-based v. warm-based ice) (Shilts 1976, 1984, 1993; DiLabio & Coker 1989; Kujansuu & Saarnisto 1990).

The effects of some of these controlling factors can be illustrated with the Hg and Sb dispersal trains. Obviously, the dispersal trains presented in Figures 4 and 5 would change in shape and length by increasing or decreasing thresholds. The purpose of this paper is not to elaborate on a methodology for defining geochemical threshold, therefore the reader is referred to Plouffe (1998) for more details on how the thresholds were defined.

The mineralization at Pinchi Lake mine crops out on a ridge 160 m above the surrounding region (920 m asl). Therefore, eroded bedrock debris enriched in Hg could have been transported in a relatively high position within the glacier. This high position would have permitted a longer distance of glacial transport. In addition, glacial erosion of the Hg mineralization may have been facilitated by the relative softness

Fig. 4. (a) Hg concentrations in the clay-sized fraction of till; (b) Hg levels in heavy mineral concentrates of till (specific gravity $> 3.3 \text{g/cm}^3$) and 0.063–0.250 mm sized fraction); and, (c) number of cinnabar grains normalized to the weight of heavy mineral concentrates in till, in the vicinity of Pinchi Lake mine (modified from Plouffe 1998). Profiles were constructed using sample sites closest to the A–B transect.

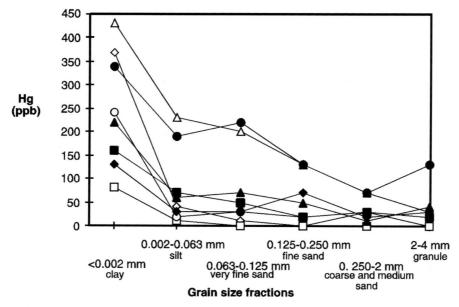

Fig. 6. Grain size v. Hg concentrations for eight till samples collected from areas located away (>20 km) from bedrock cinnabar mineralization (modified from Plouffe 1997b). See Plouffe (1997b) for the exact sample locations.

of the host rock, limestone. The greater length of the Hg dispersal train in the clay-sized fraction of till as compared to that for Sb could be related to: (1) greater geochemical contrast of Hg compared to Sb between the ore zone and the regional bedrock lithologies; and (2) the presence of secondary zones of cinnabar mineralization (potentially more abundant than stibnite mineralization) which might have produced a cumulative dispersal train. Similarly, the geochemical Hg dispersal train defined by heavy mineral concentrates is longer than the one defined by the clay-sized fraction because the ratio or contrast between the highest anomalous concentration and the background concentration is higher for heavy mineral concentrates than for the clay-sized fraction (539 times background in heavy mineral concentrates (5390/10) and 36 times background in clay (21500/600); Fig. 4a and b) (Plouffe 1998).

Physical partitioning of metals in till

Many studies have presented data on the partitioning of metals in different grain size fractions of till with the primary purpose of identifying the optimal grain size range for detecting metal enrichment in till (Shilts 1973, 1984, 1995; Nikkarinen *et al.* 1984; Sopuck *et al.* 1986; Coker *et al.* 1988, 1990; DiLabio 1988, 1995; Delaney & Fletcher 1993, 1995; Sibbick & Fletcher 1993; Cook & Fletcher 1994). In addition, these studies provided insight on the dominant mineral phases in which metals are being held in till.

Mercury

To better understand the nature of Hg in till, Plouffe (1997a) completed a partitioning experiment with till samples collected at sites both near and removed from cinnabar mineralization. Sample locations are reported in Plouffe (1997a). Results of this experiment are illustrated in Figures 6 and 7. For eight samples collected at sites at least 20 km down-ice from known cinnabar occurrences, the highest Hg concentrations are consistently present in the clay-sized fraction (<0.002 mm) which is attributed to the high adsorption capacity of clay particles for Hg (Fig. 6). Similar distributions among different grain size fractions in till have been well documented in the past for base metals, notably by Shilts (1984, p. 102, fig. 1). For four samples collected <2 km down-ice from known bedrock cinnabar occurrences, elevated Hg levels (>500 ppb) were detected in the clay-sized fraction, but the coarser grain size ranges, i.e. sand and granule (0.063–4 mm), yielded the highest Hg concentrations (Fig. 7).

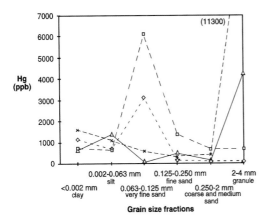

Fig. 7. Grain size v. Hg concentrations for four till samples collected a short distance (<2 km) down-ice from bedrock cinnabar occurrences located along the Pinchi Fault (modified from Plouffe 1997b). See Plouffe (1997b) for the exact sample locations.

The high Hg concentrations in the sand-sized fractions (0.063–2 mm) are attributed to the presence of sand-sized cinnabar grains derived from the mineralized bedrock (Plouffe 1998). The presence of high Hg concentrations in the granule-sized fraction (2–4 mm) is related to the presence of mineralized rock fragments. Preservation of cinnabar in near-surface till is attributed to the relatively greater stability of this mineral compared to other sulphides (Plouffe 1997a). For instance, in the upper (<0.4 m) oxidized portion of till profiles, cinnabar was found in association with goethite grains, some of which formed after the replacement of pyrite as indicated by their cubic shapes. Furthermore, because cinnabar is extremely insoluble, it is unlikely that the physical partitioning of Hg in different grain size fractions of till was affected by postglacial weathering. In other words, it is unlikely that cinnabar was substantially dissolved and re-precipitated into different grain size ranges in unoxidized till. Reproducibility of results in this partitioning experiment were optimal in samples with high Hg levels and in finer grain size fractions (<0.063 mm) (Plouffe 1997a).

The predominance of Hg in the sand-sized fraction of till near cinnabar occurrences (Fig. 7) is similar to the physical partitioning of other metals in till bound into stable minerals, which also show a large abundance in the sand-sized fraction: e.g. W in scheelite at Waverley, Nova Scotia (DiLabio 1988), lithophile elements in pyrochlore and zircon at the Strange Lake alkalic complex, Labrador (DiLabio 1995); and

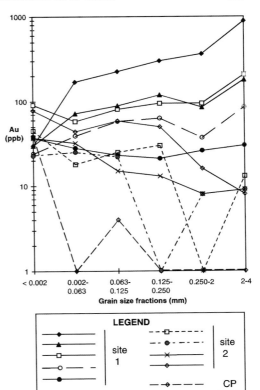

Fig. 8. Grain size v. Au concentrations for ten till samples collected near Au mineralization in bedrock. Note the log scale on the abscissa. Samples with Au concentrations below detection limit (2 ppb) are plotted at 1 ppb. Sample locations (sites 1 and 2) indicated on Figure 3; CP, Chilcotin Plateau located on Figure 1. The legend is only applicable to this figure, and not for Figures 6 and 7.

Cr in chromite near Thetford Mines (Shilts & Kettles 1990).

Gold

The partitioning of Au into different size fractions was tested with samples collected <200 m down-ice from Au mineralization in bedrock at three sites: S of Chuchi Lake ($n = 5$); 25 km N of Tchentlo Lake ($n = 4$); and, from the Fish Lake Au mineralization located on the Chilcotin Plateau ($n = 1$) S of the study area (Figs. 1 and 3). The partitioning diagram for Au is complex and does not show a constant trend with Au enrichment in a particular grain size fraction (Fig. 8). Also, there is a wide range of Au concentrations, hence the use of a logarithmic scale on the ordinate on Figure 8. For all

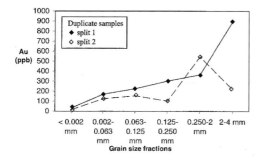

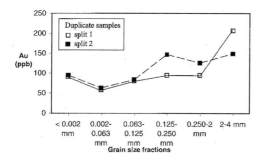

Fig. 9. Grain size v. Au concentrations of duplicate samples.

samples, concentrations > 20 ppb occur in either or both of the silt- and clay-sized fractions (< 0.002 and 0.002–0.063 mm). Several regional till geochemistry surveys in British Columbia have defined threshold Au concentrations below 20 ppb in the silt plus clay-sized fraction of till (e.g. Levson et al. 1994; Plouffe 1995b). Consequently, any one of these samples would have been considered anomalous in a regional survey. For eight samples out of ten, at least one coarser size fraction (> 0.063 mm) contains a Au concentration > 20 ppb (Fig. 8). Anomalous Au concentrations in more than one size fraction of till were noted in other studies for till collected in the immediate vicinity (less than a few hundred metres) of Au mineralization, and is attributed to the combined effects of: (1) the primary Au grain size at the bedrock or regolith source; (2) the size of re-precipitated Au in till; (3) the size of Au inclusions released following the weathering of sulphides; (4) the grain size range of glacially comminuted Au; and (5) Au adsorbed onto clay particles, oxides and hydroxides (DiLabio 1985, 1988, 1995; Coker et al. 1988; Coker & Shilts 1991; Sibbick & Fletcher 1993;

Delaney & Fletcher 1995; Sibbick & Kerr 1995). DiLabio (1985, 1988) and Sibbick & Fletcher (1993) have suggested that with increasing distance of glacial transport down-ice from Au mineralization, free Au particles and Au-bearing mineral(s) are comminuted to smaller grain size fractions. Similar to the physical partitioning patterns observed for Hg, the high Au concentrations in the coarser size fractions of till (sand and granule) near mineralization may indicate the immaturity of till with respect to its Au content. Au particles and mineralized clasts have not been glacially comminuted to finer size fractions because of the short distance of glacial transport.

The reproducibility of the partitioning study for Au was tested with two pairs of duplicate samples. Reproducibility is highest in < 0.125 mm material and decreases in the coarser size fractions (Fig. 9). This range in reproducibility likely reflects the heterogeneous distribution of Au in the coarser material (nugget effect). Using the principles presented by Clifton et al. (1969) and Harris (1982), analysing larger samples (> 30 g) would likely improve reproducibility.

Implications for mineral exploration

The presentation of the above data on the Hg dispersal train and physical partitioning in till is not intended to promote exploration for this commodity of decreasing value, but rather to demonstrate how an understanding of the mode of occurrence of a metal in till can lead to better mineral exploration practices.

As mentioned previously, the Hg dispersal train identified in the heavy mineral concentrates is longer than the one detected using the clay-sized fraction of till, because the ratio or contrast between the highest anomalous concentration and the background concentration is higher for heavy mineral concentrates than for the clay-sized fraction (Fig. 4). A longer dispersal train translates into a larger geochemical target which improves the chances of discovery in a mineral exploration program. Greater contrasts between anomalous and background concentrations are observed if the mineral counts or geochemical analyses are conducted on the size and density fraction of till where the sought commodity is dominantly present. Both metal concentrations and the anomaly:background ratio are increased by eliminating from analysis the metal poor size and density mineral fractions. In the case of Hg, heavy mineral concentrates yielded the longest dispersal train because Hg dominantly occurs in till as sand-sized cinnabar grains down-ice from cinnabar mineralization in bedrock, as indicated by the partitioning study and the microscopic observations (Plouffe 1997a).

A partitioning study, such as presented here, could easily be conducted at the onset of a property- or regional-scale exploration survey to identify the mode of occurrence of a sought commodity in till, as also proposed by DiLabio (1985). Orientation samples could be collected from the property where the exploration program is taking place or from nearby mineral occurrences similar to the exploration target.

In regional or reconnaissance till geochemical surveys in the Canadian Cordillera, sample spacing (generally on the order of 2–5 km) may exceed the length of most known Au dispersal trains (see examples and data in Plouffe 1995c; Plouffe & Jackson 1995; Levson & Giles 1997). Therefore, an isolated Au anomaly on a regional till geochemistry map could be significant from a mineral exploration point of view as it might be the only sample taken within a dispersal train and the sole indication of Au mineralization in bedrock. On the other hand, an isolated anomaly might be created by the well known 'nugget effect' (i.e. reflecting one or more Au particles heterogeneously distributed in the till and derived from a distal source or from a bedrock lithology with a Au content slightly above background level). One objective of a follow-up survey should be to differentiate between both types of anomalies. This could be achieved by re-sampling the area around the anomalous site with a dense sampling grid and conducting analyses on the same grain size fraction as the one being used in the regional survey (generally silt plus clay-sized fraction) to ensure comparability with the regional data. The size and significance of the geochemical anomaly in till can then be evaluated. In addition, knowing that high Au concentrations might be observed in the sand and granule-sized fractions of till near bedrock mineralization (Fig. 8), these size fractions could also be analysed prior to the follow-up survey. Such a practice might provide some additional information regarding the significance of the Au anomaly with little additional cost.

Conclusion

In proper drift exploration practices, an understanding of the mode of occurrence of the sought commodity in till is essential. Such information can then be used to remove metal-poor size and density fractions prior to analyses and consequently, to increase the ratio between anomalous and background metal concentrations. Data on the mode of occurrence of a sought commodity in till can partly be obtained by a simple physical partitioning test. For example, such testing confirmed that most of the Hg in till is present as sand-sized cinnabar grains down-ice from cinnabar occurrences in bedrock in the Pinchi Lake mine region of British Columbia. Consequently, the Hg dispersal train as determined from sand-sized heavy mineral concentrates is longer (24 km) than the one determined from the clay-sized fraction of till (12 km). A longer dispersal train translates into a larger exploration target which increases chances of discovery.

Physical partitioning of Au in till collected in the vicinity of Au mineralization in bedrock shows variable patterns (somewhat similar to Hg), with high Au levels in the sand and granule-sized fractions in most cases. The significance of a regional Au anomaly, as defined in the silt plus clay-sized fraction of till, could be evaluated by further determining the Au content of coarser size fractions (sand and granule) prior to conducting a follow-up survey in the field.

The first version of the manuscript was improved following the revision made by D. E. Kerr and M. B. McClenaghan. V. M. Levson and L. E. Jackson Jr. further improved the manuscript submitted to the Geological Society. Heavy mineral separations were completed by Overburden Drilling Management Ltd., Nepean, Ontario, and cinnabar grain counts by Consorminex, Gatineau, Québec. The author thanks Cominco Ltd. for providing access to their Pinchi Lake mine property. This research was funded by the Canada–British Columbia Agreement on Mineral Development and the Nechako NATMAP project. Geological Survey of Canada Contribution No. 1999193.

References

ARMSTRONG, J. E. 1949. *Fort St James map-area Cassiar and Coast District, British Columbia.* Geological Survey of Canada Memoir **252**.

ARMSTRONG, J. E. 1966. Tectonics and Hg deposits in British Columbia. *In:* GUNNING, H. C., WHITE, W. H., LITTLE, H. W., ARMSTRONG, J. E., YATES, R. G., NEY, C. S. & HOLLAND, S. S. (eds) *Tectonic history and mineral deposits of the western Cordillera.* Canadian Institute of Mining and Metallurgy, Montréal, 341–348.

BELLEFONTAINE, K., LEGUN, A. & MASSEY, N. 1995. *Northeast British Columbia mineral potential - south half: digital geological compilation (NTS 83D, E; 93F, G, H, I, J, K, N, O, P).* British Columbia Ministry of Energy, Mines and Petroleum Resources Open File **1995-24**.

BERRY, L. G., MASON, B. & DIETRICH, R. V. 1983.

Mineralogy. W.H. Freeman and Company, San Francisco.
CLIFTON, H. E., HUNTER, R. E., SWANSON, F. J. & PHILLIPS, R. L. 1969. *Sample size and meaningful gold analysis*. United States Geological Survey, Professional Paper **625-C**.
COKER, W. B. & SHILTS, W. W. 1991. Geochemical exploration for gold in glaciated terrain. *In:* FOSTER, R. P. (ed.) *Gold Metallogeny and Exploration*. Chapman and Hall, New York, 336–339.
COKER, W. B., SEXTON, A., LAWYER, I. & DUNCAN, D. 1988. Bedrock, till and soil geochemical signatures at the Beaver Dam gold deposit, Nova Scotia, Canada. *In:* MACDONALD, D. R. & MILLS, K. A. (eds) *Prospecting in Areas of Glaciated Terrain – 1988*. Canadian Institute of Mining and Metallurgy, Montréal, 241–254.
COKER, W. B., DUNN, C. E., HALL, G. E. M., RENCZ, A. N., DILABIO, R. N. W., SPIRITO, W. A. & CAMPBELL, J. E. 1990. A comparison of surficial geochemical methods of exploration for platinum group elements in Canada. *In:* BECK, L. S. & HARPER, C. T. (eds) *Modern Exploration Techniques*. Saskatchewan Geological Society, Regina, 41–53.
COOK, S. J. & FLETCHER, W. K. 1994. Platinum distribution in soil profiles of the Tulameen ultramafic complex, southern British Columbia. *Journal of Geochemical Exploration*, **51**, 161–191.
DELANEY, T. A. & FLETCHER, W. K. 1993. Size distribution of gold in some soils associated with selected gold mineralization in Canada and the United States of America. *Journal of Geochemical Exploration*, **48**, 309–327.
DELANEY, T. A. & FLETCHER, W. K. 1995. Sample representativity and gold determination in soils at the Fish Lake property, British Columbia. *In:* BOBROWSKY, P. T., SIBBICK, S. J., NEWELL, J. M. & MATYSEK, P. F. (eds) *Drift Exploration in the Canadian Cordillera*. British Columbia Ministry of Energy, Mines and Petroleum Resources, Paper **1995-2**, 135–138.
DELONG, R. C., GODWIN, C. I. & REBAGLIATI, C. M. 1991. Patterns of alteration and mineralization at the Mt. Milligan deposit. *Canadian Mining and Metallurgy Bulletin*, **84**, 88.
DILABIO, R. N. W. 1985. Gold abundances vs. grain size in weathered and unweathered till. *Current Research, Part A*. Geological Survey of Canada, Paper **85-1A**, 117–122.
DILABIO, R. N. W. 1988. Residence sites of gold, PGE and rare lithophile elements in till. *In:* MACDONALD, D. R. & MILLS, K. A. (eds) *Prospecting in Areas of Glaciated Terrain-1988*. The Canadian Institute of Mining and Metallurgy, Montréal, 121–140.
DILABIO, R. N. W. 1995. Residence sites of trace elements in oxidized till. *In:* BOBROWSKY, P. T., SIBBICK, S. J., NEWELL, J. M. & MATYSEK, P. F. (eds) *Drift Exploration in the Canadian Cordillera*. British Columbia Ministry of Energy, Mines and Petroleum Resources, Paper **1995-2**, Victoria, 139–148.
DILABIO, R. N. W. & COKER, W. B. (eds) 1989. *Drift Prospecting*. Geological Survey of Canada Paper **89-20**.
FREEZE, A. E. 1942. *Geology of Pinchi Lake, British Columbia*. PhD Thesis, Princeton University.
GARRETT, R. G. 1969. The determination of sampling and analytical errors in exploration geochemistry. *Economic Geology*, **64**, 568–574.
GABRIELSE, H., MONGER, J. W. H., WHEELER, J. O. & YORATH, C. J. 1991. Tectonic framework – part A. Morphogeological belts, tectonic assemblages and terranes. *In:* GABRIELSE, H. & YORATH, C. J. (eds) *Geology of the Cordilleran Orogen in Canada*. Geological Survey of Canada, Ottawa, **4**, 15–28.
HARRIS, J. F. 1982. Sampling and analytical requirements for effective use of geochemistry in exploration for gold. *In:* LEVINSON, A. A. (ed.) *Precious Metals in the Northern Cordillera*. Association of Exploration Geochemists, Rexdale, Ontario, 53–67.
HOLLAND, S. S. 1976. *Landforms of British Columbia, a physiographic outline*. British Columbia Department of Mines and Petroleum Resources Bulletin **48**.
JONASSON, I. R. 1970. *Mercury in the natural environment: a review of recent work*. Geological Survey of Canada Paper **70-57**.
KUJANSUU, R. & SAARNISTO, M. (eds). 1990. *Glacial Indicator Tracing*. A. A. Balkema, Rotterdam.
LEVSON, V. M. & GILES, T. R. 1997. Quaternary geology and till geochemistry studies in the Nechako and Fraser plateaus, central British Columbia (93C/1, 8, 9, 10; F/2, 3, 7; L/16; M/1). *In:* DIAKOW, L. J. & NEWELL, J. M. (eds) *Interior Plateau Geoscience Project: Summary of Geological, Geochemical and Geophysical Studies*. Geological Survey of Canada Paper **1995-2**, 121–145.
LEVSON, V. M., GILES, T. R., COOK, S. J. & JACKAMAN, W. 1994. *Till geochemistry of the Fawnie Creek map area (NTS 93 F/03)*. British Columbia Ministry of Energy, Mines and Petroleum Resources Open File **1994-18**.
LINDSAY, P. J. & SHILTS, W. W. 1995. A standard laboratory procedure for separating clay-sized detritus from unconsolidated glacial sediments and their derivatives. *In:* BOBROWSKY, P. T., SIBBICK, S. J., NEWELL, J. M. & MATYSEK, P. F. (eds) *Drift Exploration in the Canadian Cordillera*. British Columbia Ministry of Energy, Mines and Petroleum Resources Paper **1995-2**, 165–166.
MONGER, J. W. H. 1984. Cordilleran tectonics: a Canadian perspective. *Bulletin de la Société Géologique de France*, **XXVI**, 255–278.
NESBITT, B. E., MUEHLENBACHS, K. & MUROWCHICK, J. B. 1987. Genesis of Au, Sb, and Hg deposits in accreted terranes of the Canadian Cordillera. Geological Association of Canada/Mineralogical Association of Canada, Annual Meeting, Program with abstracts, Volume **12**, 76.
NIKKARINEN, M., KALLIO, E., LESTINEN, P. & ÄYRäS, M. 1984. Mode of occurrence of copper and zinc in till over three mineralized areas in Finland. *Journal of Geochemical Exploration*, **21**, 239–247.
PLOUFFE, A. 1991. Preliminary study of the Quaternary geology of the northern interior of British Colum-

bia. *Current Research, Part A*, Geological Survey of Canada Paper **91-1A**, 7–13.

PLOUFFE, A. 1994. *Surficial geology, Tezzeron Lake, British Columbia (NTS 93K/NE)*. Geological Survey of Canada Open File **2846**.

PLOUFFE, A. 1995a. Glacial dispersal of mercury from bedrock mineralization along Pinchi Fault, north central British Columbia. *Water, Air and Soil Pollution*, **80**, 1109–1112.

PLOUFFE, A. 1995b. *Geochemistry, lithology, mineralogy and visible gold grain content of till in the Manson River and Fort Fraser map areas, central British Columbia (NTS 93K and N)*. Geological Survey of Canada Open File **3194**.

PLOUFFE, A. 1995c. Drift prospecting sampling methods. *In:* BOBROWSKY, P. T., SIBBICK, S. J., NEWELL, J. M. & MATYSEK, P. F. (eds) *Drift Exploration in the Canadian Cordillera*. British Columbia Ministry of Energy, Mines and Petroleum Resources Paper **1995-2**, Victoria, 43–52.

PLOUFFE, A. 1997a. Ice flow and late glacial lakes of the Fraser Glaciation, central British Columbia. *Current Research 1997-A*. Geological Survey of Canada, 133–143.

PLOUFFE, A. 1997b. Physical partitioning of mercury in till: an example from British Columbia. *Journal of Geochemical Exploration*, **59**, 219–232.

PLOUFFE, A. 1998. Detrital transport of metals by glaciers, an example from the Pinchi Mine site, central British Columbia. *Environmental Geology*, **33**, 183–196.

PLOUFFE, A. & JACKSON, L. E. Jr. 1995. Quaternary stratigraphy and till geochemistry in the Tintina Trench, near Faro and Ross River, Yukon Territory. *In:* BOBROWSKY, P. T., SIBBICK, S. J., NEWELL, J. M. & MATYSEK, P. F. (eds) *Drift Exploration in the Canadian Cordillera*. British Columbia Ministry of Energy, Mines and Petroleum Resources Paper **1995-2**, 53–66.

SHILTS, W. W. 1973. *Glacial dispersal of rocks, minerals and trace elements in Wisconsinan till, Southeastern Quebec, Canada*. Geological Society of America, Memoir **136**.

SHILTS, W. W. 1975. Principles of geochemical exploration for sulphide deposits using shallow samples of glacial drift. *Canadian Institute of Mining Metallurgy Bulletin*, **68**, 73–80.

SHILTS, W. W. 1976. Glacial till and mineral exploration. *In:* LEGGET, R. F. (ed.) *Glacial Till: An Interdisciplinary Study*. Royal Society of Canada, Toronto, 205-224.

SHILTS, W. W. 1984. Till geochemistry in Finland and Canada. *Journal of Geochemical Exploration*, **21**, 95–117.

SHILTS, W. W. 1993. Geological Survey of Canada's contributions to understanding the composition of glacial sediments. *Canadian Journal of Earth Sciences*, **30**, 333–353.

SHILTS, W. W. 1995. Geochemical partitioning in till. *In:* BOBROWSKY, P. T., SIBBICK, S. J., NEWELL, J. M. & MATYSEK, P. F. (eds) *Drift Exploration in the Canadian Cordillera*. British Columbia Ministry of Energy, Mines and Petroleum Resources Paper **1995-2**, 149–163.

SHILTS, W. W. & KETTLES, I. M. 1990. Geochemical-mineralogical profiles through fresh and weathered till. *In:* KUJANSUU, R. & SAARNISTO, M. (eds) *Glacial Indicator Tracing*. A.A. Balkema, Rotterdam, 187–216.

SIBBICK, S. J. & FLETCHER, W. K. 1993. Distribution and behavior of gold in soils and tills at Nickel Plate Mine, southern British Columbia, Canada. *Journal of Geochemical Exploration*, **47**, 183–200.

SIBBICK, S. J. & KERR, D. E. 1995. Till geochemistry of the Mount Milligan area, north-central British Columbia; recommendations for drift exploration for porphyry Cu-Au mineralization. *In:* BOBROWSKY, P. T., SIBBICK, S. J., NEWELL, J. M. & MATYSEK, P. F. (eds) *Drift Exploration in the Canadian Cordillera*. British Columbia Ministry of Energy, Mines and Petroleum Resources Paper **1995-2**, 167–180.

SOPUCK, V. J., SCHREINER, B. T. & AVERILL, S. A. 1986. Drift prospecting for gold in the southeastern Shield of Saskatchewan, Canada. *In: Prospecting in Areas of Glaciated Terrain–1986*. Institution of Mining and Metallurgy, London 217–240.

TIPPER, H. W. 1971. Multiple glaciation in central British Columbia. *Canadian Journal of Earth Sciences*, **8**, 743–752.

Geochemical signatures around massive sulphide deposits in southern British Columbia, Canada

RAY E. LETT

British Columbia Ministry of Energy and Mines, Geological Survey Branch, P.O. Box 9320 STN PROV GOV'T, Victoria, British Columbia V8W 9N3, Canada (e-mail: Ray.Lett@gems7.gov.bc.ca)

Abstract: This paper demonstrates the application of geochemical exploration for sulphide mineralization in glaciated areas by a case history illustrating the discovery of Cu–Pb–Zn–Ag–Au massive sulphide deposits in southern British Columbia, Canada. These deposits, hosted by Palaeozoic metasedimentary and metavolcanic rocks of the Kootenay Terrane, were first detected by weakly anomalous Cu and Au values in regional stream sediment samples and subsequently confirmed by more detailed stream and soil geochemical surveys, prospecting and diamond drilling.

Till geochemistry is a very effective exploration method because there is a well developed dispersal plume of mineralized bedrock down-ice from the massive sulphide deposits. Elevated Au, Pb, Cu and As levels in till samples collected up to 8 km down-ice from the deposits are direct indicators for sulphide mineralization. Barium, Cr and Ni are pathfinders for distinguishing different types of sulphide mineralization. The relationship between the bedrock, stream sediment, stream water and till geochemistry is shown more clearly in a conceptual model. This model has a practical application to future exploration for massive sulphides in southern British Columbia by establishing criteria such as geochemical anomaly size and contrast for different sample media.

The exploration history leading to the discovery of the Rea Gold and Samatosum Au–Ag–Cu–Pb–Zn massive sulphide deposits in southern British Columbia, Canada illustrates the successful use of stream sediment and till geochemistry in discovering new mineral deposits in glaciated areas. The Homestake Au–Ag–Cu–Pb–Zn–Ba deposit W of Adams Lake had been mined intermittently since its discovery in 1898. Evidence for mineralization at Rea Gold and Samatosum, however, was first provided by elevated Cu levels in stream sediment from a nearby creek sampled in 1976 during a government regional geochemical survey (RGS) (Geological Survey of Canada 1977). The Rea Gold deposit was discovered in 1983 by A. Hilton, a prospector, who used a geochemical field test kit to detect metal anomalies in stream sediments. Follow-up surveys of these anomalies with soil sampling and prospecting along a new logging road resulted in the discovery of a hematitic gossan. Trenching, rock sampling and diamond drilling around the gossan identified the two Au–Ag–Cu–Pb–Zn massive sulphide lenses that together formed the Rea Gold deposit (Davidson & Perie 1987). The Samatosum Au–Ag–Cu–Pb–Zn massive sulphide deposit, located < 1 km NE from Rea Gold, was later discovered in 1985 by soil geochemistry, geological mapping and diamond drilling. The Rea Gold deposit proved to be sub-economic and the Samatosum deposit was mined for less than five years.

A recent regional till geochemical survey (Bobrowsky *et al.* 1997; Dixon-Warren 1998) has revealed anomalous gold, copper, silver, cadmium, arsenic and lead levels in till samples up to 8 km down-ice (SE) from the Rea Gold and Samatosum deposits. Element patterns explained by dispersal of mineralized bedrock in thick basal till. Till geochemistry is a very effective exploration technique for massive sulphides because basal till is a first product derivative of bedrock (Shilts 1975). However, the interpretation of till geochemistry can be complicated because of multiple sources of glacially transported bedrock and mineralized material. A conceptual model that links geology to stream sediment, stream water and till geochemistry for the area surrounding the Samtsosum and Rea Gold deposits is described

From: MCCLENAGHAN, M. B., BOBROWSKY, P. T., HALL, G. E. M. & COOK, S. J. (eds) 2001. *Drift Exploration in Glaciated Terrain.* Geological Society, London, Special Publications, **185**, 301–321.
0305-8719/01/$15.00 © The Geological Society of London 2001.

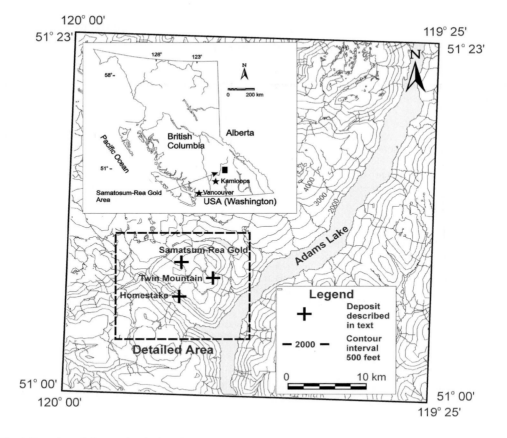

Fig. 1. Location of the geochemical study areas.

in this paper. This model is based on stream sediment, stream water and till geochemical data collected during surveys carried out in 1977, 1996 and 1997 over a 200 km² area around the deposits. Geochemical exploration criteria such as geochemical anomaly size and pathfinder elements for sulphide mineralization are defined by this model.

Location and physiography

The Samatosum, Rea Gold and Homestake deposits are located c. 100 km NE of Kamloops in southern British Columbia (BC), Canada (Fig. 1) on the edge of an undulating plateau surrounding the 600 m high Samatosum Mountain. Johnson Lake, north of the Samatosum mine, drains southwest through Johnson Creek into Sinmax Creek and then east into Adams Lake. Valley bottoms (e.g. Sinmax Creek) have scattered black cottonwood stands or have been cleared for agriculture. Hillside and plateau environments, typical of the area around Samatosum Mountain, support a dense growth of western hemlock, red cedar and Douglas fir that has been partially cleared by commercial logging. Secondary alder and lodgepole pine growth has partially regenerated in the logged areas. Dominant soil types are brunisols and podzols formed on well-drained till or organic soils in wetlands on the plateau and around lakes.

Regional geology

Regional geology has been described by Schiarriza & Preto (1987). The area around the Samatosum, Rea Gold and Homestake deposits is underlain by rocks of the Eagle Bay assemblage, which is part of the Kootenay Terrane. This assemblage comprises a Lower Palaeozoic sequence of clastic metasedimentary, carbonate and mafic metavolcanic rocks overlain by a Devonian–Mississippian succession of felsic to intermediate metavolcanic and clastic metasedimentary rocks. Structurally, the Eagle Bay assemblage forms four NW-dipping thrust sheets reflecting Jurassic-Cretaceous age deformation of the rocks. Geology of the area around the Samatosum, Rea Gold and Homestake deposits is summarized in Figure 2.

Oldest Eagle Bay rocks (Unit EBG) are Cambrian calcareous schist derived from mafic volcanic rocks. Unit EGB includes the prominent Tshinakin limestone, quartzite and conglomerate and is separated from younger Devonian–Mississippian rocks by a SW-dipping thrust fault. The Devonian–Mississippian rocks comprise Units EBF, EBA and EBS (Fig. 2). Unit EBF consists of fragmental felspathic schistose rocks derived from intermediate tuffs and volcanic breccias. Unit EBA, the host rock for the Homestake deposit, consists of chlorite–sericite quartz phyllite derived from felsic to intermediate volcanic rocks. Unit EBS consists of coarse-grained clastic metasedimentary rocks intercalated with carbonate and mafic volcanic horizons. Locally, rocks of the Eagle Bay assemblage have been intruded by Devonian orthogneiss and, regionally, by mid-Cretaceous granodiorite and quartz monzonite (e.g. the Baldy Batholith), Tertiary quartz-feldspar porphyry, basalt and lamprophyre dykes. Locally the rocks of the Eagle Bay assemblage are covered by Miocene plateau flow basalt.

Economic Geology

Eagle Bay metasedimentary and metavolcanic rocks host the Samatosum, Rea Gold and Homestake volcanogenic deposits and several other smaller mineral occurrences (e.g. Twin Mountain) in the area. The sequence of rocks along a northeast to southwest section across the Samatosum and Rea Gold mineralized zones from the deposit hanging wall to footwall comprises: (1) the Tshinikan limestone; (2) interbedded cherts and argillite (the mixed sediments); (3) mafic volcanics; and (4) the 'Mine Series' rocks (Friesen 1990), a zone of heavily altered sediments and mafic volcanic rocks, with minor felsic to intermediate volcanic rocks. The 'Mine Series' is the host rock for the Samatosum and Rea Gold deposits. Directly above the Mine Series, and forming the structural footwall of both deposits, is a thick sequence of feldspathic phyllite and schist derived from intermediate tuff and volcanic breccia (Unit EBF). The Samatosum deposit is a highly deformed quartz-vein system containing

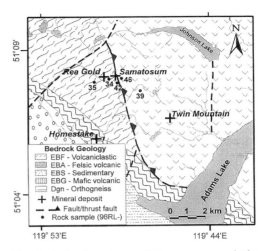

Fig. 2. Bedrock geology of the area around the Samatosum, Rea Gold area and Homestake deposits. The location of rock samples described in the text is indicated on the map (geology based on Schiarizza & Preto 1987).

massive to disseminated tetrahedrite, sphalerite, chalcopyrite and galena. Tetrahedrite is most uniformly distributed, while sphalerite, galena and chalcopyrite occur more erratically as massive lenses within the quartz vein host. Principal gangue minerals are quartz and dolomite. Alteration associated with mineralization includes sericitization, silica flooding of wallrock surrounding the orebody, dolomitization and pyrite replacement of lapilli in the mafic pyroclastic rocks. Fuchsite, a chromium-rich mica, occurs in the immediate sheared footwall portion of the ore zone.

Estimated reserves for the Samatosum deposit were 635 000 tonnes of ore grading 1.2% copper, 2.9% zinc, 1.7% lead, 1035 g/t silver and 1.9 g/t gold (Pirie 1989). The Samatosum deposit operated as an underground and open pit mine from 1988 to 1994.

At Rea Gold, the predominant minerals in two massive sulphide lenses are fine grained arsenopyrite, pyrite, sphalerite, galena, tetrahedrite, quartz and barite. The barite and sulphide minerals occur at surface beneath the original discovery gossan. Estimated reserves for the Rea Gold deposit are 376 000 tonnes of ore grading 0.33% copper, 2.3% zinc, 2.2% lead, 69.4 g t^{-1} silver and 6.1 g t^{-1} gold (The Northern Miner 1987).

Four kilometres south of the Samatosum mine is the Homestake Pb–Zn–Au–Ag–Ba deposit. Two tabular horizons of barite–pyrite–chalcopyrite–argentite–tetrahedrite–galena–sph-

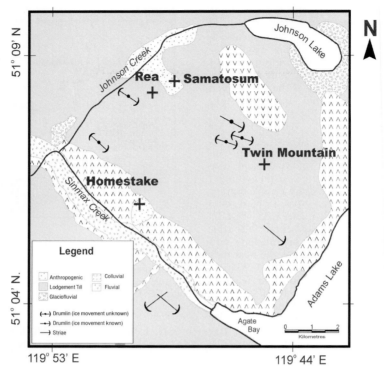

Fig. 3. Surficial geology of the area around the Samatosum, Rea Gold area and Homestake deposits (surfical geology based on Dixon-Warren 1998).

alerite lenses occur in a quartz–sericite–talc schist unit of the Eagle Bay Assemblage. The Homestake deposit contained an estimated 200 000 tonnes of ore grading 200 g t^{-1} silver, 0.58 g t^{-1} gold, 0.5% copper, 2.5 % lead and 4.0 % zinc. It was mined intermittently from 1893 until 1983 and is now abandoned (Höy 1991). The Twin Mountain prospect, located 6 km east of the Samatosum deposit, consists of galena, sphalerite, pyrite and chalcopyrite mineralization in carbonate–quartz–barite lenses hosted by a northeast-dipping zone of grey pyritic and calcareous chlorite–sericite–quartz schists (Höy 1991). The Samatosum, Rea Gold, Homestake deposits and the Twin Mountain prospect are identified on Figure 2.

Surficial geology

The surficial geology (Fig. 3) of the area east of the Samatosum–Rea Gold deposit was mapped by Dixon-Warren (1998). Surficial deposits include basal till, ablation till, glaciofluvial deposits, glaciolacustrine sediments, fluvial sediments, organic deposits and colluvium. These were deposited by the last cycle of glaciation and deglaciation (Fraser Glaciation). The plateau to the east of the Samatosum deposit is mantled by <3 m of massive, silty clay basal till deposited by a predominantly northwest to southeast ice flow (Dixon-Warren et al. 1997). Colluvial deposits occur on steep slopes and fluvial, glaciofluvial and glaciolacustrine sediments occur along valley bottoms. Eagle Bay rocks, talus and colluvium are exposed along the steep sides of the Sinmax Creek and Adams Lake vallies. Organic deposits occur locally in all types of terrain.

Previous geochemical surveys

Regional stream sediment, stream water, till and soil geochemical surveys have been carried out by government geological surveys over several thousand square kilometres surrounding the Samatosum and Rea Gold deposits. The survey methods used and the elements determined are summarized in Table 1 and are described in more detail below.

Table 1. *Summary of survey medium and analytical methods for data used in this study*

Survey Method	Analytical Methods	Elements	Reference
Stream Sediment (<177 µm) and water; 1 sample/13 km²	1. Aqua regia-AAS 2. Epithermal INAA	41 including Au, Ag, As, Cu, Co, F, Hg Mn, Mo, Ni, U, Zn	Matysek *et al.* 1991
Stream water (<45 µm filtered); 1 samples/5.5 km²	1. ICP-MS (filtered water) 2. ion electrode (F)-unfiltered water turbidimetry 3. (SO^{4-}) - unfiltered water	63 including Ca, Mg, Au, As, Ag, Cu, Cd, Pb, Zn, pH, alkalinity, SO_4	Lett *et al.* 1998a
Till (<63 µm); 1 sample/3.8 km²	1. Thermal INAA 2. Aqua regia-ultrasonic nebulization ICP-ES. 3. Lithium metaborate fusion and ICP-ES.	66 elements including Au, As, Ba, Cd, Cu, Pb, Zn, major oxides, C and S.	Bobrowsky *et al.* 1997
Till & B-horizon soil (<63 µm); 1-2 sample/km²	1. Thermal INAA 2. Aqua regia-ultrasonic nebulization ICP-ES. 3. Lithium metaborate fusion and ICP-ES	66 elements including Au, As, Ba, Cd, Cu, Pb, Se, Te, Tl, Zn, major oxides, C and S.	Lett *et al.* 1998b
Rock (<63 µm); 5 samples	1. Thermal INAA 2. Aqua regia-ultrasonic nebulizeration ICP-ES.	66 elements including Au, As, Ba, Cd, Cu, Pb, Se, Te, Tl, and Zn,	

Regional stream sediment and stream water survey

A regional stream sediment geochemical survey (RGS) carried out in 1976 by the Geological Survey of Canada and the British Columbia Ministry of Mines and Petroleum Resources (GSC–BCMMPR) over NTS map sheet 82M covered 15 400 km² around the Samatosum, Rea Gold and Homestake deposits. Stream sediment and water samples were collected from 1151 sites at an average density of one sample per 13.4 km². Between 1 and 2 kg of fine-grained sediment were taken from the active stream channel. A 250 ml unfiltered water sample was also collected from the stream. A field duplicate sediment sample was taken close to a routine sample in each batch of 20 samples collected. Sediment samples were air-dried, sieved to −80-mesh (<177 µm) and the −80-mesh fraction was ball milled. Quality control reference standards and blind duplicate samples were inserted into each batch of 20 milled samples for geochemical analysis. Sediment samples were analysed in 1977 for Cu, Co, Fe, Pb, Mn, Mo, Ni, Ag and Zn by aqua regia digestion–atomic absorption spectrometry (AAS), for Hg by aqua regia digestion–cold vapour atomic absorption spectrometry (CV–AAS) and for U by delayed neutron counting (DNC). The analyses were carried out by Chemex Laboratories (Vancouver, Canada) and by Atomic Energy of Canada, Ottawa, Canada.

The water samples collected during this survey were analysed by Chemex Laboratories for pH using a combination glass-reference electrode (GCE), for F using a specific ion-reference electrode (ION) and for U by a fluorometric method (FLU). Detection limits for these parameters are listed in Table 2.

In 1990, the sediment samples were re-analysed for an extensive suite of major and trace elements by epithermal, neutron activation (INAA) at Becquerel Laboratories (Mississauga, Canada). The elements and their detection limits are listed in Table 2. Regional geochemical survey methods, analytical techniques and results are described by Matysek *at al.* (1991).

Regional stream water survey

Another regional stream water survey, carried out in 1996 by the BC Geological Survey Branch over 1200 km² of NTS map sheets 82M/4 and 82M/5, also included samples taken from streams draining the Samatosum, Rea Gold and Homestake deposits (Sibbick *et al.* 1997;

Lett et al. 1998a). Water samples were taken from 218 sites in the survey area at an average density of one sample per 5.5 km^2. Quality control reference standards and ultra-pure water blanks (supplied by the BC Ministry of Forests Research laboratory) were inserted into each batch of 20 water samples before geochemical analysis. A sample of the water, filtered through a 0.45 μm filter and acidified with ultra-pure nitric acid, was analysed by Activation Laboratories, (Ancaster, Ontario) using inductively coupled plasma mass spectrometry (ICP-MS) for a suite of major, minor and trace elements (Table 3). An unacidified water sample was also analysed for pH and sulphate. Sampling, quality control and analytical methods are described in more detail by Lett et al. (1998a).

Till geochemical surveys

During the same year a regional till geochemical survey carried out by the BC Geological Survey Branch over 1200 square km^2 of NTS map sheets 82M/4 and 82M/5 (Bobrowsky et al. 1997) collected 535 samples at an average density of one sample per 3.8 km^2. Basal till was the preferred sample medium and 3–5 kg of material was collected at each site. The −230 mesh (< 63 μm) fraction of the till was analysed for a suite of major, minor and trace elements at Activation Laboratories (Ancaster, Ontario, Canada) by INAA and by aqua regia digestion-inductively coupled plasma emission spectroscopy (ICP-ES) at Acme Laboratories (Vancouver, Canada). Detection limits for elements determined by these methods are given in Table 4.

Table 2. *Elements and detection limits for the analysis of the < 177 μm fraction of stream sediments*

Element	Method	Detection Limit	Units
Ag	AAS	0.2	ppm
As	INAA	0.5	ppm
Au	INAA	2.0	ppb
Ba	INAA	100	ppm
Br	INAA	0.5	ppm
Ce	INAA	10	ppm
Co	AAS/INAA	2.0/5.0	ppm
Cr	INAA	5.0	ppm
Cs	INAA	0.5	ppm
Cu	AAS	2.0	ppm
Fe	AAS/INAA	0.02/0.01	%
Hf	INAA	1.0	ppm
Hg	CV-AAS	20	ppb
La	INAA	5.0	ppm
Lu	INAA	0.2	ppm
Mn	AAS	5.0	ppm
Mo	AAS/INAA	2.0/1.0	ppm
Na	INAA	0.01	%
Ni	AAS/INAA	1.0/10.0	ppm
Pb	AAS	2.0	ppm
Rb	INAA	5.0	ppm
Sb	INAA	0.1	ppm
Sc	INAA	0.5	ppm
Sm	INAA	0.5	ppm
Ta	INAA	0.5	ppm
Tb	INAA	0.5	ppm
Th	INAA	0.5	ppm
U	INAA/DNC	0.2/0.2	ppm
W	INAA	2.0	ppm
Yb	INAA	2.0	ppm
Zn	AAS	2.0	ppm
Zr	INAA	200	ppm
pH (Water)	GCE	0.1	
F (Water)	ION	20	ppb
U (Water)	FLU	0.05	ppb

Table 3. *Detection limits (ppb) for trace elements determined in stream water samples*

Element	Detection Limit	Element	Detection limit
Ag	0.02	Mo	0.02
Au	0.002	Nb	0.002
As	0.02	Nd	0.002
Ba	0.02	Ni	0.002
Bi	0.002	Pb	0.02
Br	1.0	Pd	0.02
Cd	0.002	Pt	0.002
Ce	0.002	Re	0.002
Cs	0.002	Ru	0.002
Cr	0.01	Rb	0.002
Co	0.002	Sb	0.02
Cu	0.002	Se	0.002
Dy	0.002	Sm	0.002
Er	0.002	Sr	0.002
Eu	0.002	Ta	0.002
F	20	Te	0.002
Fe	0.1	Tl	0.002
Ga	0.002	Tm	0.002
Gd	0.002	Th	0.002
Ge	0.002	Sn	0.002
Hf	0.002	W	0.002
Ho	0.002	U	0.002
I	0.02	V	0.002
La	0.002	Yb	0.002
Lu	0.002	Y	0.002
Mn	0.02	Zn	0.002

An additional 55 samples were collected in 1997 from the B-horizon soil and from basal till in a 200 km^2 area SE of the Samatosum and Rea Gold deposits as part of detailed geochemical studies (Lett et al. 1998b). These samples were prepared by the same method as the regional till samples and analysed by INAA. The prepared samples were also analysed by ICP-ES and by ultrasonic nebulization ICPU-ES following aqua regia digestion at Acme Laboratories. Five rock samples, jaw-crushed and milled to < 50 m were analysed by INAA and aqua regia (ICP-ES). The detection limits for elements determined by ultrasonic nebulization ICPU-ES are listed in Table 4.

Table 4. *Elements determined and detection limits for the analysis of the <63 μm fraction of till, B-horizon soil and <50 μm milled rock samples*

Element	Method	Detection Limit	Units
Ag	ICP-ES/ICPU-ES	0.3/0.03	ppm
Al	ICP-ES	0.01	%
As	INAA/ICP-ES/ICPU-ES	0.5/2/0.5	ppm
Au	INAA	2	ppb
Ba	INAA/ICP-ES	50/2	ppm
Bi	ICPU-ES	0.1	ppm
Br	INAA	50	ppm
Ca	ICP-ES	0.01	%
Cd	ICP-ES/ICPU-ES	0.2/0.02	ppm
Ce	INAA	3.0	ppm
Co	INAA/ICP-ES	1.0/1.0	ppm
Cr	INAA/ICP-ES	5.0/1	ppm
Cs	INAA	1.0	ppm
Cu	ICP-ES/ ICPU-ES	1.0/0.2	ppm
Fe	INAA/ICP-ES	0.02/0.01	%
Hf	INAA	1	ppm
Hg	INAA/ ICPU-ES	1000/10	ppb
La	INAA	0.5	ppm
Lu	INAA	0.05	ppm
Mg	ICP-ES	0.01	%
Mn	ICP-ES	1.0	ppm
Mo	INAA/ICP-ES/ICPU-ES	1.0/1.0/0.1	ppm
Na	INAA/ICP-ES	0.01	%
Nd	INAA	5	ppm
Ni	INAA/ICP-ES	50.0/1.0	ppm
Pb	ICP-ES/ICPU-ES	2/0.3	ppm
Rb	INAA/ICP-ES	5.0/1	ppm
Sb	INAA/ICP-ES/ICPU-ES	0.1/2/0.2	ppm
Sc	INAA	0.1	ppm
Se	ICPU-ES	0.4	ppm
Sm	INAA	0.1	ppm
Ta	INAA	0.5	ppm
Tb	INAA	0.5	ppm
Te	ICPU-ES	0.2	ppm
Th	ICPU-ES	0.2	ppm
Th	INAA	0.2	ppm
U	INAA/DNA	0.5/0.2	ppm
V	ICP-ES	2.0	ppm
W	INAA	1.0	ppm
Yb	INA	0.2	ppm
Zn	INAA/ICP-ES	2.0	ppm

Regional stream sediment and water geochemistry

Mean, median, standard deviation, maximum and anomaly threshold values (at the 95th percentile), calculated from element data for 235 stream sediment samples taken during the 1977 regional survey, are listed in Table 5. These samples represent streams draining catchment basins predominantly underlain by rocks of the Eagle Bay assemblage in NTS map sheet 82M. Results from two anomalous sediment samples, one from Johnson Creek (sample number 763043) that flows through the area north of the Samatosum and Rea Gold deposits and the other from Homestake Creek (763047) also are shown in Table 5. Only Cu (83 ppm) and Br are anomalous in the Johnson Creek sediment; Au (16 ppb) is just below the threshold for this element. Copper, Ag, Au, Ba, Sb, Pb and Zn, however, are all anomalous in stream sediment from Homestake Creek. Figures 4 and 5 show the distribution of Au and Cu values in the regional survey stream sediments. The anomalous Au and Cu values within area A highlight streams draining the Samatosum, Rea Gold and Homestake deposits. Anomalous Cu values toward the south end of Adams lake suggest additional Cu mineralization along the strike of Unit EBA to the southeast of the Homestake deposit.

The distribution of pH and Cu in stream water samples collected during the 1996 regional survey (Lett *et al.* 1998*a*) is shown in Figures 6 and 7. The pH values reveal that the creeks draining the Samatosum, Rea Gold and Homestake deposits are not acidic, suggesting that there is no oxidizing sulphide mineralization in contact with the water (Hoag & Webber 1976). Nickel is the only metal present at a significant level (1.2 ppb) in the stream water flowing from the area around Samatosum and Rea Gold. A stream water Cu anomaly (1.7 ppb) near the Samatosum deposit represents a creek draining a Cu-mineralized area north of Johnson creek (Fig. 7). One reason for a weak geochemical expression of the massive sulphide mineralization in stream water may be the effect of the acid mine drainage neutralizing plant operating at the Samatosum mine site. Neutral to alkaline water in Homestake Creek can be explained by carbonates associated with the sulphide mineralization that neutralize the acid, metal-rich mine drainage.

Stream water Ba values are typically below a 47 ppb threshold. However, Homestake creek and a cluster of streams near the Twin Mountain

Table 5. *Summary statistics for elements determined in 235 sediments from streams draining Eagle Bay rocks in NTS map sheet 82/M*

Element	Method	Mean	Median	SD	Max.	Threshold	Sample 763043	Sample 763047
Au (ppb)	INAA	8	3	25	348	18	16	110
Ag	AAS	0.14	0.10	0.27	3.6	0.2	0.1	3.6
As	INAA	11.6	5.4	20	195	54	6.5	77.2
Ba	INAA	726	690	409	4800	1544	400	4800
Br	INAA	6.8	3.4	9.2	59.2	27	30	0.5
Cr	INAA	132	110	73	460	256	130	5
Co	INAA	20	19	9	57	37	8	20
Cu	AAS	35	28	29	328	74	83	328
Fe (%)	INAA	4.2	4.1	1.59	13.0	6.90	2.1	7.3
Mn	AAS	693	520	1577	24000	1010	215	860
Mo	AAS	2.0	1.0	1.3	12	4	2	6
Ni	AAS	30	26	16	86	59	27	7
Pb	AAS	24	11	88	1060	46	9	1060
Sb	INAA	1.1	0.3	8.6	131	1.6	<1	131
Zn	AAS	95	65	184	2600	178	35	2600

All values except Au and Fe are in ppm. Element thresholds are at the 95th percentile. SD, Standard deviation.

Table 6. *Summary statistics for elements determined in 535 regional till samples from NTS 82M/4 and 82M/5*

Element	Method	Mean	Median	SD	Max.	Threshold	Sample 969172
Ag	ICP-ES	0.3	0.3	0.1	1.9	0.6	0.6
Au (ppb)	INAA	6	4	14	215	34	74
As	INAA	11.5	8.2	12	93	35.6	290
Ba	INAA	884	850	312	4600	1508	290
Cr	INAA	124	110	97	910	318	530
Co	INAA	19.4	17	11.9	66	43.1	69
Cu	ICP-ES	73	51	166	3653	405	185
Mn	ICP-ES	802	725	428	2759	1658	1212
Mo	ICP-ES	2	1	2	28	6	1
Ni	ICP-ES	63	50	83	1491	229	293
Pb	ICP-ES	27	18	32	279	91	766
Sb	INAA	0.7	0.6	0.8	13	2.3	5.3
Zn	ICP-ES	116	94	191	4168	497	607

All values except Au are in ppm. Element thresholds are at the 95th percentile. SD, Standard deviation.

occurrence have Ba levels up to 87 ppb. The elevated Ba levels in these streams can be explained by weathering of the barite associated with the sulphide mineralization.

Till geochemistry

Mean, median, standard deviation, maximum and threshold values (at the 95th percentile) calculated from element data for 535 basal till samples collected during the 1996 regional till survey (Bobrowsky *et al.* 1997) are listed in Table 6. Results of one anomalous till sample (sample number 969172) collected 200 m down-ice of the Samatosum open pit are also included in Table 6 for comparison with threshold values. The distribution of Au, As and Cu in the regional survey till samples is shown in Figures 8–10. A fan-shaped plume of anomalous Au, As and Cu values extends for almost 8 km from a point just east of the Samatosum and Rea Gold deposits eastwards almost to Adams Lake (area A in Figs 8–10). Many of these samples also contain anomalous concentrations of Ba, Pb, Ag, Co and Cr (Bobrowsky *et al.* 1997). Another cluster of anomalous Au, As and Cu values (B in Figs 8–10) occurs *c.* 20 km northwest from the Samatosum and Rea Gold deposits. These values reflect dispersal of mineralized bedrock in basal till from several small volcanogenic

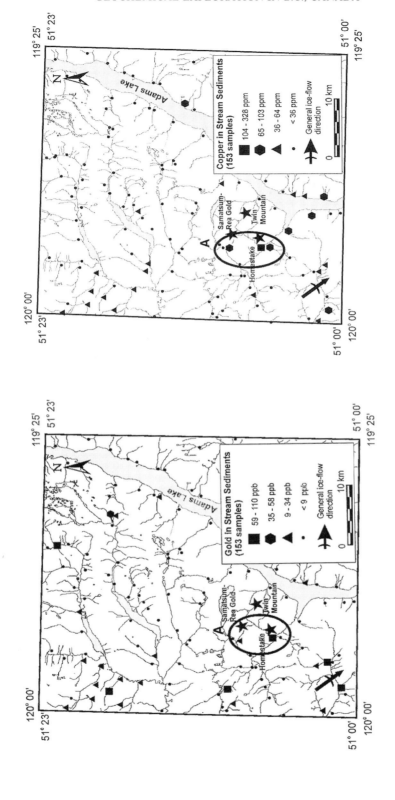

Fig. 4. Distribution of Au (ppb) in the <177 μm fraction of stream sediment samples analysed by INAA. Area 'A' outlines Au anomalies associated with the Samatosum, Rea Gold and Homestake deposits (data from Matysek et al.1991).

Fig. 5. Distribution of Cu (ppm) in the <177 μm fraction of stream sediment samples analysed by aqua regia-AAS. Area 'A' outlines Cu anomalies associated with the Samatosum, Rea Gold and Homestake deposits (data from Matysek et al. 1991).

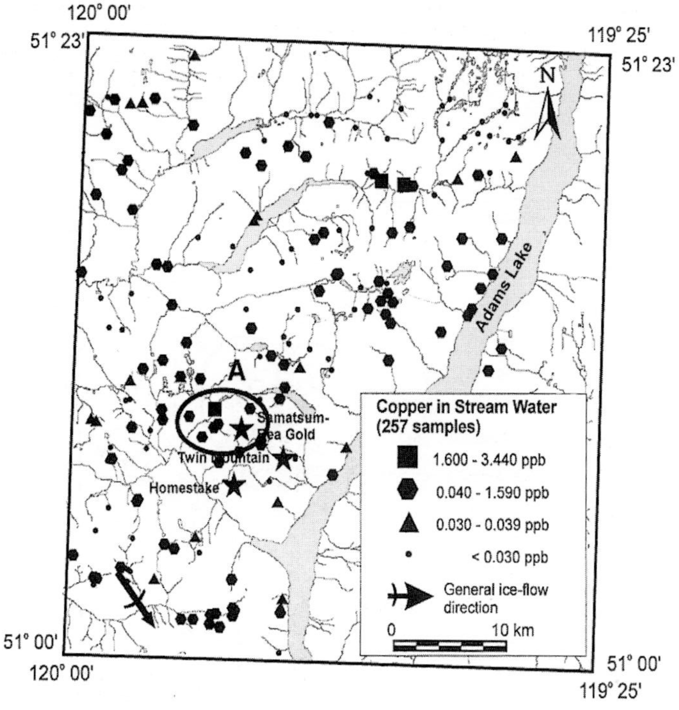

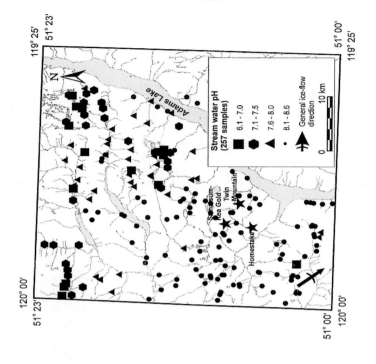

Fig. 6. Distribution of Cu (ppb) in stream water samples analysed by ICP-MS. Area 'A' outlines Cu anomalies associated with the Samatosum, Rea Gold and Homestake deposits (data from Lett et al. 1998a).

Fig. 7. Distribution of stream water pH (data from Lett et al. 1998a).

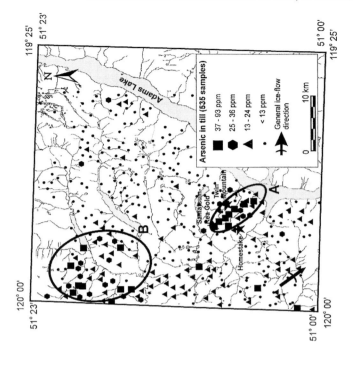

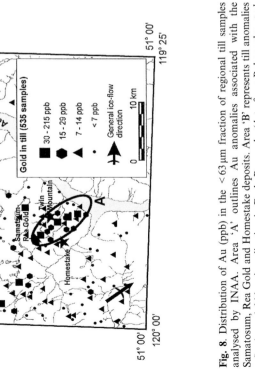

Fig. 8. Distribution of Au (ppb) in the <63 μm fraction of regional till samples analysed by INAA. Area 'A' outlines Au anomalies associated with the Samatosum, Rea Gold and Homestake deposits. Area 'B' represents till anomalies reflecting sulphide mineralization in Eagle Bay rocks (data from Bobrowsky et al. 1997).

Fig. 9. Distribution of As (ppm) in the <63 μm fraction of regional till samples analysed by INAA. Area 'A' outlines As anomalies associated with the Samatosum, Rea Gold and Homestake deposits. Area 'B' represents till anomalies reflecting small sulphide deposits in Eagle Bay Unit EBA rocks NW of the Samatosum deposit (data from Bobrowsky et al. 1997).

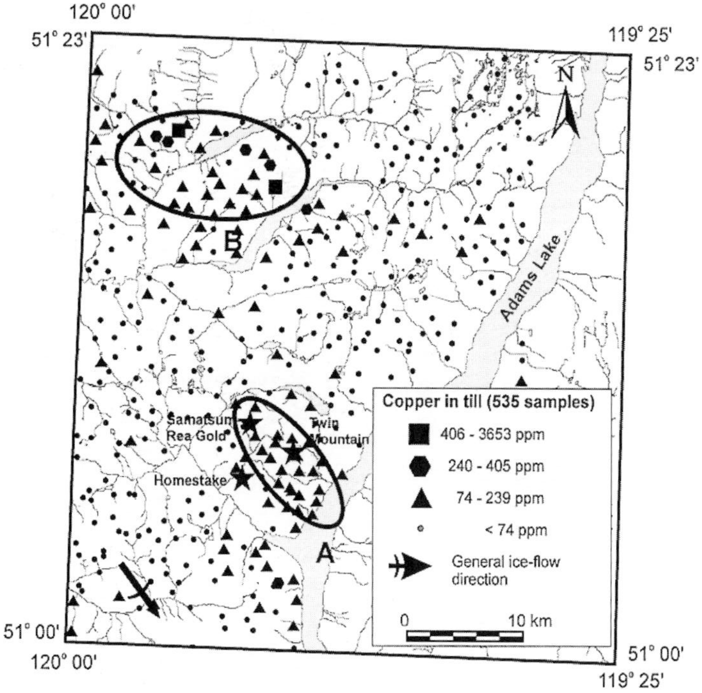

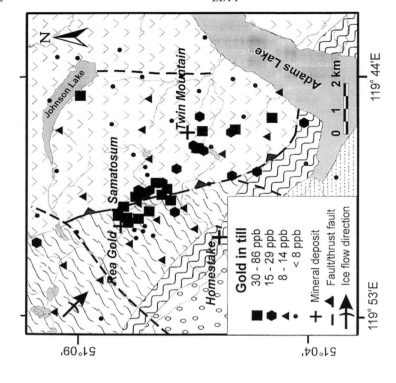

Fig. 10. Distribution of Cu (ppm) in the <63 μm fraction of regional till samples analysed by aqua regia and ICP-ES. Area 'A' outlines Au anomalies associated with the Samatosum, Rea Gold and Homestake deposits. Area 'B' represents till anomalies reflecting small sulphide deposits in Eagle Bay Unit EBA rocks northwest of the Samatosum deposit (data from Bobrowsky et al. 1997).

Fig. 11. Distribution of Au (ppb) in the <63 μm fraction of till samples by INAA around the Samatosum, Rea Gold and Homestake deposits.

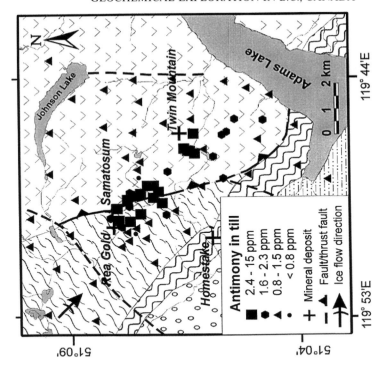

Fig. 13. Distribution of Sb (ppm) in the <63 μm fraction of till samples by INAA around the Samatosum, Rea Gold and Homestake deposits.

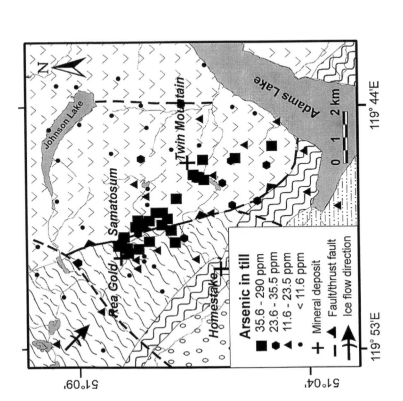

Fig. 12. Distribution of As (ppm) in the <63 μm fraction of till samples by INAA around the Samatosum, Rea Gold and Homestake deposits.

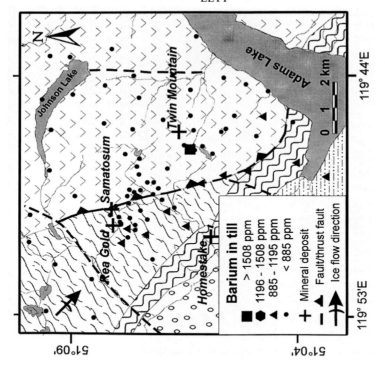

Fig. 15. Distribution of Ba (ppm) in the <63μm fraction of till samples by INAA around the Samatosum, Rea Gold and Homestake deposits.

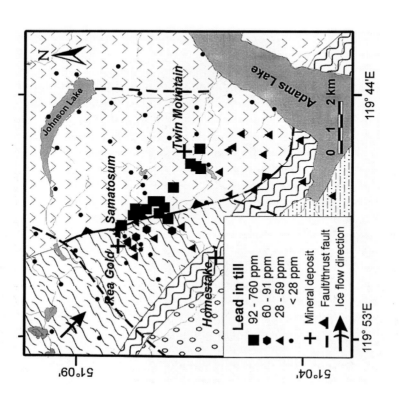

Fig. 14. Distribution of Pb (ppm) in the <63μm fraction of till samples by aqua regia and ICP-ES around the Samatosum, Rea Gold and Homestake deposits.

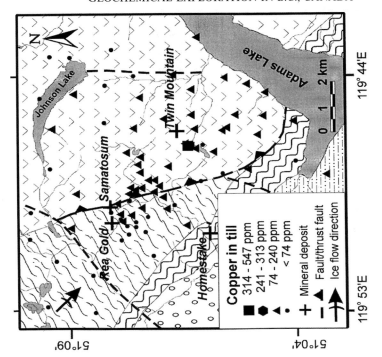

Fig. 17. Distribution of Cu (ppm) in the <63 μm fraction of till samples by aqua regia and ICP-ES around the Samatosum, Rea Gold and Homestake deposits.

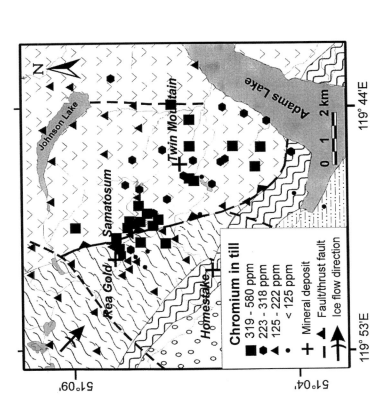

Fig. 16. Distribution of Cr (ppm) in the <63 μm fraction of till samples by INAA around the Samatosum, Rea Gold and Homestake deposits.

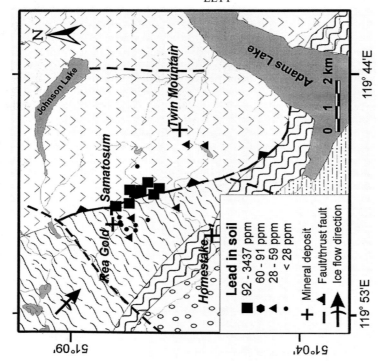

Fig. 19. Distribution of Pb (ppm) in the <63 μm fraction of B-horizon soil samples by aqua regia and ICP-ES around the Samatosum, Rea Gold and Homestake deposits.

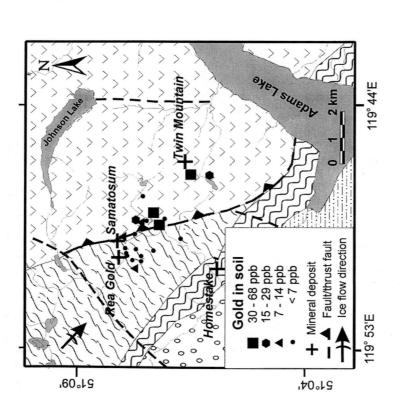

Fig. 18. Distribution of Au (ppb) in the <63 μm fraction of B-horizon soil samples by INAA around the Samatosum, Rea Gold and Homestake deposits.

Table 7. *Elements in rock samples from the area around the Samatosum and Rea Gold deposits*

Element	Method	Units	Phyllite 96RL-35	Phyllite 96RL-34	Volcanic 96RL-42	Volcanic 96RL-46	Volcanic 96RL-39
Au	INAA	ppb	6	8	13	12	< 2
Ag	ICPU-ES	ppb	147	66	46	159	38
As	INAA	ppm	28	28	24	89	4.8
Ba	INAA	ppm	1100	770	620	160	< 50
Cd	ICP-ES	ppm	0.1	0.06	0.14	0.17	0.1
Cr	INAA	ppm	160	93	390	530	450
Co	INAA	ppm	14	13	45	43	42
Cu	ICP-ES	ppm	53	125	74	99	107
Fe	INAA	%	3.09	3.08	8.14	7.32	6.44
Hg	ICPU-ES	ppb	54	28	< 10	< 10	< 10
Mg	ICP-ES	%	0.17	0.22	2.06	5.08	2.77
Mn	ICP-ES	ppm	287	1183	1034	926	877
Ni	ICP-ES	ppm	49	42	244	253	157
Pb	ICP-ES	ppm	18	7	9	15	10
Sb	INAA	ppm	0.5	0.8	1.5	1.7	1.6
Se	ICPU-ES	ppm	0.3	< 0.3	0.4	2.0	< 0.3
V	ICP-ES	ppm	9	20	66	59	117
Zn	ICP-ES	ppm	70	57	159	107	37

massive sulphide deposits hosted by Eagle Bay Unit EBA metavolcanic rocks.

Geochemical data for the 1996 regional till survey and the 1997 detailed geochemical studies (Lett *et al.* 1998*b*) have been combined to show the distribution of elements in more detail southeast of the Rea Gold and Samatosum deposits. Symbol plots for Au, As, Cu, Pb, Sb, Cr and Ba in basal till samples are shown in Figures 11–17. Immediately east of the Samatosum mine, an anomalous till sample contains 74 ppb Au, 290 ppm As, 5.3 ppm Sb, 766 ppm Pb, 186 ppm Cu and 607 ppm Zn (Table 6). These are high values compared to regional thresholds (Table 6) and are a reflection of tetrahedrite, sphalerite, chalcopyrite and galena in the Samatosum sulphide deposit. Gold values decrease down-ice from the mine to a point roughly 2 km southeast along the dispersal train and then increase again between 3 and 4 km (Fig. 11). The Au plume (> 30 ppb Au) at the 2 km point along the dispersal train is 1.5 km wide. Arsenic values (> 35.6 ppm As) also decay down-ice from the Rea Gold and Samatosum deposits, although the As dispersal train is wider than Au the 2 km point. The distribution of Sb (> 2.4 ppm Sb) in till (Fig. 13) is similar to that of Au. Lead values (> 92 ppm Pb) in Figure 14, are higher in the basal till down-ice from the Samatosum deposit than levels southeast of Rea Gold. This difference could reflect the higher Pb content in the Rea Gold deposit. Barium values (> 885 ppm) tend to be elevated in till down-ice (SE) from the Rea Gold deposit, although the highest Ba

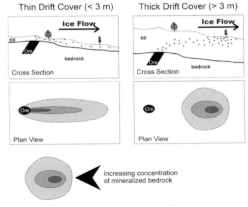

Fig. 20. Comparison of glacial dispersal patterns developed down-ice from mineralized bedrock for thick and basal thin till.

values (> 1508 ppm Ba) occur southwest from the Twin Mountain occurrence (Fig.15). The higher Ba concentrations near Rea Gold can be explained by the entrainment of barite in till from the gossan capping the massive sulphides. Although barite also is present in the Samatosum deposit, it is not exposed at the subcropping bedrock surface and was therefore not available for the advancing glacier and incorporation into the till. Chromium values (> 319 ppm Cr) are higher southeast from the Samatosum mine (Fig. 16); high Cr values, and those of Ni, V, Mg and Fe, reflect the chemistry of the mafic volcanic rock forming the Samatosum deposit hangingwall and the Cr-rich mica (fuchsite) in bedrock.

Although weakly anomalous Cu values (>74 ppm) in till extend along the down-ice dispersal train from the Rea Gold and Samastosum deposits, the highest Cu concentration (>314 ppm Cu) occurs near the Twin Mountain prospect (Fig. 17). The source for the Cu could be the chalcopyrite–barite mineralization typical of this occurrence.

Gold, As, Pb and Sb variations along the dispersal train could reflect the effect of changing till thickness or several, overlapping plumes combining material from different bedrock source areas. The distribution of Au in the B-soil horizon (>30 ppb Au) for the detailed samples shows that the highest Au values occur $c.$ 2 km down-ice from the Rea Gold and Samatosum deposits rather than at the head of the dispersal train (Fig. 18). The highest Pb in the B-soil horizon samples (3437 ppm) also occurs 2 km down-ice from the deposits (Fig. 19). The effect of different till thickness on size and shape of a typical dispersal plume is shown schematically in Figure 20. Anomalous element concentrations in the B-horizon soil may be detected almost directly above the mineralized source or could be displaced several kilometres down-ice depending on the angle of plume climb and till thickness. The relationship between the basal till and B-horizon soil Au and Pb geochemistry along the Rea Gold and Samatosum dispersal train suggests that mineralized bedrock reaches the surface 2 km down-ice from the source. Since the predicted dispersal train is <3 km, the Au, Cu and As till anomalies near the Twin Mountain occurrence may reflect additional sources of sulphide mineralization.

The relationship between till geochemistry and bedrock geochemistry is also demonstrated by element variations in five rock samples representing the host rocks at the Rea Gold and Samatosum deposits. The location of these samples is shown on Figure 2. The element concentrations listed in Table 7 are for phyllite (96RL 35 and 34), mafic volcanic rocks (96RL 42 and 46) from the Samatosum deposit hangingwall and a sample of mafic tuff (96RL 39) from an outcrop $c.$ 1 km to the east of the deposit. The phyllite (Unit EBF) has a higher Ba content whereas the mafic volcanic rocks (Unit EBG) contain more Cr, Ni, Fe, Mg and V. All samples have relatively low Pb, Sb, Cu, Zn, Ag and Au concentrations, reflecting the absence of sulphide mineralization. The As content (89 ppm) of rock sample 96RL-46 from the Samatosum hangingwall is lower than the As level (290 ppm) in a till sample (969172) just east of the deposit, suggesting concentration of As in the basal till at the head of the dispersal train. Although only a small number of rock samples have been analysed, the rock geochemistry suggests that Ba may be a pathfinder for massive sulphide mineralization (e.g. Rea Gold) in Unit EBF whereas Cr is an indicator of the Samatosum type sulphide mineralization.

Geochemical model

The relationship between surficial geochemistry, geology and sulphide mineralization is shown in Figure 21 by a model that is based on the geochemical exploration model concept proposed by Bradshaw (1975). The primary geochemical signature of the sulphide mineralization typical of Samatosum, Rea Gold and Twin Mountain is overlaid on a simplified geology block diagram. Ore indicator elements (e.g. Au, Ag, Cu, Pb, Zn), and pathfinder elements (e.g. As, Ba, Hg, Se) displayed on the bedrock layer are taken from the BC Mineral Deposit volcanogenic massive sulphide (VMS) profile (Lefebure & Ray 1995) and from results of the bedrock geochemical data described in this paper. Also shown on the bedrock layer are elements (e.g. Cr, Ni) that are enhanced along the contact between the mafic and felsic volcanic rocks. Dimensions for lithogeochemical anomalies are not shown because their actual size is unknown.

Rock and sulphide-rich debris were eroded from the bedrock surface by a Pleistocene ice-sheet advancing from the northwest and deposited in basal till down-ice as a dispersal train. This process is summarized on the surficial geology layer where the relative width and contrast of till geochemical anomalies are indicated by non-dimensional, shaded horizontal and vertical patterns. Different multi-element associations displayed on the till block diagram are intended to reflect the varying chemistry of the bedrock source for the dispersal plume. The Twin Mountain pattern is shown separated from the Rea Gold and Samatosum patterns because there are most likely different sources for the till anomalies (Yeow 1998).

Sediment and water geochemistry for elements readily weathered from bedrock and soil (e.g. Cu) and geochemically less mobile elements (e.g. Au) are displayed on the soil landscape block diagram. The soil geochemical patterns are shown displaced down-ice from the basal till patterns based on the projected relationship between till and till geochemistry shown in Figure 21. A Cu and Au stream sediment

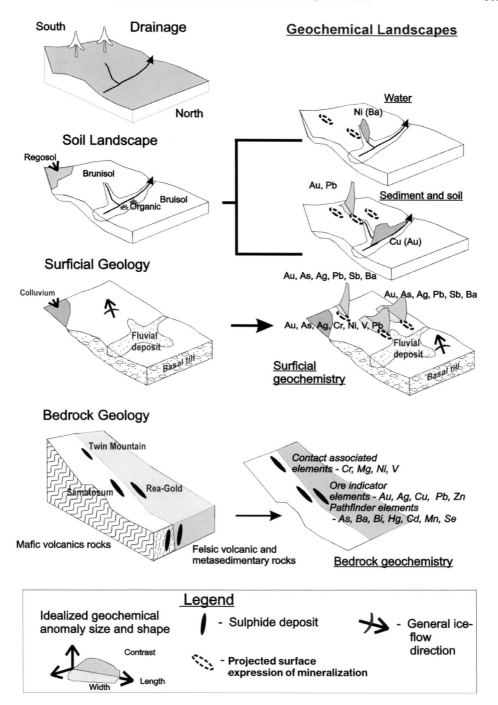

Fig. 21. A surficial geochemical model for the Samatosum and Rea Gold deposit area.

Table 8. *Geochemical exploration criteria for Samatosum and Rea Gold type massive sulphide deposits*

	Sediments	Water	Till	Soil	Remarks
Pathfinder Elements	Cu, Ba, Au, Cr, Br	Ni, Ba	Cu, Pb, Zn, Ag, Hg, Ba As, Au, Se, Cr, Ni, V, Mn, Sb	Cu, Pb, Zn, Ag, Hg, Ba As, Au, Se, Cr, Ni, V, Mn	Ba and Hg may be guides to different sulphide deposits.
Suggested Analytical Methods	INAA and aqua regia-ICP-ES or ICP-MS	ICP-MS	INAA and aqua regia-ICP-ES or ICP-MS	INAA and aqua regia-ICP-ES of ICP-MS	
Predicted Anomaly Contrast (Maximum/background)	1.5 (Cu) 1.5 (Au)	1.5 (Ni)	2(Au) to 5 (As)	2(Au) to 5 (As)	Contrast is element and sample medium dependent.
Predicted Anomaly size	<1 km along drainage	<1 km along along drainage	2 km down-ice by 1 km wide	2 km down-ice by 0.5 km wide	Anomaly size is element dependent especially for soils.

anomaly, shown in the stream draining the Samatosum and Rea Gold deposits, is based on the results of the regional geochemical survey (Geological Survey of Canada 1977). However, only Ni and Ba are weakly anomalous in stream water directly draining the two deposits (Lett *et al.* 1998*a*).

A practical application of this model is to develop geochemical exploration criteria for massive sulphide deposits similar to the Samatosum deposit. For example, weakly anomalous (95th to 98th percentile range) Cu, Au, Ba and Cr in drainage sediments collected in a regional survey (one sample per 10 to 15 km^2 scale) may indicate those watersheds likely to contain Au–Ag–Pb–Zn–Ba mineralization. However, stream water geochemistry cannot reliably detect sulphide mineralization in this area because weathering of carbonate-rich bedrock units in the Eagle Bay assemblage increases water pH and inhibits the dispersion of mobile elements (e.g. Cu, Zn). Till geochemistry can clearly detect the presence of sulphide mineralization through anomalous multi-element dispersal trains extending for up to 2 km down-ice from the mineralized bedrock source area. In Table 8 the key pathfinder elements for stream sediment, stream water and till, the predicted anomaly size and anomaly contrast have been summarized from the data presented in this paper.

Conclusions

Multi-element stream sediment and till geochemical anomalies are effective exploration methods for Pb–Zn–Ag–Cu–Au–Ba volcanogenic massive sulphide deposits in central BC. However, the application of stream water geochemistry to exploration for massive sulphides is limited because of the high dilution of elements in streams and pH control. Gold, Pb, Zn, Cu, Ag, As, and Sb in basal till are geochemical pathfinders for massive sulphides. Elevated Cr, Ni, Mg and V in till reflect underlying mafic volcanic rocks at the Samatosum sulphide deposit. Higher Ba in the till southeast of the Rea Gold deposit may be a good indicator for barite associated with the massive sulphide mineralization exposed at the bedrock-till interface.

The author very much appreciated the helpful and constructive comments on an early draft of this paper by A. Dixon-Warren. A. Yeow assisted with collection of the detailed till samples. Critical reviews of the paper by I. Thomson and S. Sibbick are greatly appreciated. A. Bichler is especially thanked for his help in preparing the figures and L. Englund and G. Lopez for carefully proof reading the text.

References

BOBROWSKY, P. T., LEBOE, E. R., DIXON-WARREN, A., LEDWON, A., MACDOUGALL, D. & SIBBICK, S. J. 1997. *Till Geochemistry of the Adams Plateau – North Barriere Lake Area (82M/4 and 5)*. British Columbia Ministry of Employment and Investment Open File **1997–9**.

BRADSHAW, P. M. D. 1975. Conceptual models in exploration geochemistry, the Canadian Cordillera and Canadian Shield. *Journal of Geochemical Exploration*, **4**, 1–213.

DAVIDSON, A. J. & PERIE, I. D. 1987. The Rea Gold massive sulphide deposit, Adams Lake, B.C.; a geochemical exploration study. *Journal of Geochemical Exploration*, **29**, 390.

DIXON-WARREN, A. 1998. *Terrain Mapping and Drift prospecting in South Central British Columbia*. Unpublished MSc Thesis, Simon Fraser University.

DIXON-WARREN, A. BOBROWSKY, P. T., LEBOE, E. R. & LEDWON. A. 1997. Eagle Bay Project: Surficial geology of the Adams Plateau (82M/4) and North Barriere Lake (82M/5) map areas. *In:* LEFEBURE, D. V., MCMILLAN, W. J., & MCARTHUR. G. (eds) *Geological Fieldwork 1996*. British Columbia Ministry of Employment and Investment, Paper **1997-1**, 405–412.

FRIESEN, R. G. 1990. Geology of the Samatosum deposit; *In:* MEYERS, R. E., WILTON, H. P. & LEGUN, A. S. (eds) *Mineral Deposits of the Southern Canadian Cordillera*. Geological Association of Canada-Mineralogical Association of Canada, Field Trip Guide Book **B2**.

GEOLOGICAL SURVEY OF CANADA. 1977. *National Uranium Reconnaissance Program Map 27-1977*. Geological Survey of Canada Open File **516**.

HOAG, R. B. Jr. & WEBBER, G. R. 1976. Significance for mineral exploration of sulphate concentrations in groundwaters. *Canadian Institute of Mining and Metallurgy*, Bulletin **69**, 86–91.

HÖY, T. 1991. Volcanogenic massive sulphide deposits in British Columbia. *In:* MCMILLAN, W. J. (ed.) *Ore Deposits, Tectonics and Metallogeny in the Canadian Cordillera*. British Columbia Ministry of Energy, Mines and Petroleum Resources, Paper **1991-4**, 89–199.

LEFEBURE, D. V. & RAY, G. *Selected British Columbia Mineral Deposit Profiles, Volume 1 – Metallic deposits*. British Columbia Ministry of Energy, Mines and Petroleum Resources Open File **1995-20**.

LETT, R. E., SIBBICK, S. & RUNNELLS, J. L. 1998a. *Regional stream water geochemistry of the Adams Lake-North Barriere Lake Area, British Columbia*. British Columbia Ministry of Energy and Mines Open File **1998-9**.

LETT, R. E., BOBROWSKY, P., CATHRO, M. & YEOW, A. 1998b. Geochemical pathfinders for massive sulphides in the southern Kootenay Terrane. *In:* LEFEBURE, D. V. & MCMILLAN, W. J. (eds) *Geological Fieldwork 1997*. British Columbia Ministry of Employment and Investment, Paper **1998-1**, 15-1–15-9.

MATYSEK, P. F., JACKAMAN, W., GRAVEL, J. L., SIBBICK, S. J. & FEULGEN, S. 1991. *British Columbia Regional Geochemical Survey, Seymour Arm (NTS 82M)*. British Columbia Ministry of Energy, Mines and Petroleum Resources RGS **33**.

PIRIE, I. 1989. The Samatosum deposit. *The Northern Miner*, **75**, 15–18.

THE NORTHERN MINER. 1987. Adit for access on Rea Gold bet at Adams Lake. *The Northern Miner*, **73**, 3.

SCHIARIZZA, P. & PRETO, V. A. 1987. *Geology of the Adams Plateau-Clearwater-Vavenby area*. British Columbia Ministry of Energy, Mines and Petroleum Resources Paper **1987-2**.

SHILTS, W. W. 1975. Principals of geochemical exploration for sulphide deposits using shallow samples of glacial drift. *Canadian Institute of Mining and Metallurgy Bulletin*, **68**, 73–80.

SIBBICK, J., RUNNELLS, J. L & LETT, R. E. 1997. Eagle Bay project: Regional hydrogeochemical survey and geochemical orientation studies (82M4/5) *In:* LEFEBURE, D. V. & MCMILLAN, W. J. (eds) *Geological Fieldwork 1997*. British Columbia Ministry of Employment and Investment, Paper **1997-1**, 423–427.

YEOW, A. 1998. *Pathfinder Elements in Soil and Till, Samatosum Mountain, British Columbia*. Unpublished BSc Thesis, University of British Columbia.

Glacial transport and secondary hydromorphic metal mobilization: examples from the southern interior of British Columbia, Canada

ROGER C. PAULEN

School of Earth and Ocean Sciences University of Victoria, Victoria, British Columbia V8W 9N3, Canada (e-mail: paulen@home.com)

Abstract: Glacial transport of trace elements was studied at seven mineral occurrences in the southern interior of British Columbia; a region where mineral exploration is hampered by the scarcity of bedrock outcrop and by a variable sediment thickness associated with the Cordilleran Ice Sheet. The till deposited in the region was, for the most part, a product of the last glacial period. A review of previous geochemical studies conducted by the mineral exploration industry provides an indication to the variable configuration of the local dispersal patterns in the area. Dispersal trains in till are short, generally < 1–2 km, rarely exceeding 10 km in length and are usually proximal to bedrock source. They are commonly ribbon-shaped and rarely exceed 1 km in width. Observed dispersal patterns suggest that drastic changes in topography might have affected basal ice velocity which increased the distance of glacial transport. In addition, the distance that separates bedrock mineralization from its surficial geochemical expression in till varies with drift thickness and topography. Several glacial dispersal trains have been modified in shape by secondary hydromorphic dispersion.

Mineral exploration in the southern interior of British Columbia is hampered by a scarcity of bedrock outcrop and by a variable sediment thickness associated with the glaciation of the Canadian Cordillera. Studies of the physical and compositional characteristics of glacial sediments in the Cordillera have for some years, provided basic models for the application of drift prospecting methodologies in this topographically rugged and geologically complex terrain (e.g. Sibbick *et al.* 1992; Kerr *et al.* 1993; Proudfoot 1993; Bobrowsky *et al.* 1995; O'Brien *et al.* 1997; Paulen *et al.* 1998a; Plouffe 1999). The methodology developed in the Canadian Shield is not necessarily applicable to the Cordillera because of the physical and geological difference between both regions.

It is important to recognize site-specific indicators when dealing with small dispersal plumes within a challenging topographical setting as is common in British Columbia. As reviewed by Bobrowsky (1999), ice flow in montane settings, is for the large part, affected by topography with the major mountain systems as the principal source areas of glaciers (Clague 1989; Bennett & Glasser 1996). This paper presents several examples of geochemical anomalies and their respective dispersal characteristics within the Cordillera. Its purpose is to illustrate the influences of topography, sample medium, and element mobility on the shape of glacial dispersal trains.

Background and setting

This paper focuses on seven sites located in south central British Columbia (Fig. 1). The sites lie within the southern part of the Shuswap Highland and the northeastern part of the Thompson Plateau. This area is within the Interior Plateau (Matthews 1986) which is characterized by moderate to high relief, and glaciated and fluvially dissected topography. Elevations range from 360 to 2380 m above sea level. Most of the area is covered by unconsolidated sediments of mixed genesis and of variable thickness, but rarely exceeding a few tens of metres. Till dominates the landscape, followed in turn by colluvial, glaciofluvial, fluvial, glaciolacustrine and organic sediments. Given the scarcity of bedrock outcrop and the abundance of glacial sediments, the region is extremely favourable for drift prospecting.

The area lies within a belt of structurally

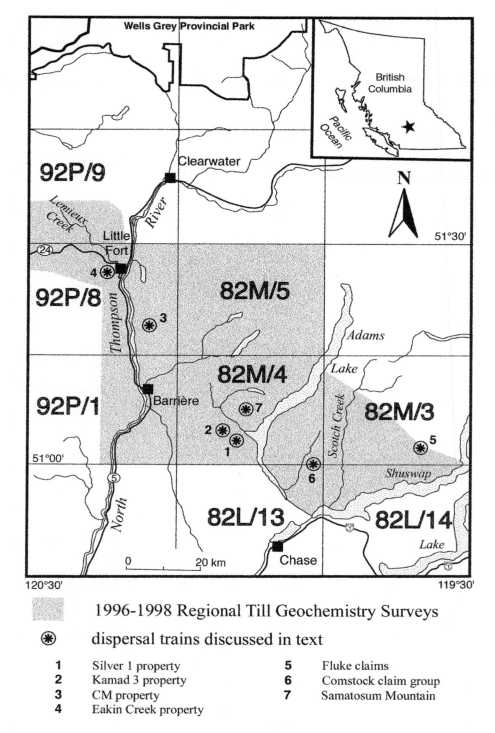

Fig. 1. Location of the Eagle Bay drift exploration project (1996–1998) study area in south central British Columbia. Numbers 1 to 7 refer to mineral occurrences and deposits.

complex, low-grade metamorphic rocks of the Intermontane Belt (Schiarizza & Preto 1987). Lower Palaeozoic to Mississippian rocks of the Eagle Bay Assemblage (Kootenay Terrane) and Permian to Devonian rocks of the Fennell Formation (Slide Mountain Terrane) underlie a major part of the area. The geology contains lithological suites hosting polymetallic precious, sedimentary exhalative and Noranda/Kuroko type volcanogenic massive sulphide (VMS) base metal deposits (Höy 1999).

Glacial history and sedimentation

The most recent affect on the present day landscape of south central British Columbia is the result of two glacial cycles, one interglacial and early Holocene erosion and sedimentation (Fulton 1975; Fulton & Smith 1978; Clague 1989; Ryder et al. 1991). However, usually only evidence from the last glacial event and various post-glacial processes are found.

At the onset of the Lake Wisconsinan Fraser Glaciation, ice build-up began in the Coast, Cariboo and Monashee Mountains. Valley glaciers formed piedmont lobes in the Interior Plateau, and eventually coalesced to form the Cordilleran Ice Sheet (Ryder et al. 1991). Ice sheet margins reached a maximum elevation between 2200 and 2400 m, burying the entire Interior Plateau beneath an ice cap by c. 19 Ka. At Fraser Glaciation maximum, regional ice flow was to the south–southeast on the Bonaparte and Adams plateaus (Tipper 1971) with deviations up to 45° (Fulton et al. 1986). This deviation was particularly noted in the eastern part of the study area, where ice from the north and west coalesced with ice flowing from the Monashee Mountains, in the east, and was subsequently directed into the Shuswap Basin.

Deglaciation of the Interior Plateau was rapid; the equilibrium line was likely to have rose considerably, reducing the area of accumulation for the Cordilleran Ice Sheet, and the ice mass decayed by downwasting and complex frontal retreat. Post glacial radiocarbon dates between 11.3 and 9.8 Ka (Clague 1980) indicate that deglaciation began about 12 Ka. Intense and vigorous erosion immediately followed deglaciation in the early Holocene. As sediment loads decreased, deposition was replaced by erosion, and water courses cut down through valley fills, leaving glaciofluvial terraces abandoned high on valley sides. Following the complete deglaciation of the region, unstable and unvegetated slopes were highly susceptible to erosion. Intense mass wasting of surface deposits on oversteepened valley slopes resulted in the deposition of colluvial fans and aprons along valley bottoms. Most post-glacial deposition occurred within the first few hundred years of deglaciation (Fulton 1967). The modern drainage pattern was established prior to 8.9 Ka (Dyck et al. 1965; Fulton 1969). Fluvial fan deposits, active talus slopes and present rivers and floodplains typify the modern sedimentation in the area.

Throughout the region, bedrock is mantled by variable amounts of massive, very poorly-sorted matrix-supported lodgement till. Deposits range in thickness from thin (<1 m) veneers to thick (>10 m) blankets. The till facies are variable, depending on local bedrock lithologies. Till thickness on the plateau is commonly <3 m, whereas valley slopes contain till thickness >10 m. Valley bottoms are likely to contain till, but are often overlain by thick glaciofluvial and glaciolacustrine sediments. Ablation till occurs in areas of hummocky terrain (Levson 2001), and deposits of ablation till occur as a thin mantle overlying basal till and/or bedrock on the higher portions of the plateaus.

Ice-flow indicators

The striation record in the study area is poor due to the lack of unweathered outcrop exposure. However, there is an abundance of sculpted landforms on the plateaus that provide regional ice-flow information during the peak of glacial activity (Fig. 2). In general, local palaeo-ice flow parallels the regional south to southeast ice flow (Fulton et al. 1986). Detailed ice flow for the study area is as follows. In the northwestern area, regional ice-flow directions are to the southeast, with ice buildup and deglacial deviations to the south in the North Thompson River valley. In the centre of the study area, the ice flowed in a southerly direction across the Adams Plateau, except in areas of variable relief where topography deflected ice flow. In the easternmost region, the landforms and striae show a south-southwesterly flow direction as ice was diverted in the Shuswap Basin. The location of southeast and southwest ice convergence is unknown due to the poor striation record, but the coalescence likely occurred near the southeastern edge of the Adams Plateau, west of Scotch Creek. Ice flow here was deflected southward, into the Shuswap Basin with a local deviation up to 45°.

Methodology

A regional till geochemistry program conducted in the area from 1996 to 1998 (Paulen et al.

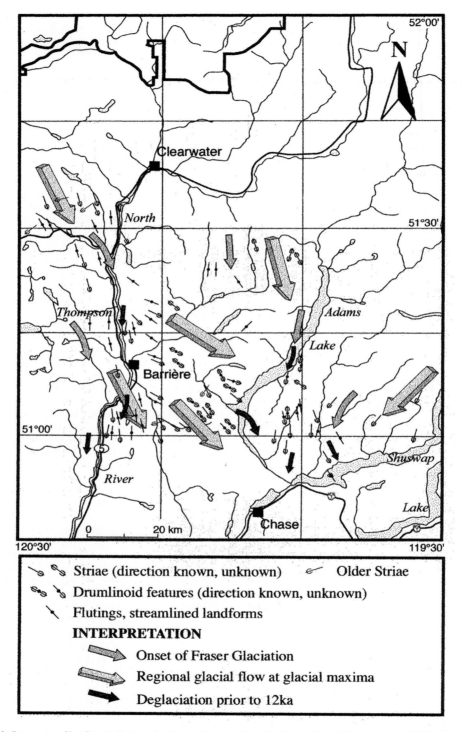

Fig. 2. Summary of ice flow indicators for the study area. Compiled from Dixon-Warren *et al.* (1997); Leboe *et al.* (1997); Paulen *et al.* (1998b, 1998c; 1999) and new field observations.

1999) prompted an exhaustive review of existing government assessment reports detailing drift prospecting activities by exploration companies in the study area. These reports provide indications of the nature of glacial dispersal and the configuration of the dispersal trains with respect to local topography. The seven case studies discussed in this paper provide insights to the dispersion and mobility of various elements and reflect both local and regional prospecting sampling programs, using various sample media. Assessment reports commonly present raw geochemical data plotted on a property grid, with little or no interpretation of the results. Soil samples were likely to be collected rapidly, with little or no attention paid to the nature of the surficial sediments. Samples were analysed for target pathfinder elements, such as Au, As, Ag, Cu, Pb and Zn. Geochemical data were often presented with arbitrarily chosen element concentration contours, commonly ignoring the problems associated with elevated background concentrations and manipulating threshold values to clarify results. In most cases, quality control data for field and analytical duplicates were either absent or ignored in the presentation of the geochemical data. Reproducibility of the analytical results gleaned from some of the assessment reports is unknown. With these limitations in mind, the following case studies were chosen. They represent the best and most reliable examples of industry geochemical survey methodology reported in the assessment files for the region.

The dispersal trains are documented here to aid the exploration industry in the interpretation of B-Horizon and C-Horizon soil sampling surveys. The contoured property maps presented here represent those commonly found in assessment reports. Glacial and hydromorphic dispersal of elements are discussed for each case study. Using simple threshold values to better define potential mineralized bedrock sources is demonstrated at the Comstock Claim group. New data from the aforementioned regional till survey also impart understanding of the relationship between glacial dispersal in areas with drastic changes in topography.

Case studies

Seven sites were examined in the study area (see Fig. 1): (1) a Co and As dispersal train at Silver 1 property; (2) a Zn dispersal train at the Kamad 3 property; (3) a Cu dispersal train at the CM property; (4) a Au dispersal train at the Cedar I to VI mineral claims (Eakin Creek property); (5)

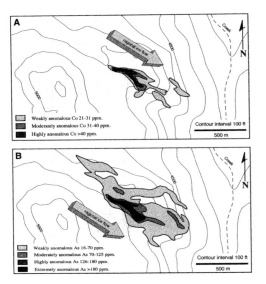

Fig. 3. Co (A) and As (B) concentrations (< 177 µm) in B-Horizon soil for the Silver 1 property (Modified from Bobrowsky et al. 1997a).

a Pb dispersal train at the Fluke claims (Crowfoot property); (6) a series of Ag dispersal trains at the Comstock claim group; and (7) As, Ag, Hg and Zn dispersal trains at Samatosum Mountain. These sites encompass several types of mineralization including porphyry Cu, Au associated with Cretaceous Plutonic emplacement, volcanogenic precious-metal enriched and base metal massive sulphides, disseminated Cu–Mo and skarn mineralization. These examples were chosen from a variety of terrains ranging from highland plateaus to steep-walled valleys. The dispersal trains were detected in various materials including basal lodgement till, ablation till and colluviated sediments. The reported dispersal trains range from 250 to 2500 m in length and are only 100 to 500 m wide, with the exception of the dispersal train at Samatosum Mountain and Rea Gold deposits which extends over 10 km.

Silver 1 property

The property is located west of Adams Lake, on the Adams Plateau in a broad saddle between two bedrock knobs. The surficial geology consists of a thick till (>2 m) with sparse rock outcrops and minor organic deposits. Drumlinoid forms are developed at higher elevations (Dixon-Warren et al. 1997). A total of 1257 B-horizon soil samples were collected by Esso

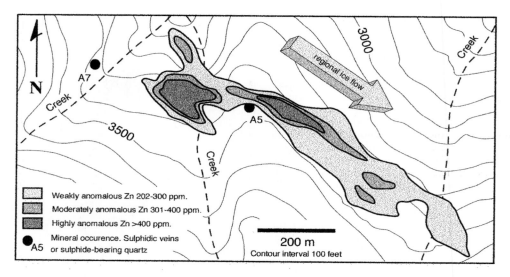

Fig. 4. Zn concentrations (< 177 μm) in B-Horizon soils for the Kamad 3 property. (Modified from Bobrowsky *et al.* 1997*a*).

Minerals Canada spaced 25 m apart on lines spaced 100 m apart (Richards 1989). Samples were subjected to multi-element analyses by aqua regia leach and inductive coupled plasma - emission spectrometry (ICP-ES). Background values were calculated by Esso Minerals to be < mean + 1 standard deviation (S.D.). Weakly anomalous, moderately anomalous and highly anomalous values were defined as mean + 1 S.D., mean + 2 S.D. and mean + 3 S.D., respectively (Richards 1989). The B-horizon soil geochemistry data display two forms of geochemical anomaly patterns.

The Co dispersal train shows a fan-shaped down-ice dispersal pattern with a sharp apex and broader fan tail extending about 600 m in length from the northwest to the southeast (Fig. 3a). The trend of this anomaly parallels the regional ice-flow direction and is interpreted to reflect clastic dispersal (cf. DiLabio 1990). As displays a less distinct dispersal pattern (Fig. 3b). The long axis of the train is parallel to the northwest to southeast regional ice flow, but the greater mobility of As in the surficial environment is evident by a secondary downslope, northeasterly component that overprints the original southeast pattern. Hydromorphic dispersion may have modified the original shape of the dispersal train creating an anomaly over 1 km long and 0.5 km wide. Co concentrations illustrate a single vector configuration parallel to ice flow, and As levels illustrate a compound two vector configuration consisting of ice-parallel plus downslope directional movements.

Kamad 3 property

This property also is located west of Adams Lake, about 2 km N–NW of the Silver 1 property and, along strike of it. The property sits on the steep upper north-facing slope of a large valley where Sinmax Creek drains into Adams Lake. Several intermittent creeks and gullies cross the property and surficial materials consist of colluviated basal till blanket overlying a thin blanket of basal till (Dixon-Warren *et al.* 1997). A total of 885 B-horizon soil samples were collected by Kamad Silver Co. Ltd. spaced 25 m apart along lines spaced 100 m apart (Marr 1989). Duplicate samples were collected at every fifteenth station. Samples were analysed by aqua regia leach and multi-element ICP-ES. Contour intervals were selected by the company at mean + 1 S.D. and mean + 2 S.D. (Marr 1989). The B-horizon samples illustrate a classic cigar-shaped dispersal plume.

The Zn dispersal train parallels the regional ice flow to the southeast. This particular anomaly is about 1 km long and 200 m wide (Fig. 4). Although the anomaly crosses a steep slope and follows the contours of the land, there is no obvious secondary downslope dispersion except at the head of the dispersal train, where it extends to the northwest down a creek valley. The anomaly was detected only 200 m down-ice from the known source, A7, with a moderately thin (<5 m) drift cover. The other source, A5, may also have contributed to the overall dispersal train; the A7 source contributed mostly to the head of the anomaly and the A5 contributed mostly to the tail.

CM property

The CM property, located 15 km N of Barriere and east of the North Thompson River, is situated on the east slope of the North Thompson River valley, directly above glaciofluvial terraces. Surficial material consists of a colluvial veneer over a till blanket (Paulen *et al.* 1998*b*). The slope is heavily gullied by post-glacial erosion. A reconnaissance survey was conducted with 25 samples of surface colluvium collected by Inco Exploration and Technical Services (Casselman 1993). Samples were analysed for trace elements by ICP-ES following an aqua regia digestion. There is no record of quality control measures nor statistical control on plotted values reported by Casselman (1993). Results indicated several ribbon-shaped dispersal trains for Cu (Fig. 5). The source remains unknown, but recent work by Lett & Jackaman (2000) indicates there may be a possible source associated with a small pluton to the northwest.

The dispersal pattern parallels the local ice-flow which was directed down the North Thompson River Valley in the early and late stages of the Fraser Glaciation. There is little or no indication of regional dispersal to the southeast during the glacial maximum. This lack of evidence of southeast ice flow could indicate that during the peak of Fraser Glaciation, this valley was filled with relatively inactive ice, as has been proposed by Ryder *et al.* (1991). The Cu anomaly is almost 2.5 km long and only 200 m wide (Fig. 5). The anomaly crosses a steep slope and follows the contours of the land with a significant secondary downslope dispersion pattern in the centre of the dispersal train which almost doubles the train's width to 400 m. The secondary hydromorphic dispersion increases the size of the target for regional exploration but may also mislead follow-up exploration. The Cu dispersal train illustrates a two-vector configuration of ice-parallel plus downslope directional transport.

Eakin Creek property

High Au concentrations have been detected at the Eakin Creek property located 8 km NW of Little Fort and directly north of Eakin Creek. The source of the Au remains unknown but is believed to be from vein deposits typical of the

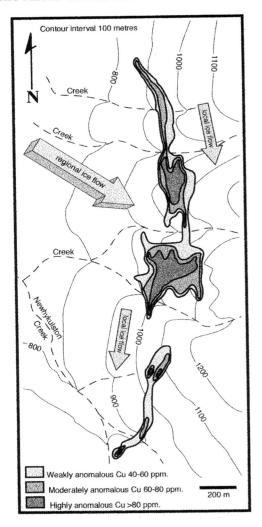

Fig. 5. Cu concentrations (<177 μm) in B-Horizon colluvium for the CM property (Modified from Paulen *et al.* 1998*a*).

local bedrock. The property straddles the eastern edge of the Bonaparte Plateau and the south-facing slope of the Eakin Creek valley. The surficial geology consists of thick till (>5 m) with a thin veneer of colluvium on the steeper portions of the property (Paulen *et al.* 1998*b*). Craven Resources Inc. collected 649 C-horizon soil samples spaced 50 m apart along lines spaced 200 m apart (Yorston & Ikona 1985). Samples were analysed by ICP-ES following an aqua regia digestion but the company did not report quality control data or statistical analysis.

The till samples illustrate two forms of

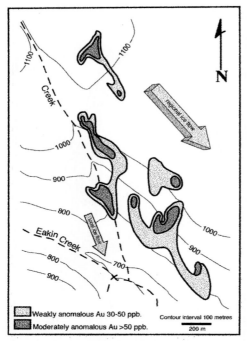

Fig. 6. Au concentrations (<177 μm) in C-horizon for the Cedar I to VI mineral claims (Eakin Creek) (Modified from Paulen *et al.* 1998a).

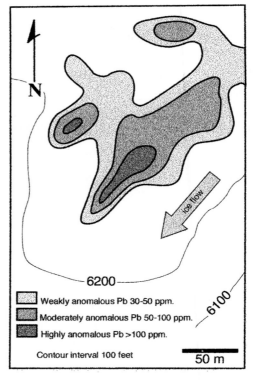

Fig. 7. Pb concentrations (<177 μm) in B-horizon soil for the Fluke claims (Crowfoot Property) (Modified from Paulen *et al.* 1999).

geochemical dispersal patterns (Fig. 6). The Au dispersal pattern shows a small fan-shaped down-ice dispersal pattern on the higher, flat ground with the axis of symmetry paralleling the regional ice flow from northwest to southeast and thus reflecting typical glacial dispersal. On the slopes north of Eakin Creek, secondary downslope dispersion has modified the original clastic dispersal fan, extending the southern boundary of the anomaly southward. The Au patterns illustrate a compound two-vector configuration consisting of glacial plus hydromorphic dispersal likely derived from several dispersed sources of Au.

Fluke claims (Crowfoot property)

This property-scale geochemical sampling program was undertaken in the flat alpine saddle on Crowfoot Mountain, north of Shuswap Lake. The property is at a high elevation (>1800 m), above the tree-line and subject to snow cover about six months of the year. An old shaft and mine workings from the 1920s and 1930s are present at the site. The surficial geology is typical of the higher peaks in the region, with bouldery ablation till deposited from downwasting of the Cordilleran Ice Sheet during early deglaciation. Drift cover is thin, commonly <1 m thick.

Although the drift cover is thin, a soil geochemical survey was conducted instead of a geophysical survey because of the conductive nature of the underlying graphitic schistose bedrock. A total of 275 B-horizon soil samples were collected by Resources Ltd., spaced 30 m apart along lines spaced 30 m apart (Allen 1977). The samples were analysed by atomic absorption spectrometry following an aqua regia digestion. There is no record of quality control data or statistical analysis of the results.

Samples taken from the B-horizon soil illustrate an amoeboid-shaped anomaly for Pb (Fig. 7) which is 250 m long and <100 m wide (Allen 1977). The anomaly parallels the regional ice flow to the south–southwest but the weakly anomalous Pb concentrations (30–50 ppm) are more extensive to the northeast. This pattern could reflect contamination from old mine workings or the possibility of higher grade bedrock mineralization near the tail of the dispersal train in an area of very thin over-

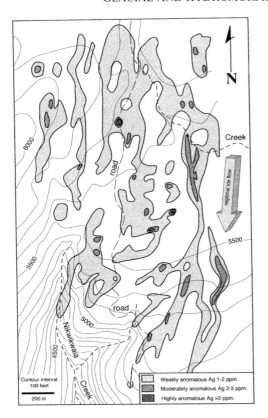 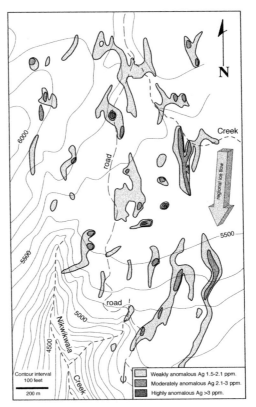

Fig. 8. Ag concentrations (<177 (μm) in B-Horizon soil from the Comstock Claim Group. Regional ice flow directions parallel bedrock strike. Several stratabound bodies of high grade massive sulphides underlie the region. Converging ice flows also modify the dispersal plumes south–southwest, which originally were emplaced south–southeast. (Modified from Wells 1987).

Fig. 9 Comstock Claim Group Ag dispersal train plotted with threshold values calculated by the author using percentile values of 90, 98 and 99.5.

burden. The possibility of post-glacial transport of Pb by surface and groundwater could produce this pattern since the drift cover is very stone-rich and thin. The dispersal pattern reflects the difficulty of interpreting geochemical patterns for soil developed on ablation till which can be locally derived or transported tens of kilometres (Bobrowsky 1999).

Comstock claim group

The Comstock claims are situated on high ground east of Nikwikwaia Creek, c. 5 km east of Adams Lake. The property is at the southern edge of the Adams Plateau. Nikwikwaia Creek and its tributaries have cut deep, steep-sided valleys into the plateau with up to 550 m relief.

The surficial geology consists of till blanket of variable thickness with minor bedrock outcrops at the ridge tops and colluvium on the steep valley slopes. A total of 1173 B-horizon soil samples were collected by Lacana Mining Corporation spaced 25 m apart on lines 25 m apart to form a grid (Wells 1987). Samples were analysed for multiple elements by ICP-ES following an aqua regia digestion. Quality control data were not reported by the company. Contour intervals were determined using mean and S.D. values. Currently, the location of mineralized bedrock source is unknown, but several dispersed bedrock sources are suspected.

B-horizon soil samples show a large number of small ribbon-shaped anomalies (Fig. 8). The background Ag values (<mean + 1 S.D.) in the area generally vary between 500 to 1500 ppb. Areas of anomalous Ag concentrations are c. 400 to 900 m in length and commonly <50 m wide. The shape of the anomalies suggests earlier glacial transport to the southeast and then a

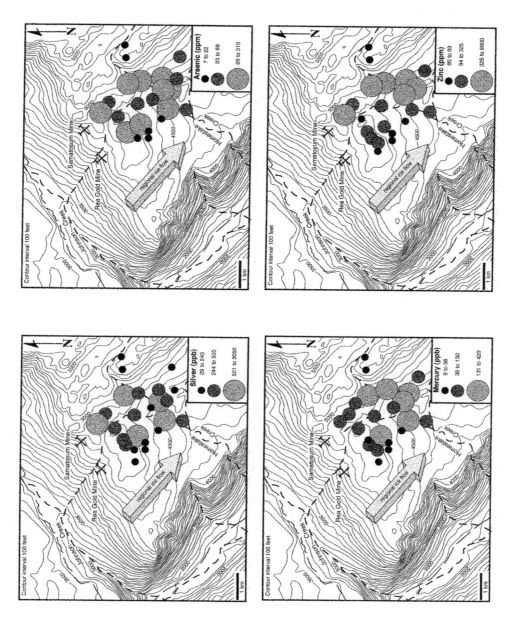

Fig. 10. Ag, As, Hg and Zn concentrations (<63 μm) in C-Horizon soils (lodgement till) at Samatosum Mountain.

shift in transport to the south–southwest. This area is within the zone of convergence of southeast and southwest flowing ice (Fig. 2). The Ag dispersal trains could represent a cumulative effect of transport to the southeast and southwest. In addition, there is the potential that hydromorphic dispersion downslope has modified the southern anomalies and transported Ag towards Nikwikwaia Creek.

Statistical analysis of the geochemical data from this property was conducted by the author to clarify the anomalies in a region characteristic of high background values to eliminate false anomalies and reduce the target area for follow-up exploration. For this property, the dataset was ranked and contours were chosen at the 90th, 98th and 99.5th percentile. The resulting map (Fig. 9) shows distinct narrow ribbon-shaped anomalies that indicate dispersal from several potential bedrock sources.

Samatosum Mountain

Samatosum Mountain is host to two past producing mines, Rea Gold and Samatosum. These mines are examples of volcanogenic Au–Cu–Pb–Zn sulphide and barite deposits hosted by felsic volcanic rocks of the Eagle Bay Assemblage (Höy 1991). The surficial geology consists of a thick (> 5 m) till blanket with minor bedrock outcrops at ridge tops (Dixon-Warren *et al.* 1997).

Regional-scale till sampling (see Fig. 1) by the British Columbia Geological Survey, has detected a major dispersal train trending southeast from the Samatosum deposit (Bobrowsky *et al.* 1997*b*). A follow-up till survey of more closely spaced samples was conducted in 1997 to illustrate how various pathfinder elements can be used to distinguish between different sources of mineralized bedrock (Lett *et al.* 1999). A total of 24 C-horizon till samples were collected from surface pits, *c.* 500 m apart.

The distribution of As, Ag, Hg and Zn in till samples reveals high concentrations forming long down-ice dispersal trains from the Samatosum and Rea deposits (Fig. 10). The anomalous zones in till are parallel to the direction of regional ice flow. Concentrations decrease sharply directly southeast of the deposit and then increase to a maximum 1.8 km down-ice (Lett *et al.* 1999). Anomalies can be traced at surface for over 10 km down-ice (Bobrowsky *et al.* 1997*b*). Topography is a likely influencing factor for hydromorphic extension of the tail of the dispersal train, where the land surface steeply drops over 900 m to Adams Lake.

Comparisons of metal concentrations in the B-horizon soil, C-horizon till and tree tissue, with respect to mineral associations and transport distances, is in progress (Lett *et al.* 1998) and may provide further insights to the extent of hydromorphic dispersion at the tail end of the dispersal train. A case study of detailed exploration geochemistry and pathfinder elements at Samatosum Mountain is presented in this volume (Lett 2001).

Discussion

As noted by Shilts (1976), the surface expression of a dispersal train depends upon basal ice conditions and local and regional ice dynamics. Authors have long recognized that dispersal trains can curve along topographical irregularities (e.g. Goldthwait 1968; Garrett 1971). Multiple tills, multiple ice advances, multiple glacial flow directions, and multiple ore sources can complicate glacial dispersal patterns. Mass movement is another dispersal agent which can add to the glacial dispersal vectors from the original bedrock source in steep-slope terrain. In addition to detrital transport, ions of Cu, Zn, As and Ag can be mobile in the surface environment and can be transported and concentrated in B-horizon soil by hydromorphic processes (Rose *et al.* 1979).

Dispersal trains in the area are thin and are classified as 'ribbons' (cf. Klassen 1997, 2001). Ribbons are streamlined along a single ice flow direction, with a width comparable to the width of the bedrock source. However, problems do arise with determining glacial transport distance in the study area because in searching for stratabound base metal deposits, the regional ice flow is often parallel to the regional bedrock strike. The question arises: are we tracing back single or multiple dispersal trains from one or more stratabound mineralized sources?

Palimpsest dispersal trains are unlikely to be found in the study area and long dispersal trains commonly coincide with drastic topographical changes, perhaps suggestive of a linear decay curve associated with a change in ice flow velocity. Observations of distributions of source components in the study area range from within < 250 m to a maximum of 10 km from source, but generally are 1–4 km down-ice from the bedrock source. This is fairly short in contrast to regional-scale dispersal trains reported in the Canadian Shield (Shilts 1976; Kaszycki *et al.* 1988; Batterson 1989) and the even larger far-travelled dispersal trains > 50 km in length reported for the dispersion of carbonate from the Hudson Bay Lowlands and heavy minerals derived from the Canadian Shield, including

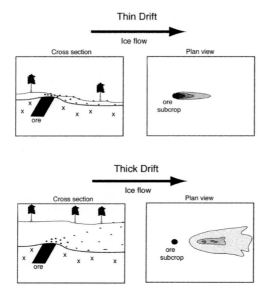

Fig. 11. Idealized geochemical signatures for a glacial dispersal train produced by a single ice-flow event. Black bodies represent ore-bearing float. Grey tones indicate relative concentration of metals in till. In areas of thin glacial drift, the anomalous material is concentrated immediately over the prospect and dispersal trains are ribbon shaped. In areas of thicker till, the geochemical signature is less concentrated at surface due to dilution and dispersal trains are larger and fan shaped. The distance between bedrock ore and its surficial expression in till depends on drift thickness and topography. Model modified from Drake (1983) and Miller (1984).

those from kimberlites (Shilts *et al.* 1979; Kaszycki & DiLabio 1986; Dyke & Morris 1988; Thorleifson & Garrett 1993; Thorleifson & Kristjansson 1993). The limited transport distance for the study area does, however, fit into the bulk of the half-distance of glacial dispersal trains, on the order of hundreds of metres to <10 km (Drake 1983; Clark 1987; Coker & DiLabio 1989; Bouchard & Salonen 1990; Finck & Stea 1995).

The examples presented here are typical surface soil and till geochemistry anomalies encountered in the southern interior of British Columbia. Generally, the surface expressions of the reported dispersal trains are small compared to those found on the Canadian Shield. However, they are still considerably larger targets than the bedrock source and can be detected with effective regional sampling. The thin drift mantling of the upland plateaus and the defined valley systems provide an excellent landscape for drift prospecting. Till in this region directly overlies the bedrock and is representative of the last glaciation. Thin till, typically seen on the upland plateaus, is usually locally derived (Bobrowsky *et al.* 1995). If there are no complications of multiple ice-flow directions to interpret, then a dispersal train should reflect only one iceflow event.

The reported dispersal trains conform to classic down-ice shapes, and are usually proximal to the bedrock source. These clastic dispersal patterns are in most cases parallel to ice flow but can be imprinted by downslope hydromorphic dispersion. Secondary dispersion occurs after glaciation, and is the result of chemical remobilization processes in the surface environment. Groundwater leaches and remobilizes certain metals, which are later adsorbed onto phyllosilicates, hydroxides or oxides (e.g. Reimann & Caritat 1998). Material affected by secondary dispersion by downslope movement can be avoided, although not eliminated, by sampling C-horizon till at greater depths.

In areas of thick drift, such as Samatosum Mountain, the till overlying and adjacent to the deposit does not reflect the underlying bedrock mineralogy until at least 1.8 km down-ice and the dispersal train is considerably longer than those in other parts of the Interior Plateau. The dispersal train is not detectable at the surface until some distance (>2 km) down-ice from the bedrock source, similar to the dispersal train postulated in Miller's (1984) model (Fig. 11). Longer dispersal trains also are found in areas with drastic differences in topography, such as at the edges of a plateau, where the dispersal train can be several kilometres in length. Longer trains here are likely to be due to a combination of a change in ice flow conditions associated with the topography and post-glacial hydromorphic dispersion.

Conclusions

Till and B-horizon soil sampling and geochemical analysis have been completed at several locations in the southern interior (Lett *et al.* 1998, 2001), which encompass several types of polymetallic mineralization. The challenges to drift prospecting in this area are the nature and structure of the mineralized bedrock. Regional ice flow directions are commonly parallel to the regional bedrock strike of the Eagle Bay Assemblage. A series of small high-grade stratabound massive and precious metal enriched sulphide lenses are known to occur within the sequences (Höy 1999). The difficulty lies in detecting these multiple lens-shaped deposits individually within till deposited by ice flow parallel to bedrock structure. The potential exists that several dispersal trains may coalesce along bedrock strike, mimicking the dispersal pattern of one larger deposit. Identifying individual trains in these regions is difficult if multiple mineralized sources are suspected. The presence of multiple mineral deposits can also supply metal-rich till over a large area, producing irregular and higher background concentrations. Applying the appropriate threshold values for contouring geochemical surveys is essential in such terrain.

The lack of bedrock exposure in some areas indicates that the proper genetic interpretation of glacial overburden is essential in delimiting and understanding potential areas of mineralization. Knowledge of ice-flow history is essential for a drift exploration program because known geochemical anomalies in till exhibit classic down-ice dispersal parallel to the regional ice flow patterns. Secondary hydromorphic redistribution of metals downslope from anomalous metal-rich till as well as mineralized bedrock has been noted in places, especially where the B-horizon soil was sampled. The distance between the underlying bedrock mineralization and its signature in the overlying surficial sediments increases with increasing till thickness. In summary, dispersal trains are small, rarely exceeding 10 km and usually very proximal to bedrock source (0–3 km). Additional research on the distribution of pathfinder elements and lithogeochemistry of source bedrock is in progress for this area.

Early drafts of the manuscript were reviewed by P. T. Bobrowsky and M. B. McClenaghan. Their critiques were invaluable and helped improve this paper. Critical reviews from A. Plouffe and B. C. Ward are greatly appreciated. Regional till sampling and analysis were carried out as a part of the Eagle Bay Project (1996–1998), funded by the British Columbia Geological Survey Branch.

References

ALLEN, G. B. 1977. *Detailed geochemical exploration program on the Fluke claims of Resourcex Limited.* British Columbia Ministry of Energy, Mines and Petroleum Resources, Assessment Report **6230**.

BATTERSON, M. J. 1989. Glacial dispersal from the Strange Lake alkalic complex, northern Labrador. *In:* DILABIO, R. N. W. & COKER, W. B. (eds) *Drift Prospecting.* Geological Survey of Canada, Paper **89-20**, 31–40.

BENNETT, M. R. & GLASSER, N. F. 1996. *Glacial Geology: Ice Sheets and Landforms.* John Wiley & Sons Ltd, New York.

BOBROWSKY, P. T. 1999. Glacial Processes and Surficial Geology. *In:* MCCLENAGHAN, M. B., COOK, S. & BOBROWSKY, P. T. (eds) *Drift Exploration in Glaciated Terrain.* Short Course Volume, 19th International Geochemical Exploration Symposium, Vancouver, Canada.

BOBROWSKY, P. T., SIBBICK, S. J., MATYSEK, P. F. & NEWELL, J. (eds) 1995. *Drift Exploration in the Canadian Cordillera.* British Columbia Ministry of Energy, Mines and Petroleum Resources Paper **1995-2**.

BOBROWSKY, P. T., LEBOE, E. R., DIXON-WARREN, A. & LEDWON, A 1997a. Eagle Bay Project; Till geochemistry of the Adams Plateau (82M/4) and North Barriere Lake (82M/5) map areas. *In:* LEFEBURE, D. V., MCMILLAN, W. J. & MCARTHUR, G. (eds) *Geological Fieldwork 1996.* British Columbia Ministry of Employment and Investment Paper **1997-1**.

BOBROWSKY, P. T., LEBOE, E. R., DIXON-WARREN, A., LEDWON, A., MACDOUGALL, D. & SIBBICK, S. J. 1997b. Till geochemistry of the Adams Plateau-North Barriere Lake Area (82M/4 and 5). British Columbia Ministry of Employment and Investment Open File **1997-9**.

BOUCHARD, M. A. & SALONEN, V-P. 1990. Boulder transport in shield areas. *In:* KUJANSUU, R. & SAARNISTO, M. (eds) *Glacial Indicator Tracing.* A. A. Balkema, Rotterdam, 87–107.

CASSELMAN, S. 1993. *Geochemical and lithogeochemical report on the CM Property, Barriere.* British Columbia Ministry of Energy, Mines and Petroleum Resources Assessment Report **23155**.

CLAGUE, J. J. 1980. *Late Quaternary geology and geochronology of British Columbia, Part 1: Radiocarbon Dates.* Geological Survey of Canada Paper **80-13**.

CLAGUE, J. J. 1989. Quaternary Geology of the Canadian Cordillera. *In:* FULTON, R. J. (ed.) *Quaternary Geology of Canada and Greenland.* Geological Survey of Canada, Geology of Canada, **1**, 15–96 (also Geological Society of America, The Geology of North America, **K-1**).

CLARK, P. U. 1987. Subglacial sediment dispersal and till composition. *Journal of Geology,* **95**, 527–541.

COKER, W. B. & DILABIO, R. N. W. 1989. Geochemical exploration in glaciated terrain: geochemical responses. *In: Proceedings of Exploration '87*: Third Decennial International Conference on Geophysical and Geochemical Exploration for Minerals and Groundwater, Ontario Geological Survey, 336–383.

DILABIO, R. N. W. 1990. Glacial Dispersal Trains. *In:* KUJANSUU, R. & SAARNISTO, M. (eds) *Glacial Indicator Tracing.* A. A. Balkema, Rotterdam, 109–122.

DIXON-WARREN, A., BOBROWSKY, P. T., LEBOE, E. R. & LEDWON, A. 1997. *Terrain geology of the Adams Plateau (82 M/4), scale 1:50 000.* British Columbia Ministry of Employment and Investment, Open File **1997-7**.

DRAKE, L. D. 1983. Ore plumes in till. *Journal of Geology,* **91**, 707–713.

DYKE, A. S. & MORRIS, T. F. 1988. Drumlin fields,

dispersal trains and ice streams in Arctic Canada. *The Canadian Geographer*, **32**, 86–90.

DYCK, W., FYLES, J. G. & BLAKE, W. Jr. 1965. *Geological Survey of Canada Radiocarbon Dates IV*. Geological Survey of Canada Paper **65-4**.

FINCK, P. W. & STEA, R. R. 1995. *The compositional development of tills overlying the South Mountain Batholith, Nova Scotia*. Nova Scotia Department of Natural Resources, Mines and Energy Branches Paper **95-1**.

FULTON, R. J. 1967. *Deglaciation studies in the Kamloops region, an area of moderate relief, British Columbia*. Geological Survey of Canada, Bulletin **154**.

FULTON, R. J. 1969. *Glacial Lake History, Southern Interior Plateau, British Columbia*. Geological Survey of Canada Paper **69-37**.

FULTON, R. J., 1975. *Quaternary geology and geomorphology, Nicola-Vernon area, British Columbia (82L W and 92 E)*. Geological Survey of Canada Memoir **380**.

FULTON, R. J. & SMITH, G. W. 1978. Late Pleistocene stratigraphy of south-central British Columbia. *Canadian Journal of Earth Sciences*, **15**, 971–980.

FULTON, R. J., ALLEY, N. F., & ACHARD, R. A. 1986. *Surficial geology, Seymour arm, British Columbia*. Geological Survey of Canada Map **1609A**, 1:250 000.

HÖY, T. 1991. Volcanogenic massive sulphide deposits in British Columbia. *In*: MCMILLAN, W. J. (ed.) *Ore Deposits, Tectonics and Metallogeny in the Canadian Cordillera*. British Columbia Ministry of Energy, Mines and Petroleum Resources, Paper **1991-4**, 89–199.

HÖY, T. 1999. Massive sulphide deposits of the Eagle Bay Assemblage, Adams Plateau, south central British Columbia (82M3, 4). *In: Geological Fieldwork 1998*. British Columbia Ministry of Energy and Mines Paper **1999-1**.

GARRETT, R. G. 1971. *The dispersion of copper and zinc in glacial overburden at the Lovem deposit, Val d'Or, Quebec*. Canadian Institute of Mining and Metallurgy, Special Volume **11**, 157–158.

GOLDTHWAIT, R. P. 1968. *Surficial geology of Wolfeboro-Winnepesaukee area, New Hampshire*. Concord, New Hampshire Department of Resources and Economic Development.

KASZYCKI, C. A. & DILABIO, R. N. W. 1986. Surficial geology and till geochemistry, Lynn Lake – Leaf Rapids region, Manitoba. *In: Current Research, Part B*. Geological Survey of Canada, Paper **86-1B**, 245–256.

KASZYCKI, C. A., SUTTNER, W. & DILABIO, R. N. W. 1988. Gold and arsenic in till, Wheatcroft Lake dispersal train, Manitoba. *In: Current Research, Part C*. Geological Survey of Canada, Paper **88-1C**, 341–351.

KERR, D. E., SIBBICK, S. J. & BELIK, G. D. 1993. Preliminary results of glacial dispersion studies on the Galaxy Property, Kamloops. *In: Geological Fieldwork 1992*. British Columbia Ministry of Energy, Mines and Petroleum Resources, Paper **1993-1**, 439–443.

KLASSEN, R. A. 1997. Glacial history and ice flow dynamics applied to drift prospecting and geochemical exploration. *In*: GUBINS, A. G. (ed.) *Proceedings of Exploration '97:* Fourth Decennial International Conference on Mineral Exploration, Toronto, 221–232.

KLASSEN, R. A. 2001. A Quaternary geological perspective on geochemical exploration in glaciated terrain. *In:* MCCLENAGHAN, M. B., BOBROWSKY, P. T., HALL, G. E. M. & COOK, S. J. (eds) *Drift Exploration in Glaciated Terrain*. Geological Society, London, Special Publications, **185**, 1–17.

LEBOE, E. R., BOBROWSKY, P. T., DIXON-WARREN, A. & LEDWON, A. 1997. *Terrain Geology Map of North Barriere Lake*. British Columbia Ministry of Employment and Investment, Open File **1997-6**.

LETT, R. E. 2001. Geochemical signatures around massive sulphide deposits in Southern British Columbia, Canada. *In:* MCCLENAGHAN, M. B., BOBROWSKY, P. T., HALL, G. E. M. & COOK, S. J. (eds) *Drift Exploration in Glaciated Terrain*. Geological Society, London, Special Publications, **185**, 301–321.

LETT, R. E. & JACKAMAN, W. 2000. Geochemical exploration techniques for plutonic-related gold deposits in southern British Columbia. *In: Geological Fieldwork 1999*. British Columbia Ministry of Energy and Mines Paper **2000-1**.

LETT, R. E., BOBROWSKY, P. T., CATHRO, M. & YEOW, A., 1998. Geochemical pathfinders for massive sulphide deposits in the southern Kootenay Terrane. *In:* LEFEBURE, D. V. & MCMILLAN, W. J. (eds) *Geological Fieldwork 1997*. British Columbia Ministry of Employment and Investment, Paper **1998-1**, 15-1–15-9.

LETT, R. E., JACKAMAN, W. & YEOW, A. 1999. Detailed Geochemical exploration techniques for base and precious metals in the Kootenay Terrane. *In: Geological Fieldwork 1998*. British Columbia Ministry of Energy and Mines, Paper **1999-1**, 297–306.

LEVSON, V. M. 2001. Regional till geochemistry and sampling techniques in glaciated shield terain: a review. *In:* MCCLENAGHAN, M. B., BOBROWSKY, P. T., HALL, G. E. M. & COOK, S. J. (eds) *Drift Exploration in Glaciated Terrain*. Geological Society, London, Special Publications, **185**, 45–68.

MARR, J. M. 1989. *1988 Fieldwork on the Kamad claims*. British Columbia Ministry of Energy, Mines and Petroleum Resources Assessment Report **18822**.

MATTHEWS, W. H. 1986. *Physiographic map of the Canadian Cordillera*. Geological Survey of Canada, Map **1701A**.

MILLER, J. K. 1984. Model for clastic indicator trains in till. *In: Prospecting in Areas of Glaciated Terrain - 1984*. Institution of Mining and Metallurgy, London, 69–77.

O'BRIEN, E. K., LEVSON, V. M. & BROSTER, B. E. 1997. *Till geochemical dispersal in central British Columbia*. British Columbia Ministry of Employment and Investment Open File **1997-12**.

PAULEN, R. C., BOBROWSKY, P. T., LITTLE, E. C., PREBBLE, A. C., & LEDWON, A. 1998a. Surficial Deposits in the Louis Creek and Chu Chua Creek Area. *In:* LEFEBURE, D. V. & MCMILLAN, W. J.

(eds) *Geological Fieldwork 1997*. British Columbia Ministry of Employment and Investment Paper **1998-1**.

PAULEN, R. C., BOBROWSKY, P. T., LITTLE, E. C., PREBBLE, A. C. & LEDWON, A. 1998b. Terrain geology of the Chu Chua Creek area, NTS 92P/8E, scale 1:50 000. British Columbia Ministry of Employment and Investment Open File **1998-3**.

PAULEN, R. C., BOBROWSKY, P. T., LITTLE, E. C., PREBBLE, A. C., & LEDWON, A. 1998c. Terrain geology of the Louis Creek area, NTS 92P/8E, scale 1:50 000. British Columbia Ministry of Employment and Investment Open File **1998-2**.

PAULEN, R. C., BOBROWSKY, P. T., LETT, R. E., BICHLER, A. J. & WINGERTER, C. 1999. Till geochemistry in the Kootenay, Slide Mountain and Quesnel Terranes. *In: Geological Fieldwork 1998*. British Columbia Ministry of Energy and Mines Paper **1999-1**, 307–319.

PLOUFFE, A. 1999. New data on till geochemistry in the northern sector of the Nechako River map area, British Columbia. *In: Current Research 1999-A*. Geological Survey of Canada, 169–178.

PROUDFOOT, D. N. 1993. Drift exploration and surficial geology of the Clusko River and Toil Mountain map sheets. *In: Geological Fieldwork 1992*. British Columbia Ministry of Energy, Mines and Petroleum Resources, Paper **1993-1**, 491–498.

REIMANN, C. & CARITAT, P. 1998. *Chemical Elements in the Environment – Factsheets for the Geochemist and Environmental Scientist*. Springer-Verlag, Berlin.

RICHARDS, B. G. 1989. *Geochemical report on the WIN group mineral claims*. British Columbia Ministry of Energy, Mines and Petroleum Resources Assessment Report **18596**.

ROSE, A. W., HAWKES, H. E. & WEBB, J. S. 1979. *Geochemistry in Mineral Exploration*. 2nd edition. A. W. Rose, Pennsylvania.

RYDER, J. M., FULTON, R. J. & CLAGUE, J. J. 1991. The Cordilleran ice sheet and the glacial geomorphology of southern and central British Columbia. *Géographie Physique et Quaternaire*, **45**, 365–377.

SCHIARIZZA, P. & PRETO, V. A. 1987. *Geology of the Adams Plateau-Clearwater-Vavenby area*. British Columbia Ministry of Energy, Mines and Petroleum Resources Paper **1987-2**.

SHILTS, W. W. 1976. Glacial Till and mineral exploration. *In:* LEGGET, R. F. (ed.) *Glacial Till: an Interdisciplinary Study*. Royal Society of Canada, Special Publication, **12**, 537–541.

SHILTS, W. W., CUNNINGHAM, C. M. & KASZYCKI, C. A. 1979. Keewatin ice sheet-reevaluation of the traditional concept to prospecting for ore deposits. *Geology*, **7**, 537–541.

SIBBICK, S. J., REBAGLIATI, C. M., COPELAND, D. J. & LETT, R. E. 1992. Soil geochemistry of the Kemess South porphyry gold-copper deposit. *In: Geological Fieldwork 1991*. British Columbia Ministry of Energy, Mines and Petroleum Resources Paper **1992-1**, 349–361.

STEA, R. R. & FINCK, P. W. 2001. An evolutionary model of glacial and till genesis in Maritime Canada. *In:* MCCLENAGHAN, M. B., BOBROWSKY, P. T., HALL, G. E. M. & COOK, S. J. (eds) *Drift Exploration in Glaciated Terrain*. Geological Society, London, Special Publications, **185**, 237–265.

THORLEIFSON, L. H. & GARRETT, R. G. 1993. *Prairie kimberlite study - till matrix geochemistry and preliminary indicator mineral data*. Geological Survey of Canada Open File **2745**.

THORLEIFSON, L. H. & KRISTJANSSON, F. J. 1993. *Quaternary geology and drift prospecting, Beardmore-Geralton area, Ontario*. Geological Survey of Canada Memoir **435**.

TIPPER, H. 1971. *Surficial Geology, Bonaparte Lake, British Columbia*. Geological Survey of Canada Map **1278A**, 1:250 000.

WELLS, R. C. 1987. *Geological, geochemical and geophysical report on the Comstock Claim Group*. British Columbia Ministry of Energy, Mines and Petroleum Resources Assessment Report **16211**.

YORSTON, R. & IKONA, C. K. 1985. *Geological Report on the Cedar I to VI mineral claims*. British Columbia Ministry of Energy, Mines and Petroleum Resources, Assessment Report **13519**.

Index

Note: Page numbers in **bold** type refer to tables; those in *italic* refer to illustrations.

Abies amabilis 155
A. balsamea 109, 155
A. lasiocarpa 155
Abitibi Greenstone Belt 31, 72, 167
   gold concentrations *207*
   gold deposits 201-224
   ice flow *202*
   location *201*
ablation till 252, 325
abrasion, clasts 7
Abukuma-type alteration 76
Adams Lake 301, 304, 307, 327, 328, 331
Adams Plateau 325, 327, 331
aillikite 87, 90
alder, *see Alnus crispa*
alnoites 90
Alnus crispa 155
alteration zones
   element concentrations 171
   indicator minerals 70
   uranium mineralization 226
amoeboid dispersal trains 9, 272, 283, 330
amphibole 79
analytical methods 36–37
anomaly evaluation 58–63
anorthosite, boulders 241
anthophyllite 75
Antigonish Highlands 244
antimony
   association with gold 139
   in dispersal trains 291–293
   in till *293*, *313*
apatite 85
Appalachia, lake sediments 131
Appalachian Ice Complex 240–241, 249, 260, 269
   and Laurentide Ice Sheet *4*, 5
Appalachian Orogen 237
Aquarius deposit 71
argentite 303
Arkhangelsk 96
Arrow Lake 55
arsenic
   association with gold 139
   in lake sediments *140*, 141, *142*
   in soils *327*, *332*
   in till 308, *311*, *313*
arsenides 74
arsenopyrite 217, 303
ashing 161
aspen, *see Populus tremuloides*
Athabasca Basin
   Landsat image *226*
   location *225*
   uranium deposits 225–235
Athabasca Group 225

Atlantic Uplands 254
Attawapiskat River 87
augers 27
Avalon Peninsula 269, 270, 277
Avalon Zone 241, 247, 255

B-horizon 23, 46, 61, 112, 334
Babine belt 47, 54, 63, 145
background concentrations 61
backhoe digging *29*, 30, 219
Baffin Island 85
Baie des Chaleurs 241
Baie Verte 277, 280–283
Bakos deposit 128
Baldy Batholith 303
balsam fir, *see Abies balsamea*
barite 75, 303, 317
barium, in till **314**
bark, biogeochemistry 156
bark ash 152
Barriere 329
basal till *47*, 48
   characteristics **49**
   copper concentrations *52*
   dispersal 51–54
   gold in 61, *62*
   massive sulphides 301
basal transport 7–9
base metals
   drilling for 31, 33
   Labrador 267
   lake sediments 128
Batchelor Lake 87
Batty Complex 88
Bay of Exploits 277
Bay of Fundy 243, 244
beach deposits, kimberlite sampling 96
Beaver River Till 251, 252, 253, 254
   clast geology *259*
   dispersal *246*
   dispersal parameters **260**
   overprinting 255
   renewal distance 258
bed roughness, and dispersal trains 6
bedrock, sculpted 54
bedrock contacts, predicted 260, *261*
bedrock geology 61
bedrock mapping 21
bedrock signature 184, 197
bedrock slopes, and ice flow 6
bedrock sources 46
   West Lawrencetown *242*
bedrock surface 22
Bell deposit 47, 63
   copper concentrations *60*

Bell Island mine 267
beryllium, Strange Lake *274*
Betts Cove 267
biogeochemical exploration 151–164
biogeochemical surveys 109, 114, 151–164
   analytical methods 161–162
   contamination 159
   rules **159**
   sampling methods 155–156
   tissue washing **160**
biogeochemical and till data 162
black spruce, *see Picea mariana* 155
'blue' ground 89
Bonaparte Plateau 325, 329
boreal forest 155
borrow pits 27
Botwood Group 277
boulder geochemistry 225–235
boulder pavements 251
boulder sampling 227–228
   illite content *227*
boulder tracing 57
boulder transport, glacial flow lines *8*
box and whisker plots 170, *171*, *184*, *191*, *192*
Bralorne Takla 288
branches, element distribution **157**
Brett River Synform 174
Brigus deposit 267
British Columbia Cordillera 47
Brook Till 241
Brown Lake Formation 275
Bruce River Group 274, 275
Buchans 267, 277
Buffalo Hills 88
Burgeo 277
Burin Peninsula 277

C-horizon 23, 25, 38, 61
Cache Creek Group 131
Cache Creek Terrane 288
Cadillac, gold deposit 201
Cadillac-Larder Lake fault 168
calcite, replacement mineral 89
Caledonia Phase 241, 251, 254, 260
Canada, Surficial Materials Map 2, *3*, 4
Canadian Cordillera 21, 45–68
   deposits 145
   drift prospecting 323
   geochemical surveys 130, 297
   map *46*
   till cover 54
Canadian Prairie
   chrome-pyrope *106*
   regional surveys 107
Canadian Shield
   drainage 129
   glaciation 21
Candle Lake 85, 107
Cape Breton Channel 243
Cape Breton trough 243
Cape Chidley 270
Capoose Batholith, gold *53*
Capoose Lake, base metals 131, 145
carbonate-rich till *10*

Cariboo Mountains 288, 290, 325
Casa Berardi 71
case study methods 58, 327–333
cassiterite 74, 77
Central Mineral Belt, Labrador 274–276
Central Volcanic Belt, Newfoundland 272
CH property 54, 55
chalcopyrite 74, 75, 77, 78, 219, 288, 308
Chaleur trough 243
Chapleau Moraine 168
Chedabucto Bay 243
chemical weathering, forested areas 23
Chignecto Phase 243-244, 254, 256
Chilcotin Group 61
Chilcotin Plateau 295
chlorite 226, 229
chrome-diopside 90
   distribution *78*
   Kirkland Lake 104
chrome-pyrope
   Canadian Prairies *106*
   Lac de Gras *105*
chromite 73, 77, 78
   high Cr-Mg 94
   Newfoundland 267
   Thetford 295
   zinc-bearing 79
chromium, in till *315*
chromium anomalies 13
   Newfoundland *278*
Chuchi Lake 290, 291, 295
Churchill Province 85
Chutanli Lake 131, 145
Cigar Lake 225, 231
cinnabar 287, 288, 291, 294, 295
clast analysis 57
clast fabric analysis 270–271, **280**, 281–282
clast uptake, Beaver River Till *252*
clasts, basal tills 48
clay content, composite boulders *229*
clinopyroxenite 79
Clisbako Lake, gold *133*, 137
Cluff Lake 225
CM property 329
Coast Mountains 47, 288, 325
Cobequid Highlands 241, 251, 252, 256
Cochrane Till 203
cohesive flows 51
colluvial fans 325
colluviated till 51, 63
comminution 7, 12, 257
composite boulder samples 227–228, *228*
   analysis 228–230
Comstock claims 331–333
concentration/distance relations 60
cones, biogeochemistry 159
copper
   Bell deposit *60*, 63
   in colluvium *329*
   Cordillera 145
   in GIS 170
   Hill-Tout Lake *133*
   in humus 177, 181
   Nak prospect 54

INDEX

Newfoundland 267
Old Fort Mountain *59*
sample statistics **171**
in soil *327*
in stream sediments *309*
in stream water *310*
in till *312*, *315*
copper anomalies, surficial deposits *173*
copper concentrations, basal till *52*
copper and zinc anomalies *172*
    correlation **189**
    and glacial dispersion 189–190
    in lake sediments 186
    lithological signatures 185, *187*, *188*, 197
    in soils 186, *194*
    in surficial deposits *185*
Coppermine River 130
Cordilleran Ice Sheet 325
coring 33, 34
Cornus stolonifera 111, 155
correlation analysis 140
Corylus cornuta 111, 155
Country Harbour 243
Counts Lakes, molybdenum *135*
crag-and-tail landforms 6, 53, 273
Cree Lake deposit 231
creep 57
Cross diatreme 85
Crowfoot Mountain 330
cryoturbation 26
cumulative frequency plots 140

Daniels Harbour 267
data collection 55, *56*
Davis Lake pluton 247
debris flow deposits
    glacigenic 50
    ice-marginal 51
    Newfoundland 282
decollement surfaces 7
deformation till 6, 48
density concentration 98
depositional environments 6
depositional processes, ice sheet margins 2
Destor-Porcupine fault 168, 201, 217, 218
diamictons 48, 51, 270, 280, 281
diamond
    drilling for 31, 33
    exploration 83–123
    kimberlite indicator minerals 70, 90
    regional surveys 103–107
diamond indicator minerals 94
Diamond Lake 102, 108
diatreme facies 88
Diavik prospect 85, 88
dickite 229
digital elevation models 183
2.5 dimensional image maps *182*, 183
diopside 77
    see also chrome–diopside
dispersal curves 60
dispersal fans 7, 249, 307, 328
dispersal profiles 20
dispersal trains 6
    in basal till 48
    case studies 327
    contacts 52
    detection 61
    diffuse 283
    elongate 22, 46, 52
    geochemical signatures *334*
    gold in 70, **71**
    ice flow *11*
    kimberlite in 83, 97, 109, 112, *113*
    length 333
    mercury in 291–293
    modelling 262
    palimpsest 7, 20, 249
    Pinchi Lake 293–294
    platinum group elements in 74
    shapes 9–12
    Strange Lake 273, 284
    surface expression 333
    three-dimensional geometry 53
    and topography 54
    see also specific forms
dissolved oxygen 127
distance/concentration plots *191*, *192*, *193*
Diversion Lake 277
Doctor's Point 153
dogwood, red, *see Cornus stolonifera*
dot maps, Swayze greenstone belt *175*
Douglas fir, *see Pseudotsuga menziesii*
drainage lakes 132–134
dravite 226, 229, 231, 232, *233*, 234
drift, carbonate-rich 9, *10*
drift prospecting 83, *239*
drill bits, contamination by 31
drill cores 33, 34, *35*
drilling, sampling from 31, *33*
drilling methods 30
drumlin tills 251–252
drumlins 6, 53, 54
    Nova Scotia 239, 241, 247
    sections *250*
    till transport 8
Dry Bones 85
Dumont sill 77

Eagle Bay, drift exploration *324*
Eagle Bay Assemblage 302, 303, 306, 325, 333
Eakin Creek property 329–330
East Kemptville 127, 143, 237, 247
East Milford Till 241
Eastern Shore 256
Eastmain deposit 126, 143
eclogites 90
Ekati diamond mine 85
element analyses
    detection limits **307**
    Newfoundland and Labrador **271**
    rock samples **317**
    stream sediments **308**
element associations 37
element concentrations
    background 61
    Kirkland Lake *111*
    Peddie kimberlite *110*

element mobilization 20
end member tills 7, 249–256, 262
end moraines, transport distances 8
Endako mine 132
Engelmann spruce, see Picea engelmannii
englacial till 251
englacial transport 9, 258
enstatite 76, 85, 90
enzyme leach 111
epidote, manganese 75
epithermal deposits 145
erosion, differential 5
erosional features 239
erosional processes, ice sheet margins 2
erosional stratigraphy 239
erratics trains 57, 252
Escuminac Ice Centre 241
Escuminac Phase 241–243, 251, 254
esker sediments 73, 95
exhalative mineralization 128
Exploits River 277

facies recognition 6
factor analysis 140
Fawnie Creek 61, 63
Fennell Formation 325
Fennoscandian Shield, glaciation 21
field data form *56*
field methods, shield terrain 21–22
fine fractions, geochemical sampling 19
Fish Lake 295
Flatwater Pond 280, 281
floods, outburst 6
Fluke claims 330
fluorine 139
fluorite 77
flutes 6, 53
forest climatic zone 21
forsterite 76, 77
Fort à la Corne 85, 88, 107, 109–111
Fort Fraser 131
Fort St James 131
fracturing, glacial 253
Francois Lake 135
franklinite 74
Fraser Basin 288
Fraser Glaciation 304, 325, 329
Fraser Lake 135
freeze-thaw reworking 257
frost action 25
fuchsite 317

G9 pyrope 94
G10 pyrope 94
gahnite 74, 75
Galaxy porphyry deposit 52
galena 217, 219, 308
Gander River 277
Garden of Eden, dispersal *248*
garnets
    kimberlite indicator minerals 73, 85
    mantle-derived 93
    plots *93*
    surface features 100–101

Gaspereau Ice Centre 241
geobotanical survey 111, *112*
geochemical analysis 36
    gold exploration 207–209
    quality control 58
geochemical anomalies 20
    evaluation 45
    and mineralization 190–195
geochemical criteria, massive sulphides **320**
geochemical dispersion, lake basins 127
geochemical distributions, characterising 170–174
geochemical exploration, Quaternary 1–17
geochemical exploration model 318, *319*
geochemical profiles *24*, *27*
geochemical sampling, for GIS 177
geochemical surveys
    property scale 131–135
    scales **21**
    southern B.C. **305**
    Swayze greenstone belt 168
geochemical variation, and partitioning 13
geographical information systems 141, 165–200
    data analysis 170–195
geological mapping 57
Germansen Landing 291
glacial deposition, modelling 238
glacial dispersal
    granite pebbles *251*
    modelling 257–260
    Newfoundland and Labrador 267–285
    processes 20
    and till thickness *317*
    zonal concept 244–249
glacial dispersal models 1, 2, 7–12
    local 58
    Maritime Canada 237–265
    vector addition 249
    zonal concept *244*
glacial erosion 6
    modelling 238
glacial flow lines, boulder transport *8*
glacial lakes 269
glacial landforms, map *3*
glacial processes, and partitioning 12–13
glacial sedimentology 6–7
glacial sediments, identification 21
glacial transport
    modelling 238
    trace elements 323–337
glaciation, Maritime Canada 239–249
glacier calving 271
glacier flow, linear 46
glaciers, polythermal 252
glaciofluvial outwash *48*
glaciofluvial sediments, kimberlite sampling 95
glaciolacustrine deposits
    characteristics 51
    sandy 95–96
    shearing 7
    size fractions 12
glaciomarine deposits
    Newfoundland 271
    shearing 7
Glenwood 277

goethite 26, 295
gold
  analytical methods 37
  anomaly patterns 221
  in basal till 61
  Capoose Batholith 53
  central British Columbia 289, 291
  Clisbako Lake 133
  concentrations 208, 217
  in dispersal trains 70, **71**
  drilling for 31, 33
  Eakin Creek 330
  lake sediment surveys 127, 128, 145
  Matheson 210
  mineralization 139
  Newfoundland 279
  partitioning 287–299
  Peterlong Lake 212
  in soil 316
  southern British Columbia 301–327
  in stream sediments 309
  in till **208**, 311, 312
  in trees **160**
  Wolf prospect 62
gold concentrations, trees **153**
gold grains
  Abitibi 205–207
  abundance 206, 211, 214, 215
  analysis 71
  laboratory methods 204
  Matheson 210
  modified 206
  pristine 205–206
  reshaped 206
  size 206–207, 207, 296
  wear 70, 72, 205, 206
gold mineralization 206, 221
gold signatures, Abitibi Greenstone Belt 201–224
gold-arsenic correlation **152**
gold-bearing formations 26
Golden Pond deposits 71, 216
  section 216
gossans 138, 301, 303
grain-size effect 257
Grand Falls 277
granite, peralkaline 273
granophile deposits 145
graphite 226
gravel, sandy 48
gravel lithology 12
gravity flow deposits 50, 51
Great Northern Peninsula 269
greenschist facies, minerals 77
greisen deposits 71, 74, 77
Grew Creek 47
grossular 77
groundwater, leaching 334
Gulf of Maine 243
Gulf of St Lawrence 241
Gullbridge 273
gyttja 126, 132, 137

Halifax 254
hand excavation 27
Harrison Lake 153
Hartlen Till 241, 251, 254, 255
harzburgite 79
hazelnut, beaked, *see Corylus cornuta*
Hazelton 131
Hazelton Group 61
heavy mineral concentrates 97, 103, 217, 291, 294
heavy mineral fractions 36, 57, 204
heavy minerals
  alteration zones **76**
  analytical methods 37, 69
  localities 70
  nickel-copper deposits **77**
  skarn and greisen deposits **76**
helicopter support 29, 135
Hemlo deposit 143
hercynite 77, 78
high strain zones, Au deposits 168
Hill-Tout Lake, copper 133, 134
Hislop gold deposit 218–219, 220
Homestake deposit 301, 302, 303
Hope Brook deposit 128, 267, 277
Houston 134
Hudson Bay
  Laurentide Ice Sheet 2
  stream sediments 96
hummocky moraine, transport distances 8
hummocky topography 51
hybrid till 7, 254–256, 262
hydromorphic dispersion 328, 335
hydrothermal alteration 73, 76

ice ablation 12
ice bed, shear stress 4
ice divide tills 252–254, 262
ice divides 2
  active 4
  evolution 6
  shifting 238
ice dynamics, variation 253
ice flow
  Abitibi Greenstone Belt 202
  indicators 325, 326
ice flow histories
  central British Columbia 288
  flowlines 240
  kimberlite exploration 96
  mapping 57
  Maritime Canada 260
  Newfoundland 276, 280
  and till formation 238
ice flow record 5–6
ice flow trends 5
ice flow velocity 2
  subglacial 2
ice rafting 51, 270
ice retreat 269
ice rises 249
ice sheet models 1, 2
ice sheets, growth and decay 5
ice streams 4, 249, 257
  deformation till 7
  landforms 6
  till 262

illite 226, 229, 231
  in composite boulders *230*, *232*
ilmenite
  Mg-rich 73, 85, 90, 101, 102, 108
  plots *91*
incompatible elements 85, 102, 277
indicator clasts 271
indicator concentration 8
indicator erratics 2, *275*
  and transport distance *7*
Indicator Lake 87
indicator minerals 69–81
inheritance, *see* till, inheritance
instrumental neutron activation analysis 37, 57, 103, 109, 128, 207, 271, 290, 306
integrated images 183
intensity-hue-saturation transforms 183
Interior Plateau 47, 51, 131, 323
Intermontane Belt 325
interpolation, in GIS 177–179
interval sampling 30, 31
Inzana Lake 291
iron ore, Labrador 267

jack pine, *see Pinus banksiana*
James Bay Lowlands 87, 96
Jericho pipe 88
Joggins 257
johannsenite 77
Johnson Creek 307
Johnson Lake 302

Kaipokok River 275
Kamad 3 property 328
kames 95
Kamloops 302
kaolin 227, 229
Kechika Trough 128, 131
Keewatin 26
Kejimikujik National Park 260
kelyphite 100
Ken deposit 132
Kennady Lake 85, 88
Kenty 168
Key Lake 128, 143, 225, 231
Kikkerk Lake 85
kimberlite indicator minerals 70, 73, 79, 90–95, *99*, 112
  geochemical methods 102–103
  local surveys 107–112
  physical features **92**
  relative abundance 102
  size range 101–102
kimberlites 33
  analytical methods 98–100
  boulders *95*
  in Canada 85
  characterisation 85
  in dispersal trains 83, *113*
  distribution *86*
  facies 88
  fields 87
  hypabyssal facies 88
  model *87*
  pebble abundance *108*
  pipe cross-sections *88*
  published surveys **84**
  root zones 88
  Siberia 114
  volcaniclastic facies 88, 110
kinoshitalite 85
Kirkland Lake 33, 73, 85, 88
  cross-section *90*
  dispersal trains 97
  erosion 89
  geochemistry 102
  gold deposits 168, 201
  local surveys 107–109
  pyrope *104*
  regional surveys 103–105
Kisseynew gneisses 76
Kitts 275
knebelite 77
komatiites 77
Kootenay Terrane 47, 302, 325
kriging 179
Kuyakuz Lake, molybdenum *134*
kyanite 75, 76
Kyle Lake 87, 96

La Manche, copper 267
La Ronge belt 72, 128, 143
Labrador, glacial history 269
Labrador and Newfoundland
  glacial dispersal 267–285
  location map *268*
Labrador tea, *see Ledum groenlandicum*
Labrador Trough 269, 272
Lac de Gras
  aerial photograph *89*
  beach deposits 96
  chrome-pyrope *105*
  dispersal train 97
  geochemistry 102
  kimberlite 83, 88
  local surveys 109
  regional surveys 105–107
Lac des Iles 128
Lac Rocher 77, 79
Ladner Creek 157
Lake Agassiz 72
Lake Melville 269
Lake Ojibway 203
lake sediments 125–149
  accumulation 126
  analytical methods 138–139
  arsenic *140*
  composition 126–127
  copper and zinc anomalies 186, 195
  field observations 137–138
  gold 145
  nickel anomalies *143*
  reconnaissance surveys 130–131
  sample site choice 136
  sampling devices 136
  sampling methods 135–138, *137*
  Shield deposits 143–145
  statistical analysis 139–141
  surveys *129*

INDEX

Swayze greenstone belt *169*
  in uranium exploration 127
Lake Timiskaming 85, 88, 102
  local surveys 107–109
lakes
  glacial 21
  profundal 126
lamproites 90
lamprophyres, ultrabasic 87, 90
landforms, streamlined 6
Larder Lake, gold deposit 201
Larder Lake-Cadillac fault zone 201
Larix occidentalis 155
Laurentian Channel 243
Laurentide Ice Sheet 2, 269, 275
  Abitibi Greenstone Belt 203
  and Appalachian Glacier Complex *4*, *5*, 241
Lawrencetown Till 241, 251, 254
  clast lithologies *256*
  overprinting 256
layered intrusions 77
leaching 111, 334
lead
  in soil *316*, *330*
  in till *314*
lead anomalies 52
leaves and needles, biogeochemistry 157
Ledum groenlandicum 155
Lemotte's Lake 277
Lewis Hills 267
lherzolites 79
limnological factors 127
lithogeochemical trends, uranium deposits 232–233
lithogeochemistry, composite boulders 230–231
lithologic signatures, in geochemical data 184–189
lithology, description 56
lithophile elements 127, 274, 295
Little Bay 267
Little Fort 329
Locker Lake Formation 225
lodgement till 6, 48, 251, 270, 281
lodgepole pine, *see Pinus contorta*
loellingite 74, 75, 77
Long Range Mountains 269
loss on ignition 139, 140
Lupin deposit 143

Mac deposit 127, 131, 145
McArthur River deposit 225, 231, 232–233
maceration 161
Mclean deposit 231
McLean Lake 269
Magdalen Shelf 241, 243, 252
magmatic sulphides 71, 128
magnesium minerals
  alteration 231
  enrichment 233
magnetic highs, Swayze greenstone belt 174
magnetic surveys 96
manganese 75
  and copper 194, *195*
Manitou Falls Formation 225, 226, 227, 230, 234
mantle, partial melting 85
map production 141

marcasite 288
marine limit
  Labrador 275
  Newfoundland 270
Maritime Canada
  bedrock geology *238*
  glacial dispersal models 237–265
  glacial evolution *240*
  glaciation 239–249
mass wasting 57, 325
massive sulphide deposits 70, 75, 128, 277, 301–321
  location map *302*
massive sulphide indicator minerals 73–79
Matachewan gold deposit 217–218, *219*
Matheson 73
  geochemical surveys 209–211
Matheson Till 203, 217
Mealy Mountains 269
megacrysts 85, 90
Meguma Zone 237, 241, 249, 251, 255, 260
Melody Lake 275
melt-out deposits 48
melt-out till 251, 257, 270
meltwater, and landforms 6
mercury
  analytical methods 37
  and grain size *294*, *295*
  lake sediments 145
  mineralization 288, 293
  partitioning 287–299
  in soils *332*
  in till 290, 291, *292*
metal mobilization, secondary 323–337
metal zonation, lake sediments 132, 134
metamorphism, Barrovian 76
metasomatism 73, 79
MgO/Al_2O_3 ratios, composite boulders *231*, 233, *234*
Michelin 275
Mine Series 303
mineral deposit types, element associations **37**
mineral exploration, and element partitioning 296–297
mineral partitioning 12, 38
mineral preservation 21
mineralization, and geochemical anomalies 190–195
mineralization zones, element concentrations 171
mineralized ground, sampling 22
Misery kimberlite 88
molybdenum 127
  Cordillera 145
  Counts Lakes *135*
  Kuyakuz Lake *134*
  lake sediments *130*, 131
  Tatin Lake *132*
Monashee Mountains 325
monticellite 85
moraines
  Athabasca Basin 226
  kimberlite sampling 95
Moran Lake 275
Moran Lake Group 274
Mount Milligan 47
Mount Peyton 277, 280
mountain hemlock, *see Tsuga mertensiana*
Mountain Lake 85, 111–112

mudboils 26, 29, 275
multiple deposits, signatures 334–335
Munro Esker 73, 95, 103
Myra Falls 47

Nadina Lake 134
nail-head striations 270
Nain plateau 273
Nain Province 74, 87, 128
Nak prospect 54
Naskaupi Lake 269
National Geochemical Reconnaissance Programme 126, 168
nearest point algorithm 189, 190
Nechako Plateau 126, 128, 130, 131, 145, 288
Nechako River
    properties 47, 54
    surveys 131
neutron activation analysis, see instrumental neutron activation analysis
Newfoundland
    glacial history 269
    ice flow histories 276
    mining 267
Newfoundland and Labrador
    glacial dispersal 267–285
    location map 268
    surficial geology 271, 272
nickel anomalies 13
    Lac de Gras 106
    lake sediments 143
    in streams 307
nickel deposit, Voisey's Bay 267
Nickel Plate 47
nickel-copper mineralization 73, 76
    indicator minerals 77–79
    Labrador 143–145
Nikwikwaia Creek 331, 333
Nipigon Diabase 74
    location map 75
Nithi Mountain 135
non-cohesive flows 51
Noranda/Kuroko deposit 325
norite 77
normal probability plots 172
normalizing data, in GIS 174, 197–198
North American Craton 288
North Star Hill 153
North Thompson River 325, 329
Northumberland Strait 241, 244
Norwegian Fjordlands 21
Notre Dame Bay 267
Nova Scotia
    cross section 243
    mineral exploration 237
nugget effect 37, 132, 179, 290, 291, 297
Nugget Pond 277
Nunavut 26, 85, 88

Old Fort Mountain, copper concentrations 59
Old Woman formation 174
olivine 77, 85, 90
Omineca Mountains 288
Ootsa Lake Group 61

orange peel texture 100–101
ore bodies, buried 22
ore-indicator elements 36
organic carbon, lake sediments 139
organic gels 126
orthopyroxene 75, 77
orthopyroxenites 79
Outokumpu 77, 79
outwash sediments 95
overburden drilling 22, 221
overconsolidation, deformation till 7
overland flows 51
overprinting, see till, overprinting
Owl Creek deposit 214–215
oxidation 26
oxidation zone 20, 22

Pacific silver fir, see Abies amabilis
palimpsest landforms 239
Pamour gold deposit 216–217, 217
paraglacial environments 51
partitioning
    and glacial processes 12–13
    minerals 12
pathfinder elements 36, 37, 102–103, 106, 208, 221, 318, 327
Peace River 85
pebble lithology 239, 255
Peddie kimberlite 88, 108, 109
    surface 91
pedogenesis 61
peralkaline granites 273, 280
percentile ranges 58, 59
percussion drills 34, 35
Percy Lake 126, 143
peridotite 73
    xenoliths 90, 94
permafrost
    active layer 22, 25
    weathering processes 25–27
perovskite 85
Peterlong Lake 206, 211–214
pH, stream water 310
phlogopite 85, 90
phyllosilicates, size fractions 58
Picea engelmannii 155
P. mariana 109, 155
Pinchi 131
Pinchi Fault 287, 288, 290
Pinchi Lake 290
Pine Cove 277
Pinus banksiana 155
P. contorta 155
P. ponderosa 155
Pipe Mine 2 body 79
pit sampling 27
placer deposits 73
plant species, selection 154–155
plant tissues
    ash yields **158**
    element concentrations **158**
    selection 156–159
platinum, Tulameen 47
platinum group elements

alloys 77
analytical methods 37
  in dispersal trains 74
  lake sediments 128
Ponderosa pine, see Pinus ponderosa
Populus tremuloides 111
porcellanite 275
porosity 114
porphyry deposits 47, 128, 135, 327
portable drills 34–36, 36
preglacial deposits 6, 12
pressure conditions 48
Prince Edward Island, erratics 241
Prince of Wales Island 10
principal component analysis 277
proportional symbol map, Swayze greenstone belt 176
provenance envelopes, West Lawrencetown 242, 262
Prunus pennsylvanica 109
Pseudotsuga menziesii 155
pyrite 26, 217, 218, 288, 303
pyrochlore 295
pyrope 73, 85, 90, 94
  abundance 103
  fractured 101
  Kirkland Lake 104
  in Munro Esker 103
  SEM images 100
pyrope-almandine garnet 94
pyroxene, omphacitic 90

quarrying, glacial 253
Quaternary, geochemical exploration 1–17
Quesnel 133
Quesnel Terrane 128
Quesnelia 288

Rabbit Lake 225
Radisson Lake 211–214
Rainy River 71, 72, 77
raised beaches 270
Rambler deposit 277
rammelsbergite 74, 77
Ranch Lake 109
Rankin Inlet 85
rapakivi granite 273, 274
rare earth elements
  Labrador 267
  Strange Lake 97, 126, 128, 145, 269
rat-tail features 270
Rea gold deposit 301, 302, 303, 317, 318
Read Lake 232, 233
reconnaissance exploration 80
regolith, size fractions 12
renewal distance 252, 253, 257
reverse circulation drills 31, 32, 32, 204
ribbon-shaped dispersal trains 9, 329, 331, 333
Rideout strain zone 174, 179
Ridge Zone 62
road cuts 27
rock hardness 257
Rocking Horse Lake 85
Rocky Mountains 47
root systems, biogeochemistry 151
rotary drill 31

rotasonic drilling 32–34, 34, 204
Rottenstone mine 153
Rouyn-Noranda 216
rutile 75, 77

St John's, copper 267
Samatosum deposit 301, 302, 303, 317, 318
  dispersal train 333
  geology 303
  location 302
  surficial geology 304
sample processing
  biogeochemistry 160–161
  flow charts 38, 97, 205
  for gold grains 204
  kimberlite exploration 97
sample sizes 96
sampling density 22, 128, 135
sampling depths 22
sampling grids 22, 55
sampling methods 21–22
  biogeochemical surveys 155–156
  composite 227
  diamond exploration 96–97
  lake sediments 135–138
  thick drift areas 30–36, 30
  thin drift 27-30, 28, 29, 221
  till geochemical surveys 55, 63, 96, 209, 271, 290
sapphirine 75
saprolite 244
scale factors, lake sediments 128–135
scale variations 1, 2, 54
scatter plots 141
scavenging, by Mn 195, 198
scheelite 74, 77, 295
Schefferville 145, 269
Scotch Creek 325
Scotian Ice Divide 243, 244, 247, 253
Scotian Phase 243, 254, 256, 260
Scotian Shelf 241
sea-level changes, Newfoundland and Labrador 269–270
seasonal variation, biogeochemical survey **154**
secondary weathering 21
sediment sampling, geochemical surveys 95–98
sedimentological data 55
sedimentology, see also glacial sedimentology
sediments
  characteristics **49**
  deformation 6
  size fractions 12
seepage lakes 134–135
selective queries, in GIS 193, 194, 198
Selwyn Basin 128
SEM images, kimberlite minerals 100
serpentine 85, 89
shearing, subglacial 7
Shebandowan Greenstone Belt 74, 78
  location map 75
Shediac Channel 243
Shield deposits, lake sediments 143–145
shield terrain, field methods 21–22
Shunsby 168, 174, 177, 179
Shuswap Basin 325

Shuswap Highland 323
Shuswap Lake 330
Siberia, kimberlites 114
significant anomalies 58
sillimanite 75, 76
silver
  in soils *331*, *332*
  in trees 157
Silver 1 property 327–328, *327*
Sinmax Creek 302, 304, 328
site duplicate samples 137
size fractions, geochemical analysis 36
skarn deposits 71, 74, 77, 327
Skeena Mountains 288
skip zones 9, 252, 262
Slave Province 85
slickensides 48
Slide Mountain terrane 325
slope wash 57
Smeaton 110
Smithers 131, 134
Smiths Cove *250*
Snap Lake 85, 88
Snow Lake 76
soil geochemistry, kimberlites 114
soil horizons 19, 22
soil profiles *23*, *25*
soils
  geochemical anomalies 61–63
  geochemical surveys 51, 54, 327
Somerset Island 85, 88
South Mountain Batholith 237, 254, 256
  bedrock geology *245*
South Mountain Ice Cap 244
spatial analysis 165
spectrometry 37, 57, 228, 271, 305
sperrylite 74, 77
spessartine 75, 77
sphalerite 74, 217, 303, 308
spinel 75, 85, 90
  discrimination 94–95
  plots *93*, *94*
Springdale 282, 283
staurolite 75, 76
  zincian 73
stibnite 55, 288, 291, 293
Stikine Terrane 128
Stony Till Plain 252
Strange Lake 97, 126, 128, 145, 267, 269
  clast distribution *273*
stream sediments
  element analyses **308**
  kimberlite sampling 96
  massive sulphides 304, **306**
stream water surveys 305, **306**
striae 21, 54, 241
striation mapping 270
striation trends, near ice divides 5
Stuart Lake 291
Sturgeon Lake 111
sub-alpine fir, *see Abies lasiocarpa*
subaerial debris flows 51
subaqueous debris flows 51
sudoite 229, 231

sulphide minerals
  gold-bearing 219
  oxidation 26
  redistribution 25
  stability 74
sulphides, *see* massive sulphides, magmatic sulphides
Superior Province 87
supraglacial till 50
surface water, percolation 51
surficial geology, Newfoundland and Labrador 271, *272*
surficial processes 51
Swayze greenstone belt 165–200
  analytical methods **169**
  bedrock geology *166*
  dot maps *175*, *177*
  lake sediments *169*
  lithogeochemical data **170**
  magnetic highs 174
  proportional symbol map *176*, 177
  surficial geology *167*
  survey methodology **168**
tabling 98
Takatoot Lake 291
Tatin Lake 127
  molybdenum *132*
Tchentlo Lake 290, 291, 295
temperate conifer forest 154–155
temperate deciduous forest 155
tephra cones, kimberlites 88
ternary images, in GIS 179, *180*, *181*, *182*, 183, *196*
Teslin Plateau 131
tetrahedrite 303, 308
textural variation 20
Tezzeron Lake 131, 291
Thetford Mines 295
thick drift areas, sampling methods 30–36
thin drift areas, sampling methods 27–30
Thompson Nickel Belt 78, 79
Thompson Plateau 323
Thuja plicata 155
Thunder Bay 128
till
  allochthonous 106
  and biogeochemical data 162
  calcareous 25
  classification 238
  immature 253
  inheritance 254, *255*, 262
  kimberlite sampling 95
  multiple sheets 226, 249
  overprinting 254–256, *255*, 262
  sampling methods 203–204
  shield-derived 12
  size fractions 109
  see also specific types, e.g. basal, lodgement
till fabric 96
till genesis, Maritime Canada 237–265
till geochemistry 19–43
  Abitibi 203, **213**
  evaluation 58–63
  field techniques 55–57
  kimberlites 102–103, 114
  laboratory methods 36–37, 57, 204

major and trace elements *107*, 108
massive sulphide deposits 301, 318
Matachewan deposit *218*
methodology 325–327
Newfoundland and Labrador 271
principles 19–21
southern B.C. 306, 307–318
survey design 55
surveys 46, 54–58, **209**, 209–219
till reworking 63, 215, 238, 249, 254
  mechanisms 256–257
till surface 22
till thickness, and glacial dispersal *317*
till wedges 257
Tilt Cove 267
Timiskaming, metasediments 217
Timmins 71, 168, 201
  till distribution *203*
tin 127, 237
  dispersal *246*, 247
Tommy prospect 145
topaz 77
Topsails granite 282
Torngat Mountains 269, 270
tourmaline 75, 77, 231
tourmalinite 153
trace elements, in exploration 168
trace metal concentrations 25
traction zone 257
transition elements 85, 102
transport distance
  boulders 227
  calculations 8
  clasts *281*, *283*
  and concentration 60
  dispersal trains 333
  indicator erratics *7*
transport distance distribution 8, *9*
transport time 72
tree tissue 109, 111
tree tops, biogeochemistry 158
trees
  element concentrations **153**, **154**, **156**
  gold concentrations **153**
Trembleur Lake 291
trenching 30
troctolite 77
trunk wood, biogeochemistry 157
Tsacha Mountain 134
Tsacha prospect 128, *144*, 145
Tshinakin limestone 303
Tsuga heterophylla 155
T. mertensiana 155
Tulameen 47
tundra climatic zone 21
tungsten 295
tunnels, subglacial 48
twigs, biogeochemistry 157
Twin Mountain 303, 317, 318

ultrabasic rocks 90, 111
ultramafic rocks
  clasts 282
  till source 13

Ungava Bay 269
uranium deposits, Athabasca Basin 225–235
uranium exploration
  drilling 31
  Labrador 267
  lake sediments in 127
uranium mineralization 26
Uranium Reconnaissance Program 127
uvarovite 77

Val d'Or, gold deposit 201
valley glacier *50*
Vanderhoof 134
variograms 177, *178*, 197
vector addition *247*, 249, 262
vegetation zones 154
vertical mixing 25
vertical profiles 12
Victoria Island 85
Victoria Lake Group 277
Vikings, iron smelting 167
visualization, in GIS 174–184, 197
Voisey's Bay 74, 143, 267
volcaniclastic facies, kimberlites 88
volcanosedimentary sulphides 70, 75–77, 128, 277

Wakami shear zone 179
water content, kimberlites 114
water sampling, massive sulphides 304–305
water table, soil horizons 23
weathering, kimberlites 88–90
weathering processes 20–21
websterite 79
wehrlite 79
Wekusko Lake 85
West Lake 277
western hemlock, see *Tsuga heterophylla*
western larch, see *Larix occidentalis*
western red cedar, see *Thuja plicata*
Westmin Lynx orebody 52
Whitesail Lake 131, 134
whole rock geochemistry 37
willemite 74
Wisconsinian
  Abitibi 203
  ice-flow history 48
  Nova Scotia 239
Wolf prospect 55, 61, *62*, 128, 131
wolframite 77
Wolverine Point Formation 225

x-ray fluorescence 37
xenocrysts 90
'yellow' ground 89
Young-Davidson gold deposit 217
yttrium 274
Yukon
  deposits 47
  lake sediments 131

zinc
  in GIS 170
  in humus 179
  Newfoundland 267

sampling statistics **171**
in soils 186, *328*, *332*
see also copper and zinc anomalies

zinc anomalies 52
zircon 90, 295
zonal concept, glacial dispersal 244–249, 262